Cases on Modern Computer Systems in Aviation

Tetiana Shmelova
National Aviation University, Ukraine

Yuliya Sikirda
National Aviation University, Ukraine

Nina Rizun
Gdansk University of Technology, Poland

Dmytro Kucherov
National Aviation University, Ukraine

A volume in the Advances in Computer and Electrical Engineering (ACEE) Book Series

IGI Global
DISSEMINATOR OF KNOWLEDGE

Published in the United States of America by
IGI Global
Engineering Science Reference (an imprint of IGI Global)
701 E. Chocolate Avenue
Hershey PA, USA 17033
Tel: 717-533-8845
Fax: 717-533-8661
E-mail: cust@igi-global.com
Web site: http://www.igi-global.com

Library of Congress Cataloging-in-Publication Data

Names: Shmelova, Tetiana, 1961- editor. | Sikirda, Yuliya, 1978- editor. | Rizun, Nina, 1968- editor. | Kucherov, Dmytro, editor.
Title: Cases on modern computer systems in aviation / Tetiana Shmelova, Yuliya Sikirda, Nina Rizun, and Dmytro Kucherov, editors.
Description: Hershey, PA : Engineering Science Reference, an imprint of IGI Global, [2019] | Includes bibliographical references and index.
Identifiers: LCCN 2018030365| ISBN 9781522575887 (hardcover) | ISBN 9781522575894 (ebook)
Subjects: LCSH: Guidance systems (Flight)--Case studies. | Digital avionics--Case studies. | Drone aircraft--Case studies.
Classification: LCC TL589.4 .C365 2019 | DDC 629.1350285--dc23 LC record available at https://lccn.loc.gov/2018030365

This book is published in the IGI Global book series Advances in Computer and Electrical Engineering (ACEE) (ISSN: 2327-039X; eISSN: 2327-0403)

British Cataloguing in Publication Data
A Cataloguing in Publication record for this book is available from the British Library.

For electronic access to this publication, please contact: eresources@igi-global.com.

Advances in Computer and Electrical Engineering (ACEE) Book Series

Srikanta Patnaik
SOA University, India

ISSN:2327-039X
EISSN:2327-0403

Mission

The fields of computer engineering and electrical engineering encompass a broad range of interdisciplinary topics allowing for expansive research developments across multiple fields. Research in these areas continues to develop and become increasingly important as computer and electrical systems have become an integral part of everyday life.

The **Advances in Computer and Electrical Engineering (ACEE) Book Series** aims to publish research on diverse topics pertaining to computer engineering and electrical engineering. **ACEE** encourages scholarly discourse on the latest applications, tools, and methodologies being implemented in the field for the design and development of computer and electrical systems.

Coverage

- Sensor Technologies
- Programming
- Analog Electronics
- Electrical Power Conversion
- Algorithms
- Chip Design
- Computer Hardware
- Applied Electromagnetics
- Microprocessor Design
- Optical Electronics

IGI Global is currently accepting manuscripts for publication within this series. To submit a proposal for a volume in this series, please contact our Acquisition Editors at Acquisitions@igi-global.com or visit: http://www.igi-global.com/publish/.

The Advances in Computer and Electrical Engineering (ACEE) Book Series (ISSN 2327-039X) is published by IGI Global, 701 E. Chocolate Avenue, Hershey, PA 17033-1240, USA, www.igi-global.com. This series is composed of titles available for purchase individually; each title is edited to be contextually exclusive from any other title within the series. For pricing and ordering information please visit http://www.igi-global.com/book-series/advances-computer-electrical-engineering/73675. Postmaster: Send all address changes to above address.

Titles in this Series

For a list of additional titles in this series, please visit: www.igi-global.com/book-series

Advanced Methodologies and Technologies in Network Architecture, Mobile Computing, and Data Analytics
Mehdi Khosrow-Pour, D.B.A. (Information Resources Management Association, USA)
Engineering Science Reference • copyright 2019 • 1857pp • H/C (ISBN: 9781522575986) • US $595.00 (our price)

Emerging Innovations in Microwave and Antenna Engineering
Jamal Zbitou (University of Hassan 1st, Morocco) and Ahmed Errkik (University of Hassan 1st, Morocco)
Engineering Science Reference • copyright 2019 • 437pp • H/C (ISBN: 9781522575399) • US $245.00 (our price)

Advanced Methodologies and Technologies in Artificial Intelligence, Computer Simulation, and Human-Computer Interaction
Mehdi Khosrow-Pour, D.B.A. (Information Resources Management Association, USA)
Engineering Science Reference • copyright 2019 • 1221pp • H/C (ISBN: 9781522573685) • US $545.00 (our price)

Optimal Power Flow Using Evolutionary Algorithms
Provas Kumar Roy (Kalyani Government Engineering College, India) and Susanta Dutta (Dr. B. C. Roy Engineering College, India)
Engineering Science Reference • copyright 2019 • 323pp • H/C (ISBN: 9781522569718) • US $195.00 (our price)

Advanced Condition Monitoring and Fault Diagnosis of Electric Machines
Muhammad Irfan (Najran University, Saudi Arabia)
Engineering Science Reference • copyright 2019 • 307pp • H/C (ISBN: 9781522569893) • US $225.00 (our price)

The Rise of Fog Computing in the Digital Era
K.G. Srinivasa (Chaudhary Brahm Prakash Government Engineering College, India) Pankaj Lathar (Chaudhary Brahm Prakash Government Engineering College, India) and G.M. Siddesh (Ramaiah Institute of Technology, India)
Engineering Science Reference • copyright 2019 • 286pp • H/C (ISBN: 9781522560708) • US $215.00 (our price)

Algorithms, Methods, and Applications in Mobile Computing and Communications
Agustinus Borgy Waluyo (Monash University, Australia)
Engineering Science Reference • copyright 2019 • 297pp • H/C (ISBN: 9781522556930) • US $205.00 (our price)

Exploring Critical Approaches of Evolutionary Computation
Muhammad Sarfraz (Kuwait University, Kuwait)
Engineering Science Reference • copyright 2019 • 390pp • H/C (ISBN: 9781522558323) • US $215.00 (our price)

IGI Global
DISSEMINATOR OF KNOWLEDGE

701 East Chocolate Avenue, Hershey, PA 17033, USA
Tel: 717-533-8845 x100 • Fax: 717-533-8661
E-Mail: cust@igi-global.com • www.igi-global.com

Table of Contents

Detailed Table of Contents

The aim of this chapter is to research and fundamentally evaluate counterfeit shrewd frameworks to recognize for outperforming human insight in the flights and its conceivable ramifications. How artificial intelligence (AI) makes current airship framework incorporates an assortment of programmed control framework that guides the flight team in route, flight administration and enlarging the security qualities of the plane, and how building aircraft engine diagnostics ontology, air traffic management, and constraint programming (CP) is useful in ATM setting. How flight security can be enhanced through the advancement and usage of mining, utilizing its outcomes and knowledge-based engineering (KBE) approach in an all-encompassing methodology for use in airship reasonable outline, is discussed. The early recognizable proof and finding of mistakes, the study of huge information and its effect on the transportation business and enhanced transit system, the agent-based mobile airline search, and booking framework using AI are shown.

In this chapter, the four layers neural network model for evaluating correctness and timeliness of decision making by the specialist of air traffic services during the pre-simulation training has been presented. The first layer (input) includes exercises that cadet/listener performs to solve a potential conflict situation; the second layer (hidden) depends physiological characteristics of cadet/listener; the third layer (hidden) takes into account the complexity of the exercise depending on the number of potential conflict situations; the fourth layer (output) is assessment of cadet/listener during performance of exercise. Neural network model also has additional inputs (bias) that including restrictions on calculating parameters. The program "Fusion" of visualization of the state of execution of an exercise by a cadet/listener has been developed. Three types of simulation training exercises for CTR (control zone), TMA (terminal control area), and CTA (control area) with different complexity have been analyzed.

The widespread use of unmanned aerial vehicles (UAVs) is currently a recognized trend. UAVs find their application in various sectors of the economy. In the chapter, based on extensive literature analysis, the possibilities of using UAVs for non-military applications are considered. The design features of various UAVs, their control features, energy requirements, and safety-related problems are considered. Particular attention is paid to public opinion related to the use of UAVs. The possibilities of using UAVs in power engineering, agriculture, for controlling traffic, for goods transporting, for controlling the means of railway transport, for first aid to people under various extreme conditions, as well as for some other applications are being explored. The UAV parameters are analyzed, which must be provided for their use in each specific case, while ensuring the minimization of the necessary financial resources.

The chapter deals with the problem of controlling the flow of information coming from a group of unmanned air vehicles by radio channel. The inevitable data losses are compensated by repetition of lost packages or reconfiguration network. Modern methods control of data flow assumes using a mechanism ARQ based on the method sliding window. The consequence of these problems is a partial or total loss of its performance manifested in a decrease in the network's starting throughput. Part of the problem of restoring the network is solved by routing mechanisms, which lead to reconfiguration of the network due to the elimination of faulty nodes. Management of computer network overload is solved by well-known routing protocols such as OSPF, IS-IS, RIP, and others. In solving the problem, the representation of the output of individual nodes of the network using the "death and reproduce" scheme was used substantially. This scheme of network operation presupposes its representation by the Markov chain and the derivation of probabilistic characteristics by solving the Kolmogorov equations.

Aviation telecommunications is one of the main means of providing guidance to civil aviation and air traffic control. Proper organization of communication is one of the main conditions for ensuring the safety and regularity of aircraft operations as well as the production activities of enterprises and civil aviation organizations. The new networks will focus on significantly improving the quality of service. The basis for their construction will form SDN networks. Therefore, the chapter analyzed the advantages and disadvantages of two SDN implementing methods. It was developed the mathematical method to assess their complex effectiveness, which considers QoS requirements of implementing service through special weights for scalability, performance, and packet delay. There were simulations of overlay networks by using soft switches to verify the adequacy of the proposed method. The results showed that the use of SDN networks more efficiently by using IP networks for large volumes of traffic and with a large amount of network equipment.

Successful business involves making decisions under uncertainty using a lot of information. Modern modeling approaches based on data science algorithms are a necessity for the effective management of business processes in aviation. Data science involves principles, processes, and techniques for understanding business processes through the analysis of data. The main goal of this chapter is to improve decision making using data science algorithms. There are sets of frequently used algorithms described in the chapter: linear, logistic regression models, decision trees as a classical example of supervised learning, and k-means and hierarchical clustering as unsupervised learning. Application of data science algorithms gives an opportunity for deep analyses and understanding of business processes in aviation, gives structuring of problems, provides systematization of business processes. Business processes modeling, based on the data science algorithms, enables us to substantiate solutions and even automate the processes of business decision making.

The information technology that is researched in the chapter provides a spatial overlay of the images received by the camera of an unmanned aerial vehicle (UAV) and raster maps of open aerial photography services. Such software helps to solve issues of actualization of maps, observation of agricultural field yields, creation of terrain photo planes, monitoring, etc. "Frames and a Map Overlay Tools" is software developed in C# using .NET4.0. All algorithms that were used during the development of the complex are described in detail, as well as the flow diagrams of the data utilities from which the complex is composed. Despite the fact that the testing of this complex has shown poorly high speed in real time, the estimates will allow the possibility of its interactive use under conditions of further refinement.

The chapter considers a description of developed control system for a group of unmanned aerial vehicles (UAV) that has a software capable to continue the flight in case of failures by using alternative control algorithms. Control system is developed on vision system by using methods of image recognition. Grouped coordinated flight of UAVs can significantly improve the performance of surveillance processes, such as reconnaissance, image recognition, aerial photography, industrial and environmental monitoring, etc. But to control a group of UAVs is a quite difficult task. In this chapter, the authors propose a model that corresponds to the principle of construction by the leading UAVs. In the case of using this model, the parameters of the system motion are determined by the direction of motion, the speed, and the acceleration of the UAVs' driving. The control system based on the methods of image recognition expands the possibilities of coordinating the group of UAVs.

The authors present a dynamic scenario-case approach to coordinated control of heterogeneous ensembles of unmanned aerial vehicles, which use coordination patterns of activity in similar situations described as spatial configurations affected by observed events. The method of obtaining deviations for approximate spatial configurations, which allows obtaining elements of the safe vehicle's trajectories. The method of qualitative safety assessment is presented. It uses a soft level topology to obtaining blurred boundaries of dynamic safety domains using fuzzy soft level sets and allows finding suitable compensations of vehicles' activity scenarios that can both keep the spatial configuration and satisfy all safety restrictions. The authors demonstrate that the proposed approach significantly reduces the computational complexity of problem solving and provides the acceptable performance.

In this chapter, the authors present the questions of aviation radioelectronic equipment operation. The structure of operation system is considered based on processes approach with adaptable control principles usage. Operation system contains processes of diagnostics and health monitoring. The authors consider the direct problem of efficiency estimation for diagnostics process, and main attention is paid to probability density function calculation for diagnostics duration. Simulation results were used for adequacy testing of these calculations. The authors also take into account the possibility of first and second kind errors presence. The inverse problem for diagnostics is defined and solved for mathematical expectation of repair time. In general case, the inverse problem can be solved for seven options of optimization.

This chapter describes two practical examples of sensors application on an unmanned aerial vehicle. The first device is a proximity sensor allowing users to measure the rotating angle of UAV's elevator. The second example discovers a measuring unit established on the UAV and processed measuring information for landing the UAV. To perform exactness control of unmanned aerial vehicles actuating mechanisms, the control system must be supplied by devices providing precision definition of values of current operation factors of those mechanisms.

Preface

The book *Cases on Modern Computer Systems in Aviation* is devoted to various applications of computer systems in aviation technology, which currently determine the effectiveness and safety of manned and unmanned aerial vehicles. These factors influence the solution of the most important economic and social tasks in society by the aviation component. However, the increase in the number of flights also leads to an increase in emergency situations in flight and on land. So for the last 6 years in the world, there have been 107 fatal airplane crashes, in which 3245 people died. In a year it is approximately 540 victims, in spite of the fact that safety issues of aviation equipment are controlled and regulated by the International Civil Aviation Organization (ICAO), as well as by the services of the countries.

An analysis of the set of dependent and independent factors of air crashes allowed us to establish that the cause of 75-80% of aviation accidents is the human factor, i.e. the level of knowledge, skills, and abilities of the crew are not adequate to the situation that arises in flight. The use of computer and computerized systems will allow controlling planes more efficiently than people do. Computers can successfully replace a person in the aviation field, where the qualified actions of the pilot, not only depends on the success of the flight but also the fate of passengers on board.

The use of computer systems will make it easier to analyze data that can change the organization of airline operations. The solution of this task will allow to reduce the time of preparatory works, to increase the efficiency of airspace management, and also to provide an individual approach to each passenger. Carrying out of diagnostic actions will allow avoiding breakdowns with accuracy to 99.5%, reducing the duration of idle time because of repair.

Computer network system is a complex mechanism, including sensors and a computer system, i.e. is a certain computing network that unites networks on different physical principles (wired and wireless) on the one hand and decides both computing and entertainment tasks on the other. These networks can be subjected to hacker attacks. One of the important applications of computer systems is cybersecurity. Today, we cannot allow a hacker to break into the aircraft control system by connecting his laptop to one of the aircraft's airborne networks and controlling the aircraft's executive systems.

Improving the controllability of the device and the level of service in this situation is possible only through the rational distribution, combination and duplication of the functions of the crew and problem-oriented "electronic assistants" that solve the problem of maximum efficiency of the control object application, effective survival, parrying failures and damages, which is solved by application intellectual technologies. The most common are the mechanisms of representation of knowledge and logical inference, agency approaches, neural network decision-making systems, training and staff training, systems for recognizing management situations.

Recently, flying robots, called UAVs, have been widely used in various fields of activity. This is associated with the minimization of risks and the cost of operation, the rapid training of maintenance personnel, more opportunities than piloted vehicles. The development of computer electronics makes it more accessible and quick to create and use a mini UAV. There are traditional areas of their application, namely, monitoring of surface space, reconnaissance and rescue operations, delivery of goods and goods. New areas of activity include the organization of communications and mobile communication networks in the territories where the organization of communications and access to Internet resources for some reason becomes an expensive or temporarily inaccessible problem. The increase in the number of such machines leads to the need for the organization and joint operation of several devices. This leads to new technical challenges for communication, management, and navigation.

The volumes of solved problems change the architecture of the computational aviation systems, they become multiprocessor. There are two main areas of application of the described systems: real-time processing of transactions (OLTP, on-line transaction processing) and creation of data warehouses for the organization of decision support systems (Data Mining, Data Warehousing, Decision Support System). The system for global corporate computing is, first of all, a centralized system with which virtually all users in the corporation work, and, accordingly, it must always be in working order. As a rule, solutions of this level are established in companies and corporations, where even short-term network downtime can lead to huge losses. Therefore, for the organization of such a system is not a suitable ordinary server with a standard architecture, it is quite suitable where there are no strict requirements for performance and downtime. These systems for global enterprise computing are characterized by such characteristics as increased performance, scalability, the minimum allowed idle time.

Chapter 1, "Artificial Intelligence Systems in Aviation," explores the possibilities of artificial intelligence (AI) tools, including software flight control systems, air traffic control, diagnostics, and others. A knowledge-oriented approach is proposed to implement the idea of a smart aircraft. Based on preliminary studies and identified errors, the study of information and its impact on the transport business, the structure of the mobile search and booking platform for airlines based on the agency approach is proposed.

Chapter 2, "Artificial Neural Network for Pre-Simulation Training of Air Traffic Controller," presents a four-layer model of a neural network to assess the correctness and timeliness of decision-making by air traffic services specialist during the training of aviation specialists. On the first layer, exercises are set for solving a potential conflict situation, the second one takes into account the physiological characteristics of the listener. The third layer takes into account the complexity of the exercise, depending on the number of potential conflict situations, on the fourth layer, the quality of the task is assessed by the listener.

Chapter 3, "Using Unmanned Aerial Vehicles to Solve Some Civil Problems," is devoted to the use of UAVs for non-military purposes. The design features of various UAVs, their control functions, energy requirements, safety problems and the ethics of their operation are considered. Parameters of UAVs that should be provided for their use in each specific case are also analyzed while minimizing the necessary financial resources.

Chapter 4, "Control of Information Stream for Group of UAVs in Conditions Lost Packages or Overloading," discusses the problem of managing the flow of information coming from a group of unmanned aerial vehicles over a radio channel. The inevitable loss of data is compensated by the repetition of lost packets or the reconfiguration network. Modern methods of flow control suggest using the ARQ mechanism based on the sliding method window. The consequence of these problems is a partial or total loss of its performance, which is manifested in a decrease in the initial network bandwidth. Part of the problem of network recovery is solved with the help of routing mechanisms that lead to a reconfiguration of the network due to the elimination of faulty nodes. Management of network congestion can be solved using known routing protocols, such as OSPF, IS-IS, RIP, etc.

Chapter 5, "Software-Defined Networking in Aviation: Prospects, Effectiveness, Challenges," discusses software-configurable computer networks, the application of which is aimed at improving the quality of service. A method for estimating their overall efficiency is shown, which takes into account the requirements of QoS to implement the service through special weights for scalability, performance and packet delay.

Modern approaches for the effective management of business processes in aviation, based on the algorithms of Data Science are presented in Chapter 6, "Data Science Tools Application for Business Processes Modelling in Aviation." The science of data includes the principles, processes, and methods of understanding business processes through data analysis. This chapter discusses linear logistic regression models, decision trees as a classic example of controlled learning and k-tools, and hierarchical clustering as uncontrolled learning.

Chapter 7, "An Information Technology of the Aerial Photo Materials Spatial Overlay on the Raster Maps: An Information Technology," explores information technology that provides spatial imposition of images obtained by an unmanned aerial vehicle (UAV) camera and raster maps of open aerial photography services. On its basis, the software has been developed that can update maps, monitor agricultural crop yields, create landscape photographic plans, monitor, etc. Algorithms and diagrams of data flow that make up the complex are also proposed. Software developed in C # using .NET4.0.

Chapter 8, "Guidance Algorithm for Unmanned Aerial Vehicles on a Basis System of Technical Viewing," deals with the management of a group of unmanned aerial vehicles (UAVs) based on software capable of controlling the flight of the apparatus in the event of failures using alternative control algorithms. A management model with a "leader" is proposed.

Chapter 9, "Information Technology of the Scenario-Case Coordinated Control of Unmanned Aerial Vehicle Teams," presents a scenario approach to the coordinated management of heterogeneous ensembles of unmanned aerial vehicles that use coordination schemes for activities in similar situations described as spatial configurations affected by observed events. The method of deriving deviations for approximate spatial configurations, allow obtaining elements of the trajectories of a safe vehicle. The method of qualitative safety assessment is presented.

In Chapter 10, "Operation of Aviation Radioelectronic Equipment," the authors present questions of the operation of aviation radio-electronic equipment using adaptive control principles. The system contains the processes of diagnosing and monitoring the state of health. The authors consider the direct task of assessing the effectiveness of the diagnostic process, and the focus is on calculating the probability density function for the duration of the diagnosis.

Chapter 11, "Information-Measuring Technologies for UAV's Application: Two Practical Examples," is devoted to the study of two sensors on an unmanned aerial vehicle. The first is the proximity sensor, which allows you to measure the angle of rotation of the UAV. The second is the measurement information for the UAV landing.

In Chapter 12, "Ensuring the Safety of UAV Flights by Means of Intellectualization of Control Systems," the authors present an analysis of the reasons for the inadequate safety of unmanned aerial vehicles (UAVs) and further justify the approach to improving safety through the intellectualization of the functioning of the control system. Intellectualization is carried out by means of computer vision. The concept of constructing algorithms for visual estimation of the position of a UAV equipped with a computer vision system is proposed.

Chapter 13, "Application of Computer Modelling in Adaptive Compensation of Interferences on Global Navigation Satellite Systems," discusses the use of global navigation satellite systems (GNSS) in an interference environment. Unlike radar and communication, where a priori information about the transmitted signal is known, in GNSS the frequency structure of the useful signal is not known in advance, which excludes the possibility of using a number of widely known methods. The section proposes a method that does not use a priori information about a useful signal and a new direct method for computing the inverse correlation matrix of interference in adaptive antennas of interference compensators.

Chapter 14, "Computer Modelling of Autonomous Satellite Navigation Characteristics on Geostationary Orbit," describes a model of the radio navigation field in which satellites off-nadir are used. This model makes it possible to assess the availability and accuracy of the characteristics of autonomous satellite navigation in space up to the geostationary orbit in order to provide the necessary research data.

Chapter 15, "Design Features of High-Performance Multiprocessor Computing Systems," analyzes the ways of developing high-performance computing systems. It is shown that a real breakthrough in the development of parallel computing technologies can be achieved by developing an additional (actually basic) level in the hierarchy of hardware capabilities of multiprocessor computing systems MPP architecture, personal computing clusters. The hardware pyramid of parallel computing technologies is studied in the form of personal computing clusters.

Chapter 16, "Realization Features of System Software of Multiprocessor Computing Systems," explores the problem of using and configuring the system software of multiprocessor computer systems. These studies also cover new ways to remotely access processor memory using RDMA technology for InfiniBand technology. It is shown that the RDMA principle, as well as the formation of a separate computer network in the data exchange environment and the implementation of VLAN mechanisms, allow data to be transferred between the nodes of the multiprocessor computer without additional buffering. This approach does not require an active operation, libraries or applications on those nodes of the system that are requested in memory.

Chapter 17, "Cybersecurity of Critical Aviation Information Systems," is dedicated to developing formalization methods for identification and security objects of critical information infrastructure (CII) in civil aviation. The analysis of modern approaches to the CII identification was carried out and this gave a possibility to determine technical weaknesses. The identification method was proposed and it gives a possibility to determine elements of CII field, mutual influences and influence on functional operations of critical aviation information system. Apart from, it was developed approach to creation of special software, which can be useful for CII elements identification and also for determination its influences on functional operations. Also some aspects of cybersecurity ensuring for identified critical aviation information system were described.

Chapter 1
Artificial Intelligence Systems in Aviation

Ramgopal Kashyap
Amity University Chhattisgarh, India

EXECUTIVE SUMMARY

The aim of this chapter is to research and fundamentally evaluate counterfeit shrewd frameworks to recognize for outperforming human insight in the flights and its conceivable ramifications. How artificial intelligence (AI) makes current airship framework incorporates an assortment of programmed control framework that guides the flight team in route, flight administration and enlarging the security qualities of the plane, and how building aircraft engine diagnostics ontology, air traffic management, and constraint programming (CP) is useful in ATM setting. How flight security can be enhanced through the advancement and usage of mining, utilizing its outcomes and knowledge-based engineering (KBE) approach in an all-encompassing methodology for use in airship reasonable outline, is discussed. The early recognizable proof and finding of mistakes, the study of huge information and its effect on the transportation business and enhanced transit system, the agent-based mobile airline search, and booking framework using AI are shown.

INTRODUCTION

This chapter will addresses challenges with Artificial Intelligence (AI) systems in aviation; it could likewise mean capacity anticipating and cautioning of approaching disappointment in computerized motor screen information. Climate estimating is somewhere else where AI will bear some significance with aeronautics. Pilots require significantly more than simply climate picture and diversionary landing strip information. Existing exploration in AI is a bit of research by each examination association, classes, and blended exchanges. Regardless of cynics, aircrew can watch inspirations to be amped up for AI which will engage planes to outline proactive and choose, because of machine learning and neural systems. Till now, all aircraft structures were to empower a pilot to rehearse power over carrier and systems. The accompanying stage is veritable essential authority endeavors. This will require significant machine learning and neural frameworks to make exceptional estimations that undertaking to 'think'

DOI: 10.4018/978-1-5225-7588-7.ch001

like a human. Thus, the point of this part is to give an outline of how the Interactive Fault Diagnosis and Isolation System (IFDIS) uses a control based ace structure made using gathering information from reports and ace appeal from specialists. The execution structure will in like manner supplant particular experts. The structure empowers the general workers to talk with the system and avoid slip-ups, wrong ends or addressing one of the specific experts. The Air Traffic Controllers (ATC) prepare by offering headings to the manufactured pilots and correspondingly for pilots to react to the ATC. The projects fuse the discourse programming made by utilizing neural systems. AI is about 'man-and-machine'; not 'man-versus machine'. The field of mechanical technology is firmly identified with AI; knowledge is required for robots to have the capacity to deal with so many errands as protest control and route. Issues to unravel incorporate possess localization, mapping what is near and movement or way arranging. Two or three trusts that human features, for instance, fake insight or a fabricated personality may be required. Starting late, the growing air development asks for, and the present airplane terminal direction help capacities are limited, contradicting interest and supplies have ended up being continuously prominent.

ATC structure is a staggering structure; the usage of framework proliferation of an aeronautics expert system is a basic research gadget. Existing proliferation instruments have two issues, one, in perspective of single focused hard to do broad scale reenactment of minute amusement; Second, the nonappearance of a straightforward controller reenactment association limits (Hwang, Kim and Tomlin, 2007) for these two request, the appropriated man-made consciousness multi-specialist advancement in air terminal direction diversion; and used Java lingo to develop a national avionics expert system in perspective of the field of the main model multi-operator common control propagation. Close by the relentless change of basic flight, the significant scale advancement of various plane terminal workplaces, the central air terminals are changing from the primary single-air terminal area to the multi-air terminal district with a particular ultimate objective to achieve the examination of the honest as far as possible assessment of the jumbled multi-air terminals terminal zone, all the more great and complete the process of mirroring game plan of the multi-plane terminals terminal region ought to be made. A Multi-agent system (MAS) is an automated structure made out of different participating shrewd Agents inside an area. Multi-operator systems can be used to deal with issues that are troublesome or unfathomable for an individual specialist or a strong structure to get it. Information may fuse some methodic, down to earth, procedural or algorithmic chase.

Air development working model of multi-air terminal zone, which relies upon the recurring pattern question of terminal zone confine and the status of research on air movement progression, was inspected by applying appropriated artificial intellectual prowess multi-operator theories and methodologies, and the working strategy for the multi-gather oral joint efforts between the aircraft, controllers, and the air terminal and so forth was pondered. The multi-specialist framework Simulation mode is created, and the specific arrangement models of some canny operators, for instance, flight operator, controller specialist and air terminal control districts specialist were presented (Ma, Tao, Zhu and Lü, 2014). In general framework and operational plan of the multi-air a terminal area air action sharp generation system was made, which relies upon the structures of the Multi-operator and finally, a foundation which is for investigating the real action conditions of the multi-air terminal locale and the affirmation of the air development keen multiplication game plan of the multi-air terminal district has be laid. The inspiration driving instrument flight procedure is to guarantee the security and improve the profitability of air movement errands in the terminal district, and flight approach design is a coherent organizing and sensible layout work for the passage and departure air courses and the framework involved them.

BACKGROUND

The flight agent holds the present flight direction depiction as a flight plan and AI assumes the profitable job in this procedure. A bit of the discretionary sources suggested for this investigation join relationship for the headway of man-made consciousness, European Association for Artificial Intelligence, Canadian Artificial Intelligence Association. The shipper contributions have furthermore been contemplated to choose the market division. The base up framework has been used to arrive at the general size of the AI in flight grandstand from the salaries of the key players.

Air Traffic Flow Management

Air Traffic Flow Management (ATFM) with its initiation, content, investigates goals and change is portrayed first. Consequent to separating the unequivocal air transportation issue inviable, the four times of the ATFM technique in the idea of activities are cleared up finally, the paper demonstrates the administrator based model with it particular methodology and the stochastic entertainment exhibiting speculation with detail. Research has been done to multi-pro based communitarian flights organizing issue. Introductory, a staggered network situated flight organizing technique is set up by strategies for a planned exertion between different air action organization units and carrier's assignments centers. The administrator based showing procedure is used to build up the inside showing and considering modules air course development organization units (Grabbe, Sridhar and Mukherjee, 2010). Research has been done on multi-authority based agreeable air courses organization issue. Introductory, a natural communitarian air courses organization framework is set up by techniques for a planned exertion between air courses movement organization units and carrier's undertakings centers. At that point, the specialist based demonstrating strategy is utilized to build the center motor and thinking modules of airships by making utilization of the point mass model and Proportional subordinate Controller. The scientists at that point built up a Multi-Agent system JADE (Sandita & Popirlan, 2015) at last, the thought will be connected to the region of constant operational control of Multi-Agent, for the proposed improvement of a model reenactment framework configuration, including practical module division, manufacture a personal computers (PCs) organize interchanges. To start with, the proposed procedure and sending of airplane kinds of exercises, based controllers strife and organization calculations; enhanced the seriousness of the contention examination calculation; proposed airship agent, the programmable inward structure of ATC robotization framework controllers agent and the agent. At long last, the model framework was tried to check, utilizing three PCs associated by means of a system. Utilize normal control imperative scene-arriving flying machine merging exchange terminal zone. Test outcomes demonstrate that three PCs were running airplane agent, Agent ATC robotization frameworks and controllers Agent can be continuous intuitive correspondences system, and more flying machine under the charge of the two controllers Agent accomplish a dynamic drop stature, settle basic clash can be given two flying machine neighboring parts.

Machine Learning and Traffic Management

For instance, what you get out depends upon the idea of what you put in, so picking the right things to measure regardless, picking the right counts and a short time later meticulously separating the results are key. That is the reason a machine learning approach in which the reasoning for the specific figure yield is joined, in like manner lessening the 'disclosure' concerns. While we have been to a great degree happy with our hidden results, we are continuing to hone the quality and accuracy of our showing with the desire that it might then be used to both predict and a short time later keep up a key separation from possible future prosperity events before they even happen following the basic accomplishment of our work with Swanwick) we've been researching diverse behavior by which we can utilize machine learning over the relationship with the introduction of a comprehensive model. It's being delivered to empower us to all the more likely appreciate the trial of rising improvement in flying surge hour gridlock and the impact it could have on our task (Jain and Kumar, 2017). This including model sees us take our development desires up to 2024 and, using bespoke computations and noteworthy data from the assignment, apply machine making sense of how to guess the potential impact of the climb in surge hour gridlock per section on delays, prosperity and nature if as far as possible and techniques for working proceeded as before by then have three examination bunches set up to separate the data the estimating and business analysis assemble looks and deferrals, the environment and airspace amass assesses the biological figures, and our safety aggregate examinations the peril to security and furthermore giving us the 'groundbreaking strategy' of what's in store, we moreover trust our comprehensive model will assist us with bearing down and explore how our advancing and prospective change endeavors could exclusively influence capability, prosperity and the earth (Shevchuk, 2010).

The ability to demonstrate fluctuating timings and augmentation for adventures, and advance fundamental authority to help meet the future regulatory targets and outside NATS, the yields could similarly be used to show the protection to the key game plan and boss in the flight business, and to our customers. For instance, judicious data of a 'does nothing' circumstance has exhibited the vital importance of modernizing the structure of airspace in the UK. AI and machine learning is emphatically something that we as a business are getting a handle on and examining, it could have a section to play in helping us thoroughly improve security, viability and lessen our biological effect. It is early days, the work will logically exhibit the potential focal points AI and machine learning has for our business, our customers and industry assistance. A year prior machine learning with empower us to predict the likelihood of potential security events, for instance, carrier level busts, or airspace infringements in our London Terminal Control task (Almubayed, Hadi and Atoum, 2015) by using bona fide data from 2015 to the present day, we use PC counts to find possible associations between prosperity events and elements, for instance, high action volume, plane terminal runway course, atmosphere conditions and that's only the tip of the iceberg. Since March 2017, we've been using this data to show seven days by week check to our partners at Swanwick Center that can alert them if our showing prescribes helpful action could keep up a key separation from a potential issue days before it might happen. The exactness of those results has been very extraordinary, running at in the region of 60 and 80 for every penny, so it's sensible there is an impetus to be had, in any case, there is still some way to deal with scattering the perspective of AI as starting from an unremarkable 'dull box'. It's an impressive measure to do with building trust, which can simply start from the precision of the yield and through the extra regard we incorporate by ace human examination of the information. While machines offer the awesome favorable position of having the ability to process and spot things in gigantic volumes of data in a way that our brains could never do, the results just really wake up with the choice of a human touch.

AI in Aviation Industry

The automated thinking in aeronautics exhibit was regarded at USD 112.3 Million out of 2017 and is most likely going to accomplish USD 2,222.5 Million by 2025, at a CAGR of 46.65% in the midst of the figure time period. The main issue driving the improvement of the AI in the flying promote fuse the use of gigantic data in the flying exchange, critical addition in capital hypotheses by flying associations, and rising gathering of cloud-based applications and organizations in the flight business ("Aviation Human Factors Related Industry News 1 sections of this area are ordered from "Flying Human Factors Industry News" and replicated with authorization of Roger Hughes.", 2015). The report goes for measuring the market size and future improvement capacity of the AI in the flying business segment in view of the offering, development, application, and geographic territory. It goes for perceiving the noteworthy market examples and components driving or constraining the advancement of the AI in the flying business part and its diverse submarkets. Additionally, the report separates conditions in the market for accomplices by perceiving the high-advancement bits of the AI in flight exhibit, intentionally profiles key market players, and thoroughly explores their market size and focus capacities in each fragment. The analyst framework used to measure and gauge the AI in the flying business area begins with getting data on key dealer salaries through discretionary research. The flight control components are given in figure 1.

AI Advance Carrier Sites' Valuing Frameworks?

Today, AI can powerfully change the look and feel of a site continuously, as voyagers draw in with it, to drastically support transformations, regardless of whether it's to offer a seat update, a more straightforward flight or exceptional offers for their trek. Also, it can do this where carriers require it most, for return guests or clients enlisted in dedication programs. As of not long ago, most carriers have been utilizing A/B testing to survey two distinct varieties of web architecture keeping in mind the end goal to enhance the client experience and increment changes. While A/B testing can be a compelling experi-

Figure 1. Flight Control Components

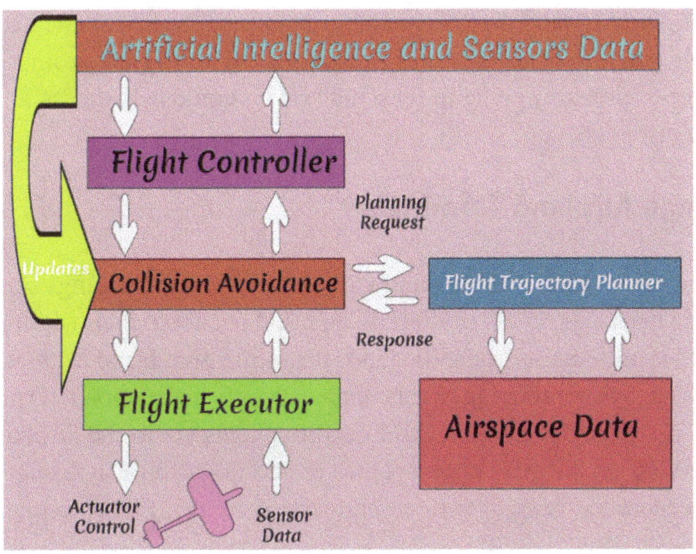

mentation apparatus, just 1 out of each 7 A/B tests brings about a positive result, making it an asset and time-serious system. Today, carriers can utilize AI to test thousands or even a huge number of plans (be it content, symbol, picture or catch shading changes) in a similar measure of time and see a 40% to half increment in transformations. AI has likewise indicated much accomplishment in anticipating the best time to purchase and offer stocks on account of its capacity to take after a total choice circle in light of watching what individuals are purchasing, arrange itself by contrasting the market with its portfolio, at that point choose and follow up on what it ought to do, purchase, hold, offer and so on. In a comparative design, AI could be utilized to choose continuously what costs it should put on flights, by watching what outside market powers are affecting expenses and costs. It can arrange itself by surveying plane fuel costs, news nourishes and traveler request and, continuously, settle on a choice regarding what cost to allot to a specific seat. It would then be able to modify that valuing as factors change after some time. How would you imagine the capacities of chatbots progressing in the following five years?

Chatbots are awesome at client benefit yet they're restricted in what they can do today. One of the huge difficulties chatbots confront is that explorers that communicate with them frequently know they are a bot and trust they have some superhuman capacity (Reynolds, 2017). This makes confusion between what benefits individuals expect and what they get. In the following 3 to 4 years, we'll see chatbots get substantially more brilliant as they begin in different types of AI to help offer items or meet client benefit needs. One illustration we could see is chatbots joining PC vision to enable individuals to find carbon copy trips at less expensive yet similarly engaging goals. For instance, if an explorer gives the chatbot a photo of a shoreline, the AI can take a gander at all the highlights of the picture and see what travel goals intently coordinate the substance, and after that show those alternatives.

Will AI Take Care of the Issue of Overbooking?

AI may not take care of the issue totally, but rather it could help essentially lessen the issue, particularly with regards to boarding travelers. The test carriers confront today is that their models and staff can't precisely anticipate what number of individuals won't appear for a flight. By dissecting past long-standing customer data and general chronicled traveler information, and changing climate designs for certain well-known courses, AI could foresee the probability that specific travelers won't show up or will swap to another flight. The AI could then surrender ground staff to-the-minute data on what number of individuals is probably going to board. It could even foresee which flyers ordinarily ask for updates or what number of workers is probably going to fly standby. This could help the issue of removing travelers from planes that have just loaded up.

How Will AI Change Airplane Terminals?

AI is as of now being utilized at air terminals and via carriers for various things. Joined Airlines has tied up with Amazon Alexa to enable clients to check in and find out about their flights (Zhang, Li & Wang, 2012). Delta Air Lines is utilizing self-benefit stands that utilization facial acknowledgment innovation to confirm client character by coordinating client countenances to international ID photographs. Different aircraft are utilizing AI (Jeffery & Rhodes, 2011) before clients get to the airplane terminal, to upsell devotion program travelers on business class updates or additional stuff by having AI naturally outline the site checkout procedure to speak to various partners of clients. Flight re-planning is the situation when a plane is flying and with the help of an AI flight planner trajectory is requested for the modified plan as given in Figure 2.

Will Pilotless Flights Turn Into a Reality?

Flights today are pretty much pilotless nearly everything except for departure and landing is controlled by locally available PCs, so it's a moderately little mechanical obstacle to overcome. There are without a doubt other, all the more squeezing reasons why flights are not completely self-ruling today, similar to the potential for digital hacking ("US military develops pilotless helicopter", 2012). Carriers should be greatly cautious and watchful when they think about the reconciliation of new advancements. It will be a while until the point when we see completely self-sufficient flights.

Are Robots More, or Less, Dependable Than People?

It depends upon the errand for general data issues where individuals can't remain mindful of the reliably propelling a moment back changes, AI is a phenomenal game plan, as per the booking cases above, impact evasion is finished with the man-made reasoning that is appeared in Figure 3.

For different issues that require more thought for human feeling, such as communicating with and serving travelers in flight, robots won't supplant people whenever soon! The aeronautics industry remains a genuine behemoth in the financial world, as it keeps on developing at an inconceivable rate year-over-year. Honestly, it experienced something of a splendid age in the region of 2009 and 2014 when the business created at a compound yearly advancement rate of around 9.5% going before accomplishing a joined estimation of $751 billion. While the business experienced a short abatement in the region of 2015 and 2016, it valued a stellar 2017 and is set to achieve record top business salaries of $824 billion preceding the present a very long time over. A considerable amount of this improvement relies upon upheld ask for, as client spending continues growing over the globe.

Correspondingly, the market has moreover bloomed with the absence of progression and new data examination propels, which have helped brands to enhance the customer experience, redo refunds and make more secure aircraft for voyaging. Colossal data and AI is starting at now controlling invigorat-

Figure 2. Flight re plan initiated with artificial intelligence

Figure 3. Collision avoidance through artificial intelligence

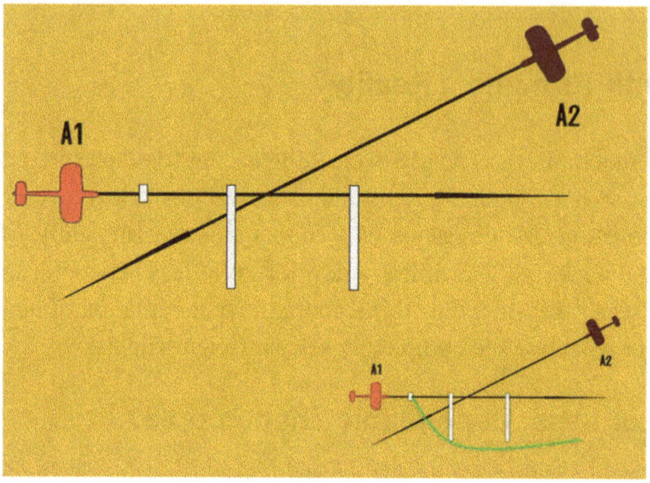

ing improvements in the flying business. The main issue driving the advancement of the AI in flight promote fuse the use of tremendous data in the plane business, enormous addition in capital hypotheses by flying associations, and rising apportionment of cloud-based applications and organizations in the air transportation industry. In this report, the AI in the flying business segment has been distributed in light of development, offering, application, and geology. Among all contributions, programming holds the greatest offer of the general AI in flight promotes. This is inferable from the progressions in AI programming for applications, for instance, observation, flight assignments, and plane terminal tasks. Among all developments, machine learning is most likely going to hold the greatest offer of the AI in air transportation promotes in the midst of the check period. Machine learning's ability to accumulate and handle immense data, close by its extended ability to perform previously boundless estimations, is invigorating the improvement of the market for machine learning.

Pitch Control of an Aircraft Using Artificial Insight

The quick progression of flying machine outline from the exceptionally restricted capacities of the Wright siblings first effectively plane to the present superior military, business and general avionics flying machine require the advancement of numerous innovations, these are streamlined features, structures, materials, drive and flight control. The advancement of programmed control framework has assumed an essential part in the development of common and military avionics. Identification of the collision time using AI is shown in Figure 4.

Present day airplane incorporates an assortment of programmed control framework that helps the flight team in route, flight administration and expanding the dependability attributes of the plane. For this circumstance, an autopilot is outlined that control the pitch of airship that can be utilized by the flight team to decrease their workload amid cruising and enable them to arrive the airship amid antagonistic climate condition ("Aircraft Pitch Control System Using LQR and Fuzzy Logic Controller", 2016). The autopilot is a component inside the flight control framework. It is a pilot alleviation system that helps with keeping up a state of mind, heading, elevation or traveling to route or landing references. Planning

Figure 4. Identification of the collision time using AI

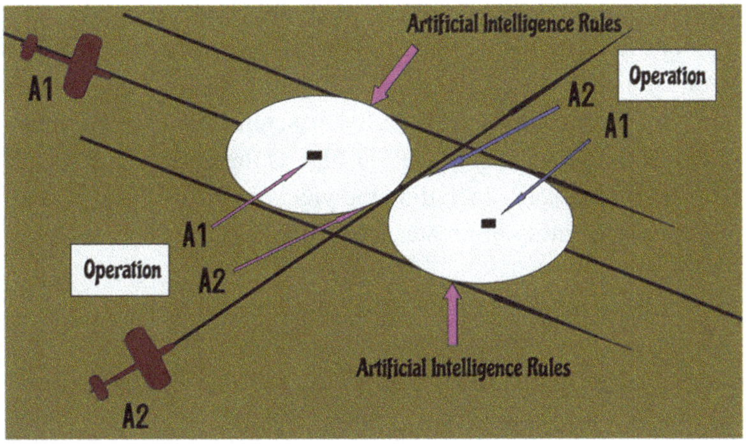

an autopilot requires control framework hypothesis foundation and information of steadiness subordinates at various heights and Mach numbers for a given plane. One of the significant issues of flight control framework is because of the blend of nonlinear flow, displaying vulnerabilities and parameter variety in describing a flying machine and its working condition. The airship movement in free flight is to a great degree entangled.

Fuzzy Logic Control (FLC)

Here is a shallow prologue to the fluffy rationale and it essential constituents as respects control of dynamic frameworks. Likewise, thought about how to construct such controllers in Simulink condition of MATLAB. Reproductions of the assembled controllers and their outcomes are moreover introduced. Fuzzy rationale controllers fall into the class of Intelligent Control Systems. A savvy control framework consolidates the procedures from the fields of Artificial Intelligence with those of control building to outline independent frameworks that can detect, reason, and plan, learn also, act in a keen way. Wise conduct is in this manner the capacity to reason, the design also, and realize which thusly expects access to learning. Counterfeit consciousness is a side-effect of the Information Technology (IT) transformation ("Speed Control of a Train using Fuzzy Logic", 2017), and is an endeavor to supplant human insight with machine knowledge. A clever control framework joins the systems from the fields of AI with those of control building to outline independent frameworks that can detect, reason, design, learn and act in an insightful way. Such a framework ought to be ready to accomplish supported wanted conduct under states of vulnerability, which incorporates.

Fuzzy PID Autopilot

A PID fuzzy controller is a controller that takes blunder, the summation of mistake and rate of progress of blunder as sources of info. Fluffy controller with three sources of info is troublesome and difficult to execute, on the grounds that it needs countless and memory. Generally, to speak to PID-FLC, it is required to outline a fluffy surmising framework with three data sources that speak to the relative, subordinate, and essential segments, and every single one of them can have up to 8 fluffy sets (Gan, Tang & Gao, 2013). Thusly, the most extreme number of required fluffy principles in any circumstance is 8x8x8 =512 rules. Yet, for this examination, just 3 fluffy sets were utilized for the run the show base, in this way the most extreme tenets it would yield would be 3x3x3=27 if three sources of info were to be executed. The PID-FLC can be developed as a parallel structure of a PD-FLC and a PI-FLC, such that the info motion for the subsidiary picks up to the PD-FLC is the control circle yield flag. The yield of the PID-FLC is framed by arithmetically including the yields of the two fuzzy control pieces, proposed by Leonid. This will lessen the quantity of greatest standards conceivable to 8x8 +8x8= 128 standards.

Enhancing Flight Wellbeing Through Information Mining

The on-screen characters in the flying business have watched that the huge measures of information gathered and put away by various on-screen characters in the field, contain concealed data that might be extremely profitable however which can't be found with purported conventional examination strategies. There are no preparatory desires about the presence of these 'chunks of learning' and that is the reason it isn't conceivable to discover them with known techniques. As such, we don't comprehend what we don't have the foggiest idea, accordingly what we are hunting down are designs that may uncover more information that could enhance well-being (Pedrycz, 2011). Information mining apparatuses and strategies give an answer to the difficulties specified. The arrangement can be viewed as twofold: other than giving an examination of printed information, information mining likewise enables it to be joined with numerical information. As the flight security information gathered incorporate both organized and account handle, the apparatuses and techniques, for example, content mining devices, for dissecting writings have just barely been created. The content mining ability seemed substantially later than information mining which is utilized for numerical information. Concerning information, the examination of which has been simple utilizing inquiries from databases and running their outcomes through devices that deliver charts, information mining gives more modern numerical investigations while looking for something that has not been predicted. Information mining helps airplane to take decision for the avoidance of collision and change the path immediately as shown in Figure 5.

Figure 5. Flight plan

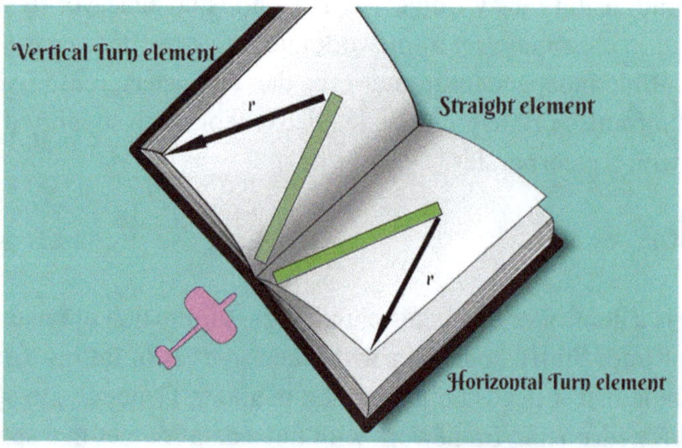

Communicated on a general level, two sorts of information are created in the carrier business the flight information recorders produce computerized information following the flight and the deviation reports composed by pilots and other workforce create printed information. As indicated by Robb, the computerized information is taken a gander at by a few elements, but since not as much work has been done on the free content information its examination needs more consideration.

RELATED RESEARCH ON AVIONICS SECURITY

A considerable measure has surely been composed about aeronautics security, in any event in the significant avionics nations, e.g. the US, the UK, Germany, and France; a considerable lot of them have their own establishments for flight security working in close contact with carriers and experts. In any case, in the region of breaking down flight security reports and applying information mining, apparatuses, and assets, look into (preceding this work) is fairly rare. In the Nordic nations, just a couple of theses concerning, or if nothing else alluding to, flight wellbeing have been distributed. In May 2015, a database seeks in Melinda (2015), the Union List of Finnish Libraries, kept running with a specific end goal to discover Finnish theses concerning flight security. The Finnish, Swedish and English words were: ilmailu (flying in Finnish), lentoturvallisuus (flight wellbeing), lentäminen (flying), lentoliikenne (air activity), flygtrafik (air movement), flygsäkerhet (flight wellbeing), flying (flying), flight wellbeing, aeronautics, avionics security. The pursuit was performed utilizing a few questions and 88 matches were created when the copies were expelled. No theories concerning flight wellbeing and nearly taking after this proposition was found. Also, regardless of the inquiry words in Finnish and Swedish, the beginning of 22 of the indexed lists originated from outside the Nordic nations (Paukkeri, García-Plaza, Fresno, Unanue & Honkela, 2012). The previous Technical University of Helsinki is currently part of Aalto University furthermore, the main place in Finland offering advanced education in aeronautics. Along these lines, a library look covering all the changed proposition composes and different reports distributed there was performed utilizing the Aalto University Library Aaltodoc seek instrument utilizing similar pursuit criteria as with Melinda. The outcome was 107 reports and just a couple of them had some association with flight security from the perspective of this investigation, which means flight tasks. Typically they had a specialized approach, for example, airship support, programming advancement for aeronautics utilize, flying machine generation, and so forth.

DECISION SUPPORT SYSTEM (DSS)

Today, chefs are influenced by expanding intricacy and vulnerability in circumstances where they need to decide, which constrains them to assign complex quantitative models which surpass the abilities of the straightforward direct models that are customarily utilized. The many-sided quality and vulnerability of the information utilized as the reason for choices keep on expanding. This is the reason the model that depicts a basic leadership circumstance should increment too with the goal that it can catch the very non-direct connections between an arrangement of factors. Quick changes command the contemporary world, making the future questionable. Because of the globalization of issues, and in addition the interrelationships between frameworks, the impact of settling on wrong strategy choices have turned out to be more genuine, conceivably having cataclysmic outcomes (Vahidov & Kersten, 2004). These vulnerabilities can be viewed as existing in all strategy making circumstances and moreover, there are a few measurements of vulnerability and also powerless understandings of various qualities. One of the fundamental issues among high hazard enterprises, similar to avionics, is the level of sureness required to limit or even boycott destructive exercises. In this way, the prudent rule has increased expanding consideration as the requirement for more helpful methodologies towards vulnerability and obliviousness in regards to administrative choices has developed (Walker, Bokelmann & Klemperer, 2003).

The choices are settled on in view of a reader's comprehension of a circumstance as being one wellspring of conceivable mistake. Another point to be said is the process by which a choice is come to. In spite of the fact that the best choice has regularly been made, a bothersome result may show up because of occasions over which the leader does not have control. In naturalistic decision making (NDM), people having space mastery in settings like flying, atomic power and seaward oil process control settle on choices in conditions that change significantly and powerfully. Likewise, the choices must be made in a restricted space of time, the objectives may struggle and the dependability of the data sources may differ. Such choices are frequently made in groups in hierarchical settings, supported by the access devices or other data assets. The intricacy of components influencing basic leadership in the data age prompts administrators to depend on advanced data examination instruments that help basic leadership in business associations yet additionally in different sorts of associations. DSS are PC based frameworks whose capacity is to help choice making exercises and forms keeping in mind the end goal to tackle semi-organized or even poorly organized issues that frequently include numerous traits, targets, and objectives. DSS manage issues which depend on the learning that is accessible. Operationally they are intelligent frameworks or on the other hand subsystems and they depend on learning and hypothesis gathered from counterfeit consciousness, database examine, scientific demonstrating, choice hypothesis, administration science, and so on. DSS is for the most part utilized for any processing application which intends to enhance the basic leadership capacity of a solitary individual or a gathering of chiefs. Due to the consistently expanding multifaceted nature of transport frameworks, it appears as a hybridization of the reproduction models is required. The new difficulties for movement administration DSS show up from the nearness of the tremendous measures of information given by the new observing and prescient frameworks (Bobek, 1992).DSS device outlines comprise of a few parts and as databases make the fundamental wellspring of information modern database administration abilities are basic. These capacities need to incorporate access to inside and outside information as well as data and learning. Effective displaying capacities are likewise required with a specific end goal to deliver helpful and justifiable handling comes about for useful choices, an effective interface is required. The interface should be easy to utilize, empowering intelligent inquiries and the revealing and showing of the outcomes in realistic shape. The examination of DSS frameworks has concentrated on enhancing the proficiency of basic leadership by building up the innovation and moving forward the adequacy of the choices made. Effective devices for building DSS developed in the mid 1990s, On-line Analytical Processing and information mining can be specified as illustrations call attention to the part of the information mining instruments as choice emotionally supportive networks, demonstrating that because of the quick improvement of information handling and the expanding multifaceted nature of the issues included, innovative answers for choice process methodologies are progressively required. From this point of view, learning disclosure in databases (Pedrycz, 2011) is featured; the part of DSS is ending up progressively basic as a huge segment in the everyday tasks of the associations. As DSS advance and develop, the need for a system that typifies hierarchical components influencing their fruitful advancement and execution turns out to be clear. The structure introduced in permits DSS analysts and in addition specialists to have the capacity to group authoritative components

(Technical, Economic, People, and Strategic) that may affect the effective execution of DSS. The system is the consequence of an amalgamation of existing examination and the individual experience of the performing artists furthermore, depends on a top to bottom investigation of the execution of an expansive scale DSS called Fuel Management System (FMS) at Delta Air Lines. The assorted variety and many-sided quality of these components make it difficult for associations to comprehend their effect on DSS usage achievement. Brookes has exhibited an improvement system for DSS in light of a model of the basic leadership process and the idea of administrative work. As indicated by this improvement system, a DSS can be separated into four capacities called: attenuation, amplification, reference, and navigation/control. Each capacity is coordinated towards a particular intellectual undertaking by the chief. Sanders and Courtney related DSS accomplishment to three expansive components:

- Choice setting (level of issue structure)
- Level of errand relationship (level of connection with others)
- Level of errand limitations (level of chief specialist and self-sufficiency)

Furthermore, they found that best administration bolster was an imperative factor in deciding the achievement of DSS. Danger recognizable proof and danger administration are the center procedures in the administration of security. This implies the components which cause or are liable to cause hurt should be recognized and comprehended. The issue of flying wellbeing extensively interests the logical field the assault made with two commandeered carriers on the twin towers in New York in 2011 has essentially expanded enthusiasm for security matters. As per an investigation the most elevated number of passing's onboard producing circumstances is the loss of the control of the flying machine on the runway, both landing, what's more, taking off, and mistaken assumptions between the pilot and the control tower, some of the time caused by a flawed information of English by one of the parties. The theoretical plan is the beginning period of a flying machine configuration process where comes about are required quickly, both logically and outwardly, with the goal that the outline can be broke down and in the end enhanced in the underlying stages. In spite of the fact that there is no need for a Computer Aided Design (CAD) display from the earliest starting point of the planned procedure, it can be an additionally preferred standpoint to have the model to get the impression and appearance. Airship setups and high-loyalty investigation devices, for example, Computational Fluid Dynamics (CFD) and Finite Element Analysis (FEA) increment the level of trust in the outlined item (Hui, Hu & Shyue, 2008) besides, this implies a consistent change into a preparatory outline is accomplished since the CAD model can continuously be made more nitty-gritty.

Aircraft Design and AI

Aircraft design is a mind-boggling process that unites diverse controls to get an all-encompassing methodology. Present day flying machine has turned out to be more costly and the time taken to fabricate has expanded extensively. Figure 6 shows a change in the reasonable plan is expected to diminish the general improvement time and cost for a flying machine.

In the applied plan, the outcomes are required quicker both scientifically and outwardly with the goal that the outline can be altered or changed at the most punctual stages. The three primary plan organizes in an airship configuration process are a conceptual outline, Preliminary plan and detail plan after the detailed outline, the airplane is checked with model testing and full creation (Wittenberg, 2001). Distinc-

Figure 6. Training systems through AI

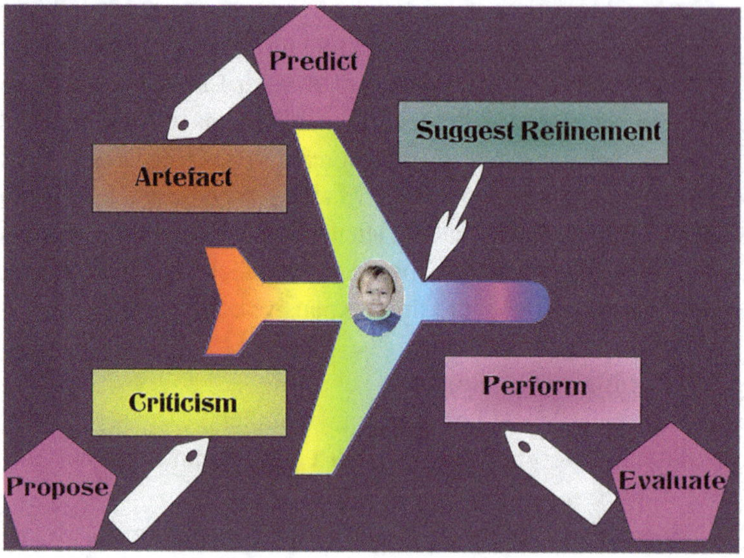

tive plans should be dissected and checked Knowledge-Based Integrated Aircraft Design, Time delay in airship ventures (Munjulury, Staack, Berry & Krus, 2015). A Handful of programming gadgets exist in the business, at universities and research centers. Some have a relationship with CAD programming, in any case, the affiliation is regularly not steady and they now and again work bi-directionally. This acquaintance was at first made to meet the necessities of the business world; be that as it may, it can moreover be associated with the setting of the present zone.

State 1, acknowledge what you know, can well be depicted as "just the same old thing new", inferring that the fundamentals and what are more the specific associations of the exercises of the affiliation are known and grasped. In this state, evaluating and declaring with dispersing is the mind-boggling works out, regardless of the way that an off-base inclination that everything is great with the world can be impelled. In State 2, in which it is fathomed what isn't known, it has been seen that movements are required remembering the true objective to help the exercises. The nearness of the issues is known yet in a demand to address them, help the examination is basic. Right when the issue zone is described, strategies like uncommonly designated or On-Line Analytical Processing (OLAP) is to a great degree effective in cognizance and estimating the game plans. The model can be particularly associated with flight prosperity and thusly to this examination.

As communicated beforehand, in the fundamental stage the business continues running as would be expected with standard enumerating. Customary uncovering about events and distinctive deviations is made, which is taken after by the examination and the reviving of headings. As indicated already, there might be an unrealistic conviction that all is great if there is no idea in regards to the nearness of an issue that may impact undertakings. In case the operational condition is protected, this sort of methodology can be seen as sufficient. In the second state, there is care around a darken (unidentified) issue and that is the reason uncovering and diverse activities ought to be created. There is basically an affirmation of weaknesses being accessible either in the errands or in the affiliation. The third state, data organization (Kashyap, 2018), is today normal to all affiliations, the more noteworthy the affiliation the more data is required and the more prominent the extent of its exercises is. The hugest state from the point of view of this examination is the fourth one in which 'the cloud isn't known. On a basic level, security factors are those that are in closest association with mischances, checking nice factors, like workforce capacities concerning pilots, air movement.

Learning Building Blocks

Main building blocks controllers are the air activity condition, flying machine capacities, climate conditions furthermore, capricious acts. Risk recognizable proof strategies are utilized as a part of a request to distinguish security factors. In present-day security examinations, diverse apparatuses and techniques are connected. The most generally utilized are measurable and inclined investigations, regularizing correlations and in addition reproduction and testing, master boards and money saving advantage examinations (Amine Chatti, 2012) all the danger recognizable proof strategies can be partitioned into two gatherings: 1) Reactive techniques 2) Proactive techniques. The main said to depend on the observing of patterns and additionally the moment examination of occasions concerning security. The last distinguish risks by breaking down the execution and capacity of frameworks with a specific end goal to find inherent dangers and potential disappointments. Cases of these are wellbeing checking and appraisal, operational security reviews, and so forth.

Information Management Process for Aviation

The larger amount of saving money association is advanced in learning yet refusal of imparting their insight into the lower level of representatives these outcomes in absence of skill obstruction in an association (Davison.et.al, 2013). Knowledge sharing activity may enhance the community devices or execute another index to share the learning. However a large portion of the association's shields themselves and their business insider facts from being unveiled (Zhang & Zhu, 2014). Information sharing should be possible with the assistance of intranet, web, GSM, sends entrances, and video conferencing and web indexes. To utilize the information suitably, appropriate inspiration and mindfulness is required. The absence of intrigue and nature are the obstructions in the appropriate utilization of information. One can discover the information from the particular one with the assistance of learning index. To store extensive estimations of data, the bank utilizes a number of SQL databases.

Manmade Brainpower in Flight Reservation Systems

Manmade brainpower is the key innovation in numerous of the present novel applications, going from keeping money frameworks that distinguish endeavored charge card misrepresentation, to phone frameworks that comprehend discourse, to programming frameworks that notice when you are having issues and offer proper counsel. These advances would not exist today without the maintained government support of principal AI explores in the course of the last three decades. The region of flight reservation frameworks is no special case to the presence of computerized reasoning. Numerous carriers have picked to strip the majority of their possessions to Global Appropriation Systems (GAS) because of which numerous frameworks are presently open to purchasers through Internet doors for lodgings, auto rental offices, and different administrations and additionally aircraft tickets (Seneff, 2002). An explorer or a movement operator can chalk out a schedule utilizing a gas which is a worldwide framework interconnecting aircraft, inns, travel operators, auto rental organizations, luxury ships and so forth. There are four noteworthy Global Distribution Systems, and they are Amadeus, Galileo, Saber, and WORLDSPAN. The Saber reservation framework is utilized by American Airlines and gloats a wise interface named Pegasus, which is a talked dialect interface, associated with Saber which enables supporters of getting a flight.

AI With Image Analysis and Biometrics

Biometrics is the science and advancement of assessing and exploring natural data in information development, biometrics escapes to progressions that measure and dismember human body characteristics, for instance, DNA, fingerprints, eye retinas and irises, voice plans, facial models and hand estimations, for approval purposes. In this examine zone of biometrics, we will center on unique mark catch, confirmation, and encryption. Biometric is a standard now that all workstations accompany biometric security choices that enable clients to store their passwords as biometric engravings and sign onto their gadgets utilizing their fingers instead of composing in passwords generally. Therapeutic pictures have turned out to be a critical wellspring of immense amounts of mind-boggling and high measurement information every now and again utilized for restorative finding, treatment appraisal, checking and arranging, and research (Kashyap, and Tiwari, 2017). Customarily, these pictures are straightforwardly translated through visual examination by a radiologist with the plan to enhance the interpretability of delineated substance, Be that as it may, this approach is unwieldy, tedious, mistake inclined and subject to weakness and diversion. In reality, it is very much recognized that such a technique requires a lot of abilities, information, and experience that may not generally be promptly accessible in this way making it infeasible. A superior inventive strategy is the utilization of computerized advances i.e. PC supported finding utilized in medicinal imaging for quick picture handling, and to supplement the conclusion of the radiologist (Kashyap, and Gautam, 2016; Juneja, and Kashyap, 2016).

The Capacity and Reprocessing of Filed Information and Images

The nonstop changes in both equipment and programming segments of therapeutic imaging innovations, for example, microarray pictures (Kashyap, and Gautam 2013), attractive reverberation imaging have seen an enormous development in current pharmaceutical (Juneja, and Kashyap, 2016). Picture division is the path toward allotting a propelled picture into various bits sets of pixels, generally called superpixels. The objective of the division is to disentangle or perhaps change the portrayal of a photograph into something that is more basic and less intricate to analyze. Picture division is by and large used to find things and purposes of restriction lines bend in pictures. All the more conclusively, picture division is the way toward naming a name to each pixel in a photograph to such an extent, to the point that pixels with a practically identical check share certain qualities. The database regard is differentiated and the biometric commitment from the end customer who has gone into the scanner and confirmation is either certified or denied one of the kind impression stages are executed as showed up in Figure 7. Along these lines, to see the use of biometrics as an astoundingly secure technique for executing security in a structure that customers' private and delicate data are being gotten to and need to keep out of the unapproved workforce to maintain a strategic distance from information misrepresentation. It has been seen from the written work that work has been done in the use of AI in-flight reservation structures and advances been used to keep up a key separation from character blackmail in portion.

Figure 7. Cased-Based reasoning

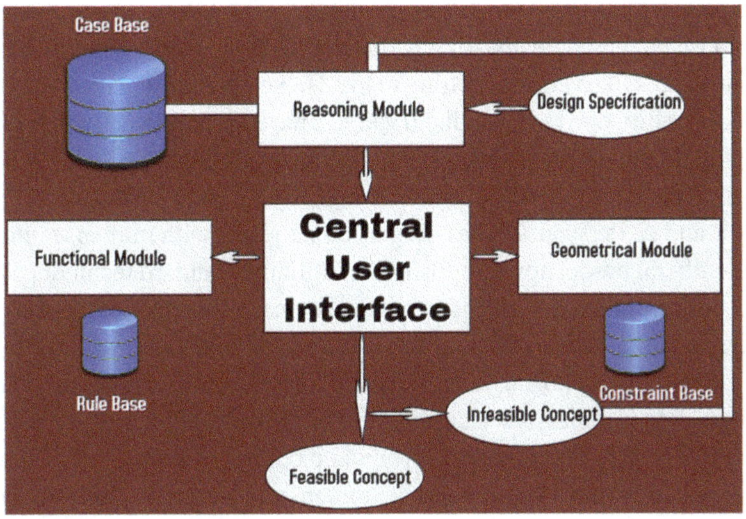

The whole above structure AI look computations, being used to play out the quest for flying machines with some understanding and besides security has been used to keep up a key separation from MasterCard theft in portion yet in the meantime the system needs information and competence in looking of airplanes where again the weight falls on the customer towards refining the interest, settling on decision in perspective of recouped comes to fruition. The system similarly has no office of looking in perspective of past comprehension or something to that effect. Grievously, diagram can't be formalized thusly. Thinking by learning can be completed with Artificial Neural Networks (ANN). An ANN includes an arrangement of centers related by methods for adaptable weights by setting up the framework with a considerable game plan of data yield consolidates, the system takes in the viable association between the data and the yield space. This sort of summed up learning can't be associated with the diagram issue in light of the way that the course of action space is pitiful and discontinuous. The third sort is best exemplified by Cased-Based Reasoning (CBR) cases are secured for a circumstance base to make a supply of issue plan mixes. Exactly when another issue is shown, CBR searches for cases with near issue depictions. Notwithstanding the way that the recuperated case, generally, does not completely fit the new issue, the recouped game plan may be a not too bad starting stage for empowering change and progression. The refinement with the other AI approaches is that CBR does not use summed up region adapting yet rather data which is locally authentic: the comprehended learning inside a case which relates an issue with an answer holds for that particular case.

The AIDA Framework

The AIDA system (AI Design of Aircraft) includes three modules and a central interface, CBR module. In this module case-based, thinking systems are completed to make a commendable beginning thought that can be adjusted in the utilitarian module. In this module manage based reasoning systems are completed to perform affectability contemplates the basic parameters of the thought. With these examinations, a handy thought is arranged. Geometrical module this module thusly models and pictures the thought. It uses incorporate based and prerequisite based showing systems. User Interface: This module handles the correspondence between the three modules (Torenbeek, 2000) the going with entries delineates the distinctive modules. For each module, existing gadgets have been used as found in Figure 8.

The nearness of advances like versatile getting ready, flowed preparing, a web of things, sensor-based structures and the accessibility of web in handheld gadgets have accomplished a time of a wide proportion of information, both made and unstructured, which is designated "Gigantic Data". The shot of managing this expansive data into basic and colossal data is being perceived by associations, affiliations, and affiliations (Kashyap, R and A. Persson, 2018a). Regardless, the test with immense information is that it is hard to oversee such a liberal proportion of information utilizing standard techniques. New contraptions, progressions, models, and systems are utilized to oversee monster information. Hadoop is an open-source structure used for preparing immense information. It is a perceivable scattered accumulating and figure condition which is utilized for anchoring and preparing for monstrous information (Kashyap, R and A. Persson, 2018b). Huge data is a massive accumulation of information which is made at an exponential rate in a wide assortment of affiliations and has wound up being difficult to oversee utilizing customary information association instrument The theory of immense information depends upon five V's: Volume: Large volume of information made each second by people, affiliations, machines, and whatnot. (Upadhyay, A., and Kashyap, R.2017) Speed: Speed at which information is being made. Collection: Various courses of action in which the information is open substance, goals, tweets, video, standardized tag, databases. Veracity: Correctness and exactness of information. Regard: Insights or data that might be conveyed by applying examination on titanic information.

Solutions With AI

As recommended by its name, the Case Based Resoning (CBR) module applies case-based thinking strategies to create a satisfactory idea from the outline details. These procedures empower the utilization of the planned encounter that is verifiably accessible in existing cases. Additionally, case-based thinking is a way to deal with learning, since the consequence of past plan sessions can be added to the case-base,

Figure 8. AI update utilizing knowledge management

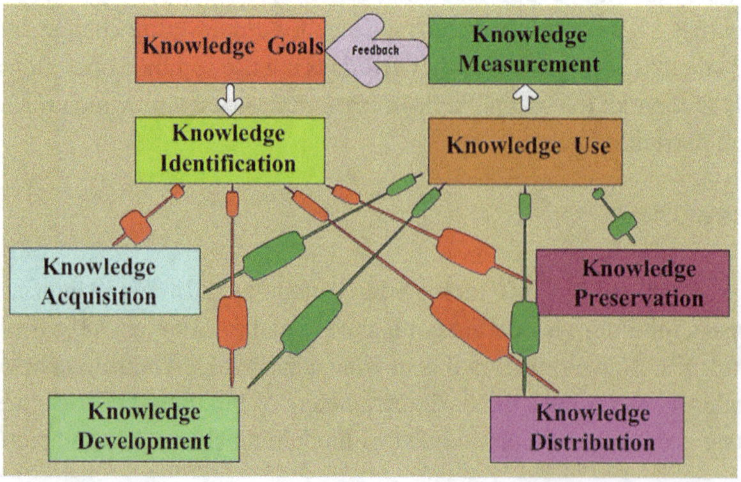

making it accessible for future outline issues. A total case-based thinking procedure can be considered as a cycle of four successive steps.

- **Retrieve:** Find cases for the situation base which looks like the issue of portrayal;
- **Reuse:** Copy case-information or join information in more cases;
- **Revise:** Evaluate the proposed arrangement; and
- **Retain:** Put effective 'educated case' for the situation base.

The issue depiction characterizes the 'new case'. In the Retrieve step, the case-base is looked for cases with information coordinating the 'new case'. The cases with most comparative information are recovered. In this progression, the coordinating procedure is generally basic. In the reuse step, information is replicated from a 'recovered case'. As a rule, the 'recovered case' does not totally coordinate the 'new case', i.e. the best coordinating case does not totally illuminate the issue. In that circumstance, the information of more 'recovered cases' can be joined. As it were, the best coordinating case is adjusted with information of other chose cases. This adjustment process requires space information and is exceptionally unpredictable. The aftereffect of the Reuse step, the 'comprehended case', is a recommended answer for the issue. It is assessed and repaired when fundamental in the revising step. The assessment procedure is regularly performed by numerical instruments. This procedure additionally requires area information, as does the repair process. The outcome is a 'tried/repaired case', or an affirmed answer for the issue. The learning perspective is actualized by including data about the affirmed answer for the case-base. The retain step handles the change from the 'tried/repaired case' into the 'learned case'.

In AIDA, just the retrieve, the reuse, and the retain steps are executed in the case-based reasoning module. The assessment in the Revise step is dealt with by the Functional module, utilizing principle based thinking procedures; see next section. The CBR module has been created in Eadocs a planning framework for composite sandwich boards. In this CBR module, the Retrieve advance, and additionally the Reuse and the Retain steps have been actualized; the revising step has been executed in another module. In flying machine plan, the cases contain information about their capacity or exhibitions, for example, the range and speed, and information about their structure or material science, for example, the weights and sizes. To empower a sort of subjective coordinating the system is made disconnected to enhance the effectiveness. The second part utilizes the system to look for cases like the predefined 'target set'. This is done online. For each piece of the objective set, the coordinating outcomes appear, and the cases are positioned as needs be. The significance of each piece of the objective set is given by priority values. Figure 8 demonstrates the AI update using knowledge management; it is difficult to alter the case truly, on account of the various associations between the helpful data and the essential data. Thusly, a system is taken after which should quick as pitiful changes as could sensibly be normal. A discretionary target set is portrayed, involving whatever is left of the conclusions that the 'best-planning' case does not satisfy, together with the most fundamental assistant and conduct data of the 'best-organizing' case. The result of the new organizing procedure will give cases that resemble the structure of the 'best-planning' case, and which resemble the utilitarian data as exhibited by the unsatisfied subtle elements. Space data is used to help the alteration technique. Ace guidelines have been accumulated which may help with focusing on the imperative data. For instance, when the 'best-organizing' case does not accomplish the foreordained speed (work), the originator should fixate on the thickness of the wing and its sweepback point (structure).

In the midst of this investigation, a couple of exercises have been instructed. Some are ordinary of all CBR structures, while others are related to the data planning method and the issues of building a system those necessities to work in all actuality. A Reliable Case-Base is Essential When the number of cases and the method for getting them were first analyzed, it created the impression that a case-base of 30-50 conflicts would have been adequately huge to start the tests and that these disputes could be handcrafted. Starting at now cleared up, both of these suppositions weren't right a result of the capriciousness of the space. As depicted in the most basic fragment of a CBR system is its library of cases (Khorasani, Jalali Aghchai, and Khorram, 2010). This was particularly substantial for ISAC. At first, the nonappearance of a change part made it vital to have a case-base with incredible breadth. Second, the multifaceted idea of the space proposed that the case-base contained heaps of cases. Finally, having a considerable measure of disputes for a circumstance base isn't adequate: every conflict needs an answer, also. Moreover, the courses of action must be understandable and must satisfy the controller. Two conditions must be respected remembering the ultimate objective to have a practical CBR structure: 1. there must be adequate cases drawn from a comparative part. If cases are from a comparable part and the case-base is used to light up conflicts on a comparable division, the chances that a similar conflict is starting at now for the circumstance base is higher. Having cases having a place with a comparable portion will diminish the multifaceted idea of the space and the degree of the case-base. 2. The responses for the conflicts that are secured for the circumstance base must be given by the controllers that generally manage that fragment. This will avoid the condition where controllers give unmistakable responses for a comparable conflict either in light of the way that they have an assorted establishment or because they use the mechanical assemblies in a sudden way. Works on being utilized in particular zones will ensure that controllers managing a comparable division will give normal courses of action. The gadget used for demonstrating the conflicts energetically affected the choice of the parameters and the courses of action of the disputes. The decision whether to use best quality level cases or noisy cases depends upon the way the case-base is secured: if the case-base is worked by hand, most noteworthy quality level cases will be used, on the other hand, if the case-base is clearly gotten from the territory, the case-base will contain more uproarious data. Some data must be entered by hand yet in an operational system each one of the data should be acquired electronically in light of the fact that the controllers will have neither time nor incline, to enter each one of the data by hand.

Various Controllers can give diverse answers for indistinguishable clash from of now stated, a controller, generally speaking, needs to get ready for more than one year on a particular zone before beginning to wear down it. This arrangement is vital to instruct the controller the favored responses for that particular fragment, yet it won't alter his tendencies and his direct when he changes territory. For example, let us consider a controller that has labored for a couple of years in the advancing toward a territory of a clamoring plane terminal where generally speaking conflicts are disentangled rapidly with a vertical move since it is the kind of move that needs the smallest watching. Exactly when this controller changes part, he will constantly be uneven and will appreciate conflicts with a vertical move. Another factor that could affect the controller's decision is the perspective towards the instruments used as a piece of the multiplication: a couple of controllers take a gander at significantly the dispute, some other don't. What's more, a couple of controllers certainly know the section used for the tests, so they are advantaged to respect to the controllers who had never watched the portion. Moreover, the segment minima got in HIPS for imagining the confined zones could change the plan given. If the division minima are excessively huge, HIPS will picture conflicts that don't exist when in doubt and the condition of the limited region will change, discrediting the courses of action. In the midst of a generation, for example, there was a 27% development in "speed" courses of action when the even division was reduced from 10 to 6 nautical miles. A response to avoid the controllers' inclinations is created a case-base containing conflicts that happened in a comparable part and to ask for their responses for controllers who tackle that section. Tendencies among controllers managing a comparative portion are less powerful in light of the fact that, having learned comparative precedents; controllers will make comparable assumptions on the conflicts. Requesting a social affair from controllers to come up with an inside and out recognized game plan, one of the fundamental choices, would take intemperate time. It could be acknowledged that the courses of action generally given by a social event of controllers managing a comparable division could be mixed by one of them, saving a huge amount of time.

DISCUSSION

CBR is better than RBS, yet with caveats, CBR diminishes the information designing issue in contrast with RBS. The claim that CBR frameworks can be executed speedier than show based frameworks is upheld by various sources. For instance, an investigation expressed that it took two weeks to build up a case-based rendition of a framework that took four months to work in lead-based shape (Watson, 1994). Likewise, and all the more as of late, engineers affirmed that a govern based framework took in excess of eight times longer to create than a case-based framework with a similar usefulness. They likewise guarantee that the support of the RBS is nonstop through the CBR framework needs no upkeep (Watson, 1994). The opportunity to successfully construct the structure that handles the information base in ISAC was short and no upkeep was essential. Adding cases to the case-base when a contention was not accurately unraveled was additionally basic. An opportunity to develop ISAC is shorter than the time that would have been important to assemble the equal lead-based framework, yet no examination between the two calculations should be possible from the perspective of the execution. Truth be told, from the accessible writing on master frameworks for ATC, it appears that the current RBS can help the controllers just in specific circumstances, however, are not solid in a general setting. Besides, their support and refresh are exceptionally troublesome using a cost work for evaluating the adequacy of an answer was considered however disposed of in light of the fact that it would have inferred constructing an entire lead-based framework as mind-boggling and costly as ISAC with the sole reason for assessing the cost. An exceptionally basic arrangement of standards (2 rules) has been utilized as a part of the progressive structure of ISAC. A few tenets are likewise utilized as a part of the adjustment step, which is exceptionally basic at this stage yet could be expanded if a more definite arrangement must be actualized. Thus, it must be said that CBR ought to be supplemented with some different frameworks, for example, RBS to construct effective applications (Bayles et al., 1993).

The knowledge engineering problem at the start of the task, a report with a few speculations on vital CBR issues (Bonzano and Cunningham, 1995) was created before having procured a profound comprehension of the issue of ATC. There were theories on the structure of the framework, on the programming dialect that could have been utilized, on the conceivable specialized and hypothetical issues and their comparing arrangements, and so forth. Some of these speculations were later uncovered to be right, while, others were definitely not. For instance, the speed of the framework in giving ongoing arrangements was viewed as one of the greatest issues toward the start, yet toward the end, it was not really. Also, it was believed that the case-base obtaining would have been one of the simplest assignments; in any case, then again it uncovered to be a standout amongst the most troublesome. These progressions are only a pointer of how complex the procedure has been. The structure of ISAC changed various circumstances. A few choices must be taken and they didn't just rely upon the CBR idea of the issue, yet additionally on its ATC nature. Besides, not just the limitations originating from the ATC space must be thought about, yet in addition the inclinations of the controllers.

CONCLUSION

How AI influences current air to dispatch system joins a variety of modified control structure that aides the flight group in course, flight organization and extending the security characteristics of the plane and how constructing airship motor diagnostics cosmology, air activity administration and imperative programming is valuable in ATM setting. How flight security can be upgraded through the progression and use of mining, using its results and Knowledge-Based Engineering (KBE) approach blend of a couple of educates in a widely inclusive strategy for use in carrier sensible blueprint is examined. The early conspicuous verification and finding of mix-ups, investigation of gigantic data and its impact on the transportation business and upgraded travel system; the agent-based mobile airline search and booking framework utilizing AI appears. The calculated plan process and recommended an outline cycle those utilizations CBR techniques to propose and adjust starting ideas, RBR-methods to dissect and assess the idea, and geometric demonstrating methods that model the idea consequently. These three strategies are actualized in three free modules, with a focal UI to interface the modules. The framework has been assessed for the calculated plan of flying machine. This application permits the disintegration of the outline item into fundamental segments. The present approach in this way depends vigorously on the disintegration of the plan item into fundamental segments.. Another part of the further research will be the execution of the different modules into one coordinated system. Since most airport regulation offices utilize practices and hardware which were created no less than 20 years back, it is normal to expect that new methodologies are required for future situations with higher airship populaces. Upgrades because of the revamping of course structures will quickly achieve an utmost in airspace and soon thereafter some central changes will be required. Most importantly, the utility of PC help will increment due additionally to expanded accuracy in anticipated directions of flying machine. The examination exhibited in this postulation was planned to research the advantages of utilizing CBR with a specific end goal to enable controllers in flying machine to compromise. The framework as it is currently is incorporated with HIPS which is installed in a particular recreation condition for assessment purposes, yet it could be in principle coordinated in any ATC device, gave that this instrument can supply ISAC with the essential information for the contention portrayal. It is our conclusion that exclusive minor changes would be expected to the structure of ISAC to be utilized as a part of an area with any ATC instrument. The presence of a solid case-base for the particular area is an alternate and more major issue.

REFERENCES

Aircraft Pitch Control System Using Lqr And Fuzzy Logic Controller. (2016). International Journal Of Advance Engineering and *Research for Development*, *3*(5). doi:10.21090/ijaerd.030564

Almubayed, A., Hadi, A., & Atoum, J. (2015). A Model for Detecting Tor Encrypted Traffic using Supervised Machine Learning. *International Journal Of Computer Network And Information Security, 7*(7), 10–23. doi:10.5815/ijcnis.2015.07.02

Amine Chatti, M. (2012). Knowledge management: A personal knowledge network perspective. *Journal of Knowledge Management, 16*(5), 829–844. doi:10.1108/13673271211262835

Aviation Human Factors Related Industry News 1Parts of this section are compiled from "Aviation Human Factors Industry News" and reproduced with permission of Roger Hughes. (2015). *Aviation Psychology and Applied Human Factors, 5*(1), 67-69. doi:10.1027/2192-0923/a000077

Bobek, S. (1992). A framework for integrating decision support systems into office information systems. *Decision Support Systems, 8*(3), 211–216. doi:10.1016/0167-9236(92)90015-H

Gan, X., Tang, X., & Gao, J. (2013). Missile Autopilot Design Based on Fuzzy Self-Tuning Immune PID Control. *Advanced Materials Research, 739*, 550–554. . doi:10.4028/www.scientific.net/AMR.739.550

Grabbe, S., Sridhar, B., & Mukherjee, A. (2010). Evaluation of an Integrated Traffic Flow Management Decision Making Approach. *Air Traffic Control Quarterly, 18*(4), 331–358. doi:10.2514/atcq.18.4.331

Hui, W., Hu, J., & Shyue, K. (2008). Role of coordinates in computational fluid dynamics. *International Journal of Computational Fluid Dynamics, 22*(1-2), 3–9. doi:10.1080/10618560701737112

Hwang, I., Kim, J., & Tomlin, C. (2007). Protocol-Based Conflict Resolution for Air Traffic Control. *Air Traffic Control Quarterly, 15*(1), 1–34. doi:10.2514/atcq.15.1.1

Jain, A., & Kumar, A. (2017). Traffic Congestion Management System: An Approach with Machine Learning. *IMS Manthan, 12*(01). doi:10.18701/imsmanthan.v12i01.10338

Jeffery, L., & Rhodes, G. (2011). Insights into the development of face recognition mechanisms revealed by face aftereffects. *British Journal of Psychology, 102*(4), 799–815. doi:10.1111/j.2044-8295.2011.02066.x PMID:21988385

Juneja, P., & Kashyap, R. (2016). Optimal approach for CT image segmentation using improved energy based method. *International Journal of Control Theory and Applications, 9*(41), 599–608.

Kashyap, R. (2018). Object boundary detection through robust active contour based method with global information. *International Journal Of Image Mining, 3*(1), 22. doi:10.1504/IJIM.2018.093008

Kashyap, R., & Gautam, P. (2013). Microarray image segmentation using improved GOGAC method. *International Journal on Computer Science and Engineering, 2*(4), 67–74.

Kashyap, R., & Gautam, P. (2015). Modified region based segmentation of medical images. *International Conference on Communication Networks (ICCN)*, 209–216. 10.1109/ICCN.2015.41

Kashyap, R., & Gautam, P. (2016). Fast level set method for segmentation of medical images. In *Proceedings of the International Conference on Informatics and Analytics (ICIA-16)*. ACM. 10.1145/2980258.2980302

Kashyap, R., & Gautam, P. (2017a). Fast Medical Image Segmentation Using Energy-Based Method. *Biometrics. Concepts, Methodologies, Tools, and Applications, 3*(1), 1017–1042. doi:10.4018/978-1-5225-0983-7.ch040

Kashyap, R., & Gautam, P. (2017b). *Fast Medical Image Segmentation Using Energy-Based Method*. In *Pattern and Data Analysis in Healthcare Settings*. Medical Information Science Reference.

Kashyap, R., Gautam, P., & Tiwari, V. (2018). Management and Monitoring Patterns and Future Scope. In Handbook of Research on Pattern Engineering System Development for Big Data Analytics. IGI Global. doi:10.4018/978-1-5225-3870-7.ch014

Kashyap, R., & Piersson, A. (2018a). Big Data Challenges and Solutions in the Medical Industries. In Handbook of Research on Pattern Engineering System Development for Big Data Analytics. IGI Global. doi:10.4018/978-1-5225-3870-7.ch001

Kashyap, R., & Piersson, A. (2018b). *Impact of Big Data on Security. In Handbook of Research on Network Forensics and Analysis Techniques* (pp. 283–299). IGI Global. doi:10.4018/978-1-5225-4100-4.ch015

Kashyap, R., & Tiwari, V. (2017). Energy-based active contour method for image segmentation. *International Journal of Electronic Healthcare*, *9*(2–3), 210–225. doi:10.1504/IJEH.2017.083165

Khorasani, A., Jalali Aghchai, A., & Khorram, A. (2010). Chatter prediction in turning process of conical workpieces by using case-based resoning (CBR) method and taguchi design of experiment. *International Journal of Advanced Manufacturing Technology*, *55*(5-8), 457–464. doi:10.100700170-010-3060-1

Ma, J., Tao, X.-P., Zhu, H.-H., & Lü, J.MA. (2014). Enhancing Multi-Agent System with Context-Awareness. *Journal of Software*, *23*(11), 2905–2922. doi:10.3724/SP.J.1001.2012.04301

Munjulury, R., Staack, I., Berry, P., & Krus, P. (2015). A knowledge-based integrated aircraft conceptual design framework. *CEAS Aeronautical Journal*, *7*(1), 95–105. doi:10.100713272-015-0174-z

Paukkeri, M., García-Plaza, A., Fresno, V., Unanue, R., & Honkela, T. (2012). Learning a taxonomy from a set of text documents. *Applied Soft Computing*, *12*(3), 1138–1148. doi:10.1016/j.asoc.2011.11.009

Pedrycz, W. (2011). Introducing WIREs Data Mining and Knowledge Discovery. *Wiley Interdisciplinary Reviews. Data Mining and Knowledge Discovery*, *1*(1), 1–1. doi:10.1002/widm.17

Pedrycz, W. (2011). Introducing WIREs Data Mining and Knowledge Discovery. *Wiley Interdisciplinary Reviews. Data Mining and Knowledge Discovery*, *1*(1), 1–1. doi:10.1002/widm.17

Reynolds, M. (2017). AI coach helps chatbots seem more human. *New Scientist*, *235*(3135), 14. doi:10.1016/S0262-4079(17)31407-0

Sandita, A., & Popirlan, C. (2015). Developing A Multi-Agent System in JADE for Information Management in Educational Competence Domains. *Procedia Economics and Finance*, *23*, 478–486. doi:10.1016/S2212-5671(15)00404-9

Seneff, S. (2002). Response planning and generation in the mercury flight reservation system. *Computer Speech & Language*, *16*(3-4), 283–312. doi:10.1016/S0885-2308(02)00011-6

Shevchuk, D. (2010). Fiber-optic intelligence systems for diagnostics contraction integrity of aircrafts. *Proceedings Of National Aviation University*, *42*(1). doi:10.18372/2306-1472.42.1816

Speed Control of a Train using Fuzzy Logic. (2017). *International Journal of Science and Research, 6*(6), 781-784. doi:10.21275/art20174396

Torenbeek, E. (2000). Aircraft design education in Europe. *Aircraft Design, 3*(4), 205–206. doi:10.1016/S1369-8869(00)00022-7

Upadhyay, A., & Kashyap, R. (2016). Fast Segmentation Methods for Medical Images. *International Journal of Computers and Applications, 156*(3), 18–23. doi:10.5120/ijca2016912399

Upadhyay, A., & Kashyap, R. (2017). Robust hybrid energy based method for accurate object boundary detection. *Journal of Advanced Research in Dynamical and Control Systems, 1*(06), 1476–1490.

US military develops pilotless helicopter. (2012). *Physics Today*. doi:10.1063/pt.5.026359

Vahidov, R., & Kersten, G. (2004). Decision station: Situating decision support systems. *Decision Support Systems, 38*(2), 283–303. doi:10.1016/S0167-9236(03)00099-X

Waoo, N., Kashyap, R., & Jaiswal, A. (2010). DNA nano array analysis using hierarchical quality threshold clustering. In *2010 2nd IEEE International Conference on Information Management and Engineering*. IEEE.

Wittenberg, H. (2001). Aircraft performance, theory and practice. *Aircraft Design, 4*(1), 79–80. doi:10.1016/S1369-8869(00)00026-4

Zhang, N., & Zhu, J. (2014). Fuzzy Analytic Hierarchy Process Method for Civil Aviation Airport Security Information Management. *Applied Mechanics and Materials, 701-702*, 40–43. . doi:10.4028/www.scientific.net/AMM.701-702.40

Zhang, T., Li, P., & Wang, S. (2012). A personal air distribution system with air terminals embedded in chair armrests on commercial airplanes. *Building and Environment, 47*, 89–99. doi:10.1016/j.buildenv.2011.04.035

KEY TERMS AND DEFINITIONS

Air Traffic Control: Air traffic control (ATC) is an administration gave by ground-based air activity controllers who coordinate airplane on the ground and through controlled airspace and can give warning administrations to the airship in non-controlled airspace. The basic role of ATC worldwide is to anticipate crashes, arrange and assist the stream of air movement, and give data and other help to pilots. In a few nations, ATC plays a security or cautious part or is worked by the military. To counteract impacts, ATC authorizes activity detachment rules, which guarantee every flying machine keeps up a base measure of purge space around it consistently. Numerous flying machines likewise have crash shirking frameworks, which give extra security by notice pilots when other airship gets excessively close.

Air Traffic Flow Management (ATFM): It is the direction of air movement keeping in mind the end goal to abstain from surpassing airplane terminal or airport regulation limit in dealing with movement consequently the elective name of Air Traffic Flow and Capacity Management (ATFCM) and to guarantee that accessible limit is utilized efficiently.

Artificial Intelligence: Artificial intelligence (AI) insight exhibited by machines, as opposed to the regular insight showed by people and different creatures. In software engineering AI investigates is characterized as the investigation of "savvy specialists": any gadget that sees its condition and takes activities that augment its risk of effectively accomplishing its goals. Colloquially, the expression "man-made brainpower" is connected when machine copies "intellectual" capacities that people connect with other human personalities, for example, "learning" and "issue solving."

Fuzzy Control System: It is a control framework in light of fluffy rationale a scientific framework that breaks down simple info esteems as far as intelligent factors that go up against constant esteems in the vicinity of 0 and 1, as opposed to established or computerized rationale, which works on discrete estimations of either 1 or 0.

Multi-Agent System: A multi-agent system is a mechanized framework made out of various collaborating shrewd operators. Multi-specialist frameworks can take care of issues that are troublesome or unimaginable for an individual operator or a solid framework to illuminate. Insight may incorporate methodic, utilitarian, procedural methodologies, algorithmic inquiry or fortification learning. Regardless of significant cover, a multi-specialist framework isn't generally the same as an operator-based model. The objective of an ABM is to look for an informative understanding of the aggregate conduct of specialists obeying straightforward guidelines, ordinarily in regular frameworks, instead of in taking care of particular use or building issues. The phrasing of ABM has a tendency to be utilized all the more frequently in the sciences, and MAS in designing and technology. Applications where multi-specialist frameworks research may convey a proper approach to incorporate online trading, fiasco response, and social structure modeling.

Chapter 2
Artificial Neural Network for Pre-Simulation Training of Air Traffic Controller

Tetiana Shmelova
National Aviation University, Ukraine

Yuliya Sikirda
National Aviation University, Ukraine

Togrul Rauf Oglu Jafarzade
National Aviation Academy, Azerbaijan

EXECUTIVE SUMMARY

In this chapter, the four layers neural network model for evaluating correctness and timeliness of decision making by the specialist of air traffic services during the pre-simulation training has been presented. The first layer (input) includes exercises that cadet/listener performs tõ solve a potential conflict situation; the second layer (hidden) depends physiological characteristics of cadet/listener; the third layer (hidden) takes into account the complexity of the exercise depending on the number of potential conflict situations; the fourth layer (output) is assessment of cadet/listener during performance of exercise. Neural network model also has additional inputs (bias) that including restrictions on calculating parameters. The program "Fusion" of visualization of the state of execution of an exercise by a cadet/listener has been developed. Three types of simulation training exercises for CTR (control zone), TMA (terminal control area), and CTA (control area) with different complexity have been analyzed.

INTRODUCTION

Statistics data (Leychenko, Malishevskiy, & Mikhalic, 2006; Allianz Global Corporate & Specialty. EMBRY-RIDDLE Aeronautical University, 2014; Aviation Accident Statistics, 2018; Statistics of the World's the Largest Aircraft Accidents for the Years 1974-2014, 2018) show us, that causality of aviation accidents didn't change over the past decade: 70-80% of accidents and disasters happened due to

DOI: 10.4018/978-1-5225-7588-7.ch002

human factor, and only 15-20% – through constructive and productive deficiencies of the aircraft. It is important to pay more attention to the training of aviation specialists because they are dealing with equipment that is becoming more complicated from year to year. Modern recommendations of International Civil Aviation Organization (ICAO) are to use information technology in aviation systems as in training as in operation (International Civil Aviation Organization [ICAO], 2007, 2008). Automation is seen as one of many resources available to the human operators, controllers and pilots alike, who retain the responsibility for management and direction of the overall Air Traffic Management (ATM) system. Additionally, unexpected or unplanned events must be a required part of planning and design when considering the systems that would replace the cognitive and adaptive capabilities of controllers or pilots. The development of training for automated systems is more difficult than for non-automated systems. One of the primary challenges in developing training for automated systems is to determine how much a trainee will need to know about the underlying technologies in order to use automation safely and efficiently. Course development based on a task analysis can be more effectiveness than traditional training development techniques.

BACKGROUND

Quality training of aviation experts, including specialists in Air Traffic Services (ATS), occupied the important part in reducing the influence of the human factor (European Organisation for the Safety of Air Navigation [Eurocontrol], 2004a). There are three types of air traffic controller (ATC) training, leading towards the issue and maintenance of an ATC license and associated unit endorsements. Initial training is the first type. ATC training phases (Figure 1) (Eurocontrol, 2004b, 2015):

1. Initial Training.
    a. Basic Training.
    b. Rating Training.
2. Unit Training.
    a. Transitional Training.
    b. Pre-On-the-Job Training (Pre-OJT).
    c. On-the-Job Training (OJT).
3. Continuation Training.
    a. Refresher Training.
    b. Conversion Training.

Simulation Training is an important part of ATC training. It is a complex of existing forms and methods of training, in which cadets / listeners through the implementation of appropriately formulated complex tasks and exercises under the guidance of the instructor develop skills and practical application of the theoretical provisions of several disciplines (Eurocontrol, 2004b, 2015). The aim of training is to improve ATC work and refinement practical skills of ATS in standard situations, potential conflict situations (PCS), in the special conditions and in the flight emergencies. The quality and number of exercises, objective evaluation of exercises influence on the effectiveness of Simulation Training.

Figure 1. Progression of ATC training
(Eurocontrol, 2015)

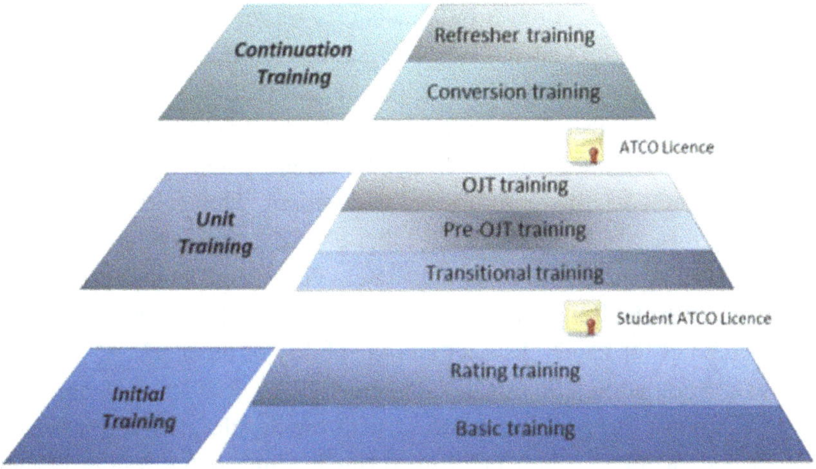

Simulation Training is a model of communication where cadet / listener, either individually or in the group receives information through a media at a rate (Eurocontrol, 2000). The combination of these elements defines the training event.

Media is the physical means by which an instructor or a training designer communicates a message. One media can use several supports we are going to define the media related to simulation but shall not attempt to make an exhaustive list of the many types of support and educational materials. It has defined the following five media:

1. Real Equipment.
2. High-fidelity Simulator (HI FI SIM).
3. Simulator (SIM).
4. Part-Task Trainer (PTT).
5. Other Training Device (OTD).

It is used any of the three rates of learning although most of the exercises are in real time:

1. Self-paced Learning.
2. Time-restricted learning.
3. Real Time.

According to the recommendations of Eurocontrol (2004b, 2015) and to optimize the effectiveness of Simulation Training, theoretical and practical training combine from the beginning of the training using the Pre-Simulation (Pre-SIMUL). The process of learning starts with getting by the cadet / listener Skill Acquisitions, then performance Part-Task Practice and continues Simulation.

Eurocontrol has defined several types of Simulation and has differentiated between pre-simulation and simulation exercises, and it has added the notion of Guided Simulation. Simulation (SIMUL) is a provision of knowledge, skills, and attitudes by means of a representation of air traffic responding to the cadet / listener action as real air traffic. Types of Simulation include Individual Simulation (IND SIMUL), Team Simulation (TEAM SIMUL) and Group Simulation (GROUP SIMUL). Simulation always includes Briefing (Brief), Debriefing (Debrief) and Tutoring. Guided Simulation (GSIMUL – Guided SIMUL) is the extensive interaction between the cadet / listener and the computer in the form of questions, feedback, comments, instructions, and assessment. This guidance assumes the existence of a theoretical model against which the cadet / listener can be compared. Guided Skills Acquisitions (GSA – Guided SA) – Skill Acquisitions (SA) with interactive assessment, comments and guidance – are actual at this time. Guided PTP (GPTP) is a Part-Task Practice (PTP), accompanied by comments, display results, assessment of the cadet / listener and the ability of feedback.

The system of ATC Simulation Training is characterized by low level of objective evaluation of exercises (Eurocontrol, 2004b, 2015). It's connected with the development of sufficient exercises at a given difficulty level. Increasing the number of exercises leading to significant growth an amount of instructional and methodical staff and requirements for their professional skills as well as time for developing the appropriate exercises of a given complexity. The most time spent on modeling of air situation in accordance with the set of objectives in the exercise, verification of exercise to meet at a given level of difficulty, verification graph to meet a planned workload and the possibility of conflict-free exercises.

In modeling of the air situation using inverse task of generating dynamic air situation at a given exercise intensity, takes into account the complexity of air traffic, presence, and quantity of PCS, flight emergencies (Ministry of Transport and Communications of Ukraine, 2007). The correctness and the timeliness are important criteria for evaluating the quality of the exercises (Eurocontrol, 2000, 2004b, 2015). Taking into account as correctness as timeliness can be done with the help of Artificial Neural Network (ANN). Automation of process of exercise verification can reduce the time for its preparation (saving up to 80% of the time).

A hybrid neural-expert system is a perspective direction of development of a neuron-information technology (Komashynskiy & Smirnov, 2002). ANN has many advantages compared to traditions and knowledge-based of diagnostic systems (Arkhangelskiy, Bohaenko, Grabowskiy, & Ryumshyn, 1999; Suzuki, 2013). It can be trained on examples, work in real-time, secure and tolerant to errors. With the help of ANN diagnoses state of the patient, performs the prediction on the stock market and the weather forecast, makes decision on granting the loan, diagnoses condition of equipment, guided operation of the engine, etc. (Komashynskiy & Smirnov, 2002; Gardner & Dorling, 1998; Goedeking, 2010). ANN is being created by the serial and parallel association of individual neurons. The neural network is grouped in two classes according to the type of connections: straight directional network (which links don't have loops) and recurrent network (with feedback connections). (Komashynskiy & Smirnov, 2002; Arkhangelskiy et al., 1999; Suzuki, 2013; Gardner & Dorling, 1998; Goedeking, 2010; Borovikov, 2008). The most common among straight directional networks are single-layer and multi-layer perceptrons, cognitron and Radial Basis Function (RBF) networks; among the recurrent network can be distinguished Hopfield, Boltzmann and Kohonen networks (Arkhangelskiy et al., 1999; Suzuki, 2013).

Using of neural networks and neuro-fuzzy systems is appropriate for solutions many problems in aviation, where it is necessary to process a large amount of fuzzy information and solving difficult formalized multiparameter nonlinear tasks (which) and problems, namely in the case of: decision making about departure and selection of an optimum alternate aerodrome landing (Kharchenko, Shmelova,

Artemenko, & Otryazhiy, 2011); choosing the optimum place of forced landing (Kharchenko, Shmelova, Sikirda, & Gerasimenko, 2011); estimation of efficiency of alternative variants of flight completion in the event of an extraordinary situation (Shmelova, Yakunina, & Yakunin, 2013); diagnostic erroneous actions of the operator of ergatic aviation system in flight emergencies (Shmelova, Yakunina, Moyseenko, & Grinchuk, 2014), etc.

It is proposed to use a neural network model of evaluation correctness and timeliness of decision making by ATC in the case of PCS during the performance of exercises in the Simulation Training through pre-training studying.

ESTIMATION OF SITUATION'S COMPLEXITY IN CASE OF POTENTIAL CONFLICT SITUATION WITH THE HELP OF FUZZY SETS METHOD

There is well-known concept of Threat and Error Management (TEM), which allows determining links between safety and operability of operator in fleeting difficult operating conditions (ICAO, 2008; Kharchenko, Chynchenko, & Raychev, 2007). Conception has descriptive character and can be used as means of diagnosis both characteristics of human efficiency and effectiveness of the system. Despite the fact that the TEM was originally developed for use in the cockpit, but it can also be used in various organizations of the aviation industry, including services for ATS.

Regarding ATC, TEM consists of three components:

- Threats;
- Errors;
- Undesirable conditions.

According to TEM, threats and errors are the routine parts of aviation activity. ATC should control undesirable conditions because it led to dangerous consequences. One of the main components of TEM is the control of undesirable conditions, and it has the same meaning as factors of threats and errors. Control of undesirable conditions is the last opportunity to avoid dangerous consequences and thus provide maintaining at a given level of aviation safety in ATS.

The concept, according to the problem of determining the timeliness of decision making in solving PCS by ATC, was developed classification stages of conflict situation evolution (Shmelova, Sikirda, Zemlyanskiy, & Danilenko, 2015).

The threat of conflict situation is the first stage of the PCS. The threat comes from the moment when the time of remaining to the conflict situation equal time, which needed to perform all elements for solving PCS, taking into account the required buffer time.

The pre-conflict situation is the second stage of the PCS. The pre-conflict situation occurs from the moment when the time of remaining until the conflict situation equal time, which needed to perform all elements to solve the PCS without buffer time. In this situation, the violation of the separation intervals is not yet come, but the probability of resolving the situation is extremely small.

The conflict situation is a stage when happens violation of the separation intervals. Since the classification shows, that for determining the timeliness in dealing with decision-making during PCS should determine when there is a transition stage to the current situation the threat of conflict and pre-conflict situation.

In order to get the quantitative indicators of the level of the situation's complexity in the developing of PCS, was used the method of fuzzy sets (Zade, 1965; Borisov, Krumberh, & Fedorov, 1990).

There are values of the linguistic variable on the scale:

- The threat of conflict;
- The pre-conflict situation;
- The conflict situation.

The minimum level of situation's complexity equal zero (0), maximum – one (1). The resulting range is divided into five intervals.

The degree of membership of linguistic variable at a certain interval defined as the ratio of answers number (where it occurs in this range) to the maximum value of this number for all intervals. There was conducted survey of 30 experts from the air traffic service training center of Flight Academy of the National Aviation University by Delphi method in two rounds. Getting results shown in Table 1.

Membership functions of the situation's complexity μ (threat of conflict, pre-conflict situation, conflict situation) in the case of PCS are in Table 2 and shown in Figure 2.

There is a system of transitions between components of TEM concept (stages of development of PCS) in the graph of states (Figure 3).

Determine the state of the exercise as (1):

Table 1. Survey results

Meaning of Membership Functions of the Complexity of the Situation in Case of PCS, μ	Interval, Units				
	0-0.2	0.2-0.4	0.4-0.6	0.6-0.8	0.8-1.0
The threat of conflict, μ_1	1	1	22	6	0
The pre-conflict situation, μ_2	0	0	8	20	2
The conflict situation, μ_3	0	0	1	12	17

Figure 2. Membership functions of the situation's complexity in case of PCS

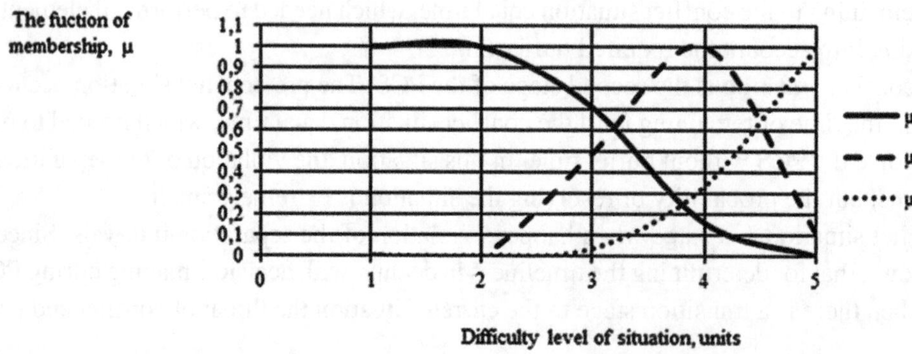

Table 2. Classification of conflict situation by the criterion of timeliness

Classification by the TEM	Classification by Timeliness	Description of Situation	Level of Complexity of Situation, Units
Threat	Threat of conflict situation	Fixed PCS which requires solving	1-2
Error	Pre-conflict situation	Parry of PCS is difficult or impossible	4
Undesirable condition	Conflict situation	Violation of separation intervals, the conflict situation has happened	5

Figure 3. Transitions between stages of PCS: T_0 – is a state, which characterized by presence of threat in PCS; T_1 – is a pre-conflict situation, happen in result of erroneous or inaction of operator (cadet / listener) ; T_2 – is a conflict situation (violation of separation intervals); t_{01} – is a transition from T_0 to T_1, characterized by time of parry of PCS and buffer time; t_{12} – is a transition from T_1 to T_2, characterized by the time of parry of PCS at stage T_1, in case of transition to stage T_2 happen critical situation; t_{10} – is a transition from T_1 to T_0, characterized by the ability of the operator to resolve the PCS; t_{21} – is a transition from T_2 to T_1, characterized by the appearance of conflict due to T_2 (violations of separation's rules), but in case of successful resolution of problem by operator during allowable time – a return to pre-conflict situation ($T_{sim} = T_2$)

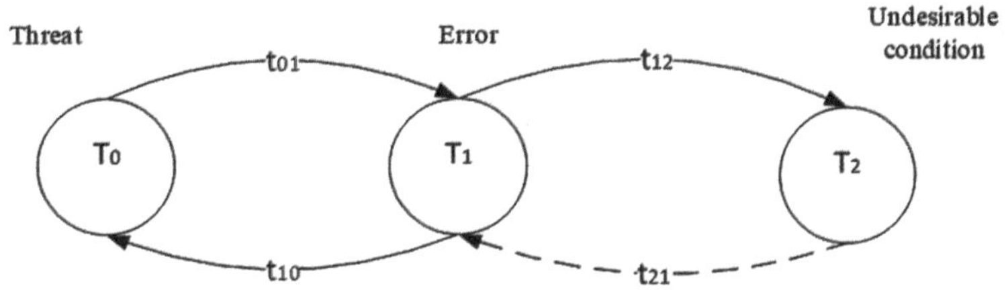

$$y = \begin{cases} 1; t_{01} + t_{12} \le T_2 \\ 0; t_{01} + t_{12} > T_2 \end{cases}. \tag{1}$$

The time to solve conflict depends on the individual characteristics of the operator (cadet / listener) (c), the numbers of PCS (n) and time of PCS (T_s) (2):

$$T_{sc}=T_c \cdot c+T_n \cdot n+T_s, \tag{2}$$

where c – is a coefficient, which defines the individual characteristics of the operator (cadet / listener); n – is an amount of PCS; T_s – is a time of developing PCS, if n=1, c=1, then $T_{sc}=T_c+T_n+T_s$.

NEURAL NETWORK MODEL OF EVALUATION CORRECTNESS AND TIMELINESS OF AIR TRAFFIC CONTROLLER'S DECISION MAKING DURING PRE-SIMULATION TRAINING

To automate the evaluation of pre-training stage of Initial Training of ATC at the stage of pre-training studying the Multilayer Perceptron Network (MPN) has been developed (Shmelova, Sikirda, Zemlyanskiy, Danilenko, & Lazorenko, 2016). It has four layers, two of which are hidden. Each neuron is characterized by the input value (dendrite) and the output value (axon), weight coefficients (synapses), threshold function. The network has additional inputs, called the Bias (offset) that takes into account additional restrictions on calculating parameters (3):

$$\sum_{i=1}^{n} w_i x_i - \theta \geq 0. \tag{3}$$

where w_i – are the weight coefficients; x_i – are the neural network inputs; $\bar{\theta}$ – is a Bias (shift).

General view of the ANN shown in Figure 4 (4):

$$\bar{Y} = f(\overline{net} - \bar{\theta}), \tag{4}$$

where f – is a non-linear function (active function); $\overline{net}$ – is a weighted sum of inputs.

Characteristics of ANN's layers:

Figure 4. General view of the ANN

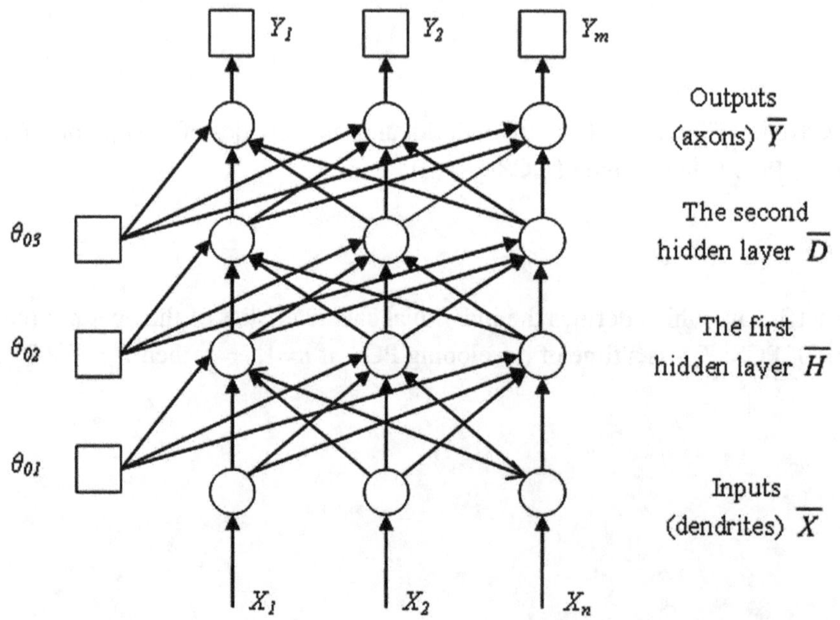

- **The First Layer (Input):** Are the exercises that perform cadets / listeners to solve PCS ($\overline{X}$);
- **The Second Layer (Hidden):** Are the physiological characteristics of cadets / listeners ($\overline{H}$);
- **The Third Layer (Hidden):** Is the complexity of the exercises, which is determined by the number of PCS ($\overline{D}$);
- **The Fourth Layer (Output):** Is an assessment of cadets / listeners during the performance of exercises ($\overline{Y}$).

Consider in more detail the topology of the neural network as an example, if three cadets / listeners (Y_1, Y_2, Y_3) perform two tasks (X_1 and X_2) (Figure 5).

The first layer (input) are the inputs x_1, x_2, ..., x_n that meet the tasks that perform by cadets / listeners to solve PCS ($\overline{X}$).

The second layer (hidden) defines the physiological characteristics of cadets / listeners ($\overline{H}$) using additional input Bias, which specifies limits on individual solving of exercises (T_{01}).

The output vector of the second layer (5):

$$\overline{H} = f(\overline{W}_1, \overline{X}) = f(\overline{net_1} - \bar{\theta}_{01}), \quad (5)$$

where $\overline{net_1} = \overline{W}_1\overline{X}$;

Figure 5. Example of ANN when three cadets / listeners perform two tasks

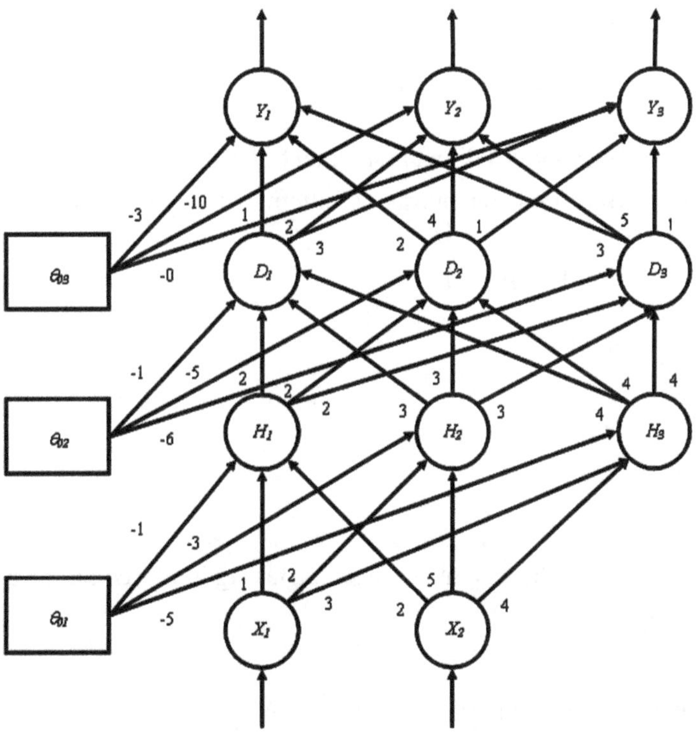

- $\overline{W}_1 = \begin{pmatrix} w_{01} & w_{11} & w_{21} \\ w_{02} & w_{12} & w_{22} \\ w_{03} & w_{13} & w_{23} \end{pmatrix}$: Are the weight coefficients, for example for studying the situation when two exercises are doing by three cadets / listeners;
- $X = \begin{pmatrix} X_1 \\ X_2 \\ 1 \end{pmatrix}$: Are the tasks that perform by cadets / listeners to solve PCS;
- $\overline{\theta}_{01}$: Is a time to solve individual training exercises.
- **The Third Layer (Hidden):** Is a complexity of the exercise, which is determined by the number of PCS ($\overline{D}$) and characterized by dynamic air situation. Auxiliary input Bias indicates the total limit of time for resolving PCS (T_{02}).

The output vector of the third layer (6):

$$\overline{D} = f(\overline{W}_2, \overline{H}) = f(\overline{net_2} - \overline{\theta}_{02}), \tag{6}$$

where $\overline{net_2} = \overline{W}_2 \overline{H}$;

- $\overline{W}_2 = \begin{pmatrix} d_{01} & d_{11} & d_{21} \\ d_{02} & d_{12} & d_{22} \\ d_{03} & d_{13} & d_{23} \end{pmatrix}$: Are the weight coefficients, which takes into account the complexity of dynamic air situation;
- $\overline{\theta}_{02}$: Is a time to solve training exercise that takes into account the complexity of the dynamic air situation.
- **The Fourth Layer (Output):** Is a direct assessment of cadet during the performance of exercises ($\overline{Y}$). Auxiliary input Bias limits the number of attempts for solving the PCS (T_{03}).

The output vector of the fourth layer (7):

$$\overline{Y} = f(\overline{W}_3, \overline{D}) = f(\overline{net_3} - \overline{\theta}_{03}), \tag{7}$$

where $\overline{net_3} = \overline{W}_3 \overline{D}$;

- $\overline{W}_3 = \begin{pmatrix} y_{01} & y_{11} & y_{21} \\ y_{02} & y_{12} & y_{22} \\ y_{03} & y_{13} & y_{23} \end{pmatrix}$: Are the weight coefficients, taking into account the quality of the exercise by the timeliness;
- $\overline{\theta}_{03}$: Are the attempts to solve the exercise.

Provides the following outputs vectors layers of neurons $\overline{H}$, $\overline{D}$, $\overline{Y}$ (8):

$$H_i D_k Y_m = \begin{cases} 1; \text{if } f(x) > 0 \\ 0; \text{if } f(x) \leq 0 \end{cases}, \tag{8}$$

where f – is a non-linear function of activation.

Consider the following set of values of weight coefficients ($\overline{W} = \overline{W}_1, \overline{W}_2, \overline{W}_3$), that take into account the performance of individual training exercises by cadet / listener depending on the physiological characteristics, the complexity of dynamic air situation, the quality of the exercise according to the timeliness:

$$H_1 = f(1x_1 + 2x_2 - 1);$$

$$H_2 = f(2x_1 + 5x_2 - 3);$$

$$H_3 = f(3x_1 + 4x_2 - 5);$$

$$D_1 = f(2d_1 + 2d_2 + 2d_3 - 1);$$

$$D_2 = f(3d_1 + 3d_2 + 3d_3 - 5);$$

$$D_3 = f(4d_1 + 4d_2 + 4d_3 - 6);$$

$$Y_1 = f(1y_1 + 3y_2 + 2y_3 - 3);$$

$$Y_2 = f(2y_1 + 4y_2 + 1y_3 - 10);$$

$$Y_3 = f(3y_1 + 5y_2 + 1y_3 - 0).$$

Present an example in vector form:

$$\begin{pmatrix} H_1 \\ H_2 \\ H_3 \\ 1 \end{pmatrix} = f\left(\begin{pmatrix} 1 & 2 & -1 \\ 2 & 5 & -3 \\ 3 & 4 & -5 \\ 0 & 0 & 1 \end{pmatrix} \cdot \begin{pmatrix} X_1 \\ X_2 \\ 1 \end{pmatrix}\right);$$

$$\begin{pmatrix} D_1 \\ D_2 \\ D_3 \\ 1 \end{pmatrix} = f\left(\begin{pmatrix} 2 & 2 & 2 & -1 \\ 3 & 3 & 3 & -5 \\ 4 & 4 & 4 & -6 \\ 0 & 0 & 0 & 1 \end{pmatrix} \cdot \begin{pmatrix} H_1 \\ H_2 \\ H_3 \\ 1 \end{pmatrix}\right);$$

$$\begin{pmatrix} Y_1 \\ Y_2 \\ Y_3 \end{pmatrix} = f\left(\begin{pmatrix} 1 & 3 & 2 & -3 \\ 2 & 4 & 1 & -10 \\ 3 & 5 & 1 & 0 \end{pmatrix} \cdot \begin{pmatrix} D_1 \\ D_2 \\ D_3 \\ 1 \end{pmatrix}\right).$$

The result of the functioning with different initial data (X = (0;0), (0;1), (1;0), (1;1)), taking into account the coefficients and conditions of performed exercises (time, attempts, characteristics of cadet / listener), are as follows (Table 3).

In general, the ANN can be represented as follows:

$$\overline{H} = f(\overline{W}_1, \overline{X});$$

$$\overline{D} = f(\overline{W}_2, \overline{H});$$

$$\overline{Y} = f(\overline{W}_3, \overline{D}).$$

From the equations we've to get the definition of fourth-layer ANN (9):

$$\overline{Y} = f(W_1 f(W_2 f(W_3(\overline{X})))), \tag{9}$$

where $\overline{X}$ – is a network input vector (exercises); $\overline{W}$ – are the coefficients of individual cadet's / listener's characteristics.

For example, for vector $\overline{H}$, which determines the physiological characteristics of cadet / listener, we have:

Table 3. Results of the functioning of the ANN

X_1	X_2	H_1	H_2	H_3	D_1	D_2	D_3	Y_1	Y_2	Y_3
0	0	0	0	0	0	0	0	0	0	0
0	1	1	1	0	1	0	0	0	1	1
1	0	0	0	0	0	0	0	0	0	0
1	1	1	1	1	1	1	1	1	0	1

- $H_1 - w_{11}, w_{21}$ – are the coefficients, which characterized by the ability of cadet / listener N°1;
- $H_2 - w_{12}, w_{22}$ – are the coefficients, which characterized by the ability of cadet / listener N°2;
- $H_3 - w_{13}, w_{23}$ – are the coefficients, which characterized by the ability of cadet / listener N°3.

Similarly, weighing coefficients that characterize the complexity of the dynamic air situation (vector $\overline{D}$) and the quality of the exercise according to the timeliness (vector $\overline{Y}$) have been taken into account.

The Table 4 shows that during performance of exercise N°2 ($X_2 = 1$) cadets / listeners N°2 and N°3 have completed the task in time (N°1 – have not complied). During solving exercise N°1 – nobody has completed the task. During simultaneously performing of two tasks ($X_1 = 1, X_2 = 1$) cadets / listeners N°1 and N°3 have completed the task, cadet / listener N°2 – have not coped with the task.

Visualization of Results of Training Exercise Performance by Air Traffic Controller

The computer program for visualization of the status of exercise performance by cadet / listener under timeliness criterion has been developed (Shmelova et al., 2016). The instructor has information about ATC's stage of solving the problem: threats, pre-conflict or conflict. Threats should be seen as a warning that it is necessary to take immediate measures to resolve the PCS. Pre-conflict stage shows that to avoid conflict is difficult or impossible. Displays information about the origin of these steps will allow those who are taught to pay attention to the necessary taking actions to resolve PCS.

To increase the effect have been proposed to duplicate the information data on those aircraft, between which is predicted PCS. For the instructor, such information will help draw the attention of the learner on the need for measures to resolve the PCS. In conducting group sessions, such information will help the instructor to know which cadet / listener cannot cope with the task.

The Institute of Air Navigation of Flight Academy of National Aviation University has developed modeling complex "Fusion", which provides the multimodal system of predicted PCS (Figure 6).

Information from the display of dynamic air situation, with regularity in one second ($t = 1$ sec), is transmitted to the objective system for storing information as to dynamic air situation. Then, considering the parameters of aircraft's movement and its relative positions can determine the type of PCS.

Figure 6. The block scheme and algorithm of advanced modeling MC Fusion

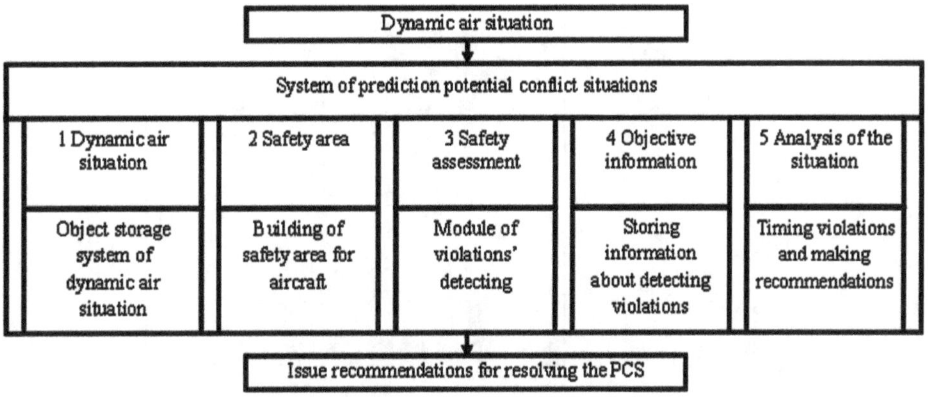

After determining the type of PCS, this information comes to the module of construction safety zone of aircraft. The security zones in MODELING COMPLEX "Fusion" are built in compliance with the regulations to maintain separation intervals (longitudinal, lateral and vertical) (ICAO, 2007). The size of the security zone depends on airspace structure and the relative position of the aircraft (for which the calculation is performed). Security zone is built along the motion vector of aircraft at each time when providing recalculation of the relative position of the aircraft. Ingestion of the aircraft in the security zone of another aircraft clearly regarded as a violation of separation intervals and recorded as the violation.

The module of detection violation's system, including the type of conflict and an active situation of security zone, defines the fact of violation of the intervals.

In cases where the system has detected a violation of safe intervals, this information comes in the system of storage information about violations. Storage information system about violations is able to keep three types of information about the conflict that took place, namely:

- Callsigns of couples aircraft, between which there was conflict;
- Time of conflict;
- Type of conflict.

The visual form of information presentation is the most appropriate form about occurrence phases of PCS for understanding by cadet / listener or instructor.

According to the developed classification, proposed to output the formula of three elements, each of which will show the number of potential conflicting situations at each stage of development. Figure 7 shows the exterior form of the output of such information.

There are elements which are colored yellow, crimson and red colors correspond to stages of conflict situation (have shown in Figure 7). The yellow element corresponds to stage "threat of conflict", crimson – "pre-conflict situation," red – "conflict situation".

Figure 7. Forms of indication: a – there are predicting four PCS (identified by yellow – the first column); b – one of the PCS has passed the stage of "pre-conflict situation" (identified by the crimson colour – the second column), c – one of the PCS has passed the stage of "conflict" (violation of separation intervals), one more of the PCS has passed the stage of "pre-conflict situation" (crimson (second column) and red (third column) colors have identified)

Indicators, which are located under the color elements, are designed to display information about the number of PCS at an appropriate stage of development. The use of the proposed indicators stages of the PCS at the workplace of future ATS specialist recommended at the stages of training, which will allow cadets / listeners gain the necessary skills to identify and resolve PCS. The similar indication at the workplace of instructor (teacher), can do the group sessions with cadets / listeners is more easier, because that will promptly identify problems in the detecting and resolution of PCS during the performance of training exercises.

Estimation of Pre-Simulation Training Tasks Complexity

The organization of the airspace over the definite area should be arranged so that it corresponds to operational and technical considerations only. In addition, aerodromes, where ATC is provided, should be designated as controlled aerodromes (Ministry of Transport of Ukraine, 2003).

States are selecting airspace classes which are appropriate to their needs. There are three zones of airspace – very important elements of ATM with individual restrictions: Control Zone (CTR), Terminal Control Area (TMA) and Control Area (CTA) (ICAO, 2007).

Control Area can be formed by TMAs of sufficient size to contain the controlled traffic around the busier aerodromes, interconnecting airways of a lateral extent, determined by the accuracy of track-keeping of aircraft operating on them, as well as the navigation means available to aircraft and their capability to exploit them; a vertical extent, covering all levels require to be provided with control service; or area-type control areas within which specific ATS routes have been defined for the purpose of flight planning and which provide for the organization of an orderly traffic (ICAO, 2007).

According to ICAO requirement, the Ukrainian ATS airspace has been classified and designated for following classes: class C and class D (Ministry of Transport of Ukraine, 2003) (Table 4).

The complexity of zone (CTR, TMA, and CTA) has been obtained with the help of expert assessments method (Beshelev & Gurvich, 1973). The experts were ATC, who operated in training course.

The algorithm of estimation of the complexity tasks in the Pre-Simulation Training with the help of expert assessments method:

1. Questionnaires for experts: m – is a number of experts, $m \geq 30$.
2. The matrix of individual preferences: R_i – is a system of preferences of i-expert, $i = \overline{1, m}$.
3. The matrix of group preferences R_j (10):

Table 4. Implementation of airspace classes in Ukrainian airspace

N°	Class of Airspace	Zones of Airspace	Restrictions
1	D	CTR	1500-2900 m
2	D (except Boryspil TMA, where class C is applied due to high traffic volumes)	TMA	D (1500-2900 m) and C (2900 m – Fl 660)
3	D and C	CTA	D (1500-2900 m) and C (2900 m – Fl 660)

$$R = R_j = R_{gr} = \frac{\sum_{i=1}^{m} R_i}{m}. \quad (10)$$

where $R = R_j = R_{gr}$ – is an experts' group opinion of the complexity of *j*-zone, $j = \overline{1, n}$; *n* – is a quality of zones.

4. Coordination of expert's opinion.
   a. Calculation of dispersion *D* (11):

$$D = \frac{\sum_{i=1}^{m} (R_{gr} - R_i)^2}{m - 1}. \quad (11)$$

   b. Calculation of square average deviation σ (12):

$$\sigma = \sqrt{D}. \quad (12)$$

   c. The coefficient of the variation ν (13):

$$\nu = \frac{\sigma}{R_{gr}} \cdot 100\%. \quad (13)$$

If $\nu_{\text{CTR, TMA, CTA}} \leq 33\%$ then the opinion is concerted and system of expert group has been obtained. If $\nu_{\text{CTR, TMA, CTA}} > 33\%$ then it is necessary to calculate Kendal's concordation coefficient *W* (14):

$$W = \frac{12S}{m^2(n^3 - n) - m\sum_{i=1}^{m} T_i}; \quad (14)$$

$$S = \sum_{j=1}^{n} \left(\sum_{i=1}^{m} R_{ij} - \overline{R} \right)^2;$$

$$T_i = \sum_{i=1}^{m} (t_i^3 - t_i);$$

$$\overline{R} = \frac{1}{n}\sum_{j=1}^{n}\left(\sum_{i=1}^{m} R_{ij}\right),$$

where S – is a generalized dispersion; t_i – is a number of the same ranks in the i-row which fixed the i-expert.

Kendal's concordation coefficient must be within the limits $0,7<W\leq1$. If $W<0,7$ it is necessary to repeat the interrogation.

5. Compare the system of preferences R_{gr} and R_i, $i = \overline{1,m}$ with the help of rating correlation coefficient R_s (Spearman's coefficient) (15):

$$R_s = 1 - \frac{6\sum_{j=1}^{n}(R_{gr} - R_i)^2}{n(n^2 - 1)}. \tag{15}$$

6. The significance of the calculations.
   a. The significance of the calculation of Kendal's coordination coefficient W, criterion $-\chi^2$ (16):

$$\chi_f^2 = \frac{S}{\frac{1}{2}m(n+1) - \frac{1}{12(n-1)}\sum_{j=1}^{n} R_j} > \chi_t^2, \tag{16}$$

where χ_f^2 – is a factual value of variable; χ_t^2 – is a table value of variable.

   b. The significance of the calculation of Spearman's coefficient R_s with using Student's t-criterion (17):

$$t_{critical} = R_s\sqrt{\frac{n-2}{1-R_s^2}} > t_{st},. \tag{17}$$

where n – is a quality of zones; t_{st} – is a tabulated value while the number of degrees of freedom $f = n-2$ and error $\alpha = 5\%$.

7. Weight coefficients w_j of j-zone complexity (18):

$$w_j = \frac{C_j}{\sum_{j=1}^{n} C_j}; \tag{18}$$

$$\sum_{j=1}^{n} w_j = 1,$$

where n – is a quality of zones; $C_j = 1 - \frac{R_j - 1}{n}$ – are the estimates, $j = \overline{1, n}$.

8. Graphical presentation of weight coefficients.

For estimation complexity of the exercises in the system of Pre-Simulation Training, it is necessary to find weight coefficients, which characterize the complexity of airspace zones. Assessments of the complexity of airspace zones with using the algorithm of estimation of the complexity tasks in the Pre-Simulation Training have been obtained (Shmelova, Lazorenko, & Bilko, 2014):

1. Questionnaires for experts – ATC with working experience.
2. The matrix of individual preferences. Evaluation of complexity of airspace zones (CTR, TMA, and CTA): R.

R_i has been obtained, $i = \overline{1, m}$, where m – is a number of experts; R_i – is a system of preferences of i-expert.

For example, $R_i = R_{iTMA} > R_{iCTA}, R_{iCTR}. 1, m$

3. The matrix of group preferences for CTR has been obtained:

$$R_{grCTR} = \frac{\sum_{i=1}^{m} R_{iCTR}}{m} = 2,64 .$$

Average value R_{gr} for TMA and CTA have the similar calculations (Table 5).

4. Concordance of expert's opinion.
   a. Calculation of dispersion D:

$$D_{!\,TR} = \frac{\sum_{i=1}^{m} (R_{grCTR} - R_{iCTR})^2}{m - 1} = 0,401099 .$$

Calculations for TMA and CTA would be the same variant (Table 5).

   b. Calculation of square average deviation σ:

$$\sigma_{CTR} = \sqrt{D_{CTR}} = 0,633324 .$$

Calculations for TMA and CTA would be the same variant (Table 5).

c. The coefficient of the variation ν:

$$\nu_{CTR} = \frac{\sigma_{CTR}}{R_{grCTR}} \cdot 100\% = 23{,}9636\% .$$

Variation for TMA and CTA have the similar calculations (Table 5).

If $\nu_{\text{CTR,TMA,CTA}} \leq 33\%$ then the opinion is concerted and the system of expert group has been obtained. For example, $R_{gr} = R_{TMA} > R_{CTR} > R_{CTA}$.

Calculations show that opinion is concerted and it is necessary to obtain weight coefficients of complexity for airspace zones.

5. Weight coefficients of complexity (Table 6):

$w_1 = w_{\text{CTR}}$ – weight coefficient for CTR zone;

$w_2 = w_{\text{TMA}}$ – weight coefficient for TMA zone;

$w_3 = w_{\text{CTA}}$ – weight coefficient for CTA zone.

6. Graphical presentation of weight coefficients of CTR, TMA and CTA (Figure 8).

Calculations have been performed with using computer program MS Excel.

Integrated estimation of tasks Q_{jl} with j-level of complexity in air traffic control for n-zones (19):

$$Q_{jl} = \sum_{j=1}^{n} \sum_{l=1}^{L} w_j q_l, \tag{19}$$

where q_l – is an estimation of the task with given complexity and type of airspace zone (CTR, TMA, and CTA); w_j – is a weight coefficient (complexity of airspace zone); l – is a level of the task.

Table 5. The matrix of group preferences

Coordination of Expert's Opinion	CTR	TMA	CTA
	x_1	x_2	x_3
R_{gr}	2.642857	1.142857	2.214286
D_i	0.401099288976	0.131868480769	0.335164576356
σ_i	0.633324	0.363137	0.578934
ν_i, %	23.9636	31.77445	26.14542

Table 6. The results of the calculation of weight coefficients

N°	Zone	R_{gr}	C_j	w_j
1	CTR	2.642857	0.453333	0.226161
2	TMA	1.142857143	0.953333	0.475786
3	CTA	2.214285714	0.596667	0.298053

Figure 8. Weight coefficients of zones' importance

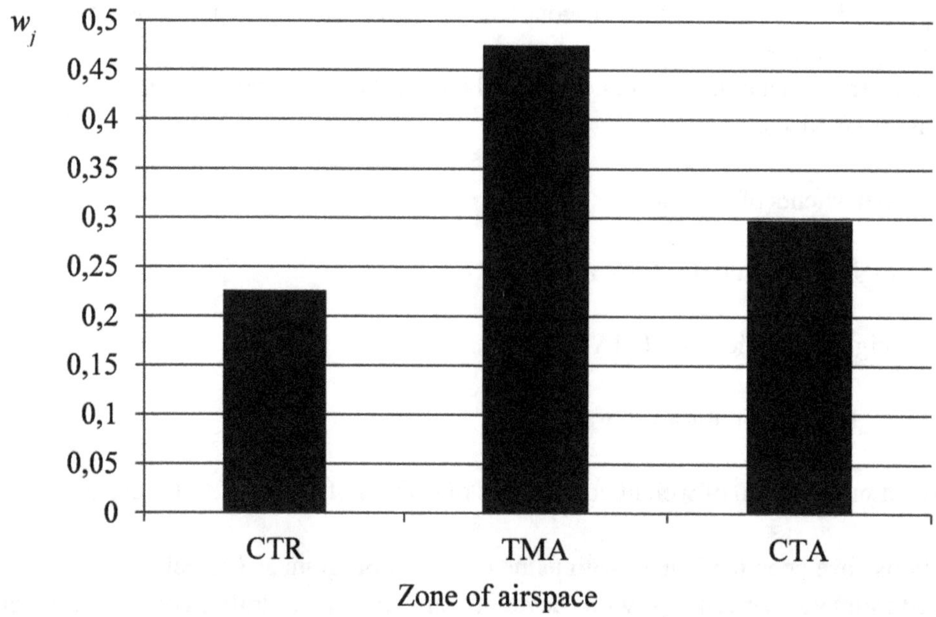

So, we can see that the most difficult zone of airspace according to the opinion of the expert is the Terminal Control Area (w_{TMA} = 0.475786).

In the future, ATC instructor would take into account this expert's opinion for the definition of a task according to difficulty.

Integrated estimation for implementation of the task with complexity characteristic of airspace zone has been calculated with the help of additive aggregation method.

For automating estimation of Pre-Simulation Training on initial stage the neural network of multi-layer type has been built (Shmelova et al., 2014). Figure 9 represents the neural network for Simulator Training of cadets / listeners with the specified number of hours and level of training. It is the multilayer perceptron with two hidden layers:

- **The First Layer:** Is a calculation of hours on theoretical training in accordance with the cadets' / listeners' knowledge evaluation;
- **The Second Layer:** Are the restrictions on the given number of hours (hidden layer);
- **The Third Layer:** Are the restrictions on passing hours (hidden layer);
- **The Fourth Layer:** Is an assessment of cadet / listener.

Threshold activation functions have been obtained according to requirements of hours and level of training marks (discipline) in compliance with the assessment criterion of tasks.

All positions of assessment criteria have been prescribed and must be used in a proper way as given below:

1. To take a duty and workplace preparation.
2. An ability to follow the prescribed standard phraseology (excepting tasks with emergency and urgency situations).
3. Coordination with adjacent ATC units and other kinds of aerodrome service provision units.
4. Handling of procedural control.
5. Handling of visual control.
6. Daily flight plan conduction.
7. Execution of traffic massages timesheets.
8. Timeliness and accuracy of decision making in ATC.
9. Compliance with safety in ATC.
10. Performing of console operations.

Some of the criteria might not be used in order to which are chosen by a supervisor before training.

During Pre-Simulation Training it is necessary to take into account that the abilities and skills of ATC are evaluating feedback.

We cannot apply general assessment criteria as they exist for the time of operation actions evaluation. However, some of them might be useful to apply. So, Pre-Simulation Training assessment criteria are:

- The retelling of the given situation;
- Phraseology to be used;

Figure 9. Neural network for Pre-Simulation Training of cadets / listeners with the specified number of hours and level of training

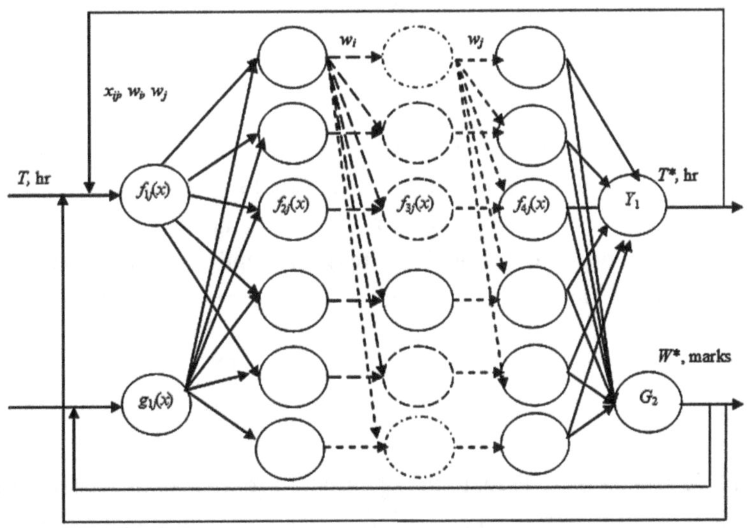

- Taking a duty and workplace preparation;
- information about traffic;
- Level change proposal;
- Vectoring and heading guidance.

The retelling of the given situation and expected phraseology is a fully individual criterion for supervisors. They must define elements which are necessary to be retail as well as to evaluate proposed phraseology.

Information about traffic must be a reflection of an ability of a student to define/detect through the plan all conflict situations. Evaluation of this element depends on the whole number of conflicts regarding the number of missed.

Level change proposal is a kind of a feedback that defines a degree of suitably given flight levels or altitudes by a student. Evaluation of this element is performed in percentage terms. Deviation of each level is equal to loss of one point.

The accuracy of a heading designation is a three degrees deviation from the desired direction. If we are out of limits we are losing one point within a criterion.

Every element has a five-point system basis. Evidently, missing all points of procedural elements by a student should be considered as non-completed element and the non-completed task correspondingly. Positively given marks are output into the mean value.

FUTURE RESEARCH DIRECTIONS

It is relevant to adapt the development artificial neural network for ATC Pre-Simulation Training of others Air Navigation System's human-operators: pilots, Unmanned Aerial Vehicles operators, flight dispatchers, rescuers, specialists of aviation security, etc. that will allow increasing the quality of their professional study.

It is our belief, that the neural network models can be used for improving the professional training of specialists in any technogenous production (hydraulic engineering, chemical and military industries, gas and oil pipelines, nuclear power plants and transport, etc.).

What's Next?

On the next steps we are planning to use neural network in the System for control and forecasting the development of emergency situations on the base of Artificial Intelligence System / Decision Support System that taking into account the influence of the professional factors (knowledge, habits, skills, experience) as well as the factors of non-professional nature (individual-psychological, psycho-physiological and socio-psychological) on the decision making process by human-operator of Air Navigation System.

CONCLUSION

On the base of the basic concepts of Threat and Error Management in air traffic control the stages of the developing conflict situation have been classified (threat – error – undesirable condition) and quantitative indicators of the complexity level at each stage using fuzzy logic have been defined. The neural network model of assessment the timeliness and correctness of the decision making by the specialist of ATS during the Pre-Simulation Training has been developed and its parameters have been obtained. The block diagram of MODELING COMPLEX "Fusion" with the ability to display phases of PCS, which simplifies the process of ATC training, as well as evaluating their actions in the performance of educational tasks by the instructor, has been presented.

Taking into account the timeliness and correctness of instructor's tasks performed during the Pre-Simulation Training with the help of using ANN will allow determining the possibility of access of specialist of ATS to Simulator Training. Multimodal system "Fusion" will give the possibility to improve the process of training of cadets / listeners through automated assessment of their actions.

With the help of methods of expert assessment, the difficulty of ATC operations (tasks) has been determined. Using the expert's opinion and criterion of weight coefficient, the hard zone for operation on Initial Training, such as simulator practice, has been defined. According to interrogation, the graph, based on preliminary calculation, has been built. The analysis of the graph has shown that TMA is in the first position according to the complexity of operation and procedure in air traffic control. The neural network for Pre-Simulation Training of cadets / listeners with the specified number of hours and level of training has been built. Automation of estimation of Pre-Simulation Training on the phase of Initial Training increases the efficiency of Simulation Training through interactive evaluation of the tasks performed by cadets / listeners. Performance of exercises is accompanying by comments, displaying results, assessments of the cadets / listeners and feedback.

REFERENCES

Allianz Global Corporate & Specialty, Embry-Riddle Aeronautical University. (2014). *Global Aviation Safety Study: A Review of 60 Years of Improvement in Aviation Safety*. Authors.

Arkhangelskiy, V., Bohaenko, I., Grabowskiy, G., & Ryumshyn, N. (1999). *Neural networks in systems of automation*. Kiev: Techniques.

Aviation Accident Statistics. (2018). *National Transportation Safety Board*. Retrieved from www.ntsb.gov/aviation/aviation.htm

Beshelev, S. D., & Gurvich, F. G. (1973). *Expert assessment*. Moscow: Science.

Borisov, A., Krumberh, O., & Fedorov, I. (1990). *Decision Making Based on Fuzzy Models: Examples of Using*. Riga: Zynatne.

Borovikov, V. P. (2008). *Neural Networks. STATISTICA Neural Networks: Methodology and Technology of Modern Data Processing* (2nd ed.). Moscow: Hotline-Telecom.

European Organisation for the Safety of Air Navigation (Eurocontrol). (2000). *Simulations Facilities for Air Traffic Control Training. HUM.ET1.ST07.3000-REP-02*. Brussels, Belgium: Author.

European Organisation for the Safety of Air Navigation (Eurocontrol). (2004a). ATM Services' Personnel (2nd ed.). ESARR 5. Belgium, Brussels: Author.

European Organisation for the Safety of Air Navigation (Eurocontrol). (2004b). *EATM Training Progression and Concepts*. Brussels, Belgium: Author.

European Organisation for the Safety of Air Navigation (Eurocontrol). (2015). *Specifications for the ATCO Common Core Content Initial Training* (2nd ed.). Brussels, Belgium: Author.

Gardner, M. W., & Dorling, S. R. (1998). Survey and Critique of Techniques for Extracting Rules from Trained Artificial Neural Networks. *Atmospheric Environment, 32*(14–15), 2627–2636. doi:10.1016/S1352-2310(97)00447-0

Goedeking, P. (2010). *Networks in Aviation: Strategies and Structures*. Springer. doi:10.1007/978-3-642-13764-8

International Civil Aviation Organization. (2007). Air Traffic Management (15th ed.). Doc. ICAO 4444-ATM/501. Montreal: Author.

International Civil Aviation Organization. (2007). Procedures for Air Navigation Services. Air Traffic Management (15th ed.). Doc. ICAO 4444-ATM/501. Montreal: Author.

International Civil Aviation Organization. (2008). *Threat and Error Management (TEM) in Air Traffic Control. Cir. ICAO 314-AN/178*. Montreal, Canada: Author.

Kharchenko, V., Chynchenko, Yu., & Raychev, S. (2007). Threat and Error Management in Air Traffic Control. *Proceedings of the National Aviation University, 3–4,* 24-29.

Kharchenko, V., Shmelova, T., Artemenko, O., & Otryazhiy, V. (2011). *A. r. Computer Program "Choosing the Pre-Flight Information and Decision Making about Departure for Automated System of Pre-Flight Information Preparation (AS PIP)"*. Kyiv: State Department of Intellectual Property.

Kharchenko, V., Shmelova, T., Sikirda, Yu., & Gerasimenko, O. (2011). *A. r. Computer Program "Optimizing the Choice of Alternative Variant of Aircraft Flight Completion in Flight Emergencies "Prompt"*. Kyiv: State Department of Intellectual Property.

Komashynskiy, V., & Smirnov, D. (2002). *Neural Networks and Its Use in Systems of Management and Communication*. Moscow: Hotline-Telecom.

Leychenko, S., Malishevskiy, A., & Mikhalic, N. (2006). *Human Factors in Aviation: monograph in two books*. Kirovograd: YMEKS.

Ministry of Transport and Communications of Ukraine. (2007). *The Rules for Issuing Certificates of Aviation Personnel in Ukraine*. Kyiv: Author.

Ministry of Transport of Ukraine. (2003). *About Approving the Flight Rules of Aircraft and Air Traffic Services in the Classified Airspace of Ukraine*. Kyiv: Author.

Shmelova, T., Lazorenko, V., & Bilko, A. (2014). Estimation of Pre-Simulating Training Tasks Complexity. *Proceedings of the National Aviation University, No., 1*(62), 17–22.

Shmelova, T., Sikirda, Yu., Zemlyanskiy, A., & Danilenko, O. (2015). Fuzzy Assessment of the Situation's Complexity during Simulation Training of Specialists in Air Traffic Services. In *Materials of International Scientific-Practical Conference "Problems of Energy Efficiency and automation in industry and agriculture"*. Kirovohrad: Kirovohrad National Technical University.

Shmelova, T., Sikirda, Yu., Zemlyanskiy, A., Danilenko, O., & Lazorenko, V. (2016). Artificial Neural Network for Air Traffic Controller's Pre-Simulator Training. *Proceedings of the National Aviation University, No., 3*(68), 13–23.

Shmelova, T., Yakunina, I., Moyseenko, V., & Grinchuk, M. (2014). *A. r. Computer Program "Network Analysis of Flight Emergency"*. Kyiv: State Intellectual Property Service of Ukraine.

Shmelova, T., Yakunina, I., & Yakunin, R. (2013). *A. r. Computer Program "Test Extraordinary Incident"*. Kyiv: State Intellectual Property Service of Ukraine.

Statistics of the World's the Largest Aircraft Accidents for the Years 1974-2014. (2018). Foringshurer Insurance. Retrieved from https://forinsurer.com/public/17/01/10/3824

Suzuki, K. (2013). *Artificial Neural Networks – Architectures and Applications*. InTech. doi:10.5772/3409

Zade, L. (1965). Fuzzy sets. *Information and Control, 8*(3), 338–353. doi:10.1016/S0019-9958(65)90241-X

Chapter 3
Using Unmanned Aerial Vehicles to Solve Some Civil Problems

Aleksander Sładkowski
Silesian University of Technology, Poland

Wojciech Kamiński
Silesian University of Technology, Poland

EXECUTIVE SUMMARY

The widespread use of unmanned aerial vehicles (UAVs) is currently a recognized trend. UAVs find their application in various sectors of the economy. In the chapter, based on extensive literature analysis, the possibilities of using UAVs for non-military applications are considered. The design features of various UAVs, their control features, energy requirements, and safety-related problems are considered. Particular attention is paid to public opinion related to the use of UAVs. The possibilities of using UAVs in power engineering, agriculture, for controlling traffic, for goods transporting, for controlling the means of railway transport, for first aid to people under various extreme conditions, as well as for some other applications are being explored. The UAV parameters are analyzed, which must be provided for their use in each specific case, while ensuring the minimization of the necessary financial resources.

BACKGROUND OF UAVS

The initial history of the development of UAVs was related to their use for military purposes. The page (History, 2018) notes that the first practical use of UAVs was in 1849 during the suppression of the Venetian insurrection by the Austrian army. In this case, unmanned balloons stuffed with shrapnel were used.

The use of winged aircraft as a drones is associated with the name of the American inventor Charles F. Kettering (Charles, 2018). Kettering designed the “aerial torpedo”, nicknamed the Kettering Bug. The sources cited above indicate different years of the creation of this invention (1910, 1914 and 1918). It was a radio-controlled bomb, created on the basis of a fairly primitive winged aircraft. Despite the fact that it was not possible to test this device in military conditions, the “Bug” is considered the first aerial missile. A total of 45 Bugs were produced, one of which is currently in the National Museum of the United States Air Force in Dayton, Ohio.

DOI: 10.4018/978-1-5225-7588-7.ch003

As the first unmanned reusable aircraft, De Havilland DH82B Queen Bee (De Havilland, 2018) can be considered. This aircraft, created for the Royal Air Force and the British Navy was used as a flying target. There were specimens with the possibility of landing on a ground surface or on a water surface. A total of about 1,000 aircraft were manufactured.

In the article (Benchoff, 2016) it is noted that the history of the emergence of the word "Drone", which is now commonly used for the name of UAVs, is most likely connected with this aircraft. In English, the words "Drone" and "Queen Bee" are used to describe the representatives of the bee family. Accordingly, we can assume that the word "Drone" has been in use since 1935.

Unguided aircraft were created and widely used during the Second World War and in the prewar years. The history of their creation and use is described in detail in the book (Everett, 2015).

Currently, drones have found numerous areas of use for non-military purposes. This chapter will be devoted to the examination of existing possibilities for using drones in various areas of human activity.

Modern Terminology

An aircraft is a flying object that can move or stay in a near-Earth atmosphere. It is used to transport people and loads (Dobryakova & Ochin, 2016). According to the definition of Unmanned Aerial Vehicle (UAV), it is an air-powered object that performs flight using aerodynamic force, without having a pilot on board and not transferring passengers. Such a device is often called a drone, which is a synonym. Due to the fact that unmanned aerial vehicles are an element of the system in which they operate, it is more accurate to describe the Unmanned Aerial System (UAS). This system, in addition to the aircraft itself, consists of other elements that are on the ground and are necessary for flights (Burdziakowski, 2016).

These elements are:

- Control station (it includes system operators, devices supporting interoperability of teams, interfaces and other elements of control);
- Communication systems between the unmanned aerial vehicle and the devices mounted on it, and the control station;
- Auxiliary equipment for servicing and transporting system components (Bujnowski, 2017).

Also introduced was the concept of Remotely-Piloted Aerial System (RPAS), which according to the decision of the International Civil Aviation Organization (ICAO) is able to meet the requirements set in civil aviation, which allows it to be allowed to fly in space air. A remotely piloted aircraft system to perform operations in non-dedicated airspace must be able to detect other airspace users and be capable of taking adaptation. In other cases, the operations of these systems must be limited to the separate airspace (Burdziakowski, 2016). In Polish regulations, the person piloting the aircraft system remotely is the operator of the unmanned aircraft, while the person who operates the aircraft model is the pilot-operator of the flying model. The division into unmanned aerial vehicles and flying models results from the purpose of their use. Structures used for sports or recreational purposes are referred to as flying models, while when used for other purposes they are called unmanned aerial vehicles. In the European Union, operations of unmanned aircraft used for civil purposes are regulated by technical criteria, which include the criterion of mass and equipment of the internal combustion engine. On this basis:

- Unmanned aerial vehicles with a weight of more than 150 kg are subject to EASA (European Aviation Safety Agency) regulation;
- Unmanned aerial vehicles of less than 150 kg weight and aircraft models are subject to the guidelines of European Union member states;
- Toys that are capable of flying but do not have an internal combustion engine are covered by Directive 2009/48/EC (Burdziakowski, 2016).

Operations with the use of unmanned recreational and sports (non-commercial) aircraft in uncontrolled airspace carried out with devices not exceeding 25kg and in sight do not require the operator to have any qualifications document and there is no need to report them in any institution. In the case of flights in sight (VLOS), however, for commercial purposes the operator's certification of the unmanned aircraft operator is required. These entitlements are obtained after passing the state examination. To receive the permission to fly out of sight (BVLOS) it is necessary to undergo theoretical and practical training (Fellner, 2015).

The Principle of Multicopter Operation

The names of multicopter correspond to the number of arms, so the device having 3 arms is a tricopter, 4 is a quadcopter, etc. Multirotors usually have a symmetrical number of arms from which their frame is built. Determining the front of such a device is necessary, however, due to later control of it. There are two configurations of simple (independent) and redundant (surplus) multicopter configurations. The first of these relies on the placement of propellers on a circle the same distance from the center of the device. Each of the rotors turns in the opposite direction to the propeller with which it is adjacent. Redundant configuration makes it possible to reduce the number of arms by half, however on each of them there are two rotors. The upper propellers rotate in one direction and the lower ones in the opposite direction. Regardless of the system used, the use of counter-rotating propellers allows the flight to stabilize completely.

Adjusting the flight altitude of such a drone depends on the pulling power of the engines. If they balance its mass, it can remain motionless. Tilt or swivel maneuvers are performed by accelerating or decelerating the rotation of individual rotors. In the quadcopter, if the two front propellers decrease the speed, and the rear one will increase, it will start to move forward, while if the reverse situation occurs, it will move backwards (Piotrowski, 2015). Currently, the number of activities that drones perform automatically increases.

The most difficult part of the flight of the multicopter is an automatic landing on a solid ground or a moving platform. In order to avoid breakdowns and potential damage, a high level of security should be maintained at this time. Most automatic landing systems use computer algorithms and other sensors such as GPS (Patruno, 2017). It is important to control the flight while approaching the unmanned aircraft for landing so that the landing takes place in the correct place, designated by the operator. When touching the multicopter to the ground, it is best to do this at the lowest possible speed to avoid potential mechanical damage to the device. Automatic landing systems use a variety of location systems, such as lasers and radio waves. It is important that these devices have the smallest possible mass, so that they do not use too much limited drone load capacity in this way (Nobahari, 2017). One of the popular multiplexes, quadcopter is shown in the Figure 1.

Figure 1. One of the popular quadcopter models
(Wang, 2016)

The Principle of the Bicopter Operation

This system is atypical; it uses motors mounted on movable arms in the transverse system (Zahorski, 2016). The appearance of the bicopter is shown in the Figure 2. Control of the bicopter's inclination takes place by symmetrical deflection of the arms with the motors mounted on them, and not, as in the case of traditional multi-rotors, by changing the thrust of the engines. To control the rotation of the bicopter around the vertical axis, the arms on which the motors are mounted lean towards the opposite sides. The revolutions of each engine are independent and their regulation allows stabilization and inclining. To control slopes, the bicopter is additionally equipped with a horizontal stabilizer with a rudder. For the synchronization of all mechanisms, a microcontroller with appropriate software and built-in accelerometers and gyroscopes is used here. The construction of the entire bicopter can be divided into three main parts: frame; housing with a stabilizer; arms with engines.

The Principle of Helicopter Operation

For the proper use of an unmanned helicopter, it is necessary to use a control system that allows you to use the construction features of the helicopter, such as the possibility of taking off and landing vertically and extended envelope loads, i.e. the factor of permissible loads depending on the speed of flight. The development of remotely controlled helicopter technology is hampered by the complexity of the flight control process. Designing a flight control system for a helicopter, in contrast to fixed wing aircraft. The helicopter body has a large inter-axial coupling, its behavior is strongly related to the atmospheric conditions in which the flight takes place, and when changing the speed of the rotor there are delays in the reaction through the body. The flight dynamics of the helicopter are also complex, which makes it difficult to model. The balancing rotor is connected to the helicopter's body by means of a flexible connector made of elastomer, which allows small propeller motions. This connector is attached to the rotor above the hinge that allows the propeller to rotate. During the start of the helicopter, the angle of attack

Figure 2. Bichopter UH-144
(Бикоптер, 2014)

of all the control blades is increased and the rotations of the main rotor are increased. It is possible to regulate the rotations, and thus the rotor thrusts, irrespective of the speed in the horizontal plane, making it possible to start or quickly change the flight altitude in place. The movement of the helicopter in the horizontal plane is carried out by changing the angle of attack of the blades and by tilting the rotor. Thanks to this, they can move both forward and backward as well as right or left. During remote control of the helicopter, the central element of the on-board system is the on-board computer cooperating with the sensors. Three types of navigational sensors are used:

- Fiber optic, using the Inertial Measurement Unit (IMU) allowing the measurement of acceleration and angular velocities;
- Global positioning system (GPS);
- Magnetic compass, informing about the direction of flight (Mettler, 1999).

The appearance of the unmanned helicopter is shown in the Figure 3.

The Principle of Fixed-Wing Aircraft Operation

Fixed-wing aircraft is an aircraft that is held in the air by generating lifting force thanks to the air flow on fixed bearing surfaces. Fixed-wing aircraft include kites, gliders, and manned airplanes, but they can also be unmanned aerial vehicles. The shape of the unmanned aerial vehicle is similar to a traditional manned aircraft, its appearance is shown in Figure 4 (Bielecki, 2015).

Due to their construction, fixed-wing aircraft can achieve higher speeds than copters, but their disadvantage is the lack of vertical launch and hovering in the air. In order to create adequate lifting force in larger unmanned fixed wing aircraft, it is necessary to use wings of considerable size. In order to facilitate the transport of fixed-wing aircrafts, it is possible to detach the wings from the body, which

Figure 3. Unmanned helicopter during the flight
(Clothier, 2015)

significantly reduces its width (Berner, 2016a). The individual elements from which the unmanned fixed-wing aircraft consists are shown in the Figure 5.

Construction of Unmanned Aerial Vehicles

The basic elements that make up the drone are:

- Frame (the most popular frames are made of carbon fiber, because it is light and durable, but it is quite fragile and expensive, often the frames are also made of aluminum, whose disadvantage is high flexibility);
- Chassis (it is designed to protect elements suspended under the drone during takeoff and landing, especially in difficult terrain);
- Propellers (usually they have two blades, although propellers with a larger number of them are also used. The direction of rotation of the rotor is also important. The propeller material is also important, those made of elastic material suppress unwanted vibrations, but need more energy. The stiff propellers better they use thrust, but they cause vibrations. The next parameter is the propeller pitch, which corresponds to the way it travels after a full rotation);
- Brushless motors (DC motors are used, there is a commutator in them, magnets are on the rotor, and the coils remain stationary. The disadvantage of these motors is the need to use a controller to adjust the motor speed);

Figure 4. The unmanned aerial vehicle S-380 Barracuda during tests
(Bielecki, 2015)

Figure 5. Construction of unmanned fixed-wing aircraft eBEE
(Berner, 2016a)

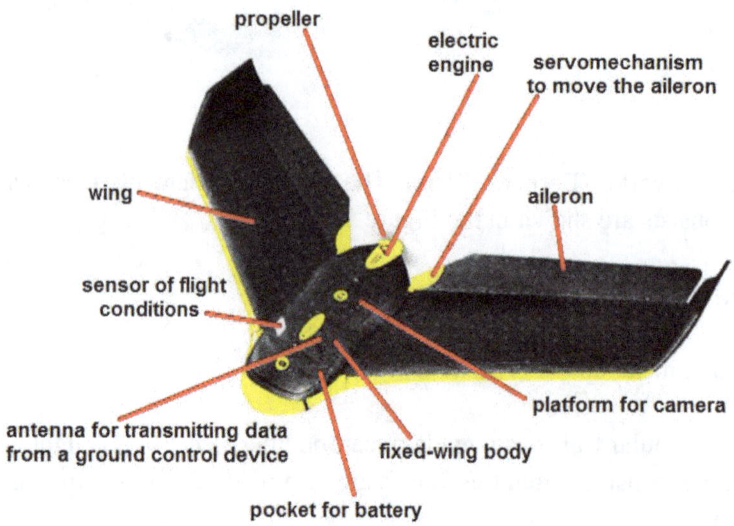

- Electronic engine speed controllers ((battery power is supplied to it, and it transfers electricity to the engine);
- A rechargeable battery (lithium-polymer batteries are used, because they have a large amount of energy in terms of their mass and volume, have a high current efficiency, provide a constant voltage value practically all the time to discharge and do not contain a liquid electrolyte that could be dangerous for the environment in the event of a drone crash) (Wyszywacz, 2016);
- Control sensors, these are:

    - GPS (position is determined by means of 31 satellites, the drone is equipped with a receiver that reads the signal from satellites and with the help of information about distances from them it is possible to determine the position of the receiver in the airspace) (Omar, 2017);
    - Compass (due to the presence of another magnetic field in every place on the ground before the start, its calibration is required, moreover, they are susceptible to magnetic field interference by ferromagnetics);
    - Height measurement sensors (the barometer measures the pressure which decreases along with the altitude, while the sonar sends the wave and measures the time in which it returns after reflection from the ground);
    - Accelerometer (it allows to determine the direction and displacement of the drone by examining its acceleration);
    - Gyroscope (it allows you to determine the angular displacement during the flight);
- Controller (it is a processor with software that receives commands from the operator and converts them to properly motors reactions);
- Remote control (it is a transmitter with an operator and a receiver mounted on an unmanned aircraft; the transmitter transmits radio control signals to the receiver);
- RC DPV (these are devices that allow you to view live images from cameras mounted on the drone, the image is sent in real time. The operator has the ability to view the image on a monitor or via video cameras);
- Gimbal (it is a device used to ensure the stability of the camera mounted on the drone, which stabilizes the image transmitted from it);
- Additional devices (e.g. storage containers for batteries, rotor covers, etc.);
- Emergency systems (the most important procedure in case of failure is Return To Home, or return to the starting place. This procedure can be implemented when the batteries are discharged to a level that only allows you to reach the starting point or in case of sudden loss of control signal from the transmitter from the operator. Then the drone rises to the appropriate altitude, then flies in a straight line to the place where it took off and lands there) (Wyszywacz, 2016).

The location of individual elements on the multicopter is shown in the Figure 6.

The device allows control of unmanned aircraft without the need to make flights in sight equipped with a screen displaying in real time the image from the camera mounted on the drone is shown in the Figure 7.

The Method of an Unmanned Aircraft Controlling

There are different types of movements that a multicopter can do. To describe the position of rigid bodies in space, three angles are used to determine the orientation of an object relative to three spatial dimensions. These angles are: rotation, tilt and deviation (Rus, 2016). They are presented in the Figure 8. The deviation angle describes how much the multicopter is moved to the side, the tilt angle shows how the drone is moved forward or backward, while the angle of rotation determines the horizontal offset.

During the control of an unmanned aerial vehicle, most of its movements are carried out using the speed control of propeller engines. The electronic engine speed regulator is able to perform rotation, deviation and tilt as well as a change in the height of the drone. For tilting and deviating, it is necessary to increase the engine revolutions on one side of the unmanned aircraft and lower the speed of the

Figure 6. Location of multicopter accessories
(Rao, 2016)

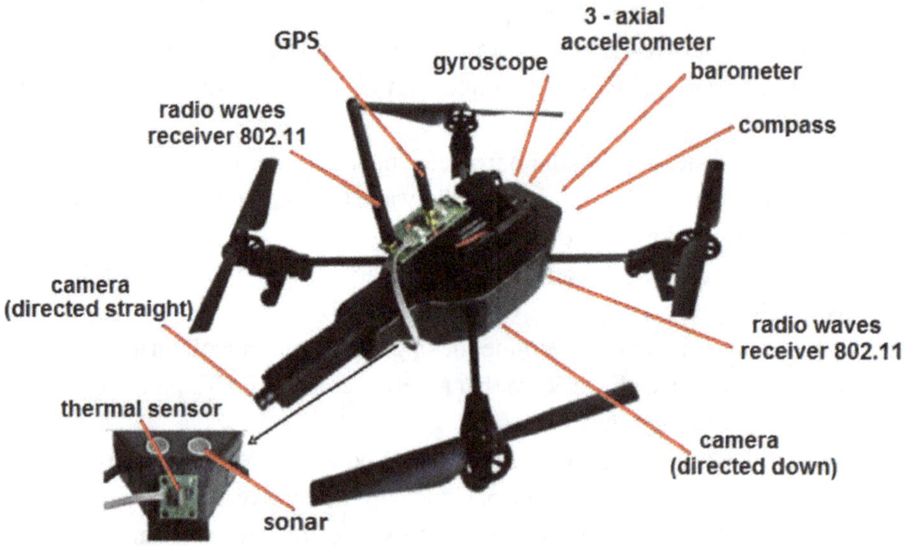

Figure 7. Device for remote control of unmanned aerial vehicles
(Wang, 2016)

engines on the other hand. To turn the quadcopter to the right, increase the engine revolutions on the left. Because of this, the left side will have a bigger driving force and it will tilt the drone to the right. A little more complicated is the performance of trading. During a normal flight, each propeller rotates counter to the adjacent rotor. This is done so that during the flight the multicopter does not rotate around its own axis. In order to rotate the drone in one direction, the speed of the pairs of propellers rotating in the opposite direction should be reduced, which will cause an imbalance of momentum and the same rotation of the unmanned aircraft. Quadcopter rotation to the right is possible by reducing the speed of the rotating pair of rotors to the left (Rus, 2016). This is shown in the Figure 9.

Figure 8. Device for remote control of unmanned aerial vehicles
(Wang, 2016)

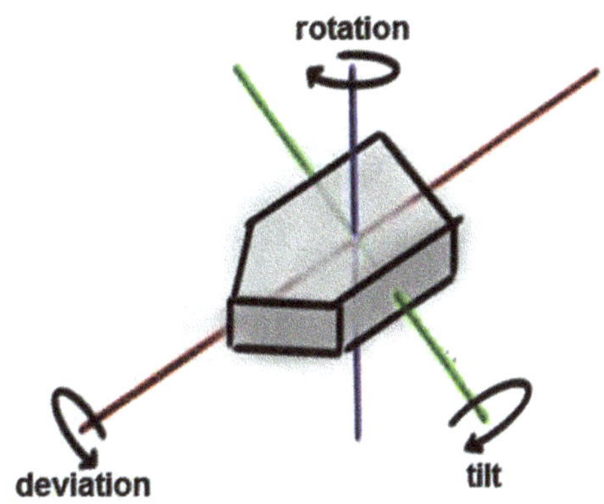

Figure 9. Quadcopter rotation to the right
(Rus, 2016)

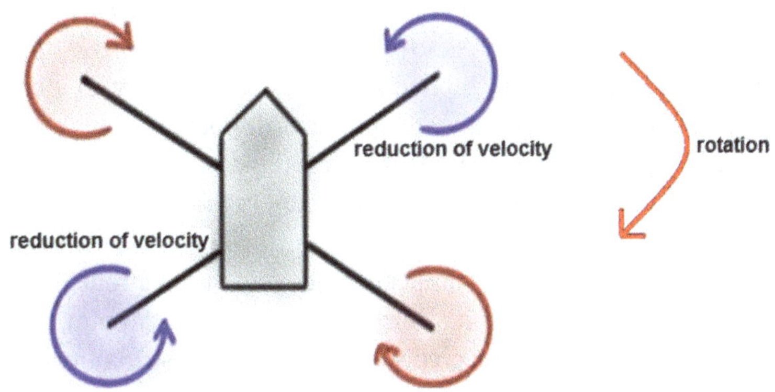

Ways of Drones Powering

Unmanned aerial vehicles most often for electricity supply, necessary for engine operation, use lithium-polymer batteries (Dudek, 2016). The battery is shown in the Figure 10.

The main disadvantage of such batteries is the limited time that the device can remain in the air. Therefore, attempts to connect this drone permanently to the power supply device have been made. The power supply was connected using a cable with a small cross-section. An unmanned aircraft powered in this way can theoretically perform operations indefinitely, but after 100 hours of flight it is recommended to descend to the ground to components control. The cable used to supply electricity also allows the transmission of data from the camera. For safety reasons, this drone has a battery that allows it to return to the starting point after interrupting the electric power supply. Another way to achieve a long flight time of unmanned aerial vehicles is to charge them during the flight by means of a laser beam sent from a device mounted on the ground in the vehicle. Yet another concept is the possibility of landing a drone on the overhead power line and picking up energy from it directly to charge the batteries (Dudek, 2016). Another way to extend the working time of an unmanned aircraft is to use a road vehicle equipped

Figure 10. Lithium-polymer battery very often used to drone power
(Piotrowski, 2015)

with a place to land and load a drone. In this case, both the drone and the road vehicle set off from the base. The road vehicle overcomes a certain road stopping several times, so that the unmanned aircraft can land, recharge the battery and start. The route in this case overcomes the drone is longer than the route of the road vehicle. It may include points that are not available for an ordinary road vehicle. After visiting all targets, both the drone and the road vehicle return to the base (Luo, 2017). A scheme of such operation is presented in the Figure 11.

Another way to eliminate the problem of limited time spent in the air by an unmanned aircraft is an automatic battery charging system that allows you to perform air operations without having to interrupt them. This system is used by iRobots's, a floor cleaning machine available on the commercial market. In the case of a discharging battery, it is automatically directed to the charging station. So far, it has not been possible to implement such a system for drones. The biggest obstacle here is the need for precise landing in the charging station, so that there is proper contact with the electrodes for their better conductivity. This requires accuracy in the navigation of the drone with respect to mm, which is very difficult outside, where there are problems with such a precise location. The most commonly used for navigating an unmanned aerial vehicle is GPS, which does not give precision so that you can use it in the charging station. A chance here is the use of wireless power transfer (WPT), which is used to charge mobile phones or other electronic devices. With it, it is possible to charge devices up to 2 meters away. The result is that using the WPT it is not necessary to land an unmanned aerial vehicle as precisely as in the case of a charging station. The possibility of using this method of charging the battery exists in the temperature range from -55°C to 120°C. The quadcopter AR Drone 2.0 was used for the research. As standard, the device is equipped with a 1000 mAh battery, while for tests it has a 500 mAh battery, so that the drone can't reach the destination without landing to the road to charge the battery. The unmanned aerial vehicle was programmed to fly to its destination. During the flight, he had to detect the discharging battery and automatically land in the place to be charged. During charging, the drone controlled the battery charge level and when it was charged it was completely started, continuing the flight to the target. This shows the possibility of using such solutions on a wider scale (Junaid, 2017).

Another way to extend the flight time of an unmanned aircraft is to use a fuel cell. It produces electricity from previously supplied fuel. The most popular fuel cells are hydrogen cells, which use hydrogen on the anode, and oxygen on the cathode. The generation of electricity occurs here under the influence of a series of chemical reactions (Gowtham, 2017). The scheme of the hydrogen fuel cell operation is shown in the Figure 12.

Figure 11. The use of a road vehicle and a drone to increase the range of the flight (Luo, 2017)

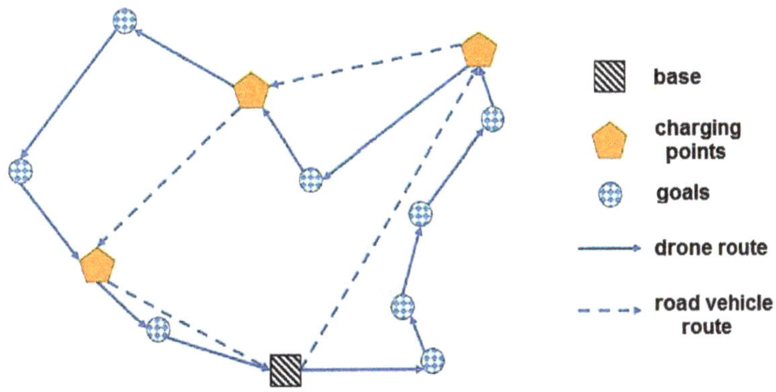

Hydrogen is stored in a special box and released as needed. On the other hand, oxygen is drawn from the air. Currently, MMC produces 2 models of drones with HyDrone 1550 and HyDrone 1800 hydrogen fuel cells that can be in the air in 2.5 and 4.5 hours in succession. Such a long time of being in the air gives a much greater possibility of using them than traditional devices powered by lithium-polymer batteries, which can stay in the air for several dozen minutes. The problem occurring when using fuel cells to power unmanned aerial vehicles is their high mass, an exemplary HPS-1800 link produced by the MMC has a mass of 9.2 kg. Another problem is the heating up of fuel cells during operation, because the drones are made mainly of plastic, which is not temperature-resistant (Keating, 2016).

The Development of Unmanned Aerial Vehicles

At the end of the 20th century, there were 76 different drone designs in Europe. At the end of 2012, this number was already 400 constructions, to reach the value of 591 in 2015. On a global scale, it already has 2115 different constructions (Merkisz, 2016a). Therefore, a large vehicle of unmanned aerial vehicles is a variety of their size and adaptation to perform a specific task (Berner, 2016a). One of the simplest devices of this type is AR.Drone produced by Parrot SA. Its cost is below 300$, it can raise a load of 278 grams and is equipped with a camera with a resolution of 1280 * 720 pixels, and the battery allows it to stay in the air up to 18 minutes. The most expensive unmanned aerial vehicles, manufactured to special order, can fly up to several hours and lift a load of several kilograms (Liu, 2015). In many places around the world, the development and use of unmanned aircraft is limited not by technical capabilities but by law. In Great Britain, a person without a pilot license can use a drone only with a weight of less than 7kg, can fly at an altitude of 122 meters, only within sight, and not allowed to fly in large clusters of people and near airports. In the United States, it is forbidden to use drones in national parks so as not to frighten animals and disturb visitors. It is also forbidden to use unmanned aerial vehicles to control hunters and fishermen (Sandbrook, 2015).

In 2014, the market of unmanned aircraft in the United States, including military applications, was worth 6.4 billion dollars (Mathews, 2015). During the summer holidays in 2015, a million drones were sold in the USA. Such an increase in the use of this type of equipment is associated with a significant drop in their prices, because one of the cheapest quadcopter from Amazon can be purchased for about 40$. It is predicted that by 2015, the industry related to the use of unmanned aircraft in the United States

Figure 12. The scheme of the hydrogen fuel cell operation
(Gowtham, 2017)

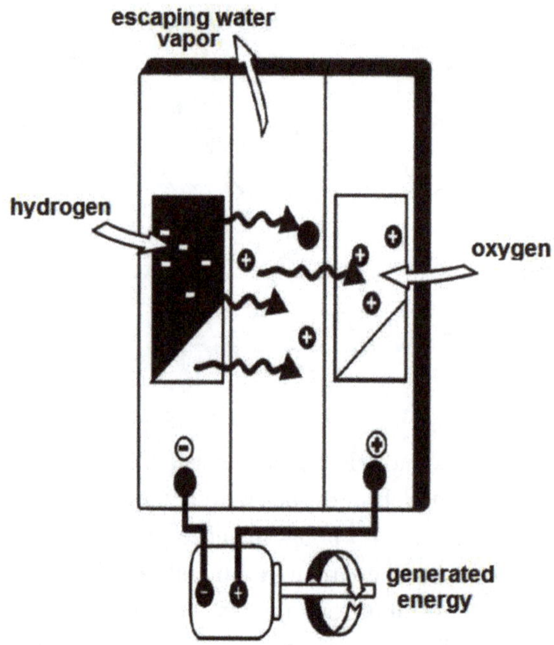

will be worth 82 billion dollars and will employ 100,000 people (Tate, 2016). Annual global revenues generated by the market related to drones in 2013 amounted to 5.5 billion EUR, while the forecast of revenues for 2018 is 6.35 billion EUR (Colomina, 2014). In the next few years it is possible to develop autonomous technology of unmanned aerial vehicles performing specific tasks without the need for operator control (Floreano, 2015). Examples of such autonomous structures are shown in the Figure 13.

Figure 13. Different designs of autonomous drone: a) winged drone equipped with cameras, b) rotocraft, due to the use of a spherical cage, it can reach narrow gaps and bounce off the walls
(Floreano, 2015)

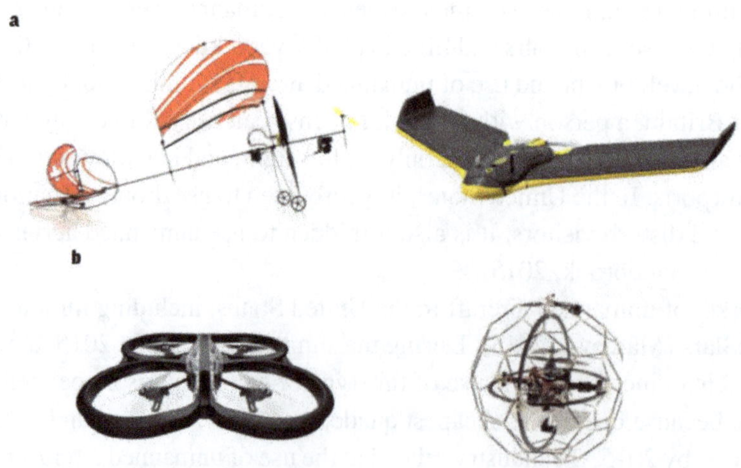

Development in this direction is currently being halted due to legal regulations both in the United States and in Europe requiring the use of a properly trained operator, and in most cases flights must be limited to the operator's line of sight. Due to the lower cost of purchase and security, unmanned aircraft with low weight become more and more popular. For example, a drone with a mass of 0.5 kg flying at 3.6 m/s has a kinetic energy of about 6.5 J which corresponds to the potential energy of an apple falling from a height of 2 meters (Floreano, 2015). With the development of unmanned aerial vehicles technology and their applications, the development of literature associated with them is related. In 2004 three articles about drone were presented at the conference of International Society for Photogrammetry and Remote Sensing in Istanbul, while in 2008 at the conference in Beijing 21 articles on the use of drones were presented. In 2012 the conference in Melbourne was an opportunity to present 50 articles about unmanned aircraft (Colomina, 2014).

Social Acceptance of the Drones Use

Unmanned aerial vehicles allow observation of many different areas. A certain threat is the ability to control using them also places inaccessible to the public, such as private areas or the interior of buildings. They can be used to collect data by state authorities like the Police, but also by unauthorized individuals (Chojnacki, 2017). The public is unaware of the possibility of unmanned aerial vehicles, which creates additional concerns about their distribution. The survey on knowledge of unmanned aircraft in the United States in March 2013 was conducted by the University of Monmouth. It was online and 2119 people took part in it. As many as 44% of respondents said that they know very little or nothing about drone. Satisfaction with the introduction of unmanned aerial vehicles for any purpose was stated by 57% of respondents, while the use of drones to ensure internal security enjoys greater social support, as 67% of respondents supported such use, even more recognition of the respondents raises the use of these devices for search and rescue, which supports 88%. The use of drones in daily use enjoys lower public support, as only 43% of respondents support this (Eyerman, 2013). These results are presented in the Figure 14. In addition to the high support of the use of unmanned aircraft, it also raised major concerns. Concern over excessive observation of both public places and private areas using drones was expressed by 75% of respondents.

Another study in Australia in Australia conducted a survey on the perception of drones by Australian public opinion. It was asked whether unmanned aircraft are more dangerous than traditional unmanned aviation, asked whether the terminology used in the description of drones affects public opinion and what concerns Australians have about the development of this technology. The results of the surveys showed that the public in Australia has a neutral attitude towards unmanned aerial vehicles, the respondents did not notice that they were both too dangerous and gave great benefits from using them. The respondents' safety compared drones to that of traditional manned air transport. The terminology used for the technical description of drone had a negligible impact on the perception of this technology. The neutral approach to public opinion drone in Australia is due to the lack of knowledge on this subject. Probably only after the development of this type of devices, implementation into various areas of life, and thus when the public knowledge about drone will increase, they formulate an opinion on unmanned aerial vehicles (Clothier, 2015).

The study of society's attitude to unmanned aircraft compared to traditional manned vessels was also carried out in Canada in 2014. In the case of road traffic monitoring, 49.5% of the respondents stated that drones are suitable for this, while the suitability of manned aircraft for this purpose was stated by

Figure 14. Survey results on the introduction of drone to various areas of life
(Eyerman, 2013)

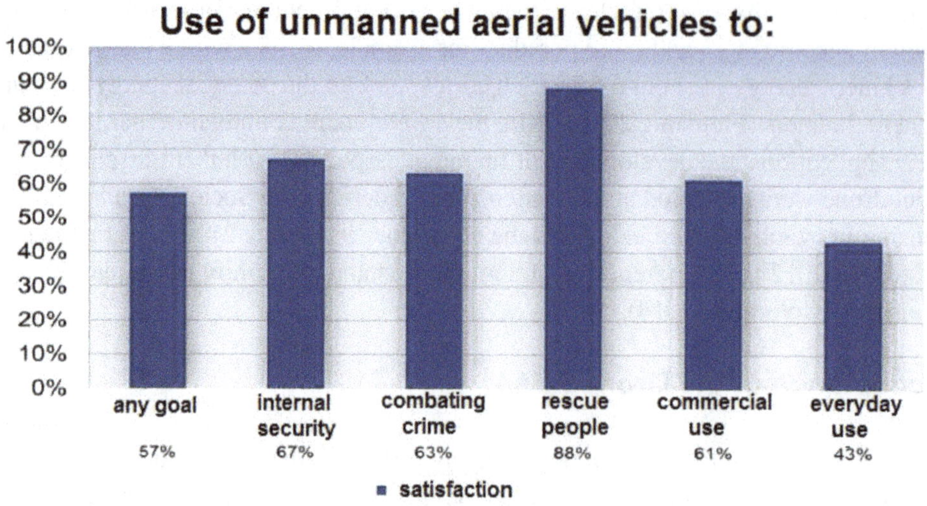

75% of respondents. The usefulness of drones for search and rescue operations was noticed by 51.6% of respondents. However, the use of traditional flying devices for this purpose positively assessed 83.5%. The situation is similar in the case of taking aerial photographs. In this case, the use of unmanned aerial vehicles was supported by 65.3% of respondents, whereas manned units - 81.3% of respondents (Saulnier, 2017). These results are presented in the Figure 15. Only in the last two applications, more than half of the respondents used drone. However, in each category, traditional manned aircraft achieved a significant advantage. This shows the lack of knowledge in the society about the possibilities of using drones. This lack of information may also be a reason for reluctance towards this mode of transport.

Figure 15. Results of a survey examining social acceptance of the use of flying crew and unmanned aircraft
(Saulnier, 2017)

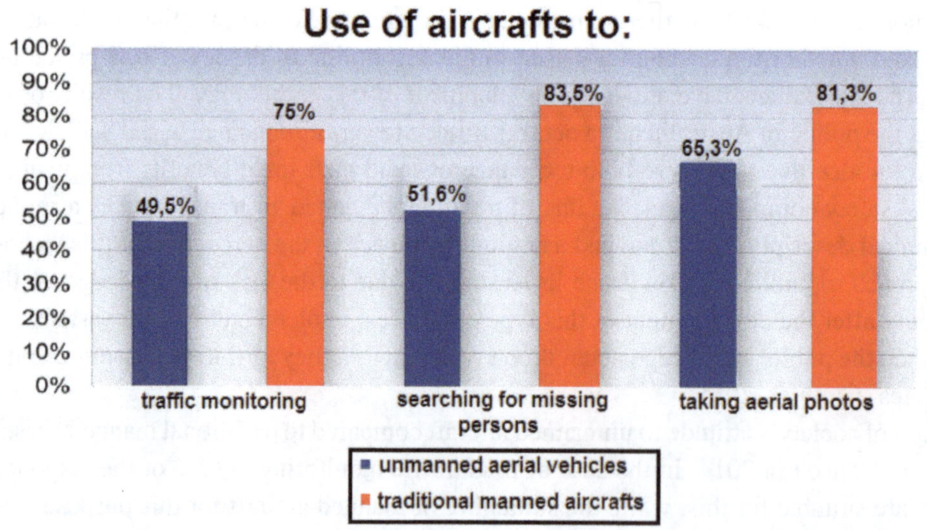

Risks Associated With the Drones Development

Drones are considered to be safer than manned aircraft because there is no pilot flying in them that could possibly be wounded. In the event of a failure, it is also assumed that they are safer than airplanes due to the smaller size and weight, thus the potential damage to the environment and to people will be lower than in the case of a plane crash. On the other hand, due to the lack of a remote control and the use of many different systems, the probability of a drone crash is much greater than a manned aircraft. On the other hand, the failure of an unmanned aerial vehicle and its fall to the ground may result in the loss of health or even the death of accidental people staying at the accident site (Sandbrook, 2015). In the event of a technical failure of the drone during the flight, the device will fall vertically downwards, which is a threat, because the operator can't move it to a safer place and the unmanned aerial vehicle may fall, for example, on infrastructure or people. The appearance of a shattered drone, which has fallen vertically down as a result of a failure, is shown in the Figure 16.

The problems and risks associated with the development of unmanned aerial vehicles can be divided into three groups. The first of these is the safety of flights and damage to both property and people that can cause a drone during a breakdown or an abnormal flight. Another group of risks is the use of civil airspace, which is limited to some extent. The third problem concerns the information that is collected by drones and looks from the point of view of everyone's privacy. Another issue is the question of responsibility for the actions taken by the operators of unmanned aircraft (Rao, 2016). The division of threats related to the implementation of unmanned aircraft to new areas of life is shown in the Figure 17.

In the case of the first group of threats, then security and freedom from fear is the basic human right that is protected by most nations. The use of drones in civil airspace has been a threat to these fundamental rights. It is connected both with technology (battery life, reliability of the drone as a whole) and with man (user errors). The real threat here is the possibility of taking control of the drone by another person, by sending a strong (but false) GPS signal. The origin of the beacons is very difficult to verify, so that in the case of damage done by a drone over which someone took control, the operator can be held responsible. Another threat is the lack of regulations controlling the cargo transported by unmanned aircraft. Currently, their use for transporting cargo is negligible, and the mass they can raise is limited, but with the development of technology it can be a bigger problem. Regarding the second group of threats, there

Figure 16. The drone crashed during the impact to the ground
(Hallsten, 2017)

Figure 17. Three groups of risks associated with the development of unmanned aerial vehicles

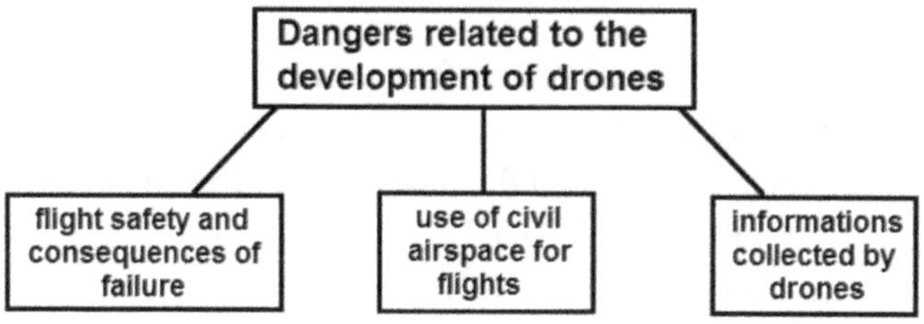

are provisions regulating in which zones and under which rigor is allowed to use drones. However, these rules are not always respected. Fortunately, now accidents for this reason have not been confirmed, while in 2014-2016 there were 900 cases of lack of certificates during flights of unmanned aircraft in places where it is required. In the case of the third group of threats, one of the problems is proper protection of information collected by the drone. This information, obtained with the help of cameras and sensors, can be very valuable. In Iraq, terrorists intercepted some of the information collected by drones belonging to the American army. Thus, if the military can't adequately secure its unmanned aerial vehicles, this problem may also affect other services, companies and individuals. Often it is even difficult to detect the leak of information (Rao, 2016).

Currently, the form of control are mainly industrial cameras located on a large part of public areas, but their location is known, they are visible and people are aware of entering the monitored area. The drone used for observation may, however, remain unnoticed, because due to the fact that it moves its location is not constant as in the case of industrial cameras and difficult to locate due to the small size and often quiet operation during the flight. This means that people observed from unmanned aircraft are not aware of this as is the case with traditional industrial cameras (Kaminski, 2013). In a public space such as a street or park, which is covered by a large number of fixed cameras, the problem of violation of privacy by drones does not occur. However, it is different in private, because here privacy is understood as a state in which no other people are observed or disturbed. The appearance of a drone on such terrain allows observation and the sound it issues can be bothersome for those around it (Rao, 2013). A drone performing a flight next to private areas may cause fear, confusion and hostility to the owners. In the event that people do not understand the use of unmanned aerial vehicles, their appearance may cause suspicion and be the reason for creating various conspiracy theories. This effect is especially possible in developing countries, where the population is not used to using technical devices. It is also possible to deliberately use drones to create fear in the observed population, so that they fear sanctions by performing illegal activities (e.g. poaching). Publications were created on this subject to use drone in Afghanistan to intimidate poachers. This application raises controversies related to ethics. The problem also occurs with the storage of data collected by unmanned aerial vehicles, because they can be intercepted or used commercially, e.g. by advertising agencies to test customer preferences (Sandbrook, 2015).

Another potential threat is the ability to fly at certain conditions without reporting this fact anywhere. This situation can be used by criminals using a drone to collect information, because it is able to remain unnoticed, which is more difficult for a man or a motor vehicle (Smith, 2015). Unmanned aerial vehicles are also used by paparazzi to take pictures of celebrities and their children also in private, which limits the right to privacy. Drones also allow reporters to observe areas where they would not get in a traditional way. Therefore, the use of these devices in journalism increases the possibilities for creating reports from events, but also carries risks due to free access to normally hard to reach places (Tate, 2016).

A big threat when flying a drone is the possibility of its being intercepted by an unauthorized person, which results in the operator losing control over it. Possible attacks to intercept unmanned aircraft can be divided into three groups: hardware attack, wireless attack and sensor fraud (Kim, 2012). Ways of intercepting control of the drone are shown in the Figure 18.

A hardware attack can take place while the person making it has direct access to the drone components. Then, the attacker can damage the data contained in the control system or install other data that interferes with the transmission of data to the operator and allows to take control. This can be done during the production, delivery, repair or storage of the drone. The wireless attack consists of remotely changing the data of the control system of an unmanned aerial vehicle. If the cipher is broken on the communication channel, the attacker will intercept the control of the drone. Another way of remote attack is to overflow the buffer in the control system, i.e. to save a larger amount of data in a certain area of memory than it is intended for. Then some of the data stored in it is lost, which causes a malfunction of the control system and the operator of the unmanned aircraft may lose control over it. This type of attack is a much greater threat than a hardware attack because it is more difficult to protect the drone from it. Attacks consisting in cheating sensors are based on falsifying data received by on-board sensors that collect data from an external environment such as a GPS sensor, sonar, etc. An attacker can send fake data via GPS channels or disrupt any of the sensors. Damage to this sensor may also result in loss of operator control over the device during flight (Kim, 2012).

Ethical Problems Related to the Development of Unmanned Aerial Vehicles

The development of unmanned aerial vehicles is associated with ethical problems, or related to morality. The actions performed by drones can be assessed only by assessing the person serving them. Here, the actions of the control person, the operation of the unmanned aircraft and the intentions of the operator are assessed. Ethical problems mainly arise when there is supervision or even surveillance, but there are also other situations. In the case of recreational flights, it is necessary to check what activities are performed during them, what their nature is and what intentions the operator has. If flights take place away from human centers, then there is no greater danger or ethical problems. However, in the case of flights in places of human population, the noise produced by a drone can be annoying for some people. They have the right to peace, for example in parks or other recreational areas (Wilson, 2014). Unmanned aerial vehicle disturbing the calm in the park depicts a Figure 19.

Figure 18. Possibilities of taking control of an unmanned aerial vehicle

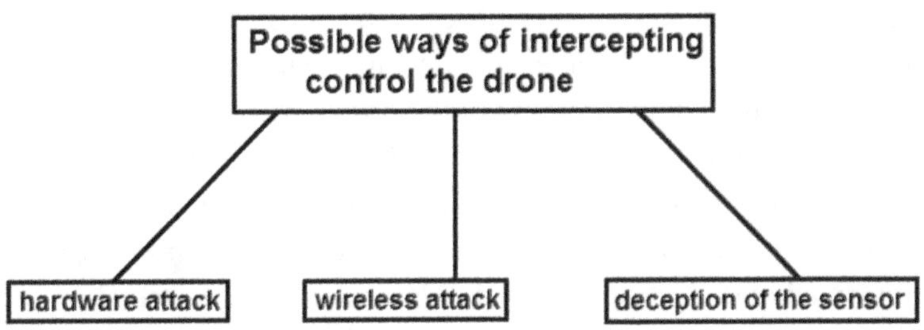

Figure 19. Dron disruptive the resting in the park
(DJI spark review, 2017)

Another problem is taking pictures. Taking artistic photos of interesting landscapes or buildings is not a problem, nor does it lead to a discussion about ethics. However, the drone can take pictures of people who do not want it, which in case of their subsequent publication may constitute a violation of privacy. An extreme case is the use of an unmanned aircraft to persecute another person. Ethical problems also involve the use of unmanned aerial vehicles in the army, the most controversial are the unmanned aircraft, which can be used to kill people. This is problematic in the assessment, because traditionally a soldier attacking the enemy himself also bears the risk of losing his life. However, when using a drone, the operator can be in a safe place and thus eliminate the opponent (Wilson, 2014).

USE OF UNMANNED AERIAL VEHICLES

The use of unmanned aerial vehicles is wider than that of manned vessels, because they can perform tasks not performed by manned structures due to adverse environmental conditions (chemical pollution, dust, radiation, too low or high temperature, too low air pressure), limitation of crew performance (mainly the length of the flight and the associated maximum duration of official duties and physiological conditions of the human body) and the technical characteristics of manned vessels currently manufactured (mainly the mass of such ships and noise occurring during operations, which eliminates them from certain zones) (Bujnowski, 2017). Most often, drones are used to observe various objects or events using a camera mounted on them. There are three phases of such observation, a conventional based on visions (observations) of object displacements, a method that eliminates the occurrence of a scale factor by scaling and compensation of the shutter and the identification of the observed system using the dependencies of the observed object relative to non-moving objects (Yoon, 2017). This scheme is shown in the Figure 20.

The use of unmanned aerial vehicles can be divided into several groups. They can be used for: sending information, transporting cargo or transferring energy. The transmission of information may be active or passive. Passive information collection means that information collected from the environment during the flight is analyzed only after landing the drone. Active transmission of information consists in sending information from the drone during his flight, usually in real time. The transport of the load may

Figure 20. Identification of objects using unmanned aircraft
(Yoon, 2017)

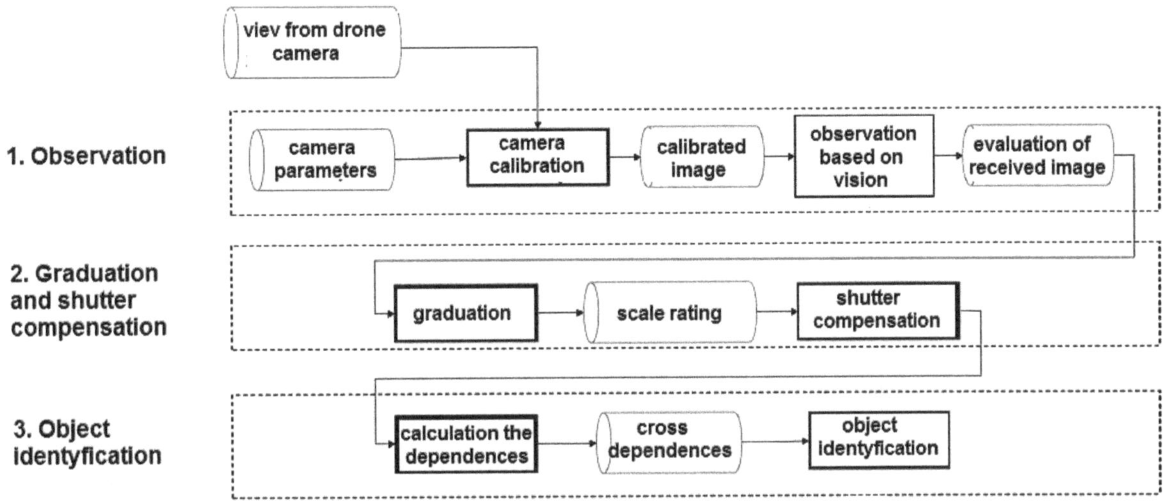

be in whole or in part. Transport in full means that the weight of the load mounted on the unmanned aircraft does not change during the flight (e.g. use of transport for shipments). Partial transport means a decrease in the weight of the load during the flight of the drone (e.g. use for spraying in agriculture, where along with the length of spraying the fertilizer mass decreases) (Lesicar, 2017). This division is shown in the Figure 21.

Some of the possible uses of unmanned aerial vehicles are shown in the diagram (Figure 22).

The Use of Drones in the Power Industry

The main task of unmanned aircraft in power engineering is to control the technical maintenance of the network, but it is also possible to apply them to network management and organization of work. When monitoring power grids (overhead), they enable the assessment of network damage after sudden weather events (storm, strong wind) and allow to check the vegetation regrowth along the line (Dudek, 2015). An unmanned aerial vehicle used for observing the power grid is shown in the Figure 23.

Figure 21. Identification of objects using unmanned aircraft
(Lesicar, 2017)

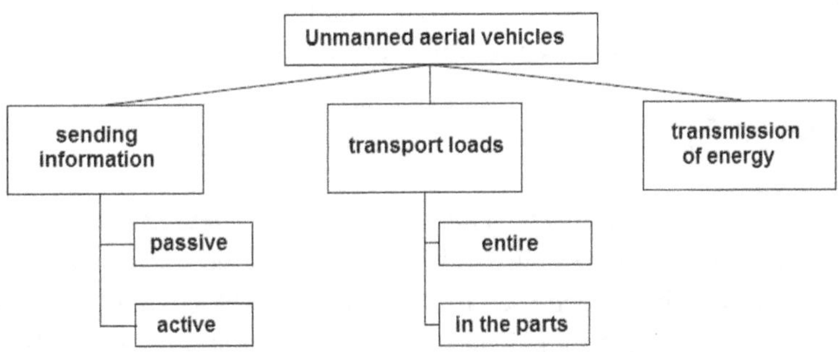

Drones are also used in various power plants: photovoltaic solar panels - for the control of mirrors, wind generators - for inspecting of turbine rotors, they can also be used in coal-fired power plant halls. The fact that unmanned aerial vehicles can rise to a height of around 200 meters is the biggest advantage in controlling wind farm systems. Attempts have also been made to use drones to work on maintaining power lines. At Energetab exhibition (2014) saw the stretching of pre-cables using an unmanned aircraft. These cables are necessary to prepare the exchange of wires on power lines (Dudek, 2015).

A frequent cause of failure of the network transmitting electricity is overheating of its individual elements caused by the transitional resistance. This resistance occurs at places of inappropriate electrical connections (e.g. damaged, worn out). An increase in resistance appears here, which causes the element to heat up, thereby generating energy losses that are converted into heat. In addition to economic losses due to overheating, a fire may occur. Thermal imaging cameras are used to locate places where the power transmission network overheats. Due to the course of such a network through inaccessible areas and the considerable height of the structure transmitting energy in a way that allows to find such damage as quickly as possible is the use of an unmanned aircraft equipped with a thermal imaging camera. An example of a system specially adapted for mounting on a drone is Wiris from Workswell. It contains a traditional camera allowing to control visible network failures and a thermal imaging camera to detect overheated areas. These devices transmit the image in real time, so that the operator receives data on the status of the network already at the time when the unmanned aircraft is in flight. On the image from the infrared camera the areas with the lowest temperature are shown in blue, while those with the maximum temperature in red. The temperature range to be tested (minimum and maximum) is selected manually and can be changed during the flight (Thermodiagnostics, 2017). The image from the thermal imaging camera showing the overheating element of the network transmitting electricity is shown in the Figure 24.

Another application of drones in the energy sector is the observation of plantations collecting solar energy. Currently, interest in this type of devices is growing as a source of renewable energy. The research was carried out using a thermovision camera mounted on an unmanned aerial vehicle for real-time observation of a solar energy plantation. These studies took place in Cordoba, Spain, where there is a plantation consisting of 104 km of parabolic and tube collectors absorbing solar radiation (Mesas-Carrascosa, 2017). The scheme of such a collector is shown in the Figure 25.

Due to the large area, the expert's traditional inspection of this plantation lasts for a week, while the less experienced person takes two weeks. Such inspection should be carried out once a month, up to once every two months. The flights were made at heights: 20, 40, 60, 80, 100 and 120 meters above the ground, moving at three speeds: 5, 7 and 10 m/s. Wind speed did not exceed 1 m/s during the tests. The results of this study have shown that it is possible to make such observations in real time using drone as a source of information on the increased absorption of solar radiation. The largest amount of information allowed to provide flights performed at the height of 100 and 120 meters. After finding the irregularities, flying over it at 20 meters allows you to provide more information about the fault. Thanks to this operation, the total time of control is reduced (Mesas-Carrascosa, 2017).

The Use of Drones in Agriculture

In agriculture, it is necessary to have an easily accessible and cheap device allowing at any time, without the need to drive the whole field with specialist vehicles, to assess the soil's richness and the state of the crop. Unmanned aerial vehicles have been used for this purpose. The flight of the drone is performed at a low altitude, which allows to collect data about the area in which it takes place. It is also important to

Figure 22. Different possibilities of using unmanned aerial vehicles

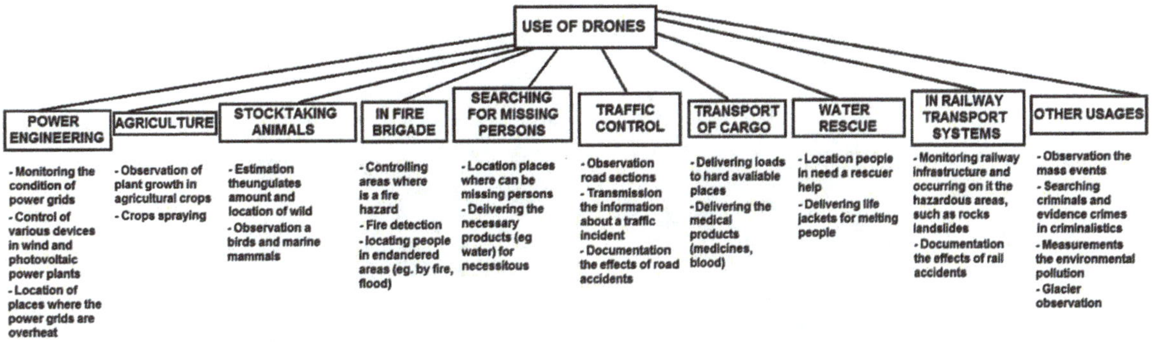

Figure 23. Drone during the observation of the state of the power grid
(Dudek, 2015)

choose the right type of unmanned aircraft, which are divided into aircraft with rotor engine (helicopters and multicopters) and fixed wing aircraft. In terms of steerability, aircraft with rotor engine have more possibilities, because they can move fully independently and can hang motionless at any place. On the other hand, fixed-wing aircraft, similar to airplanes, enable faster speeds. In precision agriculture, a higher flight speed translates into more data on the soil and crops in the same unit of time. Multispectral cameras that take pictures are used to collect data (Berner, 2016a).

Figure 24. Overheating element of the power grid
(Thermodiagnostics, 2017)

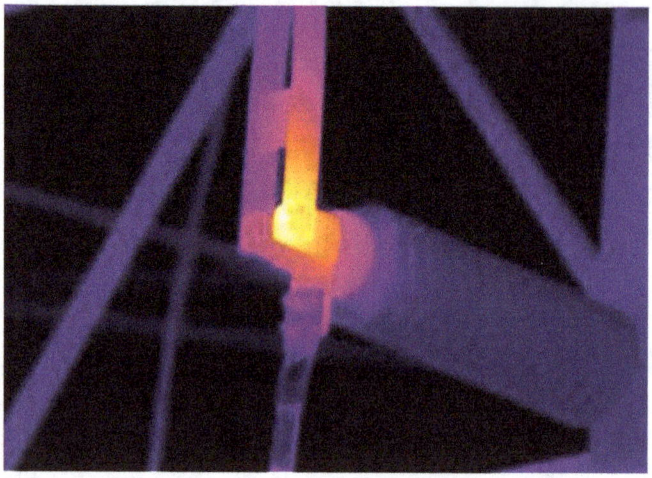

Figure 25. Scheme of a solar energy absorber
(Mesas-Carrascosa, 2017)

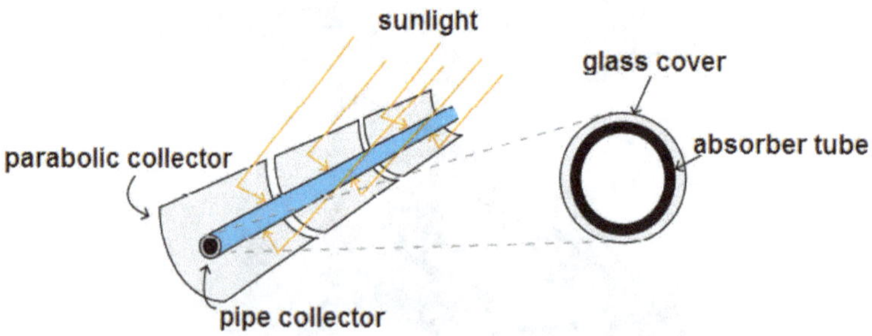

The installation of a multi-spectral camera on the drone, which collect images from many channels generalizing the primary colors, allows to assess the chlorophyll content in the green parts of the plants and locate leaf damage. Therefore, it is possible to assess in what state there are individual crops, both forest and agricultural (Tripicchio, 2015). Research of lasers was also carried out on the subject of their use for collecting data necessary for agriculture (Berner, 2016a). Pictures taken before the drones are used individually or from a certain number of images are created in computer programs of the orthomap (Kułaga, 2016). These photos allow you to assess what crops are located in your area. A comparison of several different crops is provided in the Figure 26.

In addition to traditional flat maps, three-dimensional maps of crops are made. They allow to determine the intensity of growth of individual plants by measuring their height (Chojnacki, 2017). The studies of the plant growth intensity assessment were carried out from the beginning of June to the end of July 2015, observing the winter wheat fields in Hokkaido, Japan. A Normalized Difference Vegetation Index (NDVI) was used here, which allows to determine the state of development and condition of plants using the contrast between the largest reflection in the near-infrared band and the absorption in the red band. It can take the value from -1 to 1, in areas covered with lush vegetation, this index takes

Figure 26. Various aerial photographs of fields: unhandled field, field plowed with a 25cm deep plow and a field plowed with a 50cm deep plow
(Tripicchio, 2015)

high values. In the initial stage of research, a rapid growth of wheat was recorded, with each day the value by which the plant grew was growing. After reaching the maximum value around 7-10 June 2015, daily wheat growth decreased. Wheat monitoring by means of a drone also allowed to identify places where the growth is larger or smaller. These data allow the selection of the right dose of fertilizer to reduce the dose in lush areas, and increase where the development of wheat is the smallest (Mengmeng, 2017). It was also possible to identify crop damage caused by, for example, strong wind, such damage is shown in the Figure 27.

Another example of the use of unmanned aerial vehicles in agriculture is the observation of maize cultivation. A study was conducted in which 4 different maize varieties were observed, for each of them different doses of nitrogen fertilizer were used (50, 150 and 250 N / ha were administered). This study took place in the town of Durnau in southern Bavaria, Germany, and took place on June 16, 2016. A drone equipped with a 10-megapixel camera was used to observe the crops. It was flying at 50 meters above ground level and took pictures every 5 seconds. Observations took place during such a plant development phase in which the young leaves have a lighter color than the older ones. After shooting with

Figure 27. Damaged wheat crop, the damage was detected during the test using drone
(Mengmeng, 2017)

the use of a multi-camera during processing, the contrast was increased to make the difference between these colors more visible. The obtained photos were processed in the MATLAB program, where an area marked with black representing the number of green pixels (corresponding to plants) was created (Gnadinger, 2017). This is shown in the Figure 28.

Next, traditional visual observation from ground level was performed to compare the results of aerial observation. The results showed that the observation with the use of a camera mounted on an unmanned aircraft allowed proper identification of maize development, because less than 5% of the tested plants were mistakenly assessed. This is due to the emergence of different types of weeds between the correct crops, which impede proper assessment of plant development. The use of the drone for observing the cultivation of maize allowed for its quick and low-effort evaluation with a small error (Gnadinger, 2017).

Another way to use unmanned aerial vehicles in agriculture is to use them as agricultural machines. In this case, their advantages are the lack of indentation of fields and the possibility of reaching places where it is not possible to drive a traditional agricultural tractor (e.g. wetlands). When spraying chemicals with a drone the advantage is that the remote operator is not exposed to the harmful effects of the agent. The use of unmanned aerial vehicles instead of traditional manned units enables flying at an altitude of up to 1 meter above the soil surface, and at the same time, thanks to the lower mass, it allows to reduce costs (Berner, 2017). The Figure 29 shows drone intended for spraying.

Unmanned aerial vehicles were used in China to fertilize crops with pesticides. Earlier in China, these works were done by hand, so their productivity was low and there were cases of pesticide poisoning of farmers working there. These activities reduced their labor intensity with the appearance of sprayers. Currently, unmanned aerial vehicles are increasingly used for spraying certain substances, which do not need special aerodromes and are cheaper to use than manned units. For the fertilization of crops with pesticides it was necessary to select the appropriate drone, for this purpose 2 units with gasoline and 2 electric drives were tested. Their cost ranged between 90,000-250,000 Yuan, and the distance of the flight at full load was between 15 and 30 minutes. The research was carried out in May 2016 at Anyang

Figure 28. A picture of corn growing made from a drone and a picture created in MATLAB showing the number of green pixels in the upper picture
(Gnadinger, 2017)

Figure 29. An unmanned aerial vehicle designed for spraying of agricultural crops (Berner, 2017)

City in Henan province. At that time, favorable meteorological conditions prevailed, the temperature was about 30°C, the wind speed was about 1.7 m/s, and the humidity was about 50%.

The examined fields were overgrown with wheat, 60 cm high. The tests were carried out in two parts, the uniformity of plant coverage with drops of pesticides in the area of 100 m × 100 m and the permeability of drops to the whole plant were checked. Flights were carried out at a speed of 5 m/s, at a height of 2 meters above ground level, which is a bit higher than traditional spraying using traditional agricultural machines. Two hours after the spraying by four drones were harvested from each field one kilogram of wheat for testing. They were divided into the upper, middle and lower areas. Drops of pesticides dropped from drone were smaller than during traditional spraying. Due to the higher spray height and smaller droplet size using unmanned aerial vehicles, it is more susceptible to meteorological conditions, mainly to the wind, which can direct droplets to other locations (Shilin, 2017). The arrival of drops of pesticides to various parts of the plant was made possible by filters mounted in various parts of plants. The results showed that the best result was covering the wheat plantation with pesticides an average of 65.45% by one of the drone. The worst result achieved by one of the investigated drones is 43% of the average coverage of plants with pesticides. The exact results of the coverage of individual layers by all four unmanned aerial vehicles participating in the study are shown in the Figure 30.

Another study based on the observation of winter cereal was carried out at the National Center for Precision Research in Xiao Tangshan, China, it has an area of 2 km^2 (Yue, 2017). This place is shown in the Figure 31. The studied wheat crops were divided into 3 groups, which had a different level of hydration and other doses of nitrogen fertilizer. One part was irrigated only by rainfall, the other was additionally irrigated with a normal amount (100 mm), and the third was irrigated with twice as much water (200 mm). In the case of fertilizer, the first part was covered with half of the normal fertilization level (195 kg/ha), the second with normal fertilization level (390 kg/ha) and the third with twice the normal fertilization level (780 kg / ha). Later 20 samples of plants from each part of the field were collected. Multispectral imaging made by the UHD 185 device mounted on an unmanned aerial vehicle DJI S1000 was used for the study. It has a mass of 6 kg, can stay in the air for a maximum of 30 minutes, reach a height of 50 meters above the ground and fly at speeds up to 8 m/s. It was flown on 26 April 2015 and 13 May 2015. Both of these flights included a field of 6000 m^2 and lasted 20 minutes. After the tests, the results obtained from the analysis of photos taken with the drone and the results of samples taken on the ground were compared. On the basis of these studies, plant vegetation indices and yields were calculated.

Figure 30. Covering individual parts of wheat plants by each of the drones involved in the study (Shilin, 2017)

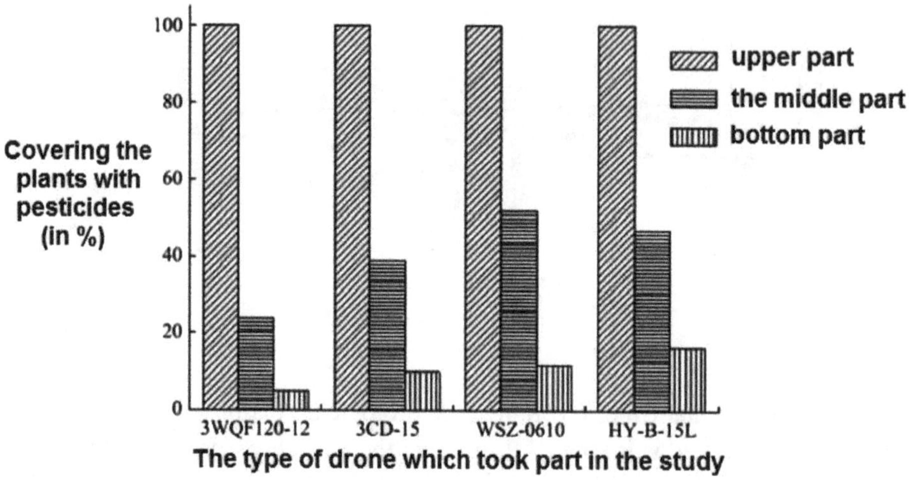

Figure 31. Location of the wheat crop studied by drones near Beijing (Yue, 2017)

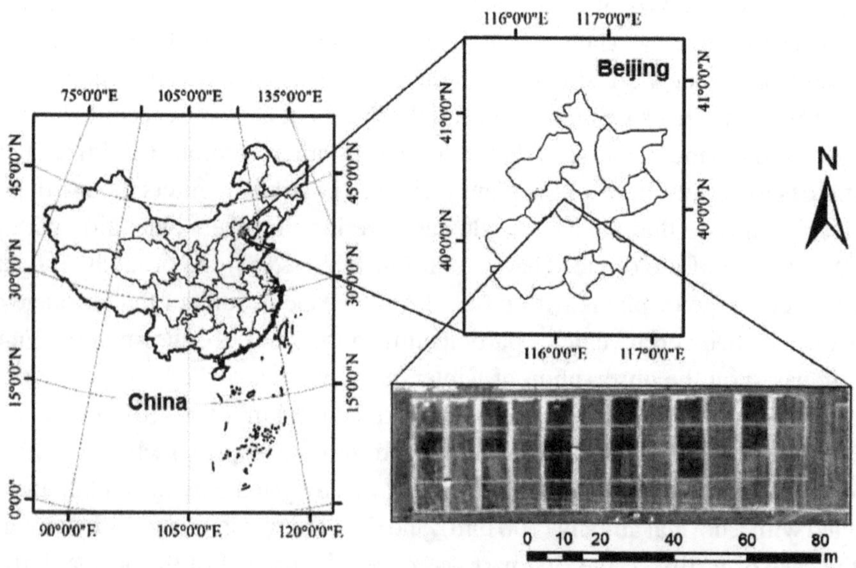

Unmanned aircrafts were also used to observe the field of consumer potatoes. The field had dimensions 450 × 200 meters and was located in the village of Reusel in the province of North Barbados in the Netherlands. These potatoes were planted on 16 April 2016. Potato cultivation secured with potassium fertilizer was examined here. This fertilization was performed on 28 June, 15 July and 9 August 2016, using a fertilizer dose of 60 kg/ha. During the fertilization of flights, 40 photos were taken. Next, polygonal views of the studied area were created from these photographs (Roosjen, 2017). This allowed to determine at each cultivation place NDVI, as shown in the Figure 32.

Figure 32. Location of the wheat crop studied by drones near Beijing
(Roosjen, 2017)

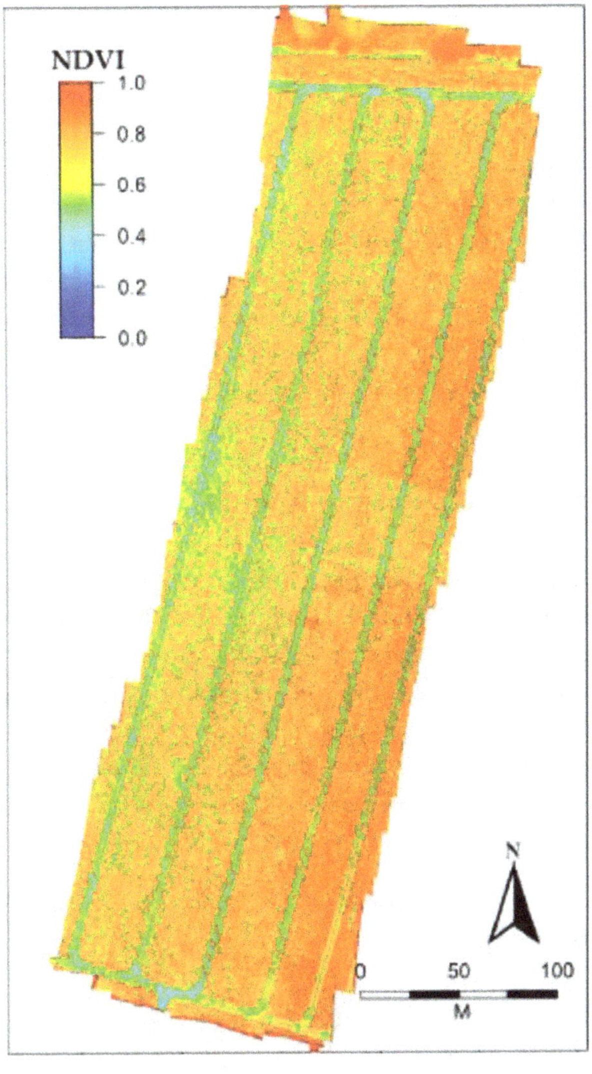

Anisotropy (dependence on the direction) of the effects of sunlight reflection on observed plants was observed. The parameters describing this anisotropy contain information on the state of potato cultivation, such as the surface of the leaves, which testify to the degree of plant growth (Roosjen, 2017).

Another use of unmanned aerial vehicles in agriculture is to use them to precision observe vine plantations. In this case, the drones proved to be competitive compared to traditional manned helicopters and satellite images, because they provide high resolution images, because of their small size have high flexibility, and at the same time their use is associated with much lower costs than using from competitive solutions. The study compared the use of manned aircraft, satellite images and unmanned aircraft to observe vegetation in vineyards. This study took place in two vineyards in the Veneto region of Italy. One of the vineyards has 5 ha and the other 50 ha. Cabernet Sauvignon, one of the best-known grape varieties used for making red wine, grows there. The spacing between the rows of vines was 2.5 meters, and between the individual plants in a row of 1.3 meters. Observations were carried out in the summer of 2012. The flights using the OktoXL multiplex were held on 18 September 2012 between 12.00 and 13.00 under a clear sky. The flight was made at a speed of 4 m/s at a height of 150 meters. The results of the study showed that in vineyards characterized by lush vegetation all of these methods of observation allowed similar identification (Matese, 2015). The comparison of images obtained from unmanned aircraft, plane and satellite is shown in the Figure 33.

Better image quality is required in vineyards with a heterogeneous development of vegetation. It was impossible to compare the costs of using all three platforms, because satellite images are a commercial product, so their prices in the commercial market were taken into account. After the analysis, it was found that the use of drone for observing the vine is profitable in the case of vineyards occupying a small area, up to a maximum of 5 hectares. In larger farms, a cheaper solution is to use satellite images, the cost of which does not change as the area of the vineyard grows. The costs of using a manned aircraft grow slightly if the vineyard is larger. However, the observation of large areas with the use of drones is difficult due to the small maximum time they stay in the air and generates additional costs (Matese, 2015).

The Use of Drones for the Inventory of Animals

An important issue in hunting is to determine the number of wild hoofed animals, because their excessive numbers in certain places may be a threat to both humans and themselves. The methods used to determine the number of herds are test runs and track traces on the snow. They are time-consuming and require a large number of people involved in these operations, and the results of their measurements are not very accurate. For this purpose, it is also possible to use unmanned aerial vehicles together with a thermal imaging camera, observing at night time, when the animals are most active, which makes them easier to detect. The tests of this method with the use of the drone were carried out in the Drawa National

Figure 33. The value of the normalized diversified vegetation coefficient obtained from photographs taken from: drone, plane and satellite
(Matese, 2015)

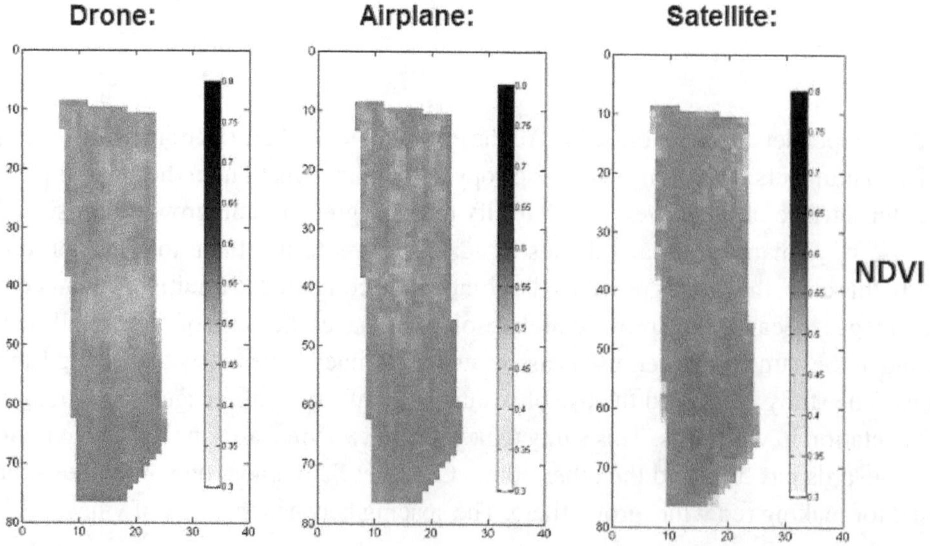

Park, conducting research at various times of the day. The flights were made at a height of about 150 meters, in the spring time, when forests overgrown with deciduous tree stands are leafless, and those with coniferous stands have incomplete short-circuits, these factors facilitated inventory from the air (Pagacz, 2016). The picture taken during the tests is shown in the Figure 34.

Studies conducted at night allowed detection of up to 5 times more wild ungulates than studies conducted at noon. Unmanned aerial vehicles allow observation of large areas in a short time, while there is no risk of loss of health or human life associated with flights at low altitudes during the night. In comparison to traditional measurement methods, the use of drones is animal-friendly, because they are not frightened or caught. It is also possible to observe animals in areas that are difficult to access, such as wetlands, swamps. The use of unmanned aerial vehicles also has some limitations. The main legal restriction is the requirement to perform VLOS flights (in sight), which means that the staff has to move frequently, which is a hindrance, but also reduces the range of observation. Flight performance is also limited by bad weather conditions (e.g. strong wind). Observations allow detection of more ungulates in deciduous forests while they are leafless. In coniferous forests the number of animals detected is lower (Pagacz, 2016).

Unmanned aerial vehicles, in addition to monitoring ungulates, can be used to observe birds, marine mammals. To observe the animals in the dry area of Savanna, a semi-automatic system was developed to detect large mammals. Traditional cameras mounted on drone were used for this purpose. The research area was a private nature reserve with 103 km2 located on the edge of the Kalahari basin in Namibia. In this area, there are about 3000 animals from 20 different species. The flights were carried out between 12 and 15 May 2014 using the Canon PowerShot S110 compact camera mounted on the SenseFly drone. It allowed to take pictures at a resolution of 3000 * 4000 pixels. 232 volunteers took part in this under-

Figure 34. Localized 2 ungulates during a night flight in a leafless deciduous forest (Pagacz, 2016)

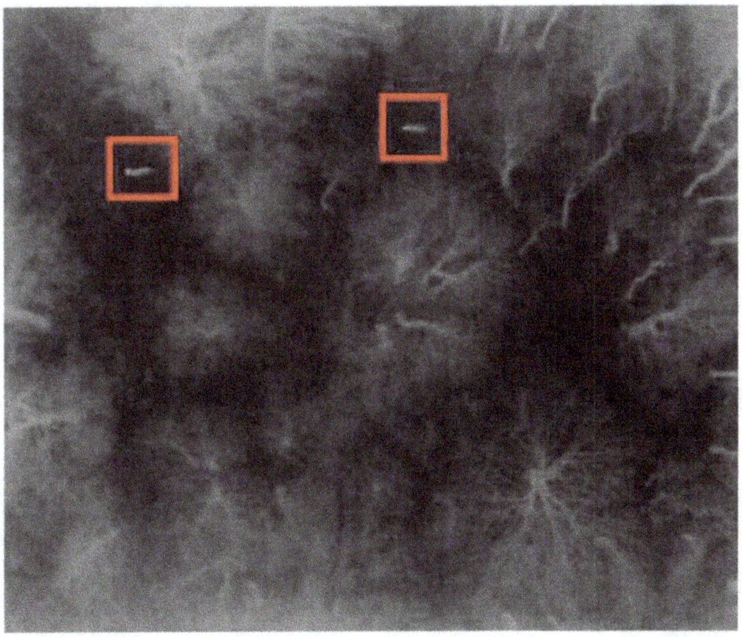

taking. They analyzed 6,500 photos and drew 7474 polygons on 645 photos showing animals. Then, these areas were merged and removed, where only one of the researchers noticed the animals. In this way, there were 976 polygons in which animals appeared (Rey, 2017). The animal detection system in this study looked as follows (Figure 35).

This method works better during the day when visibility from the camera is not difficult. To detect 176 animals during the day, 89 photos were enough, and at night to identify the same number of animals it was necessary to identify as many as 120 photos (Rey, 2017).

Another use of unmanned aerial vehicles is the location of alligator nests in the Grand Chenier Reserve in Louisiana in the United States. Traditionally, helicopters have been used for this purpose, but due to an attempt to reduce costs, drone usage tests have been carried out. Three flights lasting 25 minutes and covering an area of 28.2 ha were carried out. These flights took place on the afternoon of 23 June 2015. For this purpose, a DJI Phantom 2 Deluxe multi-rotor equipped with a camera providing real-time image transmission was used. During the study, it turned out that the optimal altitude for observation of alligator nests is 8-10 meters above the ground, then you can observe the area up to 50 meters on each side (Elsey, 2016). The picture shows an aerial photo allowing the location of an alligator nest (Figure 36).

During the first flight was managed to locate one alligator nest containing about 10-days eggs guarded by the female. During the second flight was find the old, alligator's nest, which was abandoned (Elsey, 2016).

Figure 35. Scheme of detection of animals in the study
(Rey, 2017)

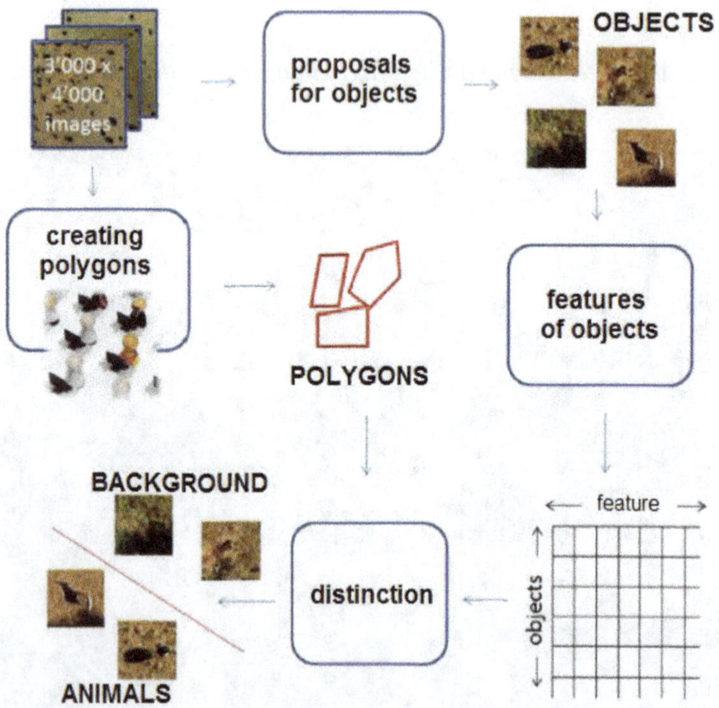

Figure 36. Photo taken from the drone, on which you can see the alligator nest (marked in yellow)
(Elsey, 2016)

For observation of large marine mammals, aerial studies from traditional manned units have long been used. However, it was decided to use drones to avoid the risk of flight and reduce costs. Such observation with the use of unmanned aircraft was carried out in one of the bays of the Indian Ocean, in the shark of the Shark Bay in Australia. The habitats of occurrence of coastal dugongs were studied here, it is a mammal adapted to water life, unable to move on land, reaching a length of 3 meters and body weight of about 400 kg. They live in the coastal zone, where the depth is about 3-4 meters. In this place, often these herbivorous mammals are exposed to attacks by predators, but being at a shallow depth, which facilitates aerial observation. The observation was made using the Drone ScanEagle from Boeing equipped with a Nikon D90 DSLR camera with a fixed focal length lens 50 mm and a polarizing filter. The image from the camera was visible to the operator during flights in real time. The research was carried out on 16-21 September 2010, 7 flights were made at three heights: 150, 230 and 300 meters above the water surface covering a total of 1.3 km2. One flight took about 25 minutes. During the observation 6243 photos were taken, out of which 627 photos were captured by coastal dugongs (Hodgson, 2013). Pictures taken during the drone flight are shown in the Figure 37.

In addition, other marine animals such as whales, dolphins and turtles have been captured (Hodgson, 2013).

The Use of Drones in the Fire Service

In some countries, unmanned aerial vehicles are used to assess the consequences of disasters or damage after natural disasters. In Poland, drones have Police, which uses them to observe areas that are difficult to access when reporting disappearances, and a fire brigades, where this equipment remains unused. However, unmanned aerial vehicles are suitable for controlling areas where there is a fire hazard (e.g. bogs, dry forest areas). As a result, they are able to detect a fire in its initial phase (Tuśnio, 2016).

Figure 37. Aerial photos of water in the bay made at the height of: a) 150 m, b) 230 m, c) 300 m and d) also at 300 m in strong wind. The localized dugongs are marked with red color
(Hodgson, 2013)

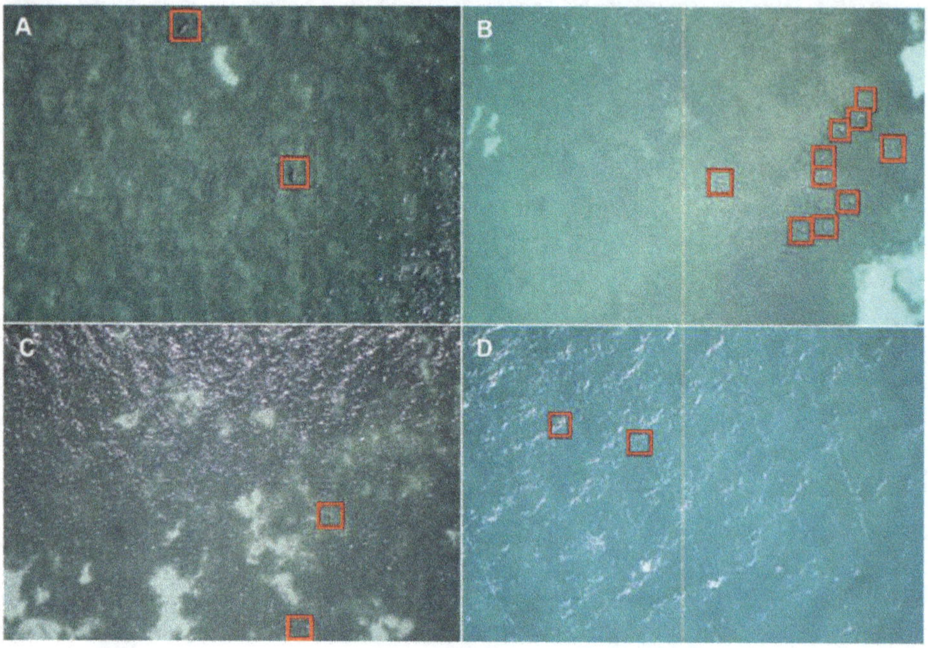

In the event of a large area fire (e.g. a forest fire), an unmanned aerial vehicle with an airplane characteristic will perform better, because it allows flying at higher altitudes and at higher speeds, which will allow you to visualize the entire fire and determine the direction of the fire spread. In the event of a point fire (e.g. a house fire) its observation with the use of a multicopter will allow you to see more details by hovering in the air. Thermovision cameras in which the drone can be equipped allow to locate the source of the fire and find those at risk inside the objects (Merkisz, 2016). The diagram of the fire detection using unmanned aircraft is shown in the Figure 38. The multicopter can also get to the burning hall and locate the source of the fire there, where toxic smoke and a large amount of carbon monoxide prevent the entry of the rescue team.

Costs of fires in areas covered with forests and other plants without significant human interference were calculated in New Zealand. There, about 20-25 large fires occur annually in areas covered with wild vegetation. This study took place in January 2014. It was assumed that every 1$ spent on fire prevention saves 3.76$ that would have to be spent on its extinguishing. Then, the value of the equipment that should be purchased to control the fire-exposed areas was summed up taking into account the purchase of unmanned aircraft, thermal imaging camera and other equipment necessary for flights. It was also taken into account that in New Zealand in forest fires and other open areas over the last 40 years, 16 people were killed, of which 4 people died in the cases of traditional manned helicopters taking part in the rescue operation. The controversy is the conversion of the value of human life into a monetary value, however, the Ministry of Transport in New Zealand, considering other accidents, e.g. road accidents, priced it in 1991 for 2 million dollars. Taking into account the current exchange rate during the study in 2014, this was converted into 3.71 million dollars. The study includes a comparison of the current situation of using only manned helicopters for observation of the terrain with three variants of implementation of drones for this purpose, at the level of 25%, 50% and 100% of observations. According to the results of the study, 6.3 million dollars a year are now being obtained by preventing fires with helicopters. In the case of replacing them with unmanned aircraft, it would be 8.6 million, 15.2 million and 17.4 million dollars, respectively (Christensen, 2015). The probability of locating a fire using a variety of unmanned aerial vehicles for observation is shown in the Figure 39. This shows that the most financially advantageous option is the complete replacement of helicopters for observing areas threatened by fires. However, due to small differences, the most likely option is the use of 50% of drone, and 50% of traditional manned units.

The Use of Drones to Search of Missing Persons

In Poland, since 2010 there has been an increase in the number of disappearances, most often accidental deaths of children and the elderly. The search area often includes areas with low population density and hard to access. Time is important in exploration efforts, as disorientation and inappropriate clothing in difficult conditions may threaten the health and even lives of missing persons. The area where the probability of finding a given person is high is to start the search (it is created on the basis of the location of the starting point, taking into account the possibility of the person moving and various scenarios). The purpose of observing the search area can be used unmanned aerial systems, because they transmit the image to the operator with the provision of accurate GPS coordinates. They can also use thermal imaging cameras, which enables searching also at night (Lorenc, 2016). The view from a thermal imaging camera mounted on an unmanned aerial vehicle during the search for missing persons in the forest area at night time is shown in the Figure 40.

Figure 38. Scheme of fire detection using a drone equipped with appropriate equipment
(Tuśnio, 2016)

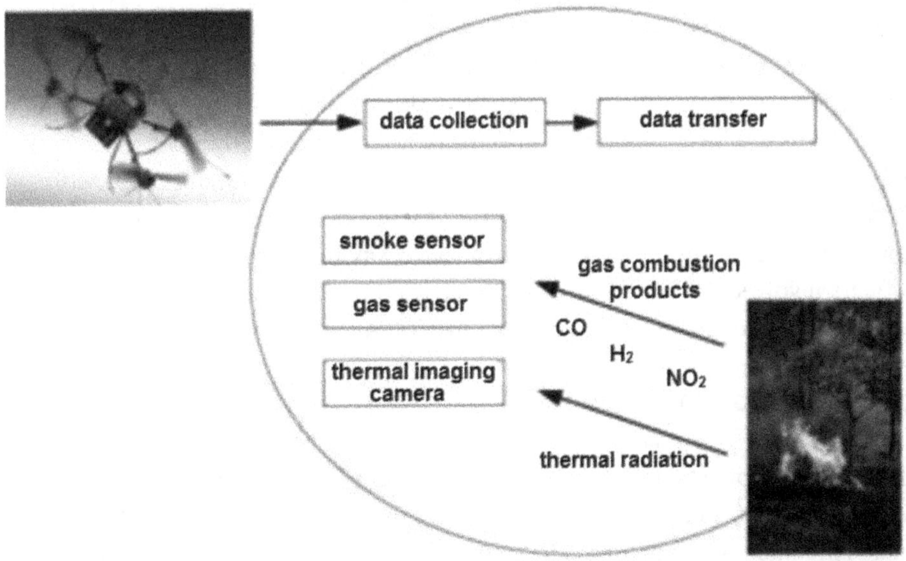

Figure 39. The probability of locating the fire and the associated cost reduction
(Christensen, 2015)

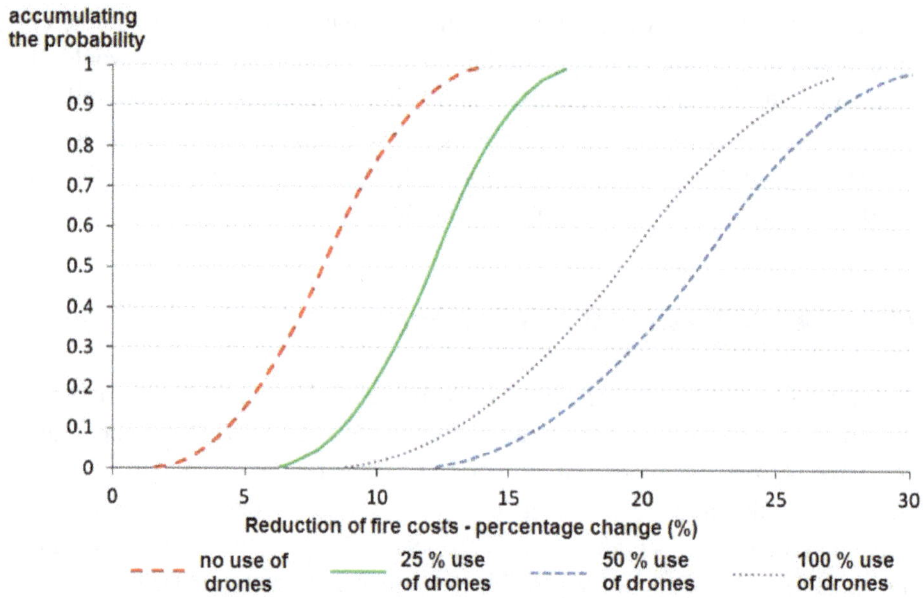

Drones can also be used to observe the condition of waters in areas threatened by flooding, because they will help to check the level of water in rivers and reservoirs, the condition of flood banks, and during the flood also the speed and direction of the flood wave. During the flood, an unmanned aerial vehicle will also allow people at risk to be located (Tuśnio, 2016). Unmanned aerial vehicles can also be used when very heavy snowfall occurs to reach places inaccessible by other means of transport. In this way, it is possible to provide medicines, food and water for people residing in the affected areas (e.g. a buried village in a mountainous area) (Guiraud, 2012). The difficult task is to determine what kind of drone will be the most suitable for the Fire Service (Merkisz, 2016).

The drone can also act as a beacon to guide the rescuers to the place where the sought-after was found. It is also possible to use an unmanned aerial system to transfer the wanted primary means (such as drinks, food, etc.) before rescuers reach it. The disadvantage of drones is the fact that you can't conduct searches using them in difficult weather conditions (Tuśnio, 2017). Due to their small size, these devices can move in hard to reach nooks and crannies (Lorenc, 2016). In London in the Hackney district, to search for a missing person, an unmanned aircraft was used by the police on 12 September 2017. The use of the drone allowed for a quick check of a large, open space, saving time and at the same time limiting the number of officers involved in the search. The device used in this case is Aeryon Skyraner, which can perform its tasks even in adverse weather conditions (Collins, 2017).

The Use of Drones to Traffic Control

In the road transport system, it is necessary to constantly observe the traffic intensity on roads, the conditions prevailing on them, and to detect the occurring difficulties. For this purpose, cameras, radars and other sensors mounted on roads are traditionally used. However, their main disadvantage is the fact that they require the deployment of a very large number of cameras and observing observers to receive full information about the area. With the help of permanent cameras, it is not possible to observe many incidents that may occur on the road network. To monitor the traffic situation, unmanned aerial systems can be used, because in addition to providing information on the current traffic volume, they can reach the place of inconvenience (e.g. accident, congestion). Road multicopters are best suited for traffic observations because they can hang still during the observation of a given area (Sosnowicz, 2016). Controlling traffic using drone should consist of four stages:

- Standard observation of a specific road section detecting vehicles moving along it, estimating their quantity,
- In the event of an unusual situation, its immediate detection and notification to the operator,
- Making a preliminary classification of an unusual situation detected,
- Adequately to the situation that occurred on the road gathering information about it and transferring it to the operator for the purpose of making appropriate decisions.

After receiving the information, the operator of the unmanned aircraft is obliged to take action. In the event of an accident, he / she should notify the Police, Ambulance, etc. During the traffic observation, it is important to detect abnormal events as soon as possible. Most often it is congestion and the location of stationary and slow moving vehicles (Kim, 2015).

The study of traffic observations using the drone was conducted in Beijing, China, on the northern part of route No. 5 with the use of the DJI Phantom 2 quadcopter equipped with a GoPro 3 camera with a resolution of 1920 × 1080 pixels transmitting the image to the operator in real time. This device can be in the air for a maximum of 25 minutes and move at a speed of 10 m / s. On the screen displaying the camera image, reference lines have been added to allow more precise control of the unmanned aerial vehicle. The real-time image sent to the operator's screen is low resolution. Better image quality is saved on the SD card and available after the observation. Observations were conducted 66 times between December 2014 and May 2015. It always started at 6:30 in the morning. The study consisted of 4 parts:

Figure 40. A view from a thermal imaging camera when searching of people missing at night
(Tuśnio, 2017)

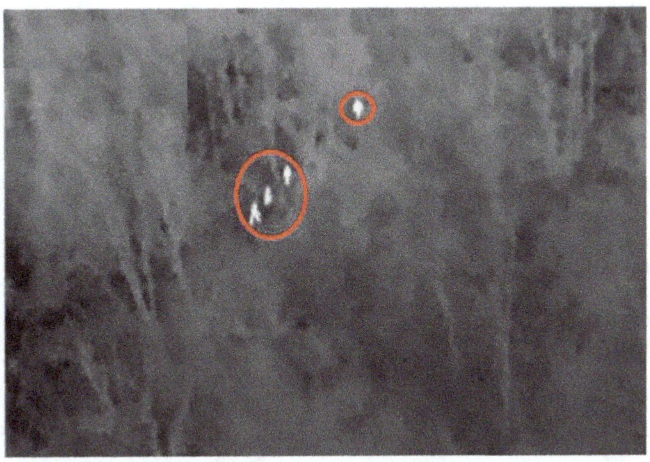

image recording, image character analysis, vehicle shape detection and vehicle tracking (Wang, 2016). The individual stages of this study were carried out on the scheme shown in the Figure 41.

A photograph taken from an unmanned aerial vehicle and used to identify road traffic is shown in the next Figure 42.

Unmanned aerial vehicles can also be used to prepare documentation after the occurrence of a traffic accident. These activities must be carried out as quickly as possible to clean the roadway as quickly as possible and restore traffic there. In addition, observation using a drone does not endanger the police officers, because they do not have to be directly on the road. This solution is used by the police in several countries, such as the USA, Canada or Italy (Sosnowicz, 2016). The picture taken from the drone during the analysis of the place of the road accident is shown in the Figure 43.

In the USA, police in Gwinnett County, Georgia, use road drones to analyze traffic accidents. It uses the DJI Inspire T600 quadcopter, which has a mass of less than 3 kg and can stay in the air for 10 minutes. Lieutenant Chris Smith claims that it takes 5 minutes to shoot photos from the area of an accident using an unmanned aircraft (Keating, 2016).

The Use of Drones for Loads Transporting

It is possible to use unmanned aerial vehicles to deliver parcels. The first time the drone was used to transport shipments was introduced by Jeff Bezos from Amazon in December 2013. Earlier, in 2007, the idea of transporting pizza and other fast food to hard-to-reach areas of the Alps in Switzerland appeared at ETH Zurich (D'Andrea, 2014). Amazon has developed a system in which drones using the handle take over the package on the packaging line and transport it to the destination, after which they leave and return automatically to the starting point (Lorenc, 2016). The unmanned aerial vehicle developed by this company presents the Figure 44.

Figure 41. Scheme of the traffic control with the use of images from the drone
(Wang, 2016)

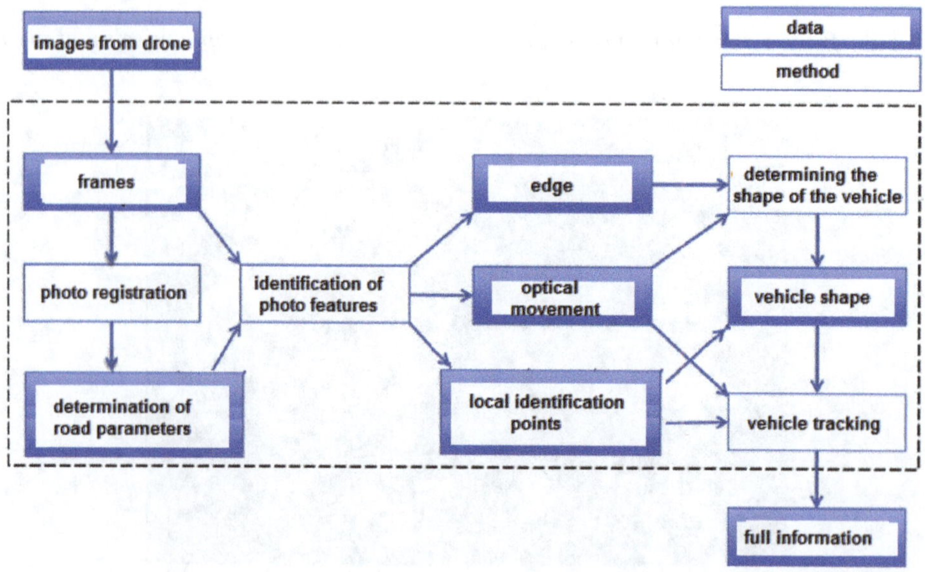

Figure 42. Photo for identifying of the road situation
(Wang, 2016)

Figure 43. Documentation of the consequences of a traffic accident with the use of unmanned aircraft
(Sosnowicz, 2016)

Ultimately, it plans to develop a technology that allows the automation of the entire delivery process from packaging to transport, so that the shipment reaches the recipient within 30 minutes (Strzelczyk, 2015). Shipments which in this way it plans to deliver using drone can be up to 2 kg and up to 10 km. It is possible to transport shipments in this way even in difficult weather conditions, with side wind up to 30 km/h (D'Andrea, 2014). Another example of the use of these devices is the use of a drone by DHL for transporting medicines to the German island Juist in the North Sea. It moves along a specially designated air zone, which allows safe flights (Strzelczyk, 2015).

The unmanned aerial vehicles used for transporting goods can be operated in automatic mode, semi-automatic mode or be controlled by the operator. In automatic mode, it is possible to plan the route earlier, and the device itself will do it thanks to navigation using GPS. The semi-automatic mode allows the operator to make changes to the previously planned route during the flight. If the GPS signal is lost, the drone will return to the starting point using the saved gyroscope and pressure sensors.

When transporting shipments, the problem is to safely move unmanned aerial vehicles in flight to avoid collisions and other problems. This problem occurs mainly in places beyond the operator's eye range. In the event of a failure, the device may fall to the ground causing damage to people, animals or material damage. In some countries, it is forbidden to operate unmanned aerial vehicles over areas where people are clustered. Thus, shipments in this way can be provided only over forested areas, unpopulated, over water reservoirs. The problem also occurs during landing at a previously planned place, because although it has been provided for it, unexpectedly, for example, a passing man may appear in it. On the other hand, the advantage is the possibility of reaching the area difficult to reach for a delivery vehicle by the drone (Berner, 2016).

Unmanned aerial vehicles can also be used in the medina to transport urgently needed drugs or blood. Some rare and expensive medicines are found only in some hospitals. Then to use them it is necessary to transport a patient from a smaller, local hospital to a large facility or transport drugs by car or helicopter. In the case of a car, the travel time is long, while the use of a helicopter is very expensive. In the case of a drone used for the transport of medicines, a short waiting time is achieved with a low cost at the same time. A similar situation occurs with rare blood groups needed for transfusions, also in this case the use of unmanned aircraft will shorten the patient's waiting time (Thiels, 2015). The appearance of an unmanned aircraft used for medical purposes is shown in the Figure 45.

Figure 44. Documentation of the consequences of a traffic accident with the use of unmanned aircraft *(Lorenc, 2016)*

Figure 45. An unmanned aerial vehicle used for transporting of medical items
(Thiels, 2015)

The Use of Drones in Water Rescue

The Water Volunteer Ambulance Service (in Polish, WOPR) supervises and protects the safety of people over water in Poland. Their main goal is to save the lives of people threatened by swimming in different water bodies as effectively as possible, and by doing so to make the number of people drowning in each year as low as possible. Unmanned aerial vehicles that can be used while searching for a drowning person on water bodies may be helpful to operations carried out by WOPR. Rescue operations carried out by the WOPR must be as fast as possible, because it may depend on the life of the melting one. Reaching an agonist with a drone is 1/3 faster than a lifeguard. This drone can be equipped with a lifebuoy, dropped to the victim and allowing him to stay on the water until the rescuer arrives.

An unmanned aerial vehicle equipped with a lifebuoy is used in Chile. In addition, he has a camera through which rescuers can constantly control the situation of the victim and throw the lifebuoy to the right place. In the case when the victim is far away from the shoreline or when the rescue operation takes place in difficult terrain (e.g. steep bank), the use of the drone will make rescuers' work easier and may even determine the success of a given action. Another drone of this type is an eight-airship with 3 lifebuoys and an additional thermal imaging camera. Thanks to the thermal imaging camera, they can also observe observation at night (Tuśnio, 2016). Such a drone is shown in the Figure 46.

Another design is the Ryptide project. It provides for the equipment of a drone with a lifebuoy, which is not inflated during the flight, but takes the air already after being dropped. This mechanism is universal and can be used in various types of multicopter. Due to the fact that the lifebuoy can be attached to any type of unmanned aircraft, this solution is economically viable, shown in the Figure 47 (Project Ryptide, 2017).

The drone intended for water rescue was also constructed in Germany. An unmanned aerial vehicle, Md4-1000, equipped with a buoy that takes on air only after being thrown into the water was used there. In addition, he had a camera that transmits the image in real time. A demonstration of the use of such a device on the Elba River for the general public was also carried out on 22 July 2016 (Drone-assisted water, 2017).

Figure 46. An unmanned aerial vehicle equipped with 3 lifebuoys
(Merkisz, 2016a)

Figure 47. Drone equipped with a Ryptide lifebuoy
(Project Ryptide, 2017)

The Use of Drones in the Transport of Defibrillators

Another field in which drones can be used is the transport of a defibrillator, or device used to stimulate the heart when it is stopped for various reasons. Computer tests of such a solution were made for both urban and rural areas. Field tests were carried out in Stockholm, in the area of nearly 6.5 thousand km2. This area is inhabited by an average of 343 people/km2, however, it is diverse, because there are rural areas with a population density below 250 people/km2 and, in the city center, population is more than 6,000 people/km2. In this zone every year, for every 46 people from 100 thousand inhabitants have place a blood circulation stop outside the hospital. During the tests, an unmanned aircraft was used that could move at speeds of up to 70 km/h and with a range of up to 10km. Three ways of delivering the defibrillator were used: dropping it from a minimum height of 25 meters, lowering the height and dropping it from a height of 3-4 meters and landing the drone with a defibrillator for disassembly and use. The study showed that in an urbanized area the drone reached the place where a defibrillator was needed in 32% of cases faster than traditional car transport. It was possible to gain about 1.5 minutes thanks to it. In rural areas, however, the drone with the defibrillator reached earlier than the traditional ambulance in 93% of cases, which allowed to save about 19 minutes. The tests also showed that the method of dropping the defibrillator from a height of several meters and the method of landing unmanned aircraft equipped with a defibrillator are applicable. Dropping a defibrillator from a height of more than 25 meters is not a practical solution, because in this way it is impossible to precisely determine the place in which it will go, and in this case also other factors like the wind decide. The results of the research have shown that using the drone to transport the defibrillator in rural areas allows to significantly reduce the waiting time for this device, which may allow saving many people (Claesson, 2016). Such a drone during the defibrillator dropping from a small height is shown in the Figure 48.

The Use of Drones as Auxiliary Units in Rail Transport Systems

There are a number of applications of unmanned aerial vehicles as devices supporting the management of rail transport systems. In Germany, the largest European carrier Deutsche Bahn (DB) uses drones for observation standing on tracks of rolling stock units, to protect them against graffiti hooligans. For this purpose, drones with a length of 1 meter are used, costing about 60 thousand Euro, which can stay in the air for up to 80 minutes and allow the speed to reach almost 120 km/h. They can be controlled automatically or by the operator. The engine of this unmanned aircraft emits only a slight noise, allowing it to remain unnoticed during the observation. The removal of graffiti from rolling stock costs Deutsche Bahn annually about 7.6 million Euro, so using drone to watch the abandoned rolling stock can reduce this cost (How drones, 2014).

The American company Union Pacific is considering using drones to increase the efficiency of maintaining its rail network. In this case, the unmanned aircraft would be able to observe the infrastructure, and if a failure was found, workers would be sent to the site. This would allow faster diagnosing of faults, due to the much higher speed of the drone compared with the walking man, while reducing costs by reducing the number of people responsible for controlling the condition of the infrastructure (How drones, 2014). The railway network in the United States has a length of over 52,000 km. It is controlled twice a week by employees throughout the area. Due to the location of part of the railway lines in areas difficult to access, away from human centers and paved roads, this control takes a lot of time. The use of unmanned aerial vehicles for her allows you to perform the controls more quickly with the simultaneous involvement of fewer employees.

Faster detection of defects can also avoid train derailments, most of which are caused by damage or poor infrastructure (Morris, 2015). Unmanned aerial vehicles are also used to carry out documentation after accidents occur. This is also the case in the case of rail accidents. Such accidents often result in a large area due to the considerable length of train sets. Due to this, their observation and documentation from the ground level is hindered. In addition, the observation of the accident area from a short distance causes the need for researchers to appear in the affected area. However, the use of helicopters generates high costs. Therefore, the most effective solution in this matter is the use of drones. They take pictures using them, use a thermal imaging camera and perform multispectral imaging (Three ways, 2017). The picture taken from the drone showing the area of the area after derailing the train is shown in the Figure 49.

Figure 48. Drone dropping the defibrillator from several meters
(Claesson, 2016)

In France, the state infrastructure manager, SNCF Réseau, conducted rock wall research in 2015 using drones. In the case where the railway line crosses rocky areas, inspections should be carried out to determine the risk of falling rock fragments on the track. Most often, the researchers reach such places on foot. The use of unmanned aerial vehicle enables quick inspection of the rock wall using the camera mounted on the drone, at the same time without exposing the hazards of the workers who are investigating and without interrupting train traffic. In France, unmanned aerial vehicles were also used to observe trees located at the track. In this way, you can locate trees threatening train traffic safety (e.g. withered, which may fall over to the track). Traditionally, such observations are carried out by patrolling the route on foot or by observing it from the driver's cab. In addition, photos taken from the drone allow you to estimate the amount of biomass that can be obtained from the planned to cut trees. These data are forwarded to the contractor performing the felling will allow him to preliminarily assess how much money he will get from the collected biomass and thus price the felling. Observations of trees intended

Figure 49. The effects of train derailment seen from the camera on an unmanned aircraft (Three ways, 2017)

for logging were carried out as part of tests in 2015 on the 200 km stretch of Dijon - Lyon. Another use of drones at SNCF Réseau is the control of the condition of the roofs of railway station buildings and other technical railway buildings. Such a control is faster and does not require special structures (e.g. scaffoldings) to be opened to allow employees to enter the roof (France: Demonstrating, 2015).

Performing observations with the use of unmanned aerial vehicles allows scientists from the Bauhaus University to use a drone to monitor the stability of a retaining wall using photogametry, that is, using photographs to determine the distance between objects. Often, train passengers have a problem with access to WIFI wireless internet, especially on mountain sections, where there is a large number of tunnels. Attempts have been made to provide access to the network using drones. However, currently, due to legal restrictions, this system is not used (Flammini, 2016).

The problem with railway accidents is the fact that large amounts of hazardous materials are transported by this type of transport (e.g. diesel oil, sulfuric acid, etc.), which during a car accident and unsealing are a major threat. A study was carried out to use an unmanned aerial vehicle equipped with an RFID reader to read information placed on a railway wagon for easier and faster identification of dangerous substances (Karoly, 2017). Schematic arrangement of the sensors on the unmanned aircraft used during the test is shown in the Figure 50.

In some cases, such as overturning a wagon carrying dangerous substances (e.g. tankers) from ground level, it is impossible to read the UN table informing about the type of transported substance. In this situation, the most quickly available information will be provided by the RFID reader placed on the drone. The problem with reading data from tables placed on wagons may also occur during their physical destruction during an accident. Then, the use of a multicopter with an RFID reader will allow you to quickly find out which substance is in the wagon and transfer this information to the appropriate emergency services. In the event that after a railway incident there is a risk of an explosion in the carriages, using a drone to gather information will help avoid the risk of losing life by rescuers. The usefulness of an RFID reader mounted on an unmanned aircraft during the determination of a substance during a railway accident depends, however, on equipping wagons with RFID tags. Currently, this system is equipped with 26,000 railway wagons in the United States, which transports 650 different types of dangerous substances.

Figure 50. Elements attached to an unmanned aircraft used in the test
(Karoly, 2017)

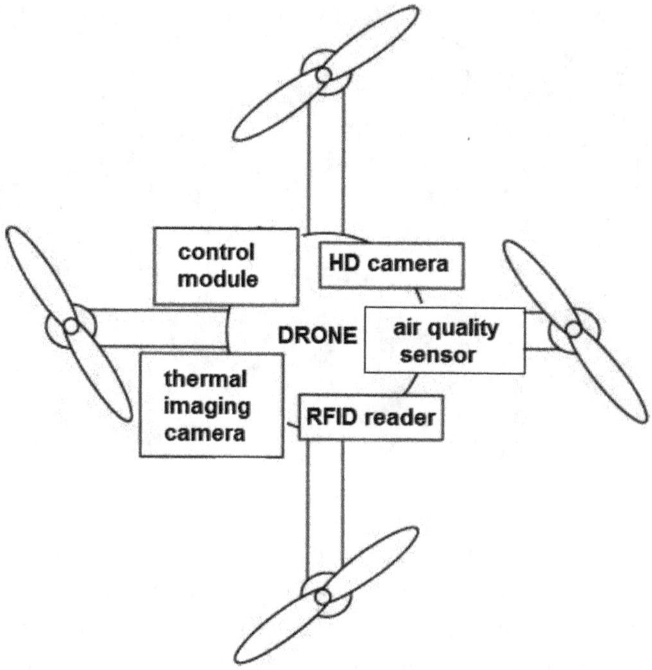

Another use of drones to increase the safety of transporting dangerous substances in rail transport is the temperature control of tank wagons using a thermal imager mounted on an unmanned aircraft (Karoly, 2017). Such a picture is shown in the Figure 51.

It is possible to use drones to almost completely monitor all railway infrastructure and railways. Communication with the unmanned aerial vehicle along the railway line is maintained using WIFI-based GSM-R wireless systems. Their range ranges from 6km to 17 km. DJI Phantom 4 unmanned aerial vehicle with a maximum speed of 72 km/h and a maximum residence time of 30 minutes was used. This allows one-time inspection of the 36 km section of the railway line (Flammini, 2016). Elements of railway infrastructure that can be controlled are shown in the Figure 52.

The scheme of how the drone monitoring the railway line works is shown in the next Figure 53.

The problem occurring on the railway network is a large number of outsiders entering the railway area. Due to the large area of the network and the course of some lines through hard-to-reach areas, it is difficult to control this area by land security forces. There is also often a conscious exposure of own life in the railway area by suicides. The number of bystanders who die under the wheels of the train is greater than the number of passengers losing their lives in railway accidents. Drones can be used to improve this situation because they give the opportunity to observe large areas in a short time (Morris, 2015). In the UK, Network Rail used drones to observe track reconstruction at a distance of about 100 meters, on the seafront in Dawlish. The track was completely destroyed there as a result of storm and storm in the winter of 2014. Later, the company started using drones to control other track works. However, it does not have unmanned aircraft and pilots, and uses subcontractors.

On the other hand, the Dutch company Pro Rail uses unmanned aerial vehicles equipped with infrared sensors. They are designed to control the moment of switching heating of turnouts in winter. Checking

Figure 51. Control of tank wagon temperature using an infrared camera mounted on a drone
(Karoly, 2017)

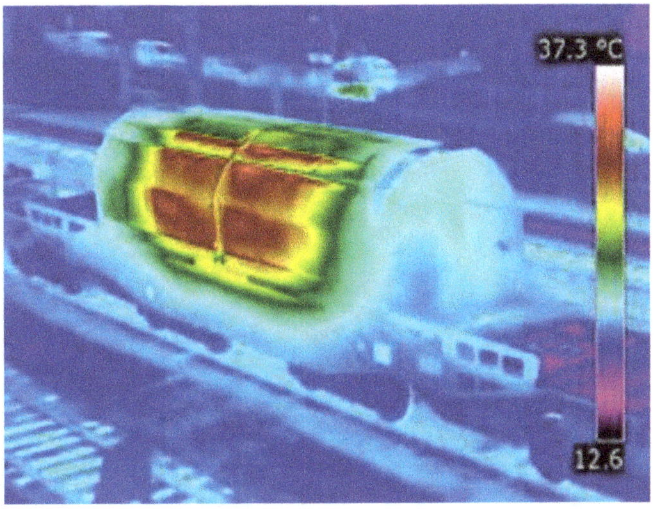

Figure 52. Possibilities of drones using to observe railway infrastructure
(Flammini, 2016)

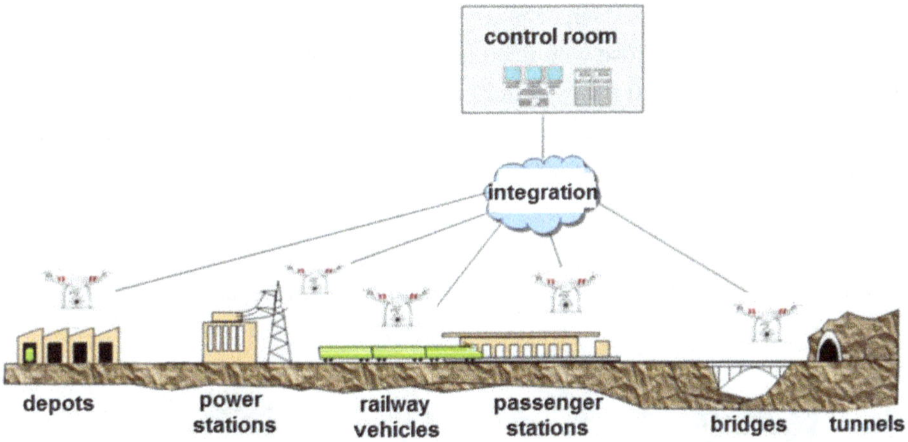

whether the moment of turning on the heating of turnouts is set correctly in the traditional way would cause the necessity to enter the track of specialists equipped with an infrared sensor. Entry of researchers to the track causes difficulties in the movement of trains and may be a threat for the investigators themselves. Therefore, the use of a drone for this purpose eliminates the occurrence of difficulties (How drones, 2014).

Other Applications of Unmanned Aerial Systems

Drones can be used to determine the number of participants and determine their change with the passage of time during a mass event. However, the limitation here is the weather conditions affecting the quality

Figure 53. Monitoring the railway line with the use of a drone
(Flammini, 2016)

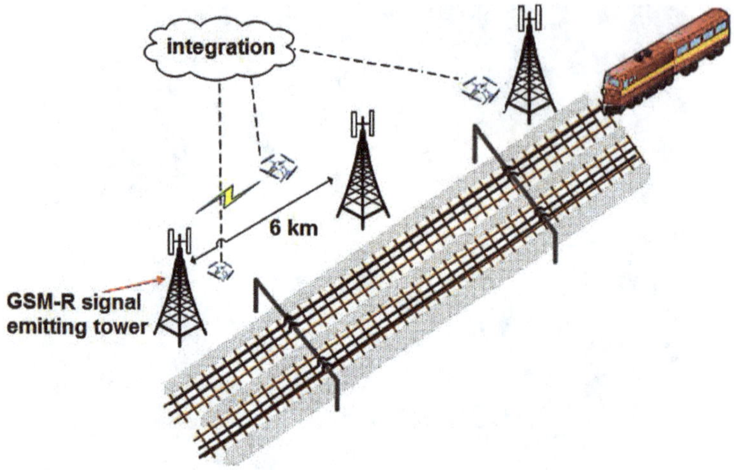

of photos and flight safety and the fact that due to human safety, do not fly directly over the place of the event, only on unpopulated area near it, which is not always possible (Smaczyński, 2015). The picture taken during the mass event research at the Poznań University of Technology is shown in the Figure 54.

In the case of organizing a mass event, the organizer is required to provide security, including the provision of medical services for those who need help from the event's participants. Pre-hospital medical help includes rescue patrols and medical help points located according to the needs depending on the size of the place where the mass event is organized. During some events a team of paramedics will suffice, while the presence of a doctor is required for gatherings over 5,000 people. For larger events,

Figure 54. Photo taken from the drone during observation of the event at the Poznan University of Technology
(Smaczyński, 2015)

an additional team of paramedics is required for every 10,000 participants. The use of unmanned aerial vehicles to observe a mass event will allow faster localization of a person in need of medical assistance. In situations threatening people's health or life, it is important to take rescue actions as soon as possible. During mass events due to the presence of a large number of people in a small area, accurate observation of the ground level is difficult. An additional obstacle may be the considerable extent of the area requiring supervision and its topographic shape. An important factor when ensuring the medical safety of a mass event is air temperature, because in hot weather often occur cases of overheating. Drones allow to achieve a much larger field of visibility than from the perspective of a person standing in a crowd. Obtained from an unmanned aircraft and sent to the operator in real time, the images can be magnified, which increases the accuracy of the observation. These devices are resistant to sunlight and high temperatures, so their use avoids exposing the rescuer to the overheating (Robakowska, 2017). In winter, when access to some mountain huts by traditional transport is difficult, it is possible to use the drone to deliver medicines or food there (Merkisz, 2016).

It is also possible to use unmanned aerial vehicles to inspect the crime scene or other such an event. In this way, you can protect the evidence before the officers get there, while leaving traces in the field intact. With their help, you can also locate hiding criminals, letting the search team avoid the danger caused by unexpected finding (Merkisz, 2016a). The picture taken from an unmanned aerial vehicle during the analysis of traces is shown in the Figure 55.

In the case of a large area where criminal offenses have been made, drones may also enable the transport of materials between crews investigating the area at distant locations. Unmanned aircrafts can also be used to observe border areas of countries by the Border Guard, which will allow to locate possible attempts to smuggle goods or attempt to enter illegal immigrants on the territory of a given country (Merkisz, 2016a).

Figure 55. Picture taken from a drone while searching for traces of crime in difficult terrain
(Merkisz, 2016a)

Another option for using unmanned aerial vehicles is the tourism industry. Traveling has been accompanied by commemorative photographs for many years. Using the drone allows you to take pictures from above, which allows you to show a wider perspective, making the photos more interesting. Tour operators may, therefore, aim to make it attractive for tourists to make commemorative photographs using unmanned aerial vehicles. Pictures from these devices can also be used to create advertising brochures or create promotional films to show your attractions (be it a natural attraction, like a waterfall or a historic object in aerial photography will look more impressive than in a traditional photo) to encourage more number of tourists to visit in a given place (Berbeka, 2016).

Unmanned aerial vehicles were also used by the organizers of the Olympic Games in Rio de Janeiro in 2016. A promotional film was created, in which aerial shots from the drone were used and depicting the Olympic park. This allowed to show the amount of buildings created especially for this event. These devices were also used during the Olympic Games in Rio de Janeiro to observe the order in the fan zones (Nadobnik, 2016).

Drones are also suitable for detecting places of illegal delivery of chemical agents to rivers or reservoirs (e.g. chlorine, sulfur), because these sites are usually located in hard-to-reach areas, where access by other means of transport is difficult or impossible (Guiraud, 2012).

Another option for using unmanned aerial vehicles is to transport them with vaccines. For this purpose, the model was made and simulations of continuous delivery of vaccines under various conditions (different terrain features, population size and traffic) were carried out using it. The results of the simulation showed that the use of drones in the transport of vaccines increased their availability from 94% to 96% while reducing transport costs compared to road transport. Depending on the simulation conditions, the transport of drone vaccines was cheaper by 0.05-0.21 $ per dose (Haidari, 2016).

Still another way to use drones is to take measurements of air pollution along roads using them. Previously, measurements of air pollutants emitted by cars were made mainly at the ground, and measurements at higher heights were made from buildings or balloons. The tests were carried out at a height of 5 to 100 meters from the ground, rising up unmanned aerial vehicle vertically upwards. The path of such a flight is shown in the Figure 56 (Villa, 2017). Test measurements of particulates made using unmanned aerial vehicles from 10:00 to 16:00 showed that at a ground level their concentration was $2*10^4$ p/cm^3, then at 40 meters this value was $3*10^3$ p/cm^3.

Drones can also be used to measure other pollutants in the air. The Chinese lowland is one of the most polluted areas in China. There is a very high ozone concentration in both urban and rural areas. Most of the pollution measurements were made at the soil level, so their vertical distribution was not known. Research using drone showed that during most flights ozone levels were higher at the top of the mixed layer than at the bottom. The highest ozone concentration was detected in the mornings in the mixed layer (Wang, 2017).

It is also possible to use unmanned aerial vehicles to assess the state of soil pollution. Very high pollution of agricultural land is a big problem in southern Italy. There is an increased risk of cancer caused by the presence of heavy metals in the soil. This contamination has both natural origin related to volcanic eruptions and is associated with industrial production. There are such elements as cadmium, chromium, copper, mercury, lead, nickel, zinc and selenium. The research area with the use of a drone covered the area around Trentola Ducenta in the province of Caserta, and its area was 4,500 m^2 (Capolupo, 2015). The location of this area is shown in the Figure 57.

Figure 56. Flight path of the drone during the test, which floated from the ground to a height of 100 meters
(Villa, 2017)

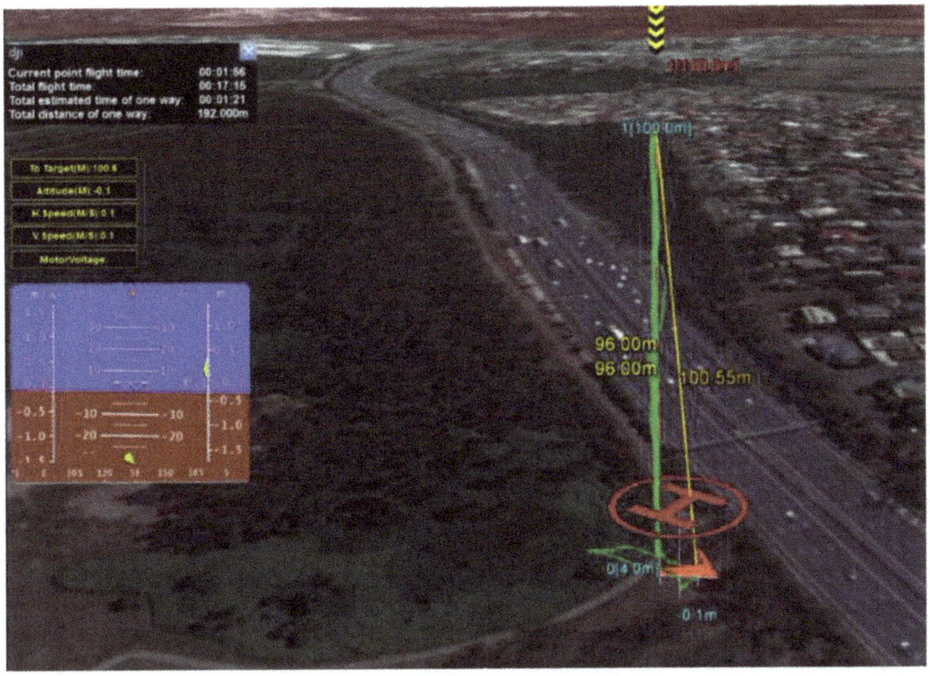

Figure 57. Location of the polluted area in Southern Italy
(Capolupo, 2015)

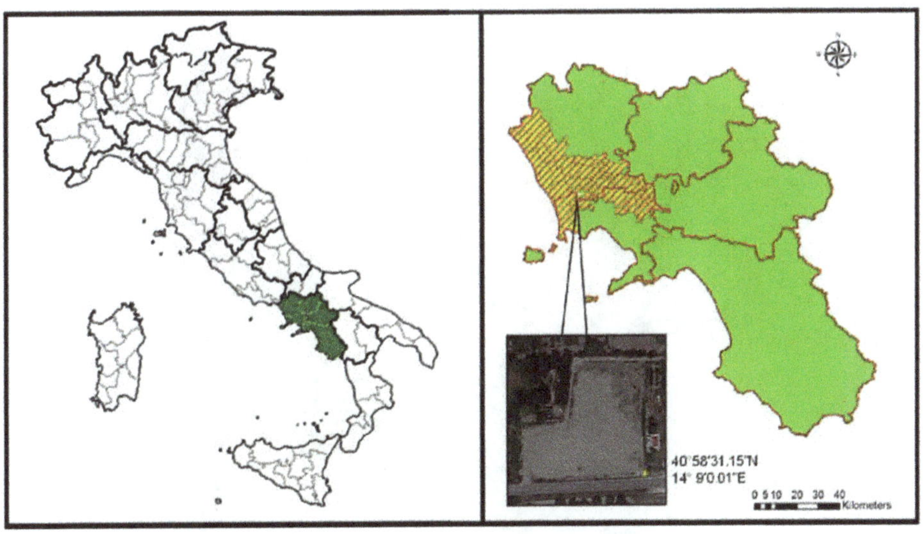

Earth samples were collected regularly from the area of 5 m * 5 m. Then, spectrometric analysis was made at the laboratory of Federico II University in Naples. Information for this analysis was collected using the DEM (Digital Elevation Model), i.e. a numerical model of the terrain was made containing its topographical height. They are created using photographs taken from unmanned aerial vehicles. The study used Canon PowerShot S100 with a resolution of 12.1 megapixels. After receiving the numerical model, its interpolation allowed to determine the level of copper in the soil. These tests allowed the detection of copper in various concentrations depending on the point under study. The maximum value was about 120 mg/kg, which is the limit value allowed by law in Italy (Capolupo, 2015). The map presenting the results of the conducted study is presented in the Figure 58.

However, the use of infrared allows using unmanned aerial vehicles to detect delaminations in elements of concrete bridge structures. It does not require physical contact with bridge elements, which means that it is not necessary to stop vehicle traffic on the tested bridge. Such a study uses a drone equipped with a thermal imaging camera. Images from such cameras should then be joined together to form thermal mosaics for the entire bridge. During the tests of this method, the mosaics were used to create a map defining various degrees of delamination of elements (Omar, 2017). Scheme of creating a thermal mosaic for a particular bridge is shown in the Figure 59.

Figure 58. A map showing the level of copper concentration in the studied area (Capolupo, 2015)

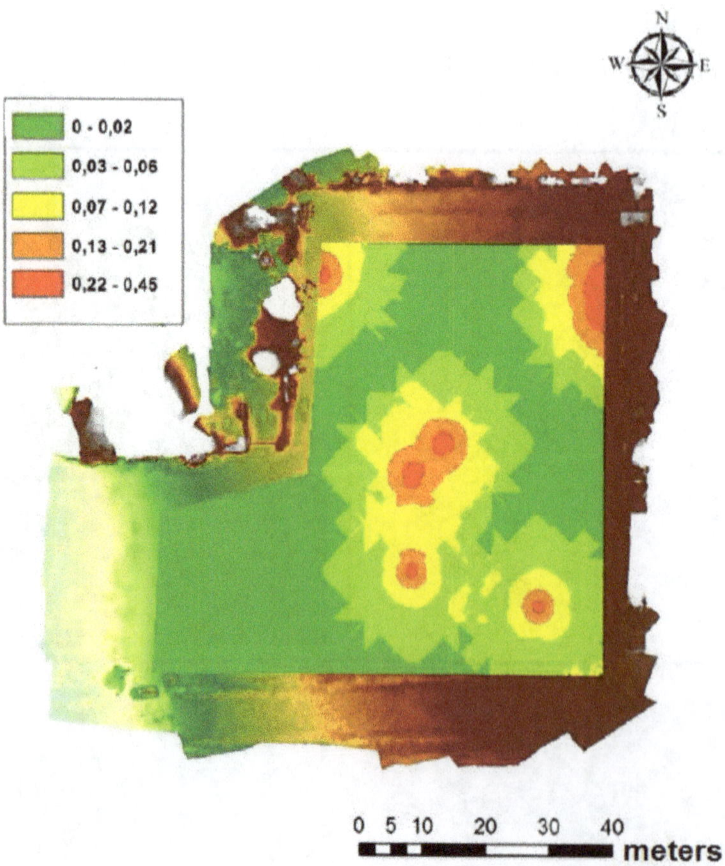

The results obtained by this method were compared by performing a non-destructive test with sound waves. The comparison showed that the method using an unmanned aircraft proved to be effective. The biggest advantage of this method is the fact that its use does not require closing the tested bridge for vehicles. As a result, objects can be checked more often in this way without causing any hindrance to traffic or closing of the examined detour site (Omar, 2017).

Another use of drones is to use them for navigation for the blind or visually impaired. Setting the goal that a person wants to get in this case must be done through voice control. The position of an unmanned aircraft for a blind person is easy to recognize, because this device gives a characteristic sound of rotating propellers. This system provides that a blind person would wear a headband with a microphone to send commands to the drone using sound. An exemplary command can be "take me home." After issuing the command, the device recognizes the location, calculates the shortest route and guides the man to the desired place. The drone while navigating is located 1-2 meters from the blind and adjusts its flight speed to its pace of movement (Avila, 2015). The use of an unmanned aircraft for navigating a blind person is shown in the Figure 60.

Unmanned aerial vehicles were also used to assess the condition of tropical forests. It can help local communities to better manage and protect them. The greatest threat here is deforestation and forest degradation. Traditionally, data on the condition of tropical forests are collected by professional scientists. However, as a test, trained volunteers and drones were used for this purpose. The data obtained in this way turned out to be just as accurate as those collected by professional scientists, but at a much lower cost. Small, light multi-rotors were used for the tests. The advantages of using drones to assess the condition of forests were both technical (the ability to take aerial photographs) and social (involvement of people from local communities) (Paneque-Galves, 2014). Tropical forests are characterized by a large diversity, as shown in the Figure 61.

Another application of drone is the journalistic industry. Unmanned aerial vehicles are a tool for collecting messages. Already in the 19th century, journalists and photographers used hot air balloons to observe different events from a higher perspective. Later, in the 20th century, planes and helicopters

Figure 59. Processes occurring when creating a bridge stress map
(Omar, 2017)

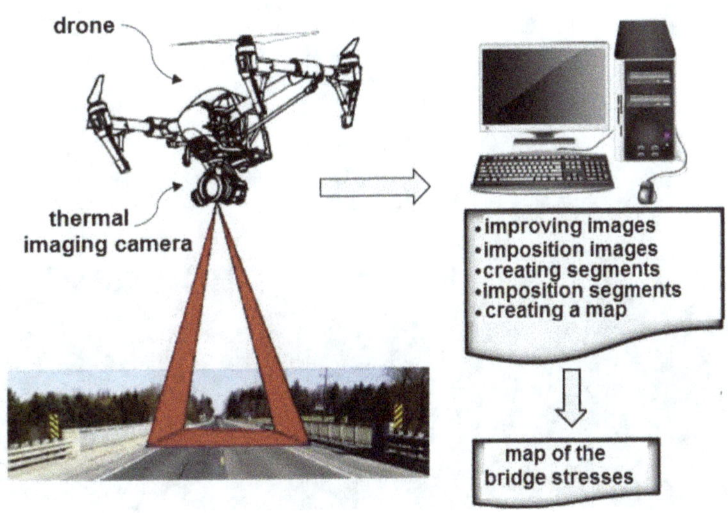

became popular for fires, natural disasters and other events. The use of unmanned aircraft, however, reduces the costs of both purchase and operation, which makes the observation of events in advance become more widely available, e.g. for journalists from the local press (Goldberg, 2014).

Drones can also be used to measure the level of radioactive radiation. In March 2011, the Fukushima Daiichi nuclear power plant disaster occurred in eastern Japan. Now, after years of forced displacement, people are starting to return to their homes. Therefore, the amount and representation of any radiological impurities had to be determined. An unmanned aircraft equipped with a laser rangefinder was able to obtain a point cloud on which a map of pollution could be imposed. These studies were carried out in May 2014 in the area of a farm comprising rice fields along with an irrigation system, the farm was located near a destroyed nuclear power plant and was abandoned. After the test, increased radiological contamination was found along the irrigation network. This is related to the flow of water, so that contaminants are transported through the water system to different places (Martin, 2016). The flight path of the drone in the contaminated area and the radiation level measured during the test are shown in the Figure 62.

The use of unmanned aerial vehicles for this purpose makes it possible to carry out measurements by an operator in the distance and does not have to be exposed directly to radiation by staying in a contaminated zone. Also in derelict areas, there are no risks to people or objects related to the flight (Martin, 2016).

Another use of drones is the observation of vertical profiles of black carbon (BC) in the lower part of the troposphere (this is the lowest layer of the atmosphere). Such research was conducted in Poland in the Wisłoka valley in the Podkarpacie region near Strzyżów on the Wisłok river (Chiliński, 2016). This place, together with its height above the sea level, is shown in the Figure 63. The aethalometer, or device allowing to measure suspended particles in the gas colloid stream, is used to measure this phenomenon. Optically it looks like smoke or haze. The research was carried out in autumn and winter during the occurrence of smog, they were carried out in two stages. In the first stage, the aethalometer was carried

Figure 60. Using a drone to navigate a blind person
(Avila, 2015)

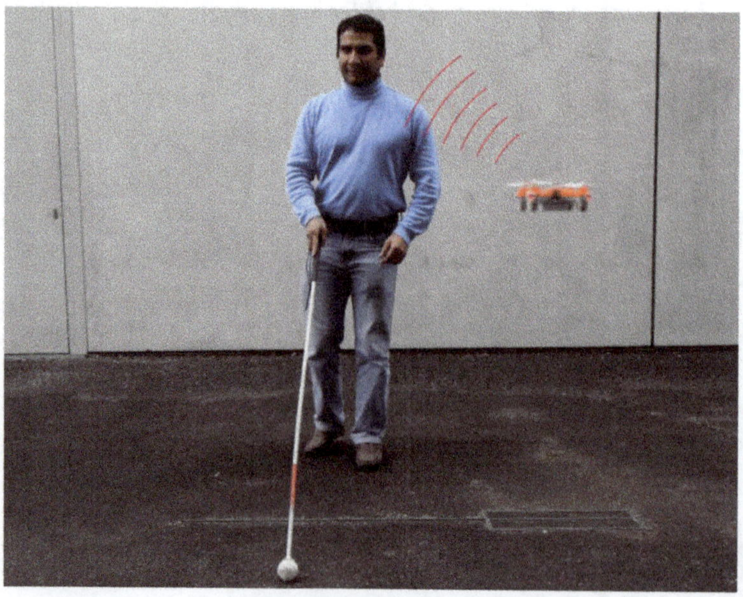

Figure 61. Aerial photographs of various forest areas: a) forest in Sabah state in Malaysia, b) Chitwan National Park in Nepal, c) palm oil plantation in Indonesia, d) recently cut piece of forest in Indonesia (Paneque-Galves, 2014)

by the investigator along the mountain basin route. In the second case, it was mounted on an unmanned aircraft, which enabled measurements up to a height of 1500 meters from the ground. Usually, the highest concentration of smog occurs at a height of up to 100 meters from ground level. The researcher's wander with an aethalometer focuses on the vertical structure of black carbon concentration. Using the drone allows you to perform various research scenarios, limited only by the wind, flight time and local legal regulations. The highest smog values occur in mountainous areas in autumn and winter, which is associated with heating of households by local residents mainly with the use of coal stoves. The highest BC value measured during the tests was over 60 mg, which is almost 60 times more than the average in this area (1.1 mg). Vertical volatility of black carbon concentration is also significant, and complex multilayer structures are often created. During the cloudless sky and weak wind, the temperature inversion takes place, which leads to vertical mixing. Because of this, smog arises in the valleys, sometimes even fog appears. However, the occurrence of a mountain breeze causes the movement of pollutants from villages to mountain valleys.

Figure 62. The route of the unmanned aerial vehicle and the results of radiation level measurements
(Martin, 2016)

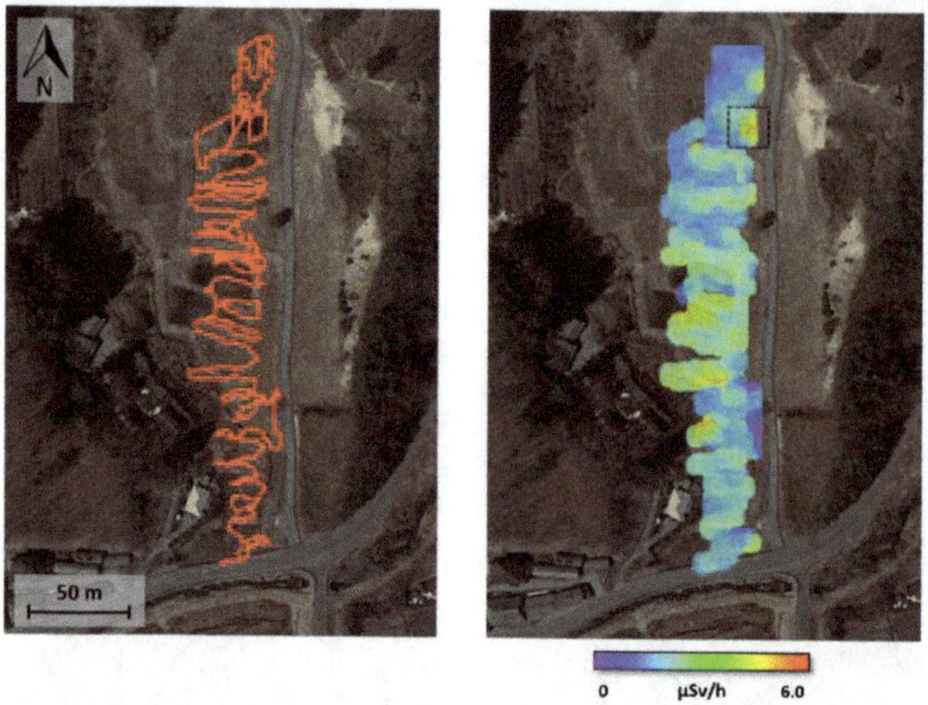

Figure 63. Place of research taking into account its height in relation to the nearest village near Strzyżów on the Wisłok River
(Chiliński, 2016)

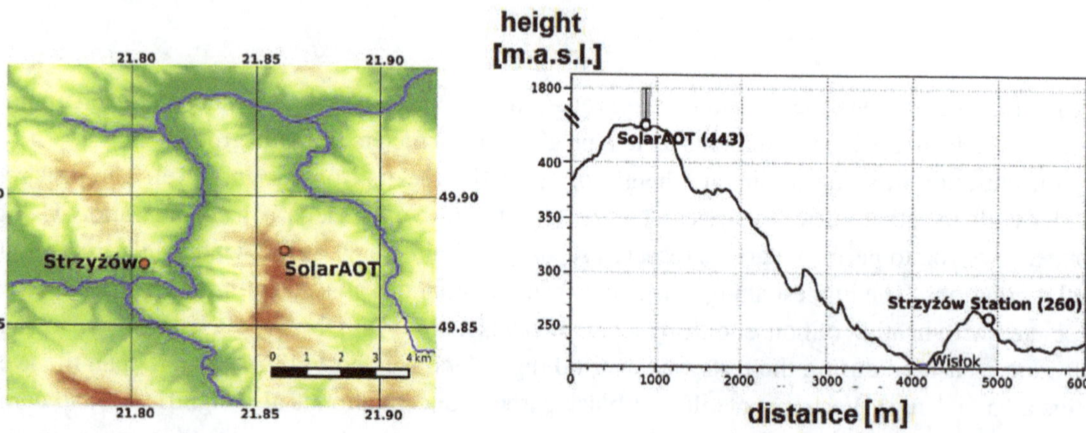

Unmanned aerial vehicles also apply when observing areas of invasive plant species. Invasive plants are non-native species that quickly spread in a new place. This has a negative impact on the ecosystem and leads to the extinction of local plant species that have grown over the area. In order to eliminate this unfavorable phenomenon, overgrowing invasive plants should be removed. The first step to this is to locate them, which is not easy in areas of considerable size. Drones can be used for this purpose. Such research was carried out by observing the areas around the city of Viana do Castelo in the north of Portugal. These areas are overgrown mainly by forests, shrubs and agricultural wasteland. An area of 2,550 ha was covered by the invasive Hakea sericea plant. It is a species of shrub originating from Australia. It reaches a height of up to 3 meters, blooms from winter to early spring. Apart from Australia, it occurs mainly in South Africa, New Zealand and in Portugal it is considered an invasive weed (Alvarez-Taboada, 2017). The area of research and plants appearing on it is shown in the Figure 64. These tests were carried out on February 15, 2012. The date of the research was chosen because of the flowering of the sought-after plant in winter, when other plants do not bloom, making Hakea sericea easier to identify. Orthophotomaps made from photographs taken with the use of unmanned aerial vehicles allowed detection of an unwanted plant in 75% of cases.

Drones were used in Canada to observe the rebirth of the forest after oil pollution. The research was carried out in 9 places in the central-western area of the province of Alberta. There are both coniferous, deciduous and mixed forests (Hird, 2017). The survey scheme in each of the areas where 5 × 5 meter surfaces were selected is shown in the Figure 65. These observations took place in the summer of 2014 and were performed for several weeks. They focused on the type of vegetation occurring in the studied area (herbs, shrubs, trees) and on the assessment of the stage of its growth. Thus, the plant height was divided into three groups: up to a height of 0.5 meters, between 0.5 and 2 meters, and above 2 meters. The research was carried out using a Panasonic Lumix GX1 mounted on an unmanned Hexacopter XL. The height of the flight above the ground was from 58 to 75 meters. After locating trees of more than 2 meters in height, both live and withered, their trunk diameter was measured with a tape. Therefore, the use of drones for this purpose as an additional element of the study allowed for a shorter time to perform preliminary vegetation observation in the studied area.

Unmanned aerial vehicles have also been used to control the level of groundwater. Important indicators characterizing groundwater are their level and depth to be obtained in order to obtain water. A study was carried out to determine the level of groundwater using orthophotomaps made of images made of drones. The study was conducted on an area of 61 ha in the northern part of the province of Alberta in Canada, about 40 km from the Peace River (Rahman, 2017). The analyzed area is shown in the Figure 66.

The Aeryton HDZoom30 camera with a resolution of 20 Mpix was used for the test, mounted on an unmanned aircraft Aeryton Skyranger. The height of the flight was 110 meters above ground level, while the speed was 4 m/s. The observations were made on 31 August 2016. On that day there were no large winds (less than 3 m/s), which could disrupt the study through vegetation movements caused by strong gusts. During the tests, a total of 851 photos were made using the drone. After the measurements with the use of orthophotomaps, 31 measurements of water wells were made to compare the results. These wells were made in the study area specifically for the purpose of comparison of results using pipes with a length of 1.5 meters and a diameter of 2.5 meters. These pipes along their entire length had holes allowing water to penetrate. This comparison allowed to determine the average error of groundwater level assessment in measurements using a drone amounting to approx. 20 cm. The biggest mistake occurred in areas densely covered with trees. On the peatlands, i.e. areas not covered with vegetation, the most accurate measurement of the groundwater level with the use of an orthophotomap has been made. Thus,

Figure 64. The studied area near the town of Viana do Castelo and vegetation occurring on it
(Alvarez-Taboada, 2017)

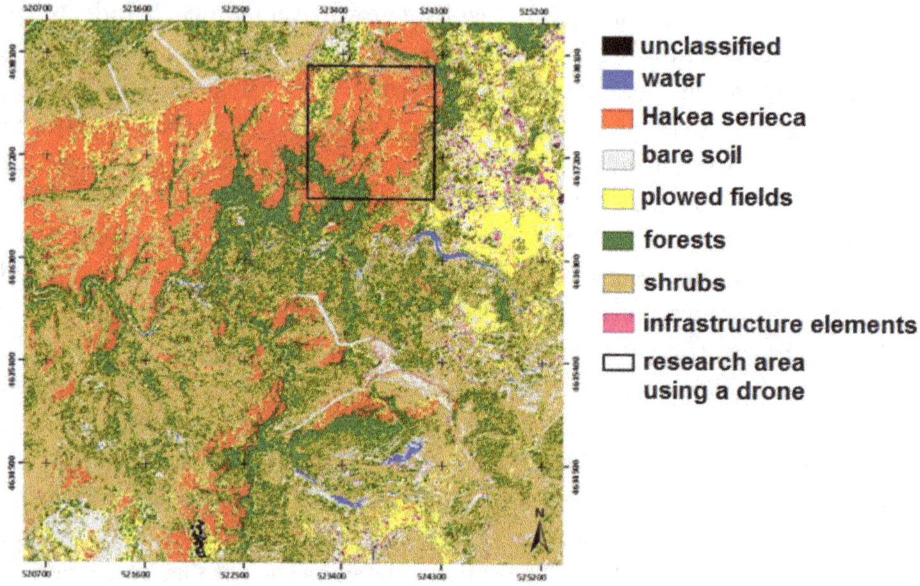

Figure 65. Scheme of activities during the examination of the selected area
(Hird, 2017)

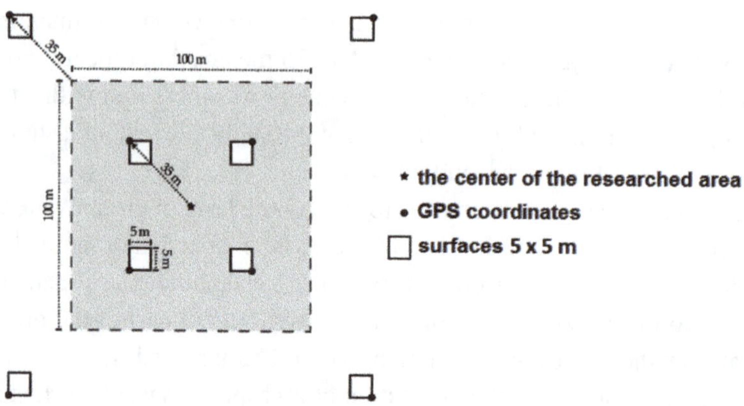

the study showed that the measurement of groundwater using photographs made from unmanned aircraft in areas heavily covered with vegetation is characterized by a measurement error, whereas this method works well in the case of groundwater level measurement on peatlands (Rahman, 2017).

Another application of unmanned aerial vehicles is the determination of the deformation of floating large diaphragm shields. Such covers are used to cover clean water reservoirs to prevent evaporation and contamination, to cover landfills so that harmful substances do not enter the environment and are used in mining (eg to cover ponds used for salt evaporation). These covers are very expensive (the cost of replacement for a new approx. 50 million $), and at the time of their discontinuation, hazardous sub-

Figure 66. Groundwater level testing area with marked measuring points
(Rahman, 2017)

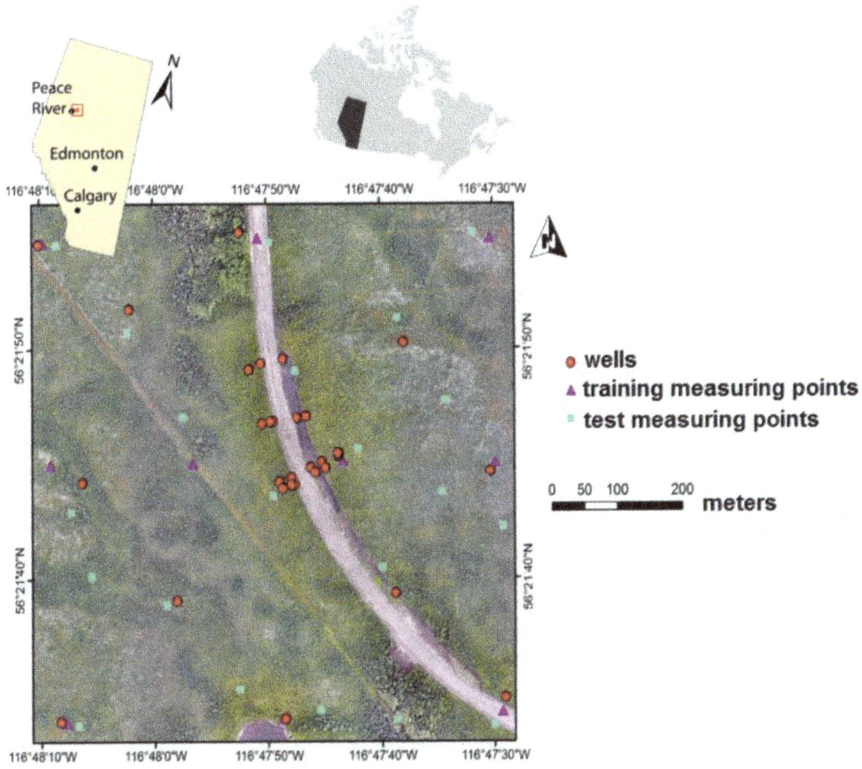

stances may enter the environment, which necessitates continuous monitoring of the condition of the shields (Ong, 2017). The design of such covers covering the landfill is shown in the Figure 67.

The study was carried out using a drone for this purpose. It took place at the wastewater treatment plant in Werribee, Melbourne, Australia. A 12mm thick cover made of polypropylene with an area of 470m × 170m is used. This cover is fixed by means of fasteners that ensure hermetic sealing. This is necessary because, during anaerobic digestion, biogas is produced here, which is used to generate electricity. The DJI E800-6 drone, which can stay in the air for approx. 30 minutes, equipped with the Olympus E-PL7, was used for the tests. Flights were made at speeds of 5, 3 and 2 m/s. The photos made with the use of unmanned aircraft were used to create a finite element model (Ong, 2017).

Drones can also be used to observe glaciers in the Himalayas. They are an important source of water after melting, and because of the difficult to access location, they have not been thoroughly investigated. Earlier, only a few field tests were performed, therefore it was decided to use remote sensing with the use of unmanned aerial vehicles. Research using this device was carried out on the tongue of the glacier located at the peak of Langtang Lirung in Nepal, close to the border with China, about 100 km north of Kathmandu (Immerzeel, 2014). The examined glacier is shown in the Figure 68.

In this area, 70% of annual precipitation occurs during the monsoon season from June to September. The examined glacier is located at an altitude of approx. 4000 m above sea level. It has a glacial language with a length of 3.5 km and a width of 500 meters. Its maximum thickness is achieved in 7234 meters near the pick. Observations were made in May and October 2013 to compare the appearance of the glacier

Figure 67. The upper picture shows the landfill covered with covers, and the lower photos show damage to the covers causing the exposition of the landfill
(Ong, 2017)

Figure 68. Tested glacier at the peak of Langtang Lirung in Nepal
(Immerzeel, 2014)

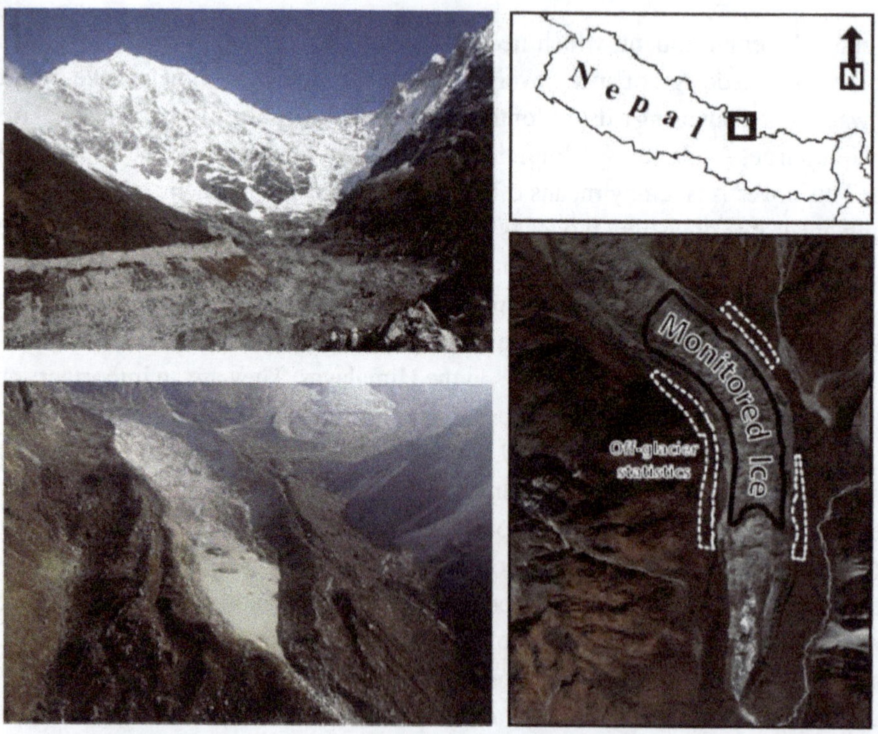

before and after the summer season in which it melts. The Swinglet CAM multikopter which can carry a weight of 0.5 kg, reaching the speed of max 36 km / h and allowing it to remain in flight for a maximum of 30 minutes, was used for the research. It was equipped with a Canon IXUS 125HS remotely operated compact camera that allows you to take 4608 × 3456 pixels. The tests used ISO sensitivity in the range of 125-250 and shutter speeds from 1/320 to 1/1200 seconds using automatic settings. Due to the limited time the drone could stay in the air during the test, several flights were made to identify the whole area of the glacier (Immerzeel, 2014). The division of the glacier into the areas observed in individual flights is shown in the Figure 69. In May 2013, a total of 7 flights for 2 days were performed (18 and 19 May), while in October there were 3 flights completed on October 22. Photos from the drone allowed to make a numerical model of the terrain representing the height of the area. This allowed to identify the mass loss by the glacier and to determine the melting speed. It was found that the glacier mass loss is small and the melting rate is very low, however, the occurrence of very high surface fluctuation of the melting rate was noticed. The most rapidly melting ice cliffs and supraglative lakes (they are formed on the surface of ice, with its melting disappear). This study showed the great potential of unmanned aircraft in glaciology (a hydrology department involved in the study of glaciers).

Another use of unmanned aircraft is monitoring landslides. The landslide research took place in the Valle Germanasca valley in the Italian part of the Alps. In March 2011, during heavy rainfall, there were several landslides, and one of them caused the rocks of the SP 170 to be blocked with rocks. This situation caused the necessity to observe the rock walls in the region. For the observation, a drone with a V-shaped rotor system was used, thanks to which the rotors do not enter the field of view of the camera (Torrero, 2015). This device is shown in the Figure 70. This multi-rotor can stay in the air for a maximum of 15 minutes. In this study, photogametry was used, so photographs taken from an unmanned aircraft had to be made at a constant speed and at one height, so that they could be applied. Flights were carried out at heights from 30 to 70 meters above the ground. In total, six flights and about 60 photos were taken. The photos taken from the drone were processed in the Agisoft Photoscan program. The use of unmanned aerial vehicles for observation made it possible to quickly take pictures of rock landslides simultaneously without the need for researchers to enter dangerous terrain [83].

Other use of drones is monitoring of dumps, the occurrence of which is a consequence of mining hard coal mining. Research using a thermal imaging camera mounted on an unmanned aircraft was carried out in the Czech Republic, near Ostrava, where there are about 50 mining dumps covering a total area of over 600 ha, of which 6 dumps show signs of thermal processes taking place in them. This activity causes hazardous substances that threaten human health to enter the environment. These dumps consist mainly of sandstone, clay, dust and coal residues, which are a flammable material, which makes them particularly dangerous. As a result of high aeration and elevated temperatures, there may be self-ignition of residual carbon on the dump. Increasing the temperature at the storage site causes exothermic reactions of the oxidation of organic substances. In the case when the energy generated from the occurring reactions is not removed, a coal gas is produced which can cause a fire. During combustion, with insufficient oxygen, substances such as sulfur oxides, nitrogen oxides and choline hydrogen are produced. The research was carried out in July and August 2017, during which the dumps of Hedvik were observed. This dump was probably founded in 1903, and its extension to the area currently occupied in the 1960s and 1970s. It covers an area of about 32 hectares, and the deposition of waste was completed on it in 1998. The first thermal activity was located on the landfill in the 1950s, it was stopped by covering the

Figure 69. The range of individual flights carried out on the glacier
(Immerzeel, 2014)

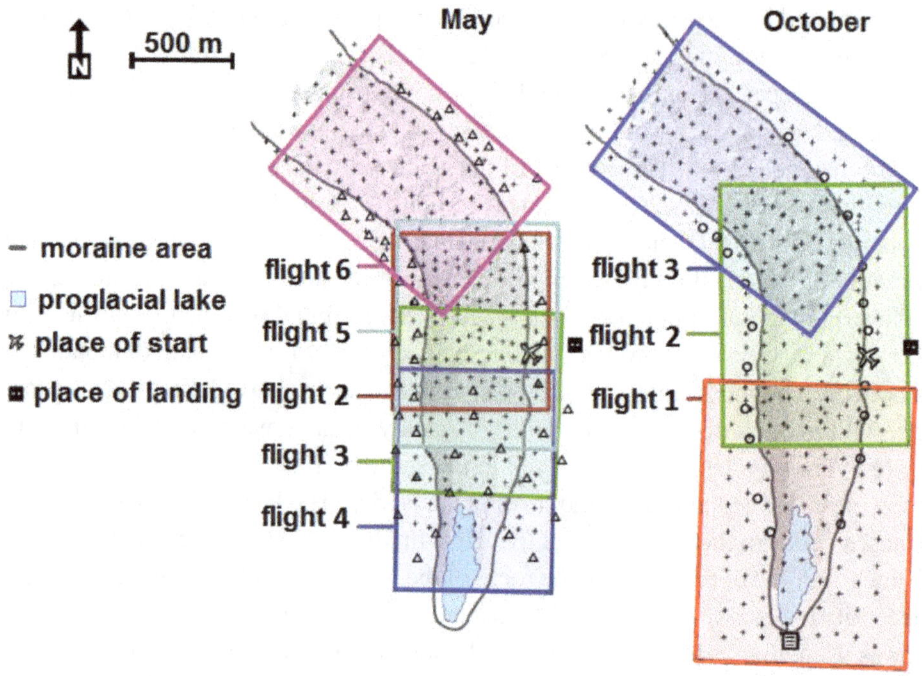

waste with an insulating layer. In the 90s of the twentieth century on this dump exceeded the temperature of 500°C in the central part covered by fire. Recent firefighting activities were undertaken here in 2006. During the research using the drone, it was found that the highest surface temperature in the investigated area of the landfill was 71.6°C. In places where the temperature was high (49.8°C and more) a coal gas emission was located (Surovka, 2017). The image from the infrared camera taken during the research in July 2017 is shown in the Figure 71.

A somewhat controversial idea, however, has long been used by the military to use armed unmanned aerial vehicles. What is disturbing here is the possibility of killing without risking the attacking army's forces. As a result of shelling carried out with the use of drones, in 2004-2014, 2,300 people were killed in Pakistan (Sałaciński, 2015). The operator of the armed aircraft may be a veteran (a member of the armed forces), in which case he may take part in combat operations. In the event that he is captured by the opposing party he will have the status of a prisoner of war. By an opponent, such an operator is treated as a military target. Drones in the army can also serve civilians, then they do not take part in battles, but only in reconnaissance missions (Kociubiński, 2016). It is also dangerous to take control of such an armed device by hackers. Loss of control and the same loss of the machine can cause the weapon to get into the wrong people. Unmanned systems flying into enemy territory are undetectable by ordinary systems used to detect flying manned units. In attack, drones may use passive means, such as smoke screens or offensive means, or rocket systems (Svaton, 2016). Another example of the negative use of drones is the possibility of using them to smuggle, for example drug smuggling (Sladek, 2013).

Figure 70. An unmanned aerial vehicle with a V-shaped rotor system used during tests
(Torrero, 2015)

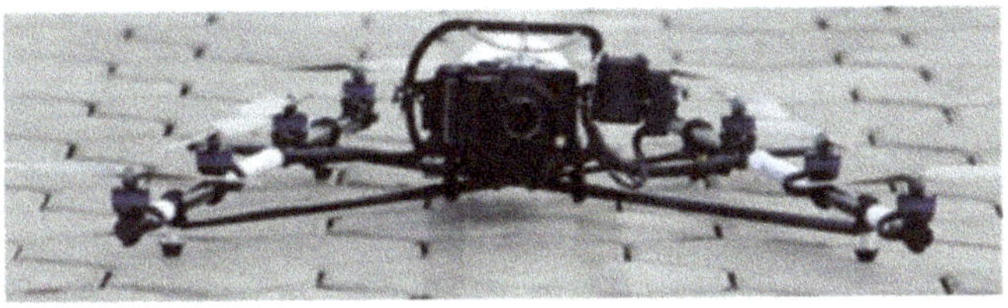

Figure 71. The area of Hedvik's dump where the temperature is high
(Surovka, 2017)

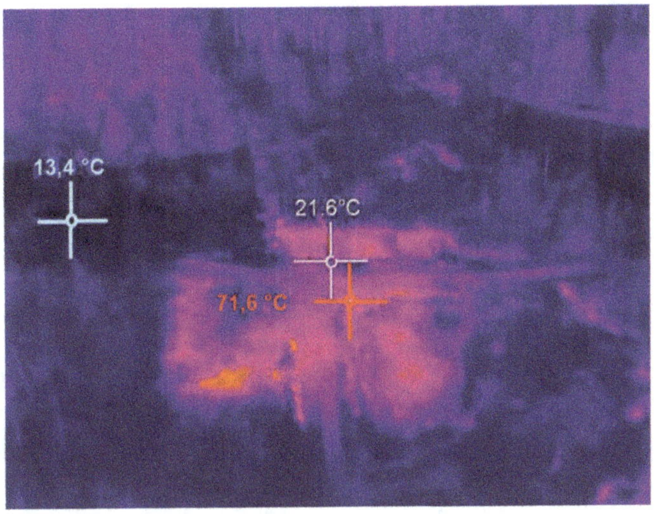

SELECTING THE DRONE FOR A PARTICULAR TASKS

There are many different uses of unmanned aerial vehicles, as shown in part 2. For each task performed by a drone, a device with different parameters will perform better, because some tasks require a long range of battery allowing for a longer period of time in the air, in others the payload is important and still others need an unmanned aircraft to stay in the needed trajectory.

Identification of Factors Important in a Specific Task

The table presents some tasks that can be performed using unmanned aerial vehicles and factors related to the construction and construction of individual drones that are important for the implementation of specific tasks. For different tasks, the weight of individual factors (C_n) associated with the design and production was assigned, taking the weights in the range from 0 to 5 (Table 1).

Determination of the Degree of Meeting the Various Criteria by the Specific Drones

The table presents the extent to which various unmanned aerial vehicles from different manufacturers meet specific factors that are important in the implementation of the drone through specific tasks. Each of the analyzed unmanned aerial vehicles was assigned parameters (P_n) with a value from 0 to 5, assuming in the case of 0 the lowest properties, and in the case of 5 properties, the highest for meeting the requirements of a given criterion (Table 2). These are expert assessments, which were obtained on the basis of a survey of 8 experts who compared the characteristics of the drones in question.

Assessing the Suitability of a Given Drone for a Specific Task

In order to assess the suitability of a given model of unmanned aerial vehicle (D) to perform a specific task, the sum of products of particular factors (C_n) important during the implementation of the drone given task and parameters (P_n) determining the degree of fulfillment of these factors by individual unmanned aerial vehicles was calculated. This action is represented by the formula

$$D = \sum_{i=1}^{n} C_i \cdot P_i . \tag{1}$$

The sum of products, where n=6, calculated on the basis of the data contained in Tables 1 and 2, allowing to choose which model of unmanned aircraft to use for a particular task is presented in Table 3.

Based on the calculations contained in the table, it is possible to determine the suitability of a given unmanned aircraft to perform a specific task. The higher the value of the product sum, the better the drone meets the criteria necessary to implement a specific task included in the table, while its smaller value means worse fulfillment of the required criteria. This sum can take values in the range from 0 to

Table 1. Weights of individual factors when performing various tasks

Using	Factors (C_n)					
	Capacity	Resistance to Wind	Maximum Speed	Flight Range	Flight Length	Stability
Monitoring of energy networks	1	3	4	5	5	5
Spraying agricultural crops	5	3	5	5	5	3
Inventory of wild animals	1	3	2	3	5	5
Fire detection	1	5	4	5	5	5
Searching for missing persons	1	5	5	5	5	3
Traffic monitoring	2	5	2	4	5	5
Shipment transport	5	5	4	5	5	4
Water rescue	4	5	5	3	5	3
Identification of railway accidents	2	4	2	3	5	5
Transport of defibrillators	4	5	5	5	5	2

Table 2. Properties of individual unmanned aerial vehicles

Dron	Parameters (P_n)					
	Capacity	Resistance to Wind	Maximum Speed	Flight Range	Flight Length	Stability
Inspire 1 DJI	1	2	2	2	2	3
md4-1000 Microdrones	3	2	1	2	3	2
R-Max Yamaha	5	4	4	5	4	5
Atlas C – Astral	2	2	3	4	4	4
Bramor C – Astral	2	3	4	4	5	4
PAM – 20	5	4	3	5	5	5
UX5 HP Trimble	3	4	3	4	4	4
AGV8A	3	2	2	2	1	3
DJI Agras MG-1	3	3	2	2	2	3

Table 3. A product that allows you to choose a drone for a given task

Using	Drone Model								
	Inspire 1 DJI	md4-1000 Micro-Drones	R-Max Yamaha	Atlas C – Astral	Bramor C – Astral	PAM – 20	UX5 HP Trimble	AGV8A	DJI Agras MG-1
Monitoring of energy networks	50	48	103	80	92	104	87	47	55
Spraying agricultural crops	50	57	117	83	96	117	94	55	63
Inventory of wild animals	42	42	85	66	76	88	73	39	47
Fire detection	54	52	111	84	98	112	95	51	61
Searching for missing persons	50	49	105	79	94	105	90	47	57
Traffic monitoring	49	51	103	76	88	106	88	48	58
Shipment transport	55	62	126	88	102	127	103	60	70
Water rescue	49	54	110	77	92	110	91	52	62
Identification of railway accidents	45	47	94	70	81	97	80	44	53
Transport of defibrillators	50	56	115	81	96	115	95	53	63

150. Green indicates the best choice. As can be seen from the table above, the drone most suitable for the task is the drone PAM – 20.

Factors Important for the Implementation of a Specific Task Including Costs

The table presents the tasks that can be Performed using unmanned aerial vehicles and factors related to the construction and construction of individual drones that are important when performing specific tasks. At the same time, costs related to the purchase and subsequent operation of drones were taken into account, because in some applications these costs have very important meanings, and for others they are not so important. For different tasks, the weight of individual factors (Cn) associated with the construction and construction was assigned, taking the weights in the range from 0 to 5 (Table 4).

The Degree of Fulfillment of Each Criterion by Specific Drones Taking Into Account the Costs

The table shows the extent to which various unmanned aerial vehicles from different manufacturers meet particular factors that are important during the implementation by means of a specific task drone. These factors include parameters related to both the purchase price of individual drone models and their operating costs. Each of the analyzed unmanned aerial vehicles was assigned parameters (Pn) with a value from 0 to 5, assuming in the case of 0 the lowest properties, and in the case of 5 properties, the highest for

Table 4. Weights of individual factors when performing various tasks

Using	Factors (C_n)							
	Capacity	Resistance to Wind	Maximum Speed	Flight Range	Flight Length	Stability	Price	Operating Cost
Monitoring of energy networks	1	3	4	5	5	5	3	4
Spraying agricultural crops	5	3	5	5	5	3	5	5
Inventory of wild animals	1	3	2	3	5	5	3	4
Fire detection	1	5	4	5	5	5	2	3
Searching for missing persons	1	5	5	5	5	3	2	2
Traffic monitoring	2	5	2	4	5	5	3	3
Shipment transport	5	5	4	5	5	4	4	5
Water rescue	4	5	5	3	5	3	3	2
Identification of railway accidents	2	4	2	3	5	5	3	2
Transport of defibrillators	4	5	5	5	5	2	1	1

meeting the requirements of a given criterion. In the case of the purchase cost and subsequent exploitation for the value of 5, the lowest cost was assumed, while in the case of 0 it is the highest cost (Table 5).

Assessment of the Suitability of a Particular Unmanned Aircraft, Including Costs

The sum of products of particular factors (Cn) valid during the implementation by the drone of a given task and parameters (Pn) determining the degree of compliance of these factors by individual unmanned aircraft calculated according to the formula (1), where n=8, necessary to assess the suitability of a given model of unmanned aircraft is presented in Table 6. Calculations was made on the basis of the weights of individual factors contained in table 4 and the degree of fulfillment of individual parameters by specific drones, which is presented in table 5. This sum may take values in the range from 0 to 200, of which values from 0 to 150 individual unmanned aerial vehicles received for fulfillment factors not related to costs, while additional values, additional from 0 to 50, are related to purchase and operating costs. Therefore, to the total sum of products, determining the suitability of a particular drone for a specific application, the costs associated with its purchase and use are included in 25%, while the remaining 75% of the product value is related to non-cost factors necessary to perform various tasks.

Based on these calculations, the suitability of a given unmanned aircraft can be determined to perform a specific task, taking into account the cost of purchasing a given drone and the costs of its subsequent operation. The higher the value of the product sum, the better the drone meets the criteria necessary to implement a specific task included in the table, while its smaller value means worse fulfillment of the required criteria. For some tasks where the price plays a significant role, these results differ significantly from those without taking into account the costs presented in Table 3. Green indicates the best choice. As can be seen from the table above, the drone PAM – 20 is the best for many tasks, but not always.

Table 5. Properties of individual unmanned aerial vehicles including costs

Dron	Parameters (P_n)							
	Capacity	Resistance to Wind	Maximum Speed	Flight Range	Flight Length	Stability	Price	Operating Cost
Inspire 1 DJI	1	2	2	2	2	3	5	5
md4-1000 Microdrones	3	2	1	2	3	2	3	2
R-Max Yamaha	5	4	4	5	4	5	0	0
Atlas C – Astral	2	2	3	4	4	4	3	2
Bramor C – Astral	2	3	4	4	5	4	2	3
PAM – 20	5	4	3	5	5	5	0	0
UX5 HP Trimble	3	4	3	4	4	4	3	2
AGV8A	3	2	2	2	1	3	4	3
DJI Agras MG-1	3	3	2	2	2	3	5	5

Table 6. Choosing a drone for a given task, including costs

Using	Drone Model								
	Inspire 1 DJI	md4-1000 Micro-Drones	R-Max Yamaha	Atlas C – Astral	Bramor C – Astral	PAM – 20	UX5 HP Trimble	AGV8A	DJI Agras MG-1
Monitoring of energy networks	85	65	103	97	110	104	104	71	90
Spraying agricultural crops	100	82	117	108	121	117	119	90	113
Inventory of wild animals	77	59	85	83	94	88	90	63	82
Fire detection	79	64	111	96	111	112	107	68	86
Searching for missing persons	70	59	105	89	104	105	100	61	77
Traffic monitoring	79	66	103	91	103	106	103	69	88
Shipment transport	100	84	126	110	125	127	125	91	115
Water rescue	74	67	110	90	104	110	104	70	87
Identification of railway accidents	70	60	94	83	93	97	93	62	78
Transport of defibrillators	60	61	115	86	101	115	100	60	73

CONCLUSION AND FURTHER STUDIES

In recent years, there has been a rapid development of unmanned aerial vehicles. It involves certain risks and emerging new threats, but it also gives new opportunities to observe hard-to-reach places and allows to reduce costs by replacing expensive, manned flying units. Different drone characteristics are required for different applications, so another application requires the use of another type of drone from another manufacturer. Therefore, Table 1 shows what factors are important for the various applications of unmanned aerial vehicles. Different models of drones also have different parameters. For several different models selected, these parameters are presented in Table 2. After adding the products of the required factors for specific applications and parameters that various unmanned aerial vehicles have, it is possible to determine which model will be suitable for a given use, as shown in Table 3. Additionally, important factors were taken into account specific tasks related to costs (table 4) and the degree of parameter fulfillment also related to costs (table 5). On this basis, the usefulness of individual drones was assessed taking into account the costs, which is included in Table 6. The values obtained allowed to conclude that:

1. Large unmanned aerial vehicles with both internal combustion or electric propulsion are suitable for most applications.
2. Small drones to a lesser degree meet the high criterion level of requirements, mainly due to the limited time spent in the air and the limited mass that they can raise.

3. Taking into account purchase costs and subsequent exploitation, large unmanned aerial vehicles for many applications that don't require a large payload and a significant time of being in the air give way to small and cheap multicopter.

The scope of drones for non-military purposes is expanding every year and it is even difficult to find any area of the economy in which the use of drones is impossible. One can imagine that in connection with huge transport problems of large megacities in the very near future there will be air taxis. The design of such aircraft is already at the testing stage. And this means that the next step will be the use of unmanned aerial vehicles for this purpose.

At present, due to climate change, forest fires have increased. The use of drones for the localization and identification of fires has been described above. But the use of drones to extinguish them is an urgent need of the present day.

These are only a few possible examples of the use of drones in the future, but it is obvious that in the future drones will get elements of artificial intelligence for their control, which will significantly expand from the field of use.

REFERENCES

Бикоптер UH-144 из Halo. (2014). Available at: http://www.parkflyer.ru/static/user_files/2014/5/9/1.1402320683936.jpg

История развития дронов. (2018). Available at: https://dronomania.ru/faq/istoriya-razvitiya-dronov.html

Alvarez-Taboada, F., Paredes, C., & Julian-Pelaz, J. (2017). Mapping of the Invasive Species Hakea sericea Using Unmanned Aerial Vehicles (UAV) and WorldViev-2 Imagery and an Object-Oriented Approach. *Remote Sensing*, *9*(5), 913–931. doi:10.3390/rs9090913

Avila, M., Funk, M., & Henze, N. (2015). Drone Navigator: Using Drones for Navigating Visually Impaired Presons. *Assets*, *15*(10), 1–2.

Benchoff, B. (2016). *A Brief History of 'Drone'*. Available at: https://hackaday.com/2016/09/26/a-brief-history-of-drone/

Berbeka, J. (2016). Fotografia powietrzna jako innowacyjne narzędzie współtworzenia przeżyć turystów. *Prace naukowe Uniwersytetu Ekonomicznego we Wrocławiu, 473(30),* 69-76.

Berner, B., & Chojnacki, J. (2016). Wykorzystanie dronów do transportu towarów [The use of drones for goods transporting]. *Autobusy*, *R17*(6), 233–235.

Berner, B., & Chojnacki, J. (2016a). Wykorzystanie dronów w rolnictwie precyzyjnym [The use of drones in precision farming]. *Technika Rolnicza i Leśna*, *3*, 19–21.

Berner, B., & Chojnacki, J. (2017). Zastosowanie bezzałogowych statków powietrznych do opryskiwania upraw rolniczych [The use of unmanned aerial vehicles for spraying agricultural crops]. *Technika Rolnicza Ogrodnicza Leśna*, *2*, 23–25.

Bielecki, K. (2015). *Zastosowanie dronów do inspekcji sieci ciepłowniczych* [Application of drones for inspection of heating networks]. Available at: http://www.inzynierbudownictwa.pl/technika,materialy_i_technologie,artykul,zastosowanie_dronow_do_inspekcji_sieci_cieplowniczych,8704

Bujnowski, M. (2017). Integracja bezzałogowych statków powietrznych z unijnym systemem lotnictwa cywilnego [Integration of unmanned aerial vehicles into the EU civil aviation system]. *Kwartalnik Antymonopolowy i Regulacyjny*, *6*(2), 74–97.

Burdziakowski, P. (2016). Przegląd budowy i funkcjonalności współczesnych bezzałogowych statków powietrznych do celów fotogrametrycznych [Review of the construction and functionality of modern unmanned aerial vehicles for photogrammetric purposes]. *Biuletyn WAT*, *65*(4), 69–91. doi:10.5604/12345865.1228956

Capolupo, A., Pindozzi, S., Okello, C., Fiorentino, N., & Boccia, L. (2015). Photogrammetry for environmental monitoring: The use of drones and hydrological models for detection of soil contaminated by copper. *The Science of the Total Environment*, *514*(9), 298–306. doi:10.1016/j.scitotenv.2015.01.109 PMID:25668282

Charles F. Kettering. (2018). Available at: https://en.wikipedia.org/wiki/Charles_F._Kettering

Chiliński, M. T., Markowicz, K. M., & Markowicz, J. (2016). Observation of vertical variability of black carbon concentration in lower troposphere on campaigns in Poland. *Atmospheric Environment*, *137*(4), 155–170. doi:10.1016/j.atmosenv.2016.04.020

Chojnacki, J. (2017). Status operatorów bezzałogowych statków powietrznych w świetle międzynarodowego prawa humanitarnego [Status of operators of unmanned aircraft in the light of international humanitarian law]. *Witelona w Legnicy*, *22*(1), 137–147.

Christensen, B. R. (2015). Use of UAV or remotely piloted aircraft and forward-looking infrared in forest, rural and wildland fire management: Evaluation using simple economic analysis. *New Zealand Journal of Forestry Science*, *45*(16), 1–9.

Claesson, A., Fredman, D., Svensson, L., Ringh, M., Hollenberg, J., Nordberg, P., ... Ban, Y. (2016). Unmanned aerial vehicles (drones) in out-of-hospital-cardiac-arrest. *Resuscitation and Emergency Medicine*, *24*(124), 1–9. PMID:27729058

Clothier, R. A., Greer, D. A., Greer, D. G., & Mehta, A. M. (2015). Risk Perception and the Public Acceptance of Drones. *Risk Analysis*, *35*(6), 1167–1183. doi:10.1111/risa.12330 PMID:25689883

Collins, K. (2017). *London police deploy drones to search for missing people*. Available at: https://www.cnet.com/news/london-police-deploy-drones-to-search-for-missing-people/

Colomina, I., & Molina, P. (2014). Unmanned aerial systems for photogrammetry and remote sensing: A review. *ISPRS Journal of Photogrammetry and Remote Sensing*, *92*(4), 79–97. doi:10.1016/j.isprsjprs.2014.02.013

D'Andrea, R. (2014). Guest Editorial Can Drones Deliver? *IEEE Transactions on Automation Science and Engineering*, *11*(3), 647–648. doi:10.1109/TASE.2014.2326952

De Havilland DH82B Queen Bee. (2018). Available at: https://www.dehavillandmuseum.co.uk/aircraft/de-havilland-dh82b-queen-bee/

DJI spark review: a smart selfie drone that's fun and frustrating. (2017). Available at: http://planforcrisis.com/dji-spark-review-a-smart-selfie-drone-thats-fun-and-frustrating/

Dobryakova, L., & Ochin, E. (2016). W sprawie klasyfikacji systemów transportowych w tym bezzałogowych [Regarding the classification of transport systems, including unmanned vehicles]. *Autobusy, R17*(6), 1680–1687.

Drone-assisted water rescue demo. (2017). Available at: https://www.airmedandrescue.com/video/1581

Dudek, B., & Czapaj-Atłas, R. (2016). Rozwój dronów i ich zastosowań w energetyce [The development of drones and their applications in the energy sector]. *Energetyka (Warszawa), 1*, 21–26.

Elsey, R. M., & Trosclair, P. L. III. (2016). The Use of an Unmanned Aerial Vehicle to Locate Alligator Nests. *Southeastern Naturalist (Steuben, ME), 15*(1), 76–82. doi:10.1656/058.015.0106

Everett, H. R. (2015). *Unmanned Systems of World Wars I and II*. MIT Press.

Eyerman, J., Letterman, C., Pitts, W., Holloway, J., Hinke, K., Schanzer, D., ... Kaydos-Daniels, S. C. (2013). *Unmanned Aircraft and the Human Element: Public Perceptions and First Responder Concerns. Research Brief, Post-Project Report*. Institute for Homeland Security Solutions.

Fellner, R. (2015). Bezpieczeństwo i bezzałogowe statki powietrzne w dyskursie naukowo-prawniczym [Safety and unmanned aerial vehicles in the scientific-legal discourse]. *Seria Transport, 95*, 197–212.

Flammini, F., Pragliola, C., & Smarra, G. (2016). Railway Infrastructure Monitoring by Drones. *2016 International Conference on Electrical Systems for Aircraft, Railway, Ship Propulsion and Road Vehicles & International Transportation Electrification Conference (ESARS-ITEC)*, 1-6.

Floreano, D., & Wood, R. (2015). Science, technology and the future of small autonomous drones. *Nature, 521*(5), 460–466. doi:10.1038/nature14542 PMID:26017445

France: Demonstrating the use of drones for railway maintenance and service. (2015). Available at: https://uic.org/com/uic-e-news/464/article/france-demonstrating-the-use-of?page=iframe_enews

Gnadinger, F., & Schmidhalter, U. (2017). Digital Counts of Maize Plants by Unmanned Aerial Vehicles (UAVs). *Remote Sensing, 9*(5), 544–559. doi:10.3390/rs9060544

Goldberg, D., Corcoran, M., & Picard, R. G. (2014). Remotely Piloted Aircraft Systems & Journalism. *Reuters Institute for the Study of Journalism, 6*, 1–41.

Gowtham, S., Nazeer, A., & Saranya, G. (2017). Survillence Drones imparted with Optimized Cloud Server Techniques. IEEE, 17, 1-3.

Guiraud, M., & Nachescu, M. (2012). Drones as a tool of sustainable development - consequences of their usage and new forms of risk management. *Economic and Environmental Studies, 12*(4), 467–479.

Haidari, L. A., Brown, S. T., Ferguson, M., Bancroft, E., Spiker, M., Wilcox, A., ... Lee, B. Y. (2016). The economic and operational value of using drones to transport vaccines. *Vaccine*, *34*(24), 4062–4067. doi:10.1016/j.vaccine.2016.06.022 PMID:27340098

Hallsten, J. (2017). *The essential guide to building your Amazon drone delivery pad.* Available at: http://www.hallsteninnovations.com/drone-delivery-pad/

Hird, J. N., Montaghi, A., McDermid, G. J., Kariyeva, J., Moorman, B. J., Nielsen, S. E., & McIntosh, A. C. S. (2017). Use of Unmanned Aerial Vehicles for Monitoring Recovery of Forest Vegetation on Petroleum Well Sites. *Remote Sensing*, *9*(5), 413–433. doi:10.3390/rs9050413

Hodgson, A., Kelly, N., & Peel, D. (2013). Unmanned Aerial Vehicles (UAV) for Surveying Marine Fauna: A Dugong Case Study. *PLoS One*, *8*(11), 1–15. doi:10.1371/journal.pone.0079556 PMID:24223967

How drones are already being deployed by railways around the world. (2014). Available at: https://www.smartrailworld.com/how_drones_are_already-being-used-by-railways-around-the-world

Immerzeel, W. W., Kraaijenbrink, P. D. A., Shea, J. M., Shrestha, A. B., Pellicciotti, F., Bierkens, M. F. P., & de Jong, S. M. (2014). High-resolution monitoring of Himalayan glacier dynamics using unmanned aerial vehicles. *Remote Sensing of Environment*, *150*(5), 93–103. doi:10.1016/j.rse.2014.04.025

Junaid, A. B., Konoiko, A., Zweiri, Y., Sahinkaya, M. N., & Seneviratne, L. (2017). Autonomus Wireless Self-Charging for Multi-Rotor Unmanned Aerial Vehicles. *Energies*, *10*(6), 803–817. doi:10.3390/en10060803

Kaminski, M. E. (2013). Drone Federalism: Civilian Drones and the Things They Carry. *California Law Review Circuit*, *4*(5), 57–74.

Karoly, G., Leizer, K., & Tokody, D. (2017). Radiofrequency identification by using drones in railway accidents and disaster situations. *Interdisciplinary Description of Complex Systems*, *15*(2), 114–132.

Keating, M. (2016). Georgia country police use drones to investigate serious traffic wrecks. *American City & Country.* Available at: http://americancityandcounty.com/public-safety/georgia-county-police-use-drones-investigate-serious-traffic-wrecks-related-video

Kim, A., Wampler, B., Goppert, J., Hwang, I., & Aldridge, H. (2012) Cyber Attack Vulnerabilities Analysis for Unmanned Aerial Vehicles. American Institute of Aeronautics and Astronautics. doi:10.2514/6.2012-2438

Kim, N. V., & Chevonenkis, M. A. (2015). Situation Control of Unmanned Aerial Vehicles for Road Traffic Monitoring. *Modern Applied Science*, *9*(5), 1–13.

Kociubiński, J. (2016). *Rozwój bezzałogowych statków powietrznych w świetle prawa do prywatności na gruncie Europejskiej Konwencji o Ochronie Praw Człowieka.* Available at: www.eumed.net.rev.rednma.33.koc.html

Kułaga, Ł. (2016). Używanie dronów w celu zwalczania terroryzmu w świetle 'ius ad bellum' [The use of drones to combat terrorism in the light of 'ius ad bellum']. *Zeszyty Prawnicze*, *16*(2), 141–162. doi:10.21697/zp.2016.16.2.07

Lesicar, J.C. (n.d.). Conceptual approach to utilization of the unmanned aerial vehicles in diverse activities. *Interdisciplinary Description of Complex Systems, 2*(15), 144-148.

Liu, Z., Li, Z., Loannis, R., Liu, B., Fu, X., & Ren, Q. (2015). Rise of Mini-Drones: Applications and Issues. *Pamco '15. Proceedings of the 2015 Workshop of Privacy-Aware Mobile Computing*, 7-12.

Lorenc, W., Chodnicki, M., Kordowski, P., & Nowakowski, M. (2016). Drony transportowe – nowa era transportu towarów i ludzi [Transport drones - a new era of transporting goods and people]. *Autobusy, R17*(6), 627–631.

Luo, Z., Liu, Z., & Shi, J. (2017). A Two-Echelon Cooperated Routing Problem for a Ground Vehicle and Its Carried Unmanned Aerial Vehicle. *Sensors (Basel), 17*(5), 1144–1161. doi:10.339017051144 PMID:28513552

Martin, P. G., Kwong, S., Smith, N. T., Ymashiki, Y., Payton, O. D., Russell-Pavier, F. S., ... Scott, T. B. (2016). 3D unmanned aerial vehicle radiation mapping for assessing contaminant distribution and mobility. *International Journal of Applied Earth Observation and Geoinformation, 52*(6), 12–19. doi:10.1016/j.jag.2016.05.007

Matese, A., Toscano, P., Di Gennaro, S. F., Genesio, L., Vaccani, F. P., Primicerio, J., ... Gioli, B. (2015). Intercomparison of UAV, Aircraft and Satellite Remote Sensing Platforms for Precision Viticulture. *Remote Sensing, 7*(3), 2971–2990. doi:10.3390/rs70302971

Mathews, B. D. (2015). Potential Tort Liability for Personal Use of Drone Aircraft. *St. Mary's Law Journal, 46*, 573–602.

Mengmeng, D., Noboru, N., Atsushi, I., & Yukinori, S. (2017). Multi-temporal monitoring of wheat growth by using images from satellite and unmanned aerial vehicle. *International Journal of Agriculture and Biology, 10*(5), 1–14.

Merkisz, J., & Nykaza, A. (2016). Perspektywy rozwoju i wykorzystania bezzałogowych statków powietrznych w służbach ratowniczych [Prospects for the development and use of unmanned aerial vehicles in emergency services]. *Autobusy, R17*(6), 291–296.

Merkisz, J., & Nykaza, A. (2016a). Zastosowanie bezzałogowych statków powietrznych w kryminalistyce rozpoznawczej i wykrywawczej [The use of unmanned aerial vehicles in forensic and detection criminalistics]. *Autobusy, R17*(6), 297–301.

Mesas-Carrascosa, F. J., Santano, D. V., Porras, F. P., Merono-Larriva, J. E., & Garcia-Ferrer, A. (2017). The Development of an Open Hardware and Software System Onboard Unmanned Aerial Vehicles to Monitor Concentrated Solar Power Plants. *Sensors (Basel), 17*(6), 1329–1343. doi:10.339017061329 PMID:28594353

Mettler, B., Tischler, M. B., & Kande, T. (1999). *System Identification of Small-Size Unmanned Helicopter Dynamics.* American Helicopter Society 55th Forum, Montreal, Quebec, Canada.

Morris, D. Z. (2015). *Why BNSF Railway is using drones to inspect thousands of miles of rail lines.* Available at: http://fortune.com/2015/05/29/bnsf-drone-program/

Nadobnik, J. (2016). The use of drones in organizing the Olympic Games. *Handel Wewnętrzny*, *365*(6), 288–299.

Nobahari, H., & Mohammadkarimi, H. (2017). Application of model aided inertial navigation for precise altimetry of Unmanned Aerial Vehicles in ground proximity. *Aerospace Science and Technology*, *69*(7), 650–658. doi:10.1016/j.ast.2017.07.033

Omar, T., & Nehdi, M. L. (2017). Remote sensing of concrete bridge decks using unmanned aerial vehicle infrared thermography. *Automation in Construction*, *83*(6), 360–371. doi:10.1016/j.autcon.2017.06.024

Ong, W. H., Chiu, W. K., Kuen, T., & Kodikara, J. (2017). Determination of the State of Strain of Large Floating Covers Using Unmanned Aerial Vehicle (UAV) Aided Photogrammetry. *Sensors (Basel)*, *17*(6), 1731–1755. doi:10.339017081731 PMID:28788081

Pagacz, S. & Witczuk, J. (2016). Wykorzystanie samolotów bezzałogowych i termowizji do nocnej inwentaryzacji zwierząt kopytnych. *Studia i materiały CEPL w Rogowie, 49A*(4), 50-57.

Paneque-Galves, J., McCall, M. K., Napoletano, B. M., Wich, S. A., & Koh, L. P. (2014). Small Drones for Community-Based Forest Monitoring: An Assessment of Their Feasibility and Potential in Tropical Areas. *Forests*, *5*(6), 1481–1507. doi:10.3390/f5061481

Patruno, C., Nitti, M., Stella, E., & D'Orazino, T. (2017). Helipad detection for accurate UAV pose estimation by means of a visual sensor. *International Journal of Advanced Robotic Systems*, *5*, 1–15.

Piotrowski, P., Witkowski, T., & Piotrowski, R. (2015). Bezzałogowa zdalnie sterowana jednostka latająca [Unmanned remote controlled flying unit. Measurements Robotics Automation]. *Pomiary Automatyka Robotyka*, *R19*(1), 49–55. doi:10.14313/PAR_215/49

Plaza, J. (2017). *Will Hydrogen Fuel Cells Help Drones Stay in the Air?* Available at: https://www.expouav.com/news/latest/hydrogen-fuel-cells-drones/

Project Ryptide - A life Saving Drone Accessory. (2017). Available at: https://www.kickstarter.com/projects/flying-robots/project-ryptide-a-life-saving-drone-accessory

Rahman, M. M., McDermid, G. J., Strack, M., & Lovitt, J. (2017). A New Method to Map Groundwater Table on Peatlands Using Unmanned Aerial Vehicles. *Remote Sensing*, *9*(5), 1057–1071. doi:10.3390/rs9101057

Rao, B., Gopi, A. G., & Maione, R. (2016). The social impact of commercial drones. *Technology in Society*, *45*(5), 83–90. doi:10.1016/j.techsoc.2016.02.009

Rey, N., Volpi, M., Joost, S., & Tuia, D. (2017). Detecting animals in African Savanna with UAVs and the crowds. *Remote Sensing of Environment*, *200*(8), 341–351. doi:10.1016/j.rse.2017.08.026

Robakowska, M., Tyranska-Fobke, A., Nowak, J., Slezak, D., Zuratynski, P., Robakowski, P., ... Ladny, J. R. (2017). The use of drones during mass events. *Disaster and Emergency Medicine Journal*, *2*(3), 129–134.

Roosjen, P. P. J., Suomalainen, J. M., Bartholomeus, H. M., Kooistra, L., & Clevers, J. G. P. W. (2017). Mapping Reflectance Anisotropy of a Potato Canopy Using Aerial Images Acquired with an Unmanned Aerial Vehicle. *Remote Sensing*, *9*(5), 417–440. doi:10.3390/rs9050417

Rus, C., & Patrascoiu, N. (2016). Technical and legal aspects on the use of drones. *Annals of the University of Petroscani*, *18*(6), 69–78.

Sałaciński, T. (2015). *Pojazdy bezzałogowe - nowe wyzwanie dla materiałów wybuchowych. Przegląd.* Available at: http://yadda.icm.edu.pl/yadda/element/bwmeta1.element.baztech-244768f7-5471-4098-b4b9-298a2916e671/c/Salacinski1_t_6.pdf

Sandbrook, C. (2015). The social implications of using drones for biodiversity conservation. *Ambio*, *44*(4), 636–647. doi:10.100713280-015-0714-0 PMID:26508350

Saulnier, A., & Thompson, S. N. (2017). Police UAV use: Institutional realities and public perceptions. *Policing: An International Journal of Police Strategies & Management*, *39*(4), 680–693. doi:10.1108/PIJPSM-11-2015-0136

Shilin, W., Jianli, S., Xiongkui, H., Le, S., Xiaonan, W., Changling, W., ... Yun, L. (2017). Performances evaluation of four typical unmanned aerial vehicles used for pesticide application in China. *International Journal of Agriculture and Biology*, *10*(4), 22–31.

Sladek, J. (2013). Drones and their use in geovedical research. *Nové trendy v geovedách, 088UK(4),* 1-19.

Smaczyński, M. (2015). Wizualizacja dynamiki zmian liczby uczestników imprezy masowej z wykorzystaniem dronów. *Badania Fizjograficzne - Geografia Fizyczna, 66*(6), 157-172.

Smith, K. W. (2015). Drone Technology: Benefits, Risks, and Legal Considerations. *Seattle Journal of Environmental Law*, *5*(1), 290–302.

Sosnowicz, K. (2016). Dronem do wypadku [With drone to an accident]. *Geodeta Magazyn Informacyjny*, *257*(10), 44–47.

Strzelczyk, P., & Macek-Kamińska, K. (2015). Kontroler lotu dla bezzałogowych obiektów latających [Flight controller for unmanned aerial vehicles. Measurements Robotics Automation]. *Pomiary Automatyka Robotyka*, *R19*(4), 69–73. doi:10.14313/PAR_218/69

Surovka, D., Pertile, E., Dombek, V., Vastyl, M., & Leher, V. Monitoring of Thermal and Gas Activities in Mining Dump Hedvika, Czech Republic. *IOP Conference Series: Earth and Environmental Science*, *92*(1), 1-6. 10.1088/1755-1315/92/1/012060

Svaton, M. (2016). *Low-cost implementation of Differential GPS using Arduino* (PhD thesis). Prague: Czech Technical University in Prague.

Tate, A. (2016). Miley Cyrus and the Attack of the Drones: The Right of Publicity and Tabloid Use of Unmanned Aerial Vehicles. *Texas Review of Entertainment and Sports Law*, *17*(1), 73–99.

Thermodiagnostics in the power engineering sector. (2017). Available at: https://www.drone-thermal-camera.com/drone-uav-thermography-inspection-highvoltage/

Thiels, C. A., Aho, J. M., Zietlow, S. P., & Jenkins, D. H. (2015). Use of Unmanned Aerial Vehicles for Medical Product Transport. *Air Medical Journal*, *34*(2), 104–108. doi:10.1016/j.amj.2014.10.011 PMID:25733117

Three ways that drones could change the railway industry. (2017). Available at: http://railwayinnovation.com/3-ways-that-drones-could-change-the-railway-industry/

Torrero, L., Seoli, L., Molino, A., Giordan, D., Manconi, A., Allasia, P., & Baldo, M. (2015). The Use of Micro-UAV to Monitor Active Landslide Scenarios. *Engineering Geology for Society and Territory*, *5*, 701–704.

Tripicchio, P., Satler, M., Stefanini, G., Dabisias, G., Ruffaldi, E., & Avizzano, C. A. (2015). Towards Smart Farming and Sustainable Agriculture with Drones. *Computers & Society*, *29*, 140–143.

Tuśnio, N., Tuśnio, J., Wolny, P., & Nowak, A. (2016). Bezzałogowe statki powietrzne w działaniach Państwowej Straży Pożarnej - propozycja dedykowana Państwowej Straży Pożarnej [Unmanned aerial vehicles in the activities of the State Fire Service - a proposal dedicated to the State Fire Service]. *Zeszyty Naukowe SGPS*, *58*(2), 105–122.

Tuśnio, N., & Wolny, P. (2017). Nowoczesne narzędzia i sprzęt wykorzystywane do poszukiwań osób zaginionych [Modern tools and equipment used to search for missing persons]. *Zeszyty Naukowe SGPS*, *61*(1), 7–23.

Villa, T. F., Jayaratne, E. R., Gonzalez, L. F., & Morawska, L. (2017). Determination of the vertical profile of participle number concentration adjacent to a motorway using an unmanned aerial vehicle. *Environmental Pollution*, *230*(6), 134–142. doi:10.1016/j.envpol.2017.06.033 PMID:28649041

Wang, L., Chen, F., & Yin, H. (2016). Detecting and tracking vehicles in traffic by unmanned aerial vehicles. *Automation in Construction*, *72*(5), 294–308. doi:10.1016/j.autcon.2016.05.008

Wang, R., Xu, X., Jia, S., Ma, R., Ran, L., Deng, Z., ... Ma, Z. (2017). Lower tropospheric distributions of O3 and aerosol over Raoyang a rural site in the North China Plain. *Atmospheric Chemistry and Physics*, *17*(3), 3891–3903. doi:10.5194/acp-17-3891-2017

Wilson, R. L. (2014). Ethical Issues with Use of Drone Aircraft. *IEEE Explore*, *56*, 1–4.

Wyszywacz, W. (2016). *Drony, budowa, loty, przepisy*. Łódź: Wydawnictwo Poligraf.

Yoon, H., Hoskere, V., Park, J., & Spencer, B. Jr. (2017). Cross-Correlation-Based Structural System Identification Using Unmanned Aerial Vehicles. *Sensors (Basel)*, *17*(9), 1–12. doi:10.339017092075 PMID:28891985

Yue, J., Yang, G., Li, C., Li, Z., Wang, Y., Feng, H., & Xu, B. (2017). Estimation of Winter Wheat Above-Ground Biomass Using Unmanned Aerial Vehicle-Based Snapshot Hyperspectral Sensor and Crop Height Improved Models. *Remote Sensing*, *9*(5), 708–727. doi:10.3390/rs9070708

Zahorski, T., & Żurek, Ł. (2016). Wykorzystanie druku przestrzennego w budowie bezzałogowego statku powietrznego [The use of 3D printing in the construction of unmanned aircraft]. *Autobusy*, *R17*(12), 1495–1499.

KEY TERMS AND DEFINITIONS

BVLOS: Beyond visual line of sight – permission to pilot the drone out of sight.

Drone: The most common name of UAVs, which began to be used in 1935.

EASA: European Aviation Safety Agency.

GPS: Global positioning system.

ICAO: International Civil Aviation Organization.

IMU: Inertial measurement unit.

NDVI: Normalized difference vegetation index.

RPAS: Remotely-piloted aerial system.

UAV: Unmanned aerial vehicles – generic name, can be of different types (multicopter, bicopter, quadcopter, etc.).

VLOS: Visual line of sight – permission to fly with drones in sight.

WOPR: Water Volunteer Ambulance Service (Polish).

WPT: Wireless power transfer.

Chapter 4
Control of Information Stream for Group of UAVs in Conditions Lost Packages or Overloading

Dmytro Kucherov
National Aviation University, Ukraine

Igor Ogirko
Kazimierz Pulaski University of Technology and Humanities in Radom, Poland

Olga Ogirko
State University of Internal Affairs, Ukraine

EXECUTIVE SUMMARY

The chapter deals with the problem of controlling the flow of information coming from a group of unmanned air vehicles by radio channel. The inevitable data losses are compensated by repetition of lost packages or reconfiguration network. Modern methods control of data flow assumes using a mechanism ARQ based on the method sliding window. The consequence of these problems is a partial or total loss of its performance manifested in a decrease in the network's starting throughput. Part of the problem of restoring the network is solved by routing mechanisms, which lead to reconfiguration of the network due to the elimination of faulty nodes. Management of computer network overload is solved by well-known routing protocols such as OSPF, IS-IS, RIP, and others. In solving the problem, the representation of the output of individual nodes of the network using the "death and reproduce" scheme was used substantially. This scheme of network operation presupposes its representation by the Markov chain and the derivation of probabilistic characteristics by solving the Kolmogorov equations.

DOI: 10.4018/978-1-5225-7588-7.ch004

INTRODUCTION

Today the number of researchers takes attention to the collective management of unmanned machines based on radio remote control. It is the task of executing the works with the risk to human life, the need to fulfill the tasks in constraints of time, a long period carries out routine work. Application tasks of collective control include search and monitoring operations, extinguishing fires in large areas of the earth's surface and other (Dang & Horn, 2015 & Nathan et al, 2005), combat and antiterrorist operations (Karpenko, 2010). More recently, theoretical studies of the problems of application, challenges, and research problems related to the network use of UAV appeared (Mozaffari et al, 2018).

When managing the behavior of the group UAVs, it needs to transfer large amounts of information to the ground control station. To simplify the processing of information, the information is transmitted in small packets. The movement of aircraft, equipment failures, jamming of the radio signals, some data packets may be lost or, for example, there may be no confirmation of its receipt.

Some of the authors offer methods of the networks actions planning of UAVs, which based on discipline schedule, i.e. the order of processing of the transmitted packages (Kaur, 2011 & Niyato, 2005; Issariyakul, 2006; Bezruk et al, 2011; Gong et al, 2018). Nevertheless, an important task becomes developing algorithms compensate for lost packets or overloading. A natural approach to solving this problem is to resend lost packets to the point of reception and processing this information or overcoming this overloading. Then we have a network, and alone UAV the node of this network.

Network overload is one of the main problems that users of computer networks occasionally encounter. This problem causes a decrease in the bandwidth of the network, an increase in the passage time or loss of packets. It is this phenomenon that results in the termination of some network services, such as VoIP, interactive applications, chat, access to remote resources, and others. Lately, when there is an exponential growth of networks, this problem during their operation becomes the acuter.

One of the overload factors is excess buffering of the transmission channel (Gettys, 2011; Arefin & Amin, 2010). A buffer is required for data transmission over a communication line if the sender and the recipient have different processing pace. In this case, the buffer delays the transmission of packets at the time of acceptance and initial processing by the recipient. Filling the buffer leads to loss of packets transmitted. This phenomenon can be observed in routers, wireless access points, bridges, gateways, satellite devices.

Exclusion of overload is solved in several ways. You can achieve buffer overload by controlling queues and methods of prediction. Overload control methods control overload after it occurs, while prediction methods eliminate overload by controlling the transmission rate of the network on the network, enabling overloading and preventing it in typical "bottlenecks" of the network. The main tool of network operating systems for preventing overloads in Cisco is the use of algorithms of Weighted Random Early Detection (WRED) (Cisco, 2004).

In the duplex mode of switching ports control it is possible to implement a feedback mechanism that is introduced for Ethernet networks with IEEE 802.3x specification. The mechanism is implemented by introducing a sub-level of MAC level control, which introduces a time-stamping parameter for other nodes. The time is measured at 512-bit intervals of a specific Ethernet implementation, the range of possible stopping options is in the range 0 – 65535. After the stop time is completed, the transmission is restored.

Overloading can be eliminated if bandwidth reservation is introduced based on binary methods. In this case, the user QoS applications provided a portion of the throughput of the channel, the other part is reserved for other users, which is carried out through the use of the logical connection. But this happens at the hardware level, which introduces priorities to specific user applications, such as video conferencing. Therefore, the entire network of equipment in the transmission channel must support this technology, but this is not always the case.

In mobile networks, a typical solution to overload is the network reconfiguration. The paper analyzes the possibility of recovering throughput of a computer network, generating traffic in different conditions of network connections, determining the configuration with less load, establishing the relationship between the fall of traffic and its restoration due to the reservation.

The structure of chapter as follows: the next section contains some relative research in this problem and problem statement, Section 3 – 6 presented some results respect to lost packages in different schemes of transmission and their simulation, Section 7 – 9 present reconfiguration of a system based on graph approach if it is in overloading condition, main results of chapter present Section 10.

BACKGROUND AND PROBLEM STATEMENT

The computer network usually operates with different data rates. In this case, we can meet the overloading problem. The problem of overloading takes place not only in Ethernet networks but also in mobile networks. The presence of many agents of the mobile network results in its overload, which to a large extent is solved by routing task under the condition of changing the topology of the network. Since there is no single effective routing algorithm for computer networks, therefore, dynamic traffic distribution is proposed to solve based on routing problems in the existing agent system to determining related agents (Kucherov, 2016).

Using UAV equipped with the appropriate kit is one of the solutions of the problem of mobile networks congestion (Rohde & Wietfeld, 2012). The question of overloading a mobile network in conditions of interference of intentional and unintentional origin is solved by the authors of work (Kucherov & Kozub, 2015). An efficient mobile communication network is created by reconfiguring the directional diagrams of the base station antennas. The requirements for a topology with the best noise immunity of cellular communication are put forward.

One of the recent works in this direction (Lastovchenko et al, 2009) proposes a method for constructing optimization models of a computer network, which are designed to perform tasks by placing nodes of the network structure using the spreadsheet environment. The nodes of the network are divided into servers and clients linked to them. The configuration of the graph of such a network is created by solving the maximizing problem based on nonlinear optimization.

For optimizing the data flow when tracking multiple objects requires an iterative approach and therefore time costs (Zhang et al, 2008). A method of tracking the window in controlling the flow of data in the local network formed by mobile agents is proposed by the authors (Issariyakul, 2006). An analysis of the above approach shows the effectiveness of using an adaptive observation window based on a flow control mechanism known as Automatic Repetition Query (ARQ). The established properties of the control mechanism allow determining the optimal size of the window of observation and packets for data transmission over the network.

A comparative analysis of the wireless network topologies for the UAV group showed the existence of a number of problems that must be solved for stable and reliable network operation (Gupta, 2015). The method of coordination in case of loss of communication for the reconfiguration of the network topology can be used in the management of autonomous UAVs (Grancharova, 2013). The methodology for evaluating the quality of the reconfigurable topology of Wi-Fi Asynchronous Transfer Mode (WATM) networks based on the reliability of their functioning is given in (Lastovchenko et al 2009). An approach developed by the author of the paper (Gorbunov, 2006) is based on step-by-step monitoring of the thresholds of non-repudiation and restoration of the computer network. The next step is determined by the ratio of the growth rate of readiness to increase the cost of its support. In (Fortz & Throup, 2002) the bandwidth of the multisensory system, which is based on the network principle of action, is considered.

The adequacy of network reconfiguration in the conditions of natural and artificial obstacles is analyzed according to the scheme "destruction-reproduction" in (Ventcel, 1988). It should be noted that in most studies, for example (Frenkel et al, 2013), network elements are assumed to be non-recoverable, which, of course, simplifies the task, but does not always correspond to practice.

The purpose of the paper is to evaluate the reliability of traffic of a computer network, which operates in conditions of overload due to artificial and natural constraints, which leads to a change in the configuration of the system.

We will consider a computer network, the nodes of which are able to process information and exchange data, and in addition to changing the configuration of the system. The network management method corresponds to the «client-server» architecture.

The network has a definite topology, which is described by a weighted graph $G = (V, E)$, in which V is a set of the nodes, the number of nodes $V(G) = N$, and E are weighted edges, the number of edges $G(E) = M$. Each edge (i, j) corresponds to the number $w_{ij}>0$, which is called the weight of the edge (i, j), $i = 1..N, j = 1..M$. In the case when $(i, j) \notin G$, $w_{ij} = \infty$.

The network has two specific vertices designated S (source) and T (terminal). It is assumed that for any vertex $V \neq S$ there exists a path leading from S to T and passing through V. Vertexes are considered to be absolutely trustworthy and cannot be denied, but edges can refuse. The weight of the edge w_{ij} characterizes the performance of the channel. If it fails, the performance drops and the weight w_{ij} decreases. Restoration begins at the time of edge failure. The functions of fail-free operation $F_E(t)$ and restoration $W_E(t)$ (distribution functions) of the edge E_{ij} are given. The functions $F_E(t)$ are assumed to be absolutely continuous, i.e. there are density $f_E(t)$.

The set of edges $E_1, ..., E_M$ is called a cut if their simultaneous malfunction entails a complete network failure (the path from s to t is violated). A cut is called minimal if no edge can be removed from it so that the network goes into a failure state. The network can be considered as a series-parallel connection of arcs. In parallel connection of arcs, they belong to one cut.

If $S_1, ..., S_T$ is a cut set, then the network productivity at the moment t is determined by

$$P\left(V'\left(t\right)\right) = \min_{1 \le j \le T} \sum_{E \in S_j} c_j \left[1 - V_E\left(t\right)\right], \tag{1}$$

where $V'\left(t\right) = \left(V_E\left(t\right), E \in G\right)$, $V_E(t)$=0, if the edge E is in working order at the moment t, and $V_E(t)$, if the edge is on recovery. In accordance with formula (1), the performance is determined by the minimum amount throughput of vertices in working state across all cuts. At each point in time, the system requires a certain level of performance. If at some point the performance becomes less than the availability of level, then a functional failure occurs.

We introduce a Markovian process ξ having a finite set of states $I = \{1, \ldots, n\}$ and given by the probabilities P_{ij} of the transition of the embedded Markovian chain and the conditional functions $B_{ij}(t)$ of the time distribution between the transitions.

The function $B_{ij}(t)$ is assumed to be absolutely continuous; there are densities $b_{ij}(t)$. It is also assumed that all states form one class of essential states, and $\xi(0) = 1$ with probability 1. If $\xi(t) = j$, then the level of required capacity is equal $\varsigma(j)$, i.e. a random process $\varsigma(\xi(t))$ specifies the performance required from the system at the time t. It is assumed that $\varsigma(j)$ is a monotonically increasing function, $\varsigma(j)>0$. The system operates in normal mode, if $P(V'(t)) > \varsigma(\xi(t))$. The moment t of its refusal is defined as $t_r = \inf \{t: P(V'(t)) > \varsigma(\xi(t))\}$.

Our aim of the study is to estimate the probability $Q(t)$ of a functional failure from a degree of network congestion on a long interval, i.e. $t \to \infty$.

PACKETS TRANSMISSION

Let have a group from n UAVs, which gives information to the operator's remote control by radio link. Figure 1 shows the typical use of UAVs in the monitoring task. Because of the obstacles on the ground, the presence of monitoring noise the transferring information to control point may come by the chain, in which the intermediate UAVs are repeaters.

Operator and UAV exchange this information as usually by packages. The package is some amount of binary information, organized in a certain way that named protocol.

All UAVs transmit information to the ground control station in asynchronous mode; therefore, the input stream in the processing system has the form

$$I = I_{UAV_1} + I_{UAV_2} + \ldots + I_{UAV_n}, \tag{2}$$

where I_{UAV_i} is the data flow from i of the source and $i = \overline{1, n}$. Here the problems of auto identification, authentication don't solve. The packages from all UAVs are considered as whole one stream.

Figure 1. A typical application of the group of UAVs.

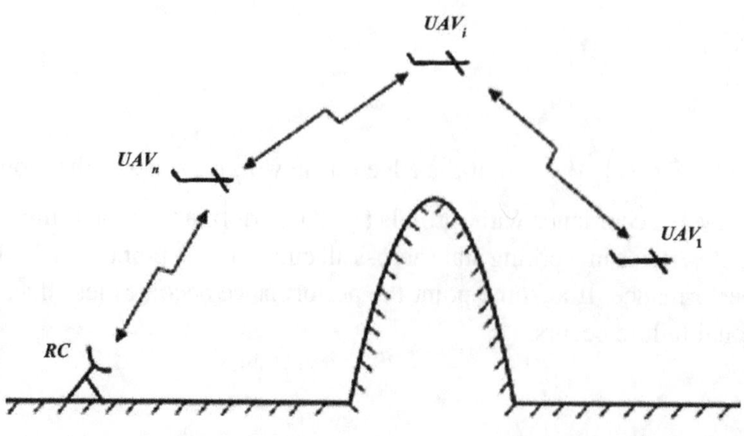

The main problem in this transferring data is motion UAVs, mismatch processing speeds in reception and transmission points, buffer overload is due to data retrieval on a low speed, re-transmission data to another address, and errors during transmission data. Consequently, it becomes necessary to control the flow of data.

There are some methods to control the flow of data. Among them stop and wait, return to the M steps, selective rejection. The last two approaches are known as a mechanism automatic repetition query (ARQ) [8]. The first scheme works as follows: if the sender sent the package, the receiver sends the confirmation it readies to get the next package. In this case, we have errors of two types. One type of error can turn out because of transmission package and another – in time receiving confirmation. In accordance to the second scheme is entering a "sliding window" for transmission M packages. Error in package leads to the need to repeat their and all the subsequent packets transmitted in this window. In the case, selective rejection is repeated only the packages that have been damaged and the packages which have the waiting time are expired.

Assume that the system consists of a source, which transmits packages of fixed length Tpac, and the receiver, which confirm its decision sent to the request. Time data transmission in one direction is Tdir. Considered that the time duration of packages processing and transmission confirmation is very small, so they can be neglected, and then the time required transferring one packet

$$T = T_{pac} + 2T_{dir}. \tag{3}$$

If necessary, N-time to repeat transmission one package until successful reception that this time increases to a value

$$T = N(T_{pac} + 2T_{dir}). \tag{4}$$

Also is given the probability of damage to the package p. Let consider different schemes of the packets' transmission.

THE SCHEME OF STOP AND WAIT

As known, throughput is a metric characteristic that showing the relation limiting the number of passing units (information items, the volume) per unit of time through a channel. In our case throughput is value C, which calculated such as

$$C = \frac{T_{pac}}{T} = \frac{T_{pac}}{T_{pac} + 2T_{dir}}. \tag{5}$$

If we set $\bar{t} = T_{dir} / T_{pac}$ that (5) we can write in the form

$$C = \frac{1}{1+2\overline{t}}. \tag{6}$$

The throughput C in (6) is normalized and takes values in the interval [0; 1]. If $\overline{t} < 1$ then C→1, that corresponds to a high productivity channel, and if $\overline{t} > 1$ then C→0, that corresponds to low productivity channel.

If a transmission error occurs and the information is the necessary repetition of data being sent, the throughput channel deteriorates in N-time

$$C(\overline{t}) = \frac{1}{N(1+2\overline{t})}. \tag{7}$$

In the general case N in (7) is a random value, which for a long transmission interval determined by the probability

$$N = \sum_{k=1}^{\infty} kp^{k-1}(1-p), \tag{8}$$

where k – the number of repetitions transmission and p<1. In accordance with [17, formula (21.2-39)] record (8) transformed to

$$N = \frac{1}{1-p}, \tag{9}$$

then (7) write in the form

$$C(\overline{t}) = \frac{1-p}{1+2\overline{t}}. \tag{10}$$

Formula (10) is setting loading ability radio channel in the scheme with stop and wait.

ARQ SCHEME

Let's begin to consider its effect when errors transmission is absent. For the M-packages defining the size of the sliding window, there are two variants for transferring information.

Case 1: The size sliding window M more than the time allowed for transmission of information, i.e. M $\geq 1+2\overline{t}$. In this case, the channel is not overloaded and has the best performance. This statement allows to set the numerator (6) limit the transmission time value, then the throughput is equal to C = 1.

Case 2: The transmission time of the channel is limited to the size of the window, i.e. $M < 1 + 2\bar{t}$.
Here the numerator equals M and (6) can be written as

$$C(\bar{t}) = \begin{cases} 1, if\ M \geq 1 + 2\bar{t}, \\ M/(1 + 2\bar{t}), if\ M < 1 + 2\bar{t}. \end{cases} \tag{11}$$

Transmission errors are deteriorated throughput that in the case with selective rejection when returning on N steps for single package occurs in virtue of (9), (10), takes the form

$$C(\bar{t}) = \begin{cases} 1 - p, if\ M \geq 1 + 2\bar{t}, \\ M(1 - p)/(1 + 2\bar{t}), if\ M < 1 + 2\bar{t}. \end{cases} \tag{12}$$

In the scheme of return on N steps retransmit L packages. Here, the number of return N steps is a function of the number of transmitted packages

$$N = \sum_{k=1}^{\infty} f(k) p^{k-1} (1 - p). \tag{13}$$

In (13) f (k) is the total number of retransmitted packages and a lost packet transmitted k times. f (k) can be represented as (Stallings, 2002)

$$f(k) = 1 + (k - 1)L. \tag{14}$$

After the substitution of (14) to (8) is obtained by the number of the repeated packages

$$N = \frac{1 - p + Lp}{1 - p}. \tag{15}$$

For the derivation of (15) essentially used the results of (21.2-38) and (21.2-39) in (Korn and Korn, 1968).

Substituting (15) into (7) and considering the cases 1 and 2 relatives to the size M = L of the sliding window, we obtained throughput for ARQ-mechanism that based of return on N steps

$$C(\bar{t}) = \begin{cases} \dfrac{1 - p}{1 - p + Mp}, if\ M \geq 1 + 2\bar{t}, \\ \dfrac{M(1 - p)}{(1 + 2\bar{t})(1 - p + Mp)}, if\ M < 1 + 2\bar{t}. \end{cases} \tag{16}$$

Expressions (12) and (16) correspond to throughput data flow control mechanism with ARQ. It is noteworthy that for M = 1, both expressions degenerates to scheme stop and wait (9).

Expression (6), (11), (12), (16) prove the following.

Theorem: If the size M of the window satisfies the condition $M \geq 1 + 2\overline{t}$, then the channel is effectively used and $C(\overline{t}) = C_{\max}$.

Corollary: The number of bits l of the sliding window is determined by the expression

$$l \geq 1 + \log_2 T_{pac} - \log_2 T_{dir}. \quad (17)$$

Formula (17) follows from the theorem and the value $\overline{t} = T_{dir} / T_{pac}$ and

$$M = 2^l - 1. \quad (18)$$

SIMULATION PACKET TRANSMISSION

We validate the analytical results obtained and analyze them. In all study estimates throughput, we assume that the probability of errors in the transmission of the data packages is p = 10-3.

The value M is selected from the condition (18), where l - the number of bits of the transmitted package. Selected the numbers of 1, 7, 127, wherein the number M1 = 1 corresponds to a scheme stop and wait, M2 = 7 applies to packets of small size (3 bits) and M3 = 127 (7 bits) is typically used in high-speed wide-area networks.

The value of the number a=$\overline{t}$ is determined by the time of the transmission and time spread packages to the point of control. Considering different data rates and different transfer size package the values $\overline{t}$ are in interval 0 to 1000.

Figure 2-6 shows the results of numerical modeling throughput for these initial data. Thus, in Figure 2, 3 is shown throughput selective rejection scheme and based on return on window sizes M1, M2, M3. Figure 4 and 5 are graphs of the bandwidth for different schemes of the window size M2 in comparison with the scheme stop and wait. So, Figure 4 correspondence for window size M2, and Figure 5 windows for size M3. Summarizes the results is shown in Figure 6 that allows us to understand the correspondence between the various schemes of flow control schemes that are considered.

These figures show that if the effective load channel for scheme stop and wait for correspondence level is a = 1, then return and selective rejection schemes this is approximately equal to the number of M.

This analysis leads to the conclusion that the greatest throughput schemes are based on the approach of creating a sliding window. There is a direct relationship between the window size and bandwidth. The larger the window size, the higher the throughput is. By comparison two schemes creating a “sliding window” the best characteristics has the selective rejection scheme.

Figure 2. Function C (a) for stop and wait and selective rejection with sizes M1, M2, M3 sliding window.

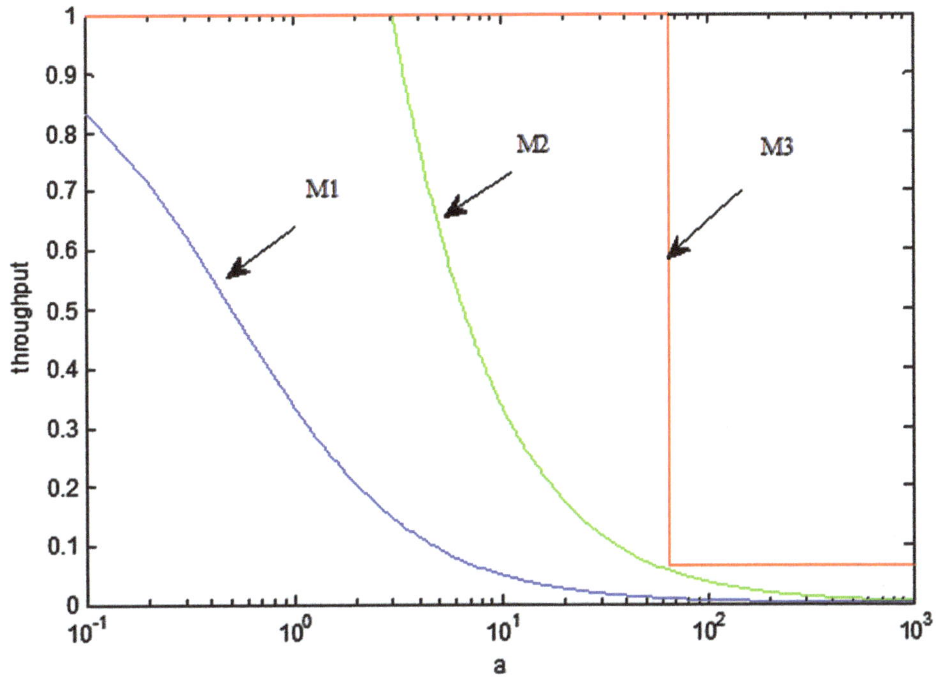

Figure 3. Function C (a) for stop and wait and scheme ARQ based on return on M1, M2, M3 steps.

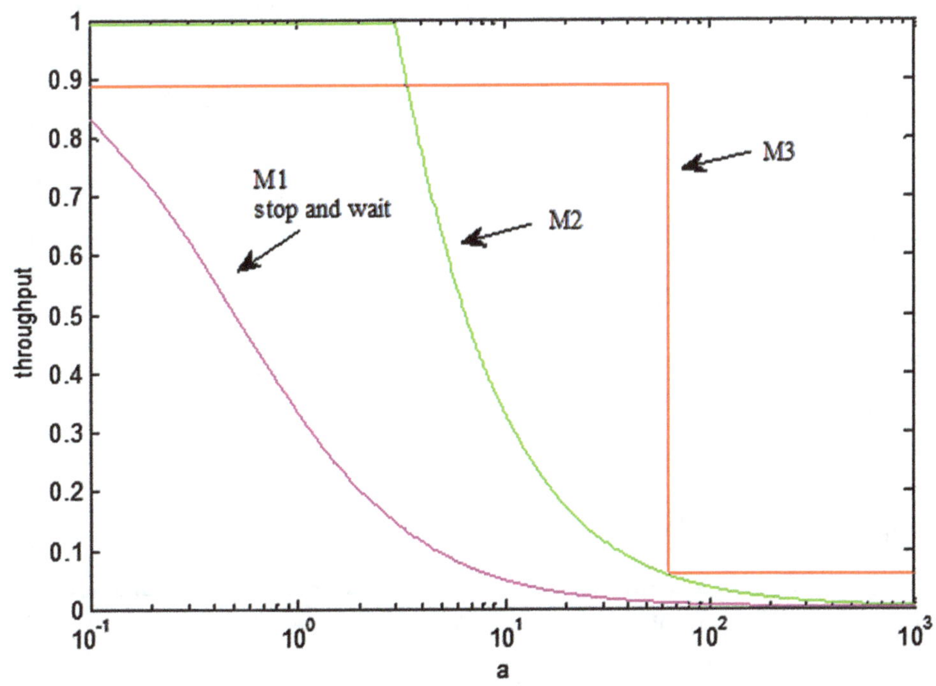

Figure 4. Function C (a) for stop and wait and scheme ARQ based on return on M2 steps.

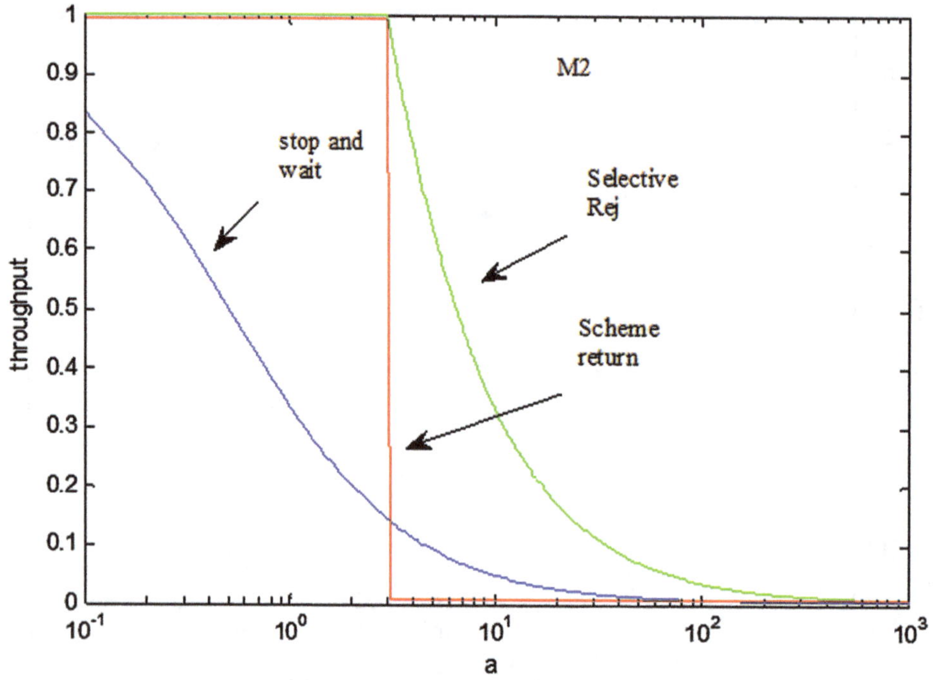

Figure 5. Function C (a) for stop and wait and scheme ARQ based on return on M3 steps.

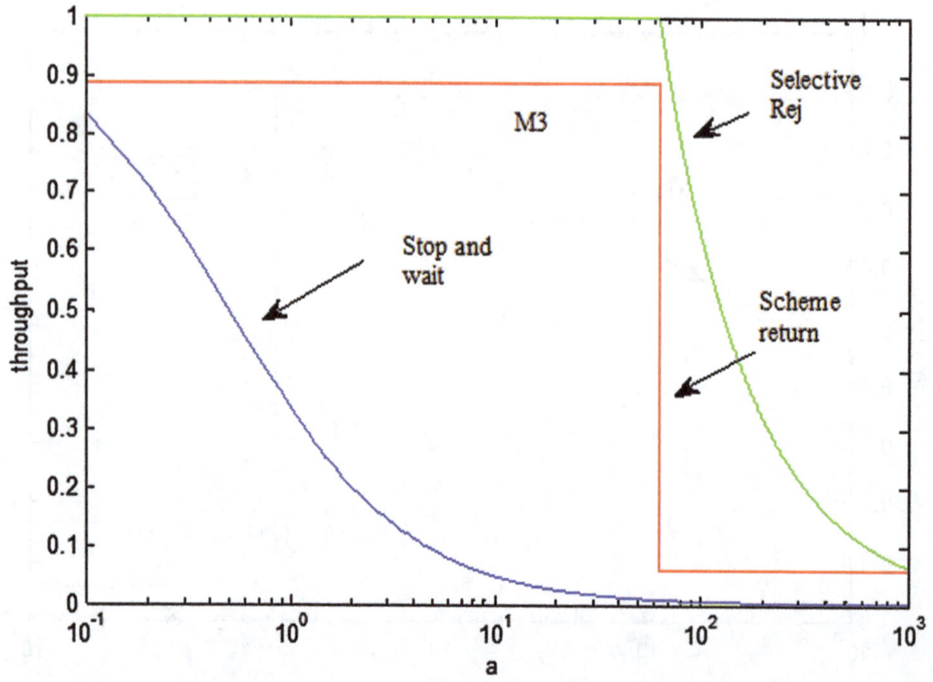

Figure 6. Function C (a) for stop and wait and schemes ARQ based on return and selective rejection of M1, M2, M3 steps.

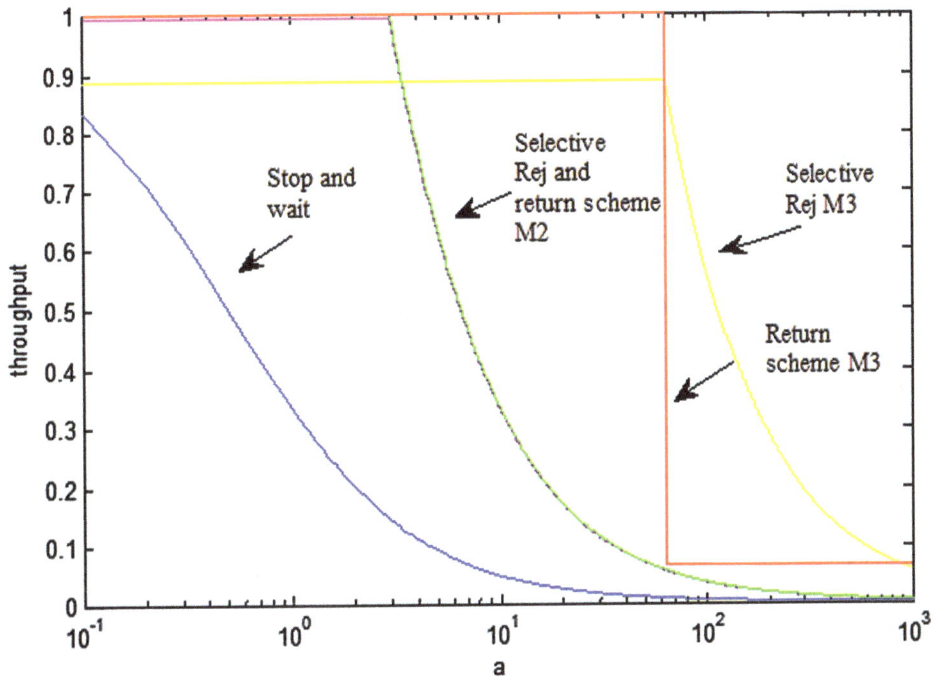

THE SYSTEM GRAPH APPROACH

Data transmission on the network takes place according to the rules that are governed by the applicable protocols, according to which the address, size, and structure of the packet being transmitted are determined by the speed of data transmission.

The address can be individualized when the message is sent to only one agent, group when the message is sent to each agent in the group, or to a certain subgroup of agents. This method of addressing is basic to addressing in computer networks. The data transmission takes place from the source address to the receiver's address.

If the number of nodes is significant, then there is a need to pass information through transit nodes. However, it becomes necessary to lay a route to ensure a minimum of time for the delivery of information achieved by a minimum of transit nodes or the presence of channels with high bandwidth and reliability of communication lines. It goes without saying that a small increase in the number of transit nodes with high bandwidth is overwhelming over the minimum number of nodes with low bandwidth in terms of the time of transmission. Reliable communication channels are characterized by minimal loss of transmitted information.

Several sub-streams can pass through a node; they are distinguished by the address of the destination. It is clear that in order to determine the route that would provide equal time flows of different volumes of data, it is necessary to take into account the transmission speed of the individual lines and to enable the distributed of sub-streams and their assembly. The switching of nodes for the transmission of sub-streams is carried out by multiplexing of free channels. This task consists in determining the route of information passing to the endpoint. Distinguish between static and dynamic routes. A static route is given once or at a certain schedule and does not change within a certain time. Dynamic routes are calculated by appropriate algorithms depending on the topology and status of the information network. These include algorithms for finding the shortest distance in width, Dijkstra, Bellman-Ford (Stallings, 2002).

THE DIJKSTRA'S ALGORITHM

The most common algorithm is the Dijkstra's algorithm; it is considered to be the procedure for finding the shortest path on a weighted oriented graph. The algorithm is used by OSPF and IS-IS routing protocols in IP networks (Fortz and Throup, 2002). The edges of the graph have a weight wij such that

$$w_{ij} = \begin{cases} 0, \text{ if } i = j, \text{ is the same vertex,} \\ w > 0, \text{ if } \mid i - j \mid < 2 \text{ i.e. neighbour vertex,} \\ \infty, \text{ else.} \end{cases} \tag{19}$$

The path length at each step k from the vertex s is determined by the rule

$$L(k) = \min_{V}[L(k), L(V) + w_{ij}], \tag{20}$$

where $L(k)$ is path length after k step. Rule (20) does not allow passage along the arcs of graph G with a high weight. Thus, the set of vertices in the graph G is an ordered sequence of connected nodes that contains the shortest path from the vertex s to k. This path is shown in Figure 7 by the arrows

At each scheduled time point i on the router comes the total flow of information intended to be transmitted to each router j. This general stream defines a routing table that can be calculated by the matrix P (t) of the dimension N×M with the zero main diagonal

Figure 7. The structure of the weighted graph.

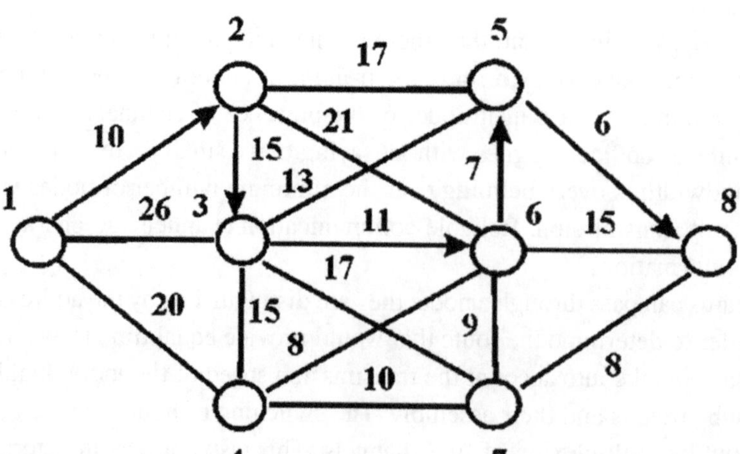

$$P(t) = \sum_{i,j} c_{ij}(t). \tag{21}$$

In formula (21) the information streams associated with the corresponding IP addresses in the routing table, and cij is bandwidth. Cisco routers define weights according to the formula [20]

$$w_{ij} = \frac{10^8}{c_{ij}(t)} \tag{22}$$

Since the algorithm (20) is iterative, the number of iterations is determined by the number of vertices of the graph, so the time complexity of the algorithm O(N). Within each iteration, a new passage takes place taking into account the new (j + 1) vertex. In this case, the peaks with the greatest weight are released, and the length of the path with new vertices is renewed, the best result is remembered. This is the same as the number of vertices. The overall performance of the algorithm is estimated to be O (N2). Thus, the Dijkstra's algorithm is resource-intensive, but due to the knowledge of the network topology and the path to the desired node, the router always finds an alternative path to the desired network node in the event of problems in any node of the specified path.

Resetting capacity should take into account the network load. To control the router is supplemented by means of load measurement, which will create a similar (2) matrix of loads C = | cij |. Then the backup reservation algorithm can be written as

$$\overline{c}_{ij}(t) = \begin{cases} c_{ij} + \Delta, \text{ if } c_{ij} < x_{ij}, \\ c_{ij,} \text{ if else,} \end{cases} \tag{23}$$

where Δ is the fraction that compensates for the overload.

ESTIMATED SYSTEM RECONFIGURATION

By reconfiguration of the information system, we will understand the change in the structure of the system, which refers to the size or its topology. The feature of reconfiguration is the restructuring of the structure and topology of the system to eliminate overloads and failures in the system. It is believed that the network carries out its functions, while exchanges between nodes are carried out. The examples of a reconfigurable system are shown in Figure 8, 9.

Thus, due to overloads or crashes, the topology of a local full network is changed to the "bus".

In accordance with the general idea of a computer network (1), it is a system with a limited number of possible states. Therefore, its behavior can be modeled using a mathematical apparatus for analyzing Markov chains.

We will assume that in the process of operation, a computer network consisting of a finite set of elements N, exchanges information between all the elements in the OSPF or RIP protocols, which corresponds to the conditions for the normal functioning of the system. The initial state of the network is denoted by S1. If for some accidental reason the elements of the network are starting to fail, or infor-

Figure 8. The structure of the information system in graph form.

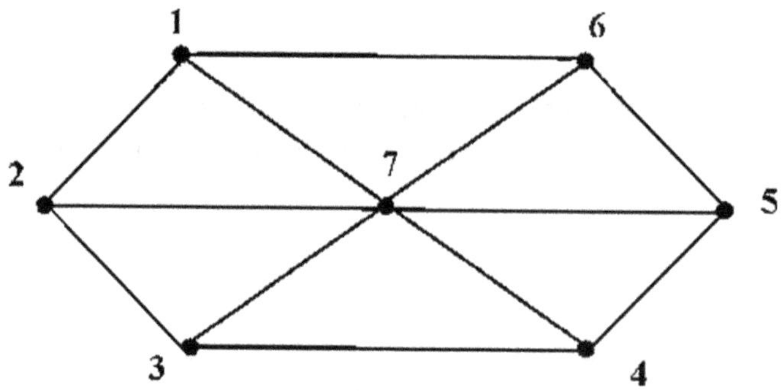

Figure 9. Changing the structure of the information system as a result of a loss of communication.

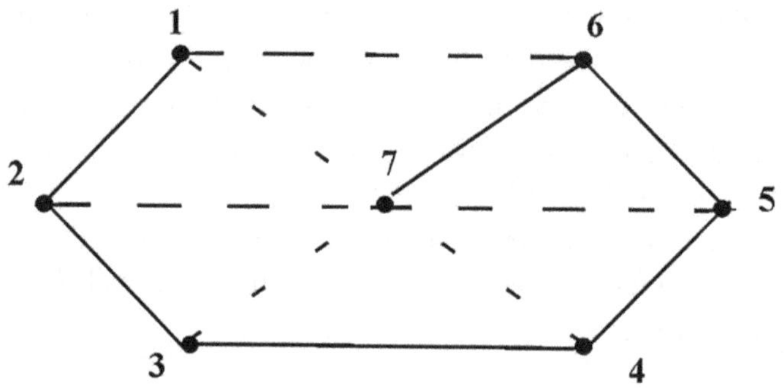

mation exchange may be lost, the system is moving to another state. We will assume that the elements of the system are not overloaded simultaneously, but one by one, therefore successive transitions from state S1 to states S2, S3, ..., Si are performed at certain intervals of time Δt, i stands for state number. A sequential set of states of the computer network and transitions between them forms the Markov chain. Since the chain is consistent, the operation of the system can be filed in the form of a scheme "death-propagation" (Ventcel, 1988).

Let the computer network consist of n nodes, therefore, according to the approach to the "death-multiplication" scheme, we introduce the states Si, i = 1..(n + 1), where S1 is the state of the network, which corresponds to the functioning of all nodes without overload. Under the influence of external and internal factors, the throughput of channels in the network with fixed intensity deteriorates λ, which is associated with weighting factors wij in (6). In this case, the transition to the state occurs, when the nodes gradually lose the packets, successively one after the other. Thus, the transition to the state S2 occurs when no works the node number 1, and so on, and therefore Sn + q – the computer network has ceased to perform tasks in connection with the failure (n – q) nodes. The system can also take measures to increase the throughput, which occurs with the intensity of $\mu > \lambda$. At the same time, the successive transitions from the states of Si to the Si-1 state occur with intensity μ. We find the probabilities qi of finding a computer network in each of the finite states Si and analyze them. The probabilities of finding a system in the final states are based on the formulas [20]

$$q_i = \frac{q_0}{i!}\frac{\prod_{k=1}^{i}\lambda^k}{\prod_{k=1}^{i}\mu^k},\ i \neq 0,\ q_0 = \left(1 + \sum_{i=1}^{n}\frac{1}{i!}\frac{\prod_{k=1}^{i}\lambda^k}{\prod_{k=1}^{i}\mu^k}\right)^{-1} \tag{24}$$

Example: Let us consider this problem for the case network as in Figure 8 where n = 5, λ = 0.5, and μ1 = 0.8, μ2 = 1.6, μ3 = 2.4. The results of the calculation by the formulas (7) are shown in Figure 10. The analysis of the figure shows that the less the μ, the better the system is handling the packets, the more likely it is to be in a state with minimal delay and less probability of being in the state with the greatest delays.

As can be seen from the figure, if μ> (2..5) λ, the system becomes more sensitive to loads of different types. The failure of two or more nodes becomes critical for a network of this type.

CONCLUSION

Exchange information with the group UAV is done by radio channel "UAV- point control". The movement of UAV and equipment imperfections causes errors in reception information or overloading. Troubleshooting the received information is achieved by repetition of data packages or reconfiguration network is called flow control.

Figure 10. Degradation of the computer network in conditions of intense loads.

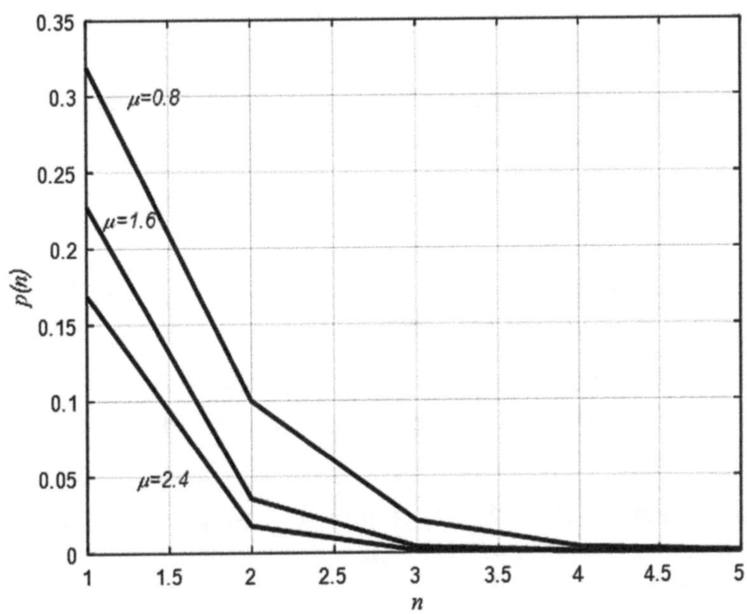

In this chapter analyzes the known methods of flow control that are focused on re-processing of the lost information. These are methods to stop and wait, to repeat the last N packages and selective rejection. The absence of errors in the received information is estimated by ARQ based on a sliding window. In the investigation, the indicator of bandwidth is essentially used.

Also, based on the analysis of protocols and routing algorithms, it has been established that the efficiency of a computer network is determined by the possibility of its operation in conditions of overloads and failures, which is the result of excessive buffering of the system. One of the effective ways to reduce the impact of overloads on the network is to reserve the bandwidth of the channels and to compensate for its share in the channels that are most exposed. According to the analysis of the network under the scheme of "death-propagation", it is established that the action of the system functions if the ratio of the intensity of overload to the intensity of the increase in throughput does not exceed the value of 0,2 ... 0,5.

Further research on the functioning of the computer network is planned to focus on analyzing its dynamic properties.

REFERENCES

Arefin Md, T., & Amin Md, R. (2010). Congestion Control and Buffering Technique for Video Streaming over IP. *International Journal of Latest Trends in Computing., 1*(2), 133–137.

Bezruk, V., Chebotareva, D., Ivanenko, M., & Jo, S. (2014). Multicriteria Optimization in Planning of Mobile Communication Networks. *Proceeding of the 20th International Conference on Microwaves, Radar and Wireless Communications*, 633–639. 10.1109/MIKON.2014.6899974

Cisco. (2004). *Internetworking technologies*. Cisco Press.

Dang, A. D., & Horn, J. (2015). Formation Control of Leader-Following UAVs to Track a Moving Target in a Dynamic Environment. *Journal of Automation and Control Engineering, 1*, 1–8.

Fortz, B., & Throup, M. (2002). Optimizing OSPF / IS-IS weights in a changing world. *IEEE Journal on Selected Areas in Communications, 6*, 1–31.

Frenkel, I. B., Karagrigoriou, A., Lisnianski, A., & Kleyner, A. V. (2013). *Applied reliability engineering and risk analysis: probabilistic models and statistical inference*. Wiley. doi:10.1002/9781118701881

Gettys, J. (2011). Bufferbloat: Dark Buffers in the Internet. Internet Computing. *IEEE Computer Society, 3*(15), 96–95.

Gorbunov, I. E. (2006). Methodology of analysis and synthesis of reconfigurable topologies of mobile communication networks. *Mathematical Machines and Systems, 2*, 48–59.

Grancharova, A., Grøtli, E. I., & Johansen, T. A. (2013). Rotary-Wing UAVs Trajectory Planning by Distributed Linear MPC with Reconfigurable Communication Network Topologies. *IFAC Proceedings Volumes, 46*(27), 198 – 205.

Gupta, L., Jain, R., & Vaszkun, G. (2015). Survey of Important Issues in UAV Communication Networks. *IEEE Communications Surveys and Tutorials*, 1 – 32.

Issariyakul, T., & Hossain, E. (2006). Channel-quality-based opportunistic scheduling with ARQ in multi-rate wireless networks: Modeling and analysis. *IEEE Transactions on Wireless Communications, 4*(5), 796–806. doi:10.1109/TWC.2006.1618929

Karpenko, A. V. (2010). *Complex «LEER» with Unmanned Aerial Vehicle «ORLAN-10»*. Retrieved from http://bastion-opk.ru/orlan-10/

Kaur, P., Singh, D., Singh, G., & Singh, N. (2011). Analysis, comparison and performance evaluation of BNP scheduling algorithms in parallel processing. *International Journal of Information Technology and Knowledge Management, 1*(4), 279–284.

Korn, G. A., & Korn, T. M. (1968). *Mathematical handbook*. McGraw-Hill Book Company.

Kucherov, D. P. (2016). Reconfiguration multisensory system in conditions of impact of destabilizing factors. *Sensor Electronics and Microsystem Technologies, 2*(13), 101–112. doi:10.18524/1815-7459.2016.2.73659

Kucherov, D. P., & Kozub, A. N. (2015). Control System Objects with Multiple Stream of Information. *Proceedings IEEE 3rd International Conference "Actual Problems of Unmanned Aerial Vehicles Developments (APUAVD)"*, 290–293. 10.1109/APUAVD.2015.7346623

Kuzmichov, A.I., & Dodonov, E.O. (2017). Optimization models of network structure. *Registration, Storage, and Processing of Data, 2*, 24 – 35.

Lastovchenko, M. M., Zubareva, E. E., & Sachenko, V. O. (2009). Method for analyzing the efficiency of the reconfiguration of the topology of wireless multi-server networks with increased noise immunity. *USM, 6*, 79–86.

Mozaffari, M., Saad, W., Bennis, M., Nam, Y.-H., & Debbah, M. (2018). *A Tutorial on UAVs for Wireless Networks: Applications, Challenges, and Open Problems*. Retrieved from https://arxiv.org/abs/1803.00680

Nathan, S., Janusz, M., Milind, T., Paul, S., Lewis, J. P., & Kasinadhuni, N. (2005). *The Future of Disaster Response: Humans Working with Multiagent Teams Using DEFACTO*. Published Articles & Papers. Paper 41. Retrieved from http://research.create.usc.edu/published_papers/41

Niyato, D. (2005). Analysis of fair scheduling and connection admission control in differentiated services wireless networks. *Proceedings of the IEEE International Conference on Communications*, 3137–3141. 10.1109/ICC.2005.1494984

Rohde, S., & Wietfeld, C. (2012). Interference Aware Positioning of Aerial Relays for Cell Overload and Outage Compensation. *Proceedings IEEE Vehicular Technology Conference (VTC Fall)*, 1 – 5. 10.1109/VTCFall.2012.6399121

Stallings, W. (2002). *High-Speed Networks and Internets. Performance and Quality of Service*. Prentice Hall PTR.

Ventcel, E. S. (1988). *Issledovanie operacii: zadachi, principi, metodologia. Moscow*. Nauka.

Zhang, L., Li, Y., & Nevatia, R. (2008). Global Data Association for Multi-Object Tracking Using Network Flows. *Proceedings IEEE Conference on Computer Vision and Pattern Recognition*, 1 – 8. 10.1109/CVPR.2008.4587584

ADDITIONAL READING

Cooper, R. B. (1981). *Introduction to Queueing Theory*. Elsevier North Holland.

Cormen, T. H., Leiserson, C. E., Rivest, R. L., & Stein, C. (2002). *Introduction to Algorithms*. McGraw-Hill Book Company.

Dijkstra, E. W. (1959). A Note on Two Problems in Connexion with Graph. *Numerische Mathematik*, *1*(1), 269–271. doi:10.1007/BF01386390

Jensen, P. A., & Barnes, J. W. (1980). *Network Flow Programming*. John Wiley & Sons.

Taha, A. H. (2003). *Operations Research: An Introduction*. Pearson Education Inc.

KEY TERMS AND DEFINITIONS

ARQ: Automatic repeat request, also known as automatic repeat query, is an error-control method for data transmission that uses acknowledgements (messages sent by the receiver indicating that it has correctly received a packet) and timeouts (specified periods of time allowed to elapse before an acknowledgment is to be received) to achieve reliable data transmission over an unreliable service.

IS-IS: This is a routing protocol designed to move information efficiently within a computer network, a group of physically connected computers or similar devices.

MAC: This is the abbreviation of media access control, it is the open systems interconnection basic reference model layer.

OSPF: This is the abbreviation of the open shortest path first, it is a routing protocol for internet protocol (IP) networks.

Stop and Wait: A scheme telecommunications to send information between two connected devices. It ensures that information is not lost due to dropped packets and that packets are received in the correct order.

Weighted Random Early Detection: The active queue management algorithm for managing router overflows with the ability to prevent overload.

Wi-Fi Asynchronous Transfer Mode: These are the standards for carriage of a complete range of user traffic, including voice, data, and video signals.

Chapter 5
Software-Defined Networking in Aviation:
Prospects, Effectiveness, Challenges

Roman Odarchenko
National Aviation University, Ukraine

EXECUTIVE SUMMARY

Aviation telecommunications is one of the main means of providing guidance to civil aviation and air traffic control. Proper organization of communication is one of the main conditions for ensuring the safety and regularity of aircraft operations as well as the production activities of enterprises and civil aviation organizations. The new networks will focus on significantly improving the quality of service. The basis for their construction will form SDN networks. Therefore, the chapter analyzed the advantages and disadvantages of two SDN implementing methods. It was developed the mathematical method to assess their complex effectiveness, which considers QoS requirements of implementing service through special weights for scalability, performance, and packet delay. There were simulations of overlay networks by using soft switches to verify the adequacy of the proposed method. The results showed that the use of SDN networks more efficiently by using IP networks for large volumes of traffic and with a large amount of network equipment.

INTRODUCTION

Aviation telecommunications is one of the main means of providing guidance to civil aviation and air traffic control. Proper organization of communication is one of the main conditions for ensuring the safety and regularity of aircraft operations, as well as the production activities of enterprises and civil aviation organizations. The telecommunication of civil aviation is a set of centers, communication stations, terminals, various telecommunications equipment, interconnected by telecommunication networks (ICAO, 1999). Aviation telecommunications performs the following main functions: the transfer of instructions, orders and various types of messages to ensure the safety and regularity of air traffic, messages at all stages of the flight; the interaction of air traffic control centers (points) in the process

DOI: 10.4018/978-1-5225-7588-7.ch005

of air traffic control, planning and organization of flights; operational interaction between the services of airlines; transfer of administrative-management and production information; data transfer of various applications of civil aviation. Basic requirements for civil aviation telecommunications: timeliness of communication; reliability and continuity of communication; ensuring the required information transfer rate; securing the necessary secrecy in the transmission of information; maximum efficiency and economy of telecommunication operation.

One of the main parts of these modern aviation telecommunication systems is computer networks. Modern networks are not without disadvantages, such as: the complexity of network management, the high cost of network equipment, etc. The main drawbacks of modern networks, which are used, including in aviation, are:

1. *Difficulties in managing.* Many corporations use shared networks for voice, data and video transmission. Although existing networks can provide differentiated QoS levels for each type of traffic, resource allocation in many cases requires manual control. Administrators must configure each manufacturer's device individually and configure parameters such as bandwidth and QoS based on sessions and applications, since the static nature of the networks does not allow dynamic adaptation when changing traffic, applications, and user queries.
2. *Difficulties in scaling.* Existing network technologies include many different sets of protocols, designed for reliable communication between hosts at arbitrary distances, different channel throughputs and topologies. The industry has developed a set of network protocols that provide higher performance, reliability and security. Protocols solve a specific problem. This generates one of the main constraints - the complexity. For example, to add or remove a particular device, network administrators should make adjustments in multiple switches, routers, firewalls, and more. The topology of the network, manufacturers and models of switches, software versions should also be taken into consideration. Such complexity leads to relative static networks, since administrators seek to minimize the risks brought by the service.
3. *The need for highly skilled specialists.* The disadvantages of the networks described above result in significant difficulties in configuring a variety of proprietary network equipment, since the interfaces for interacting with them and the sets of commands are different. It is clear that this puts much higher requirements for the qualification of the administrators.
4. *High cost of the network.* Modern IP networks require a variety of hardware (routers, switches, and various middleboxes - network equipment that transforms, filters, checks, or otherwise manipulates traffic for purposes other than routing. Examples of middleboxes: firewalls, load balancers, IDS / IPS, etc.). This equipment is quite expensive (especially for large networks), which leads to high costs for building and scaling the network. Large networks require significant network maintenance due to the need to attract qualified specialists for their configuration, management and control.

These and many other shortcomings have led the world of telecommunications to the need to transpose the modern network organization concept. As a result of this transforming process new class of networks appeared – Software-Defined Networks.

BACKGROUND

Software-defined networking (SDN) is a revolutionary network architecture that separates out network control functions from the underlying equipment and is an increasing trend to help enterprises build more manageable data centers where big data processing emerges as an important part of applications (Qin and others, 2017; ONF, 2012). So, the main trend of network development is the transition from networks of the past and the existing generation, to the next generation networks (Matias, 2014; Haleplidis, 2015), where it is planned to use the SDN concept. Each generation comes out with the best parameters, but traditional networks design become ineffective in the dynamic environments (Haleplidis, 2014). In (Xie and others, 2012) it is noted that SDN-networks, have advantages over traditional networks, but have many disadvantages, the main of which is a very complex and long transition to a new architecture (Skoldstrom, 2015; Heller and others, 2012; Haleplidis, 2014). Therefore, before making a choice, it is necessary to understand, in what cases it is really necessary (Xie, 2012). When deploying SDN in real networks, large networks are always partitioned into several smaller ones due to numerous reasons: privacy, scalability, incremental deployment, security and so on (Heller, 2012; Porras, 2012;). So, as we can see from the above analysis, there are several problems, which it is necessary to solve immediately. They relate to the use cases, deployment scenarios, effectiveness evaluation etc. That's why the main purpose of this chapter is to create novel SDN effectiveness evaluation method, which will help to compare different architectures, which are planned to be deployed.

INTRODUCTION TO SDN

SDN (Software-Defined Networking) is a relatively new technology in telecommunications, designed to address shortcomings of traditional IP (Internet Protocol) networks (Tootoonchian, 2010). SDN is today one of the most promising technologies in the field of computer networks (Smieliansky, 2012). The SDN model has a number of advantages over traditional networks, among which developers allocate the following according (Orlov, 2014; Cherniak, 2012): increased network equipment efficiency by 25-30%; a decreased operating cost of networks; vroviding users with the ability to create programmatically new services and quickly upload them to network equipment. SDN is better suited to the requirements of critical information infrastructure networks, which include networks for aviation applications.

The concept of SDN is based on the following principles (Smieliansky, 2012):

- Control plane and data plane separation. The SDN divides information traffic and traffic management processes.
- An unified supplier-independent interface between the control level and the data transmission level.
- Logically centralized network management, carried out with the help of a controller.
- Programmability of the network. The purpose of SDN is to apply software applications that will affect the entire network. These applications can enhance network reliability by providing new security features, improving traffic routing and prioritizing, leading to better service quality etc.
- Virtualization of the network physical resources.

For a better understanding of the SDN concept, we must consider the architecture of traditional IP-networks and determine their differences.

In the modern IP network (pictured in Figure 1) routers and switches, that functioning, are complex network devices using a large number of protocols. In this case, their main function is the switching of communication channels. To perform this function, routers use routing protocols such as RIP (routing information protocol), OSPF (Open Shortest Path First), IS-IS (Intermediate System to Intermediate System) - for routing inside autonomous systems and BGP (Border Gateway Protocol) - for routing from / to other stand-alone systems. The switches use the STP protocol to function in networks with nodes and virtual local area networks (VLANs) for traffic isolation. The whole functionality is performed by each of the devices themselves, while none of the devices have a detailed view of the network, but only connects to its direct neighbors. Therefore, between the neighbors is constantly generated service traffic on the same channels of communication, which throw is also transmitted information traffic.

The SDN concept involves the separation of traffic and control planes (Smieliansky, 2012). Simplified architecture is depicted in Figure 2. In SDN networks, the entire hardware intelligence (control plane) is transferred to a single control center. Thus, SDN-switches are simple devices which main functions are switching and data transmission (data plane function). The work of the controller can be partially or fully automated with software applications that independently manage network resources. The controller takes all decisions based on the full view of the network and applies them simultaneously to all network devices. For example, when a large traffic flow arrives, the routing rules of which the switch does not know, the controller makes decisions once for all devices and in a traditional IP network each device must rout separately. The signal information is sent to it by individual channels, isolated (often physically) from data channels. All this improves network performance and increases the speed of data processing, especially for large traffic volumes.

The realization of the SDN concept in practice will allow network owners, including aviation, to obtain equipment independent control over the entire network from a single location, which greatly simplifies its operation. Equally important, the configuration of the network is much simpler, and administrators will not have to enter hundreds of lines of code separately for different switches or routers. The network characteristics can be quickly changed in real time, respectively, the timing of the introduction of new applications and services will be significantly reduced.

Two main methods are used to implement SDN-networks (Orlov, 2012): - SDN implementation based on special switches (OpenFlow protocol). In this network construction, switches are used as a network equipment, which use the OpenFlow protocol for interaction with the controller. The OpenFlow device consists of at least three components (Smieliansky, 2012):

Figure 1. Simpe IP network architecture

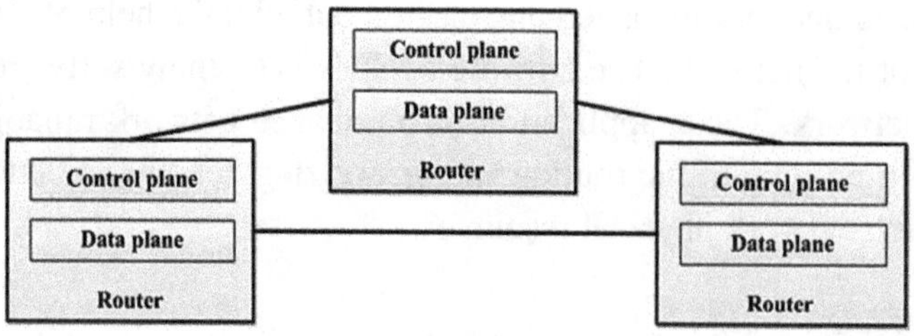

Figure 2. Simple SDN architecture

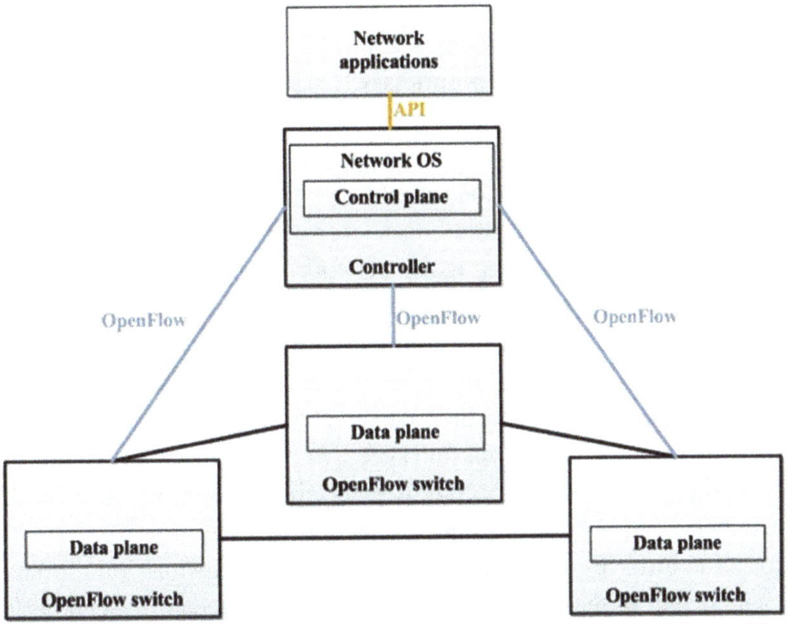

- Flow table;
- Secure channel;
- Own OpenFlow protocol.

Other equipment, such as IP-switches and routers, is not used on a "pure" SDN network.

- *Realization of SDN on the basis of virtual switches on technology Overlay.* Such SDN network can be constructed over an ordinary network, with a virtual switch (eg Open vSwitch) running on the hypervisor connected to the physical switch or router that supports the OpenFlow protocol (Kim, 2011). All traffic coming to the physical device is forwarded to the software switch where it is routing. After that, the traffic is sent to the desired destination from the physical switch.

SDN Possibilities

SDN provides a number of features, most of which can not be realized by using traditional architecture:

- Centralized network resources management. In a traditional architecture, the control plane, which includes the routing process, and the data layer, which is responsible for forwarding packets from one interface to another, are combined in one device. The concept of SDN involves transferring control functions to the central controller, thus replacing the traditional distributed routing model with a centralized one. Advantages of centralized management: the whole network configuration is stored in one place; it is convenient to manage the entire network at once; it is easy to complete the network with new equipment.

- Management unification. Switch designers can freely implement the internal structure of switches, but the package lookup procedure and semantics of instructions must be identical for everyone.
- The connection between network devices and the SDN controller is provided through network protocols that can be both open and proprietary. Open protocols based on standards allow you to interact with different manufacturers of network equipment among themselves, increasing the choice for the customer.
- API applications Programming. The concept of SDN allows converting the design and configuration of the network into the programming task. Accordingly, the network management process, which includes the creation of routes, is the network programming as a whole. Until now, the configuration of network devices was mainly through the command line or the Web interface. But these tools are limited by the shell of the programming offered by the manufacturer.
- Equipment capabilities expansion. Due to the correct flow tables' setup, the same OpenFlow switch operates simultaneously as a traditional switch, like a router and as a firewall.
- Routing in SDN. Increasing the transmission speed. SDN simplifies the process of creating routes. In modern networks, the working topology is determined jointly by all devices, in the SDN - it implements a network simulation program with given parameters.
- The separation of control levels (Control Plane) and data transmission eliminates switches and routers from a large part of the calculated load. All that they need is to send packets from one port to another as quickly as possible, according to the routing table provided by the network controller. The controller instead of routing for each packet, as provided in the traditional network, decides once and then passes all the same type of the packets flow to route ready until the network status changes or the nature of the traffic. Instead of calculating the route for each packet, the principle of data streams is used, the routing of which does not need to make changes. This means that they will be processed at the highest possible speed. There is also an increase in the channel utilization rate due to more rational use of equipment.
- Virtual SDN. When network virtualization refers to the isolation of network traffic - grouping (multiplexing) multiple data streams with different characteristics within the same logical network that can share a single physical network with other networks or logical network cuts (network slices). Each such cut can use its addressing, its routing algorithms, quality service management, etc. On the basis of the physical network, we can construct several logical networks that will be isolated from each other. This allows to conduct experiments on the network without interrupting the work of the main network. Due to this, we can test new network settings before putting them into operation.
- Security features. SDN technology has some security issues that will be described below. Nevertheless, it allows to implement new security services that will work at the controller level and will quickly change the network configuration in case of external attacks or internal security policy violations. One example of security products is the OpenFlowSec.org open source project for designing solutions that allow to implement some security services while using SDN. Similar security features can be implemented or for other controllers, or be portrayed on them (opennetworking.org/, 2014). On the one hand, the SDN opens up new targets for attackers and abuses, and on the other hand provides new opportunities for creating information security services.
- Dynamic reconfiguration. Consider, for example, a network reconfiguration in the case of a "drop" of the channel between some nodes in the network. In the traditional model, the network will move to a new state because the router will tell its neighbors about this event. Routers will be engaged

in the development of new routes, and information about the new topology will be distributed to neighboring routers. It will just take a while. In case of the control center use, the calculation of the new topology is performed on the basis of knowledge of the entire network. We can also specify the required topology of the next state. Since the creation of a new topology is a computational process, that can be executed much faster.

DISADVANTAGES OF SDN CONCEPT

The SDN architecture also has its drawbacks. For effective SDN network implementation, it is necessary to understand its weaknesses in order to develop methods for their elimination. The study of the main drawbacks of the concept is given below.

1. Problems related to the drawbacks of technology (Korgov, 2013):
    a. Stopping the network when the controller fails.
    b. Error ability in application programming.
    c. Delay when the controller receives information about downtime or channel loading.
    d. The complexity of network building.
2. Problems related to the novelty of the technology.

Out of the controller management. The most obvious drawback of SDN stems from its centrality. When the controller fails, the whole network will stop. This may be caused by external interference (since the controller will be the most vulnerable point in the network and can be the target of targeted attacks) or internal failures (physical damage, lack of equipment).

This problem can be solved fairly easily:

1. Backup the controller. The spare controller must continuously work in the "idle" mode for the possibility of replacing the main controller. Another option is if the functionality is shared between the controllers. If the controllers are geographically spaced, each one will be responsible for the part of the network, the delay to which is less. At the same time, each controller must be prepared to accept all functionality in case of failure.
2. The building of the hybrid network "SDN + Traditional Network". Then, in the event of a failure of the central control element, the switches go into the offline mode of calculating the routes and begin to work on the old technology - though slower, but without breaking the connection.
3. Using new security software that protects the controller from external attacks. Examples of such tools are given above.
4. Sometimes a combination of these methods is used.

Error in programming API applications. Error in the programming of the controller can cause serious problems throughout the network it serves. Example: Several simultaneously running programs will compete for network resources, with a high load, there may be a situation where one program monopolizes them, which will cause other applications to stop functioning.

Possible solutions:

- Use only verified applications and their combinations.
- Conduct testing of new applications on a virtual network before they are put into operation.

Delay when the controller receives information about the failure or load of the channel. The IP protocol is aimed at ensuring that routers themselves monitor the download and performance of the device-linked channels, and in the SDN, the state of the network follows the controller, which receives data on the failure of the channels and their load "from the side", that is, there is a delay in receiving information and making a decision. This does not help to increase reliability.

The complexity of building a network that has more than one owner due to the inability to divide the functions of one controller among themselves. Two operators will not be able to share the functions of one controller among themselves, so a reliable software-configurable network can only be built if the entire network has one owner. Note, that complexity occurs only if one logical controller is used.

The possible solution is the virtualization and isolation of the traffic of two VLANs. Such a network, however, is more complex for management and debugging, due to the need to apply algorithms for such a network resource distribution that meets the needs of both ISP operators.

- Non-technological problems. The need to rebuild existing networks.
- Necessity of re-training of network administrators.
- A small number of "pure" OpenFlow switches on the market. For the most part, OpenFlow is presented in network equipment only as an additional functionality. This leads to an increase in equipment, and therefore, the introduction of a software-configurable network requires more money.
- A small number of programs for an open source software controller.

All these problems will be removing over time, although companies with high financial capabilities and skilled specialists can solve them now.

SDN IN AVIATION USAGE PERSPECTIVES

In order to assess the future of technology development, it is possible to track on its rapid progress over the past years.

The first commercial project in the area of SDN executed in 2007 by the company Nicira (Semenov, 2014). Nicira has developed its own Networking Virtualization Platform (NVP). As a result of several investment rounds, in July 2012 the company purchased VMware for $ 1.26 billion, which marked the beginning of the formation of the SDN solutions market (Semenov, 2014, Wanderer, 2014).

It should be noted that a number of manufacturers already ready to sell their own solutions in the field of SDN. For example, Cisco Systems, in addition to launching Nexus and Catalyst 35XX switches, able to work on traditional networks and SDNs, announced the Open Network Environment (ONE) platform specifically designed to support SDN solutions. In addition, the company announced the development of a pilot version of the software for controllers, as well as a pilot version of the agent OpenFlow to collect information about the work of network infrastructures SDN. Juniper Networks added the OpenFlow option to the JunOS SDK operating system, and in June announced the implementation of this technology in the series of EX and MX series switches. NEC, Pronto and Marvell offer switches that implement only the OpenFlow protocol, and IBM has released the IBM System Networking Programmable Network Controller as a software application on an OpenFlow-based Linux platform. HP implements the HP Virtual Application Networks strategy, which includes the release of controller, applications, and SDN-based services and solutions, while Brocade introduced the first SDN-enabled products, including the Brocade VDX 8770 switchboard. The Intel team, which demonstrated on IDF is a solution for the SDN switch and OpenFlow based Linux software.

In April 2012, Urs Holz, senior vice president of Google's technical infrastructure, said that the company had translated the entire G-Scale internal network for the exchange of traffic between Google's data centers around the world on SDN, by making OpenFlow switches on its own, since existing analogs on the market were. At that moment, the company is not available. Google OpenFlow switches can scale up to hundreds of 10-Gigabit Ethernet ports (Wanderer, 2014).

Already SDN was gaining momentum in networks with large volumes of traffic. Different companies may be interested in SDN due to the benefits they provide. Let's list the main areas where SDN is currently used:

- Data centers. Data centers are interested in increasing the speed of routing, which is especially noticeable for large volumes of traffic (opennetworking.org, 2014). This is achieved by streaming routes rather than packets. Configuring with the API allows for flexibility in managing network resources.
- "Cloud" data centers.Cloud services are interested in creating scalable shared-use systems, automatic resource allocation, which is easier to implement with SDN.
- Corporate networks. SDN technology helps to realize virtual workplaces for large corporations, to automate the work of a private "cloud".
- ISP Providers. The interest of providers is a flexible provision of services, easier centralized management and policy-based analytics to optimize and monetize the services provided.
- Scientific institutions. Such organizations are interested in the possibility of conducting experiments over existing networks, without damaging their work.

MANAGEMENT LEVEL CREATING APPROACHES

The controller, as a centralized control, encompasses many functions (S. Schmid and J. Suomela, 2014). The main ones are: management of network elements, resources and applications; construction of the network topology and continuous monitoring of its conditions; data transmission. Initially, the management-level architecture provided for the presence of only one controller (Figure 3).

But with the growth of the network, there was a need to make changes to the existing architecture. Increasing the management level scalability has been achieved through the introduction of additional controllers, as well as changes in the management level structure - the formation of decentralized and hierarchical management level structures. The solution is to develop a distributed SDN architecture such as Hyperflow (Ganjali, 2010), Onix (Koponen et all, 2010), and Kandoo (Yeganeh, 2012).

Hyperflow is a decentralized structure, where all controllers have equal rights and functions. Hyperflow provides scalability, while the network remains logically centralized: each Hyperflow controller has information about the entire network and serves the needs of switches that are in close proximity, which minimizes the time to create a new thread.. In addition, Hyperflow does not require any changes to the OpenFlow standard and requires minor changes to existing applications. Hyperflow guarantees data

Figure 3. Centralized architecture

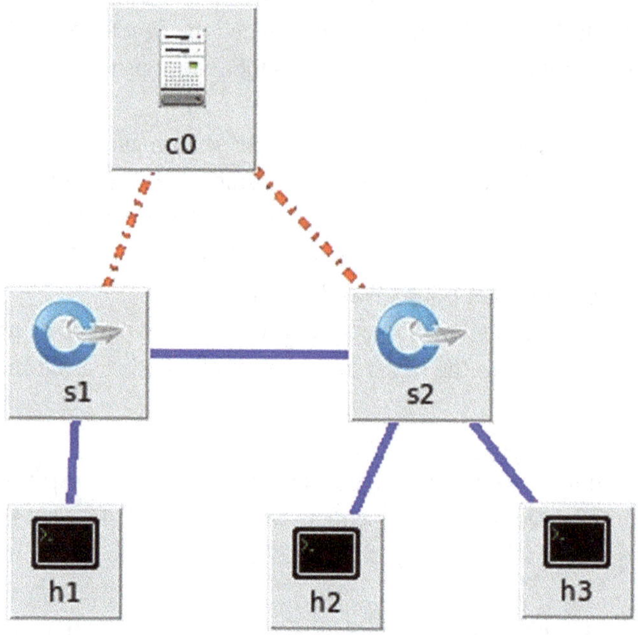

transfer without loops. Hyperflow is one of the first distributed control planes for OpenFlow. A similar design is used in FlowVisor, which has other purposes. FlowVisor allows you to use multiple controllers on the network, cutting the network resources and deleting controls of each cut to one controller (Fig. 4).

An alternative design is to save the state of the controller in a distributed data warehouse and enable the local cache on individual controllers. For example, if a solution (for example, a path setting) can be made for many threads after accessing a local cache, inevitably some streams require removal from remote controllers, resulting in increased maintenance time. In addition, this design requires modification of applications to maintain a state in a distributed data warehouse. In contrast, Hyperflow actively announces the change of status of all other controllers, thereby providing controllers with up-to-date information for locally servicing all threads.

Onyx is a decentralized network structure that is distributed among multiple controllers. Each controller has information about its local area network (Fig. 5).

There are four components on the network controlled by Onix, and they have different roles:

- Physical infrastructure. It includes a network, which consists of switches and routers, as well as any other networks that supports an interface that allows Onix to view and memorize how to manage the behavior of an item. These network elements do not require any software other than those recommended to support the operation of this equipment.
- The connection between the physical network equipment and Onix goes into the communication infrastructure. The communication infrastructure should support two-way communication between Onix instances and switches.
- Onyx is a distributed system that works in a cluster from one or more physical servers. Onyx provides logic for managing software access to the network.

Figure 4. Decentralized architecture (global)

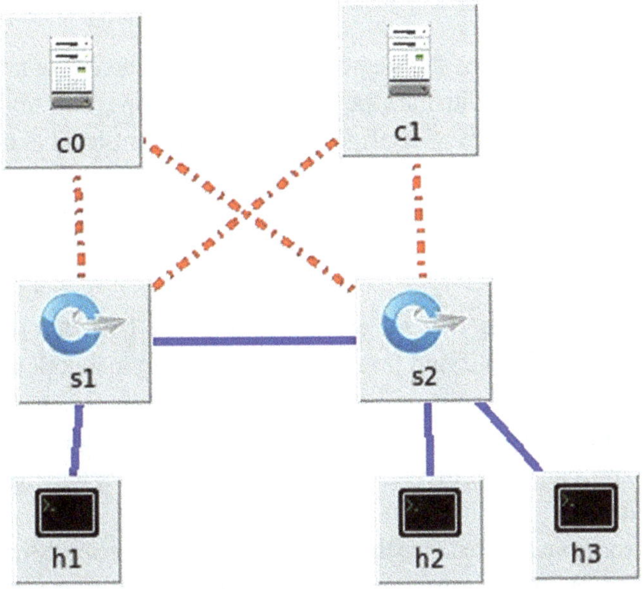

Figure 5. Decentralized architecture (local)

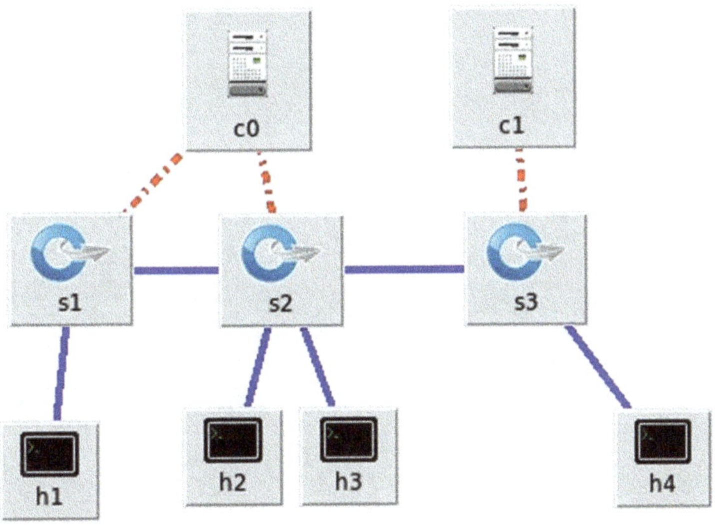

- The logic of network management is implemented at the top of the Onix API. This control logic determines the desired behavior of the network.

Kandoo is a hierarchical structure (Fig. 6). There are two classes of controllers: the root controller and the local controller. Local controllers have limited functionality and do not contain information about the entire network topology. A local controller manages one or more switching speakers' proxy switches for root controllers. And the root controller has information about the whole network. The root controller manages all local controllers and can run applications that require network access. Local controllers can be implemented directly in OpenFlow switches. Local controllers can be linearly scaled with the number of switches.

Kandoo does not apply applications that require knowledge of the entire network state. The implementation of Kandoo fully complies with the OpenFlow specifications. The advantage of Kandoo is that it gives network operators the freedom to configure the management level and its functionality, based on the management application characteristics.

SDN Efficiency Estimation

Designing a network is a challenging task for every operator. The first step is to understand common networking requirements. After identifying these requirements, key network characteristics can be selected that meet these requirements.

Networking devices must reflect the goals, characteristics, and policies of the service provider in which they operate. Two primary goals drive networking design and implementation: Application availability and Cost of ownership.

A well-designed network can help balance these objectives. When properly implemented, the network infrastructure can optimize application availability and allow the cost-effective use of existing network resources.

So, starting to design or optimize the network, or, for example, choosing the best route, main requirements have to be known for the new applications, which we want to implement.

Figure 6. Hierarhical structure

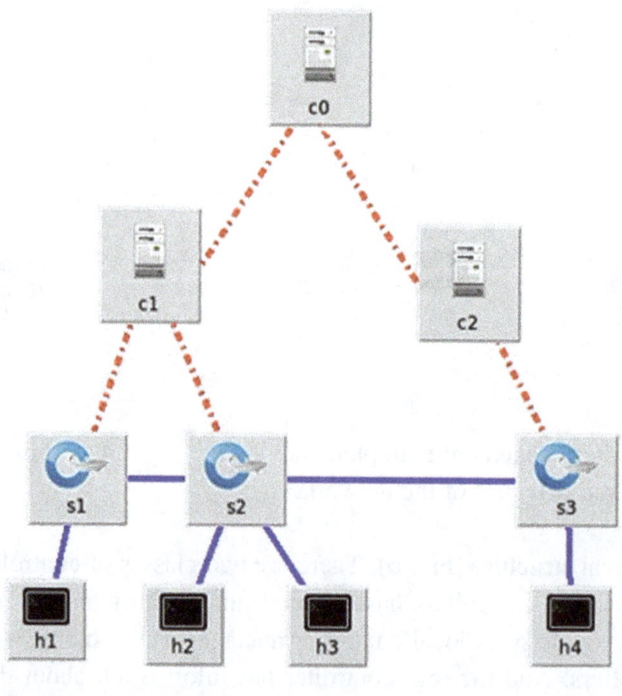

Figure 7. New application implementation process

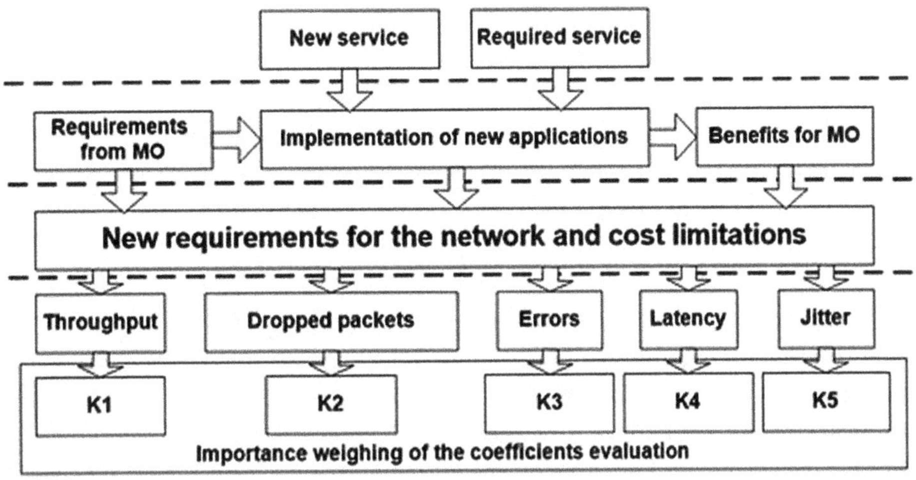

So, as shown in Figure 7, implementing the new service, network provider has to check main characteristics of the network for compliance with the minimum threshold for each implemented application. Some characteristics will be more valuable and disparity of some is not critical for application. Thus, the weights K1 – K5 will be formed, that will be used in future calculations of the total efficiency.

As part of the task of SDN networks effective functioning first must be assessed such characteristics as performance, latency and scalability. These characteristics depend on the number of managed switches and their connectivity with the controller, the intensity of receiving requests, as well as the time query processing of the controller. Thus, the efficiency of the SDN networks depends on the controller adapt to the increasing intensity of coming requests from switches and capabilities to ensure the quality of service by increasing the scale of the network. Thus, network scalability evaluation, based on the concept of SDN, delay and productivity, is an important task in the design of new or expansion of existing network architecture (tutorialspoint.com, 2014). So, the main task of this work is SDN network efficiency evaluation method development, which will allow to calculate main characteristics of SDN (latency, performance, scalability) for each type of the network in different cases.

Determining the efficiency of entire SDN networks, or parts thereof, for a particular application will be kept to the definition of complex criteria using the analytic hierarchy process for each network operator. First, as shown in Fig. 8, will be conducted definition of priorities for different criteria.

The figure shows the hierarchy in which the default priorities of elements are considered equal, that all four criteria are of equal importance from the standpoint of goal, and priorities of all alternatives are equal in all criteria. In other words, in this example alternatives are indistinguishable. Thus, the amount of elements priority at any level, is one.

Global alternatives priorities regarding the aim are computed by multiplying the local priority of each alternative on each criterion priority and summation for all criteria.

Acceptation of priorities solutions can be either based on objective data (including optimization methods and probabilistic and statistic models) or based on the views of specialist (experts). In tasks of feasibility analysis always uses a variety of expert estimate methods.

Figure 8. The hierarchical structure with priorities

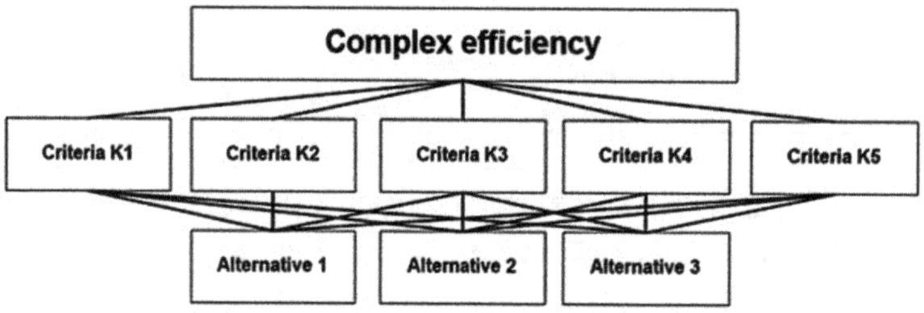

To assess the complex effectiveness, the use of following simpler model is proposed (Figure 9).

In this model (Figure 9) complex efficiency is defined as the weighted sum of scalability, network delays and its performance. Thus, further will develop mathematical tools and conduct, where it's necessary, modeling of SDN networks to determine the complex efficiency.

To compare the quality of data transmission the following parameters are using: bandwidth, packet transmission delay, the jitter and packet loss rate. The delay is the most important indicator of quality, so it is useful to compare the average packet delay in SDN/OpenFlow, SDN/Overlay and IP networks.

To evaluate the average packet delay in the Overlay SDN network use of environment was made for modeling SDN networks – Mininet, and graphic editor for it – MiniEdit. The Mininet is a program that provides emulation for SDN network operation, allows to create a realistic virtual network and conducts its setting and research via command line.

Let us calculate how long delivery of packets takes for each type of network. IP, SDN/OpenFlow, SDN/Overlay network topologies are shown in Figure 10.

In IP-networks, general packet delivery time is the amount of delay per channel and processing time in each switch:

$t_{IP}(p) = m \bullet t_{link} + n \bullet p \bullet t_{switch}$ – for IP-switches.

In SDN/OpenFlow-networks to calculate the total time of packet delivery to the sum of delay per channel and processing time in each switch is also added delivery time of information about the first packet to controller and vice versa, as well as deciding the time of the controller:

$t_{SDN\_OFS}(p) = m \bullet t_{link} + n \bullet p \bullet t_{OFS} + 2 \bullet t_{link} + t_{ctrl}$ – for OpenFlow switches.

Figure 9. The estimate model of SDN networks complex efficiency

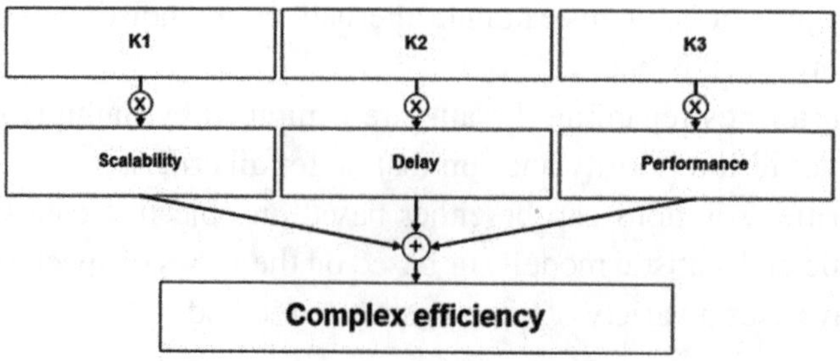

Figure 10. Researched network topologies:a – simplified IP network topology; b – simplified SDN/ OpenFlow network topology; c – simplified SDN/Overlay network topology

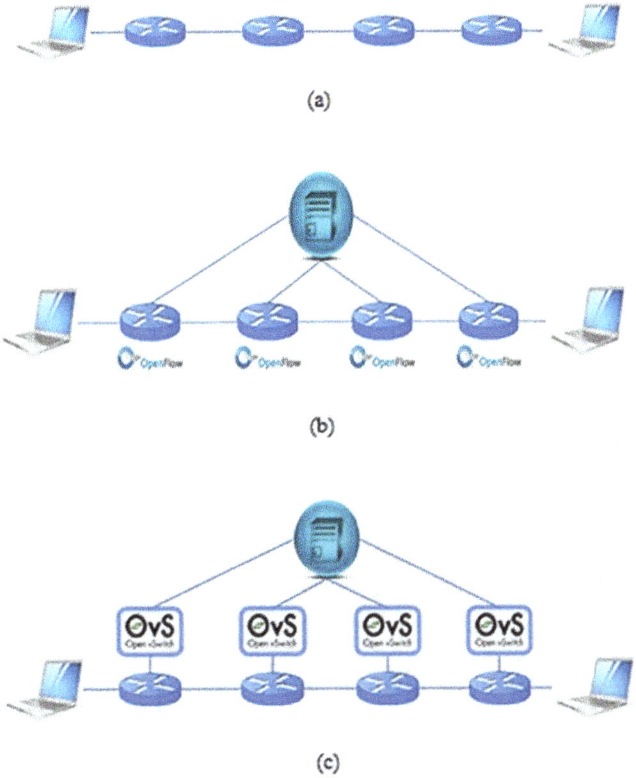

In SDN/Overlay-networks, the total delivery time of packets is calculated so as for SDN/OpenFlow networks, but instead of OpenFlow-switch processing time is using software Open vSwitch processing time, which takes into account packet delivery time from the physical to the virtual switch and vice versa:

$t_{SDN\_OVS}(p) = m \bullet t_{link} + n \bullet p \bullet t_{OVS} + 2 \bullet t_{link} + t_{ctrl}$ – for Open vSwitch switches

Total packet delivers time should be divided by the number of packets to calculate the average packet delay:

$t_{average\_IP}(p) = (m \bullet t_{link} + n \bullet p \bullet t_{switch})/p$ – for IP-switches;

$t_{average\_SDN\_OFS}(p) = (m \bullet t_{link} + n \bullet p \bullet t_{OFS} + 2 \bullet t_{link} + t_{ctrl})/p$ – for OpenFlow switches;

$t_{average\_SDN\_OVS}(p) = (m \bullet t_{link} + n \bullet p \bullet t_{OVS} + 2 \bullet t_{link} + t_{ctrl})/p$ – for Open vSwitch switches.

Next step is to construct dependence of the total packet delivery time and average delay against the number of packets in the data stream. Graphs of dependence for the three types of networks shown in Figures 11 and 12. Graphs are built for the number of packets from 1 to 1000. The red solid line – for

Figure 11. Total time of data transmission against the number of packets (when n = 4)

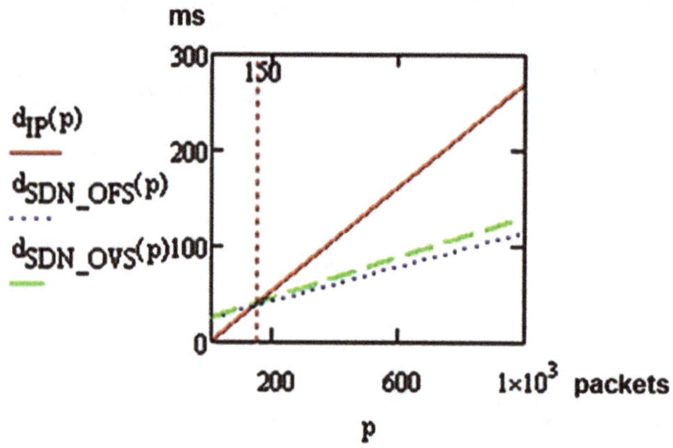

Figure 12. The average packet delay against the number of packets (when n = 4)

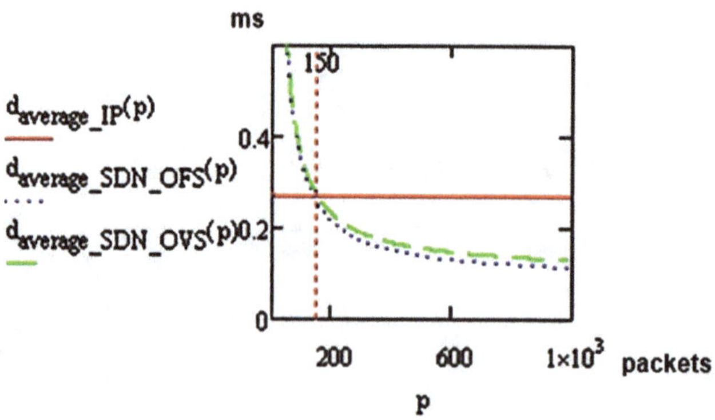

IP-network, the green dashed line – for SDN/Overlay network, lines with blue dots – for SDN/OpenFlow network.

From the dependence graphs, we can conclude that the average delay in SDN network decreases with increasing number of packets in the stream. The number of packets in the flow is not affected on average delay in traditional networks. Thus, the hypothesis is confirmed that SDN is efficient to use in networks with a large volume of traffic. Thus, the network with OpenFlow-switches are more efficient than a network built on Overlay.

Then calculate how delay changes in the network depending on the number of network equipment (switches). For the purity calculations, we will take the number of packets when the average delay is about the same for all networks. As seen from the graphs, the lines intersect roughly at the mark of 150 packets. So, fix the number of packets $p = 150$.

Total time of data transmission depending on the number of switches expressed by features:

$t_{IP}(n) = (n+1) \cdot t_{link} + n \cdot p \cdot t_{switch}$ – for IP-switches;

$t_{SDN\_OFS}(n) = (n+1) \cdot t_{link} + n \cdot p \cdot t_{OFS} + 2 \cdot t_{link} + t_{ctrl}$ – for OpenFlow switches;

$t_{SDN\_OVS}(n) = (n+1) \bullet t_{link} + n \bullet p \bullet t_{OVS} + 2 \bullet t_{link} + t_{ctrl}$ – for Open vSwitch switches.

The average packet delay depending on the number of switches expressed by features:

$t_{average\_IP}(n) = ((n+1) \bullet_{tlink} + n \bullet p \bullet t_{switch})/p$ – for IP-switches;

$t_{average\_SDN\_OFS}(n) = ((n+1) \bullet t_{link} + n \bullet p \bullet t_{OFS} + 2 \bullet t_{link} + t_{ctrl})/p$ – for OpenFlow switches;

$t_{average\_SDN\_OVS}(n) = ((n+1) \bullet t_{link} + n \bullet p \bullet t_{OVS} + 2 \bullet t_{link} + t_{ctrl})/p$ – for Open vSwitch switches.

Construct dependence of the total packet delivery time and the average delay against the number of switches in the network. The dependence graphs for the three types of networks are shown in Figures 13 and14. Graphs are built for the number of switches from 1 to 16.

From the above graphs, we conclude that the average delay in SDN network increases with the number of switches slower than in a traditional network. At 16 switches, the gain in delay is more than in 2 times. So, SDN efficient to use in networks with a large number of network equipment. Network with OpenFlow-switches are more efficient than network built on Overlay.

At the management level of SDN performance F(N) can be defined as follows (Yeganeh, A. Tootoonchian, and Y. Ganjali, 2013):

$$F(N) = \varphi(N) \cdot \frac{T(N)}{C(N)}$$

where

N: The number of network nodes;
φ(N): The capacity of management level in network requests processing;
T(N): The average response time of each request;
C(N): The cost of management level deployment.

Figure 13. Total time of data transmission against the number of switches (p = 150)

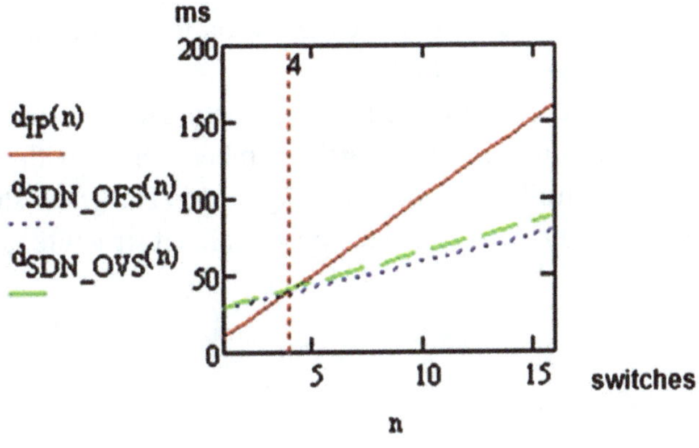

Figure 14. The average packet delay against the number of switches (p = 150)

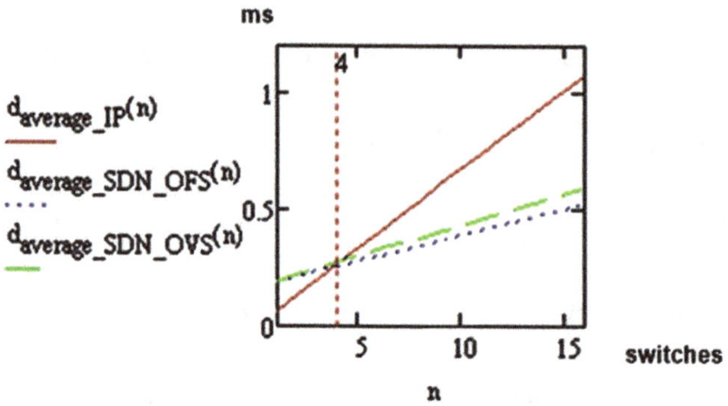

Scalability in SDN management level, when the size of the network varies from N_2 to N_1, is defined as (Yeganeh, A. Tootoonchian, and Y. Ganjali, 2013):

$$\psi(N_2, N_1) = \frac{F(N_2)}{F(N_1)}$$

where

$F(N_2)$: Management level performance in processing requests from N_2 network elements;
$F(N_1)$: Management level performance in processing requests from N_1 network elements.

Evaluation of SDN network scalability appropriate to carry out for the three main architectures. Depending on controllers' connectivity management level structure can be roughly divided into three types: centralized, decentralized and hierarchical:

Scalability of management level depends on the type of structure that will be used. For different types of structures are given following dependencies for the average response time upon request (Table 1) (Yeganeh, A. Tootoonchian, and Y. Ganjali, 2013; Schmid and J. Suomela, 2013):

Based on the above presented formulas we calculated scalability for SDN management level (Table 2) (Yeganeh, A. Tootoonchian, and Y. Ganjali, 2013; Schmid and J. Suomela, 2013).

According to the formulas from table 2, it is possible to construct graphs of scalability of different types of management levels structures and compare their scalability. The local decentralized structure and hierarchical structure have the best scalability.

The scalability of the centralized structure, the global decentralized structure, the local decentralized structure and the hierarchical structure, depending on the number of hosts with the number of controllers 6 and the mean distance to the controller 2, is shown in Figure 15. The graph shows that the centralized architecture has the worst scalability, while the local decentralized structure and hierarchical structure are the best.

Table 1. Average response time upon request in SDN network

SDN Network Structure	Average Response Time Upon Request
Centralized structure	$E\left(T_c\left(N\right)\right)=\frac{1}{\mu_c-\lambda_c}$
Decentralized structure	$E\left(T_{D,l}\right)=\frac{L_{D,l}\cdot m_D}{N\cdot\left(N-1\right)\cdot\lambda}=\frac{1+\frac{N\cdot\left(m_D-1\right)}{\left(N-1\right)\cdot m_D}\cdot d_D}{\mu_{D,l}-\lambda_{D,l}}$
Hierarchical structure	$E\left(T_H\right)=\frac{\frac{N-\frac{N}{m_H}}{N-1}}{\mu_{H,r}-\lambda_{H,r}}+\frac{\ln\left(\frac{N-\frac{N}{m_H}}{N-1}\cdot d_H+1\right)+1}{\mu_{H,l}-\lambda_{H,l}}$

where

λ: The average number of flow requests per second from one host to another;

μ_c : The average receipt rate of the flow request;

λ_c : The average processing rate of flow requests by controller;

N: The number of network nodes;

m_D : The number of controllers in a decentralized structure;

d_D : The average distance to the controller in a decentralized structure;

$L_{D,l}$: The queue length for each controller in a decentralized structure;

$\lambda_{D,l}$: The receipt rate of the flow initiation request for each controller in a decentralized structure;

$\mu_{D,l}$: The average service rate of controller in a decentralized structure;

m_H : The number of controllers in a hierarchical structure;

d_H : The average distance to the controller in a hierarchical structure;

$\lambda_{H,r}$: The receipt rate of the flow initiation request for each controller in a hierarchical structure;

$\mu_{H,r}$: The average service rate of controller in a hierarchical structure.

SDN Modeling

In this section, we will conduct experiments to explore different approaches for building an SDN management layer architectures, set up an HTTP server on one of the hosts, and connect from another host to the server. Also we will investigate the messages exchanged by hosts on the network.

- **Centralized Management Level Structure:** In the MiniEdit software environment (mininet.org, 2014) SDN simulation model has been developed. The simulation model consists of a controller (c0), three switches (s1, s2, s3), seven hosts (h1, h2, h3, h4, h5, h6, h7). The first four hosts are

Table 2. Scalability for SDN management level

SDN Network Structure	Scalability for SDN Management Level
Centralized structure	$\psi_c\left(N_1, N_2\right) \approx \dfrac{K - N_2^4 \cdot \lambda}{K - N_1^4 \cdot \lambda}$
Global decentralized structure	$\psi_{D,g}\left(N_1, N_2\right) \approx \dfrac{K \cdot m_D - N_2^4 \cdot \lambda}{K \cdot m_D - N_1^4 \cdot \lambda}$
Local decentralized structure	$\psi_{D,l}\left(N_1, N_2\right) \approx \dfrac{K \cdot m_D^4 - N_2^4\left(d_D \cdot m_D - d_D + m_D\right)\lambda}{K \cdot m_D^4 - N_1^4 \cdot \lambda\left(d_D \cdot m_D - d_D + m_D\right)\lambda}$
Hierarchical structure	$\psi_H\left(N_1, N_2\right) \approx \dfrac{K \cdot m_H^4 - N_2^4\left(d_H \cdot m_H - d_H + m_H\right)\lambda}{K \cdot m_H^4 - N_1^4\left(d_H \cdot m_H - d_H + m_H\right)\lambda}$

Figure 15. SDN scalability fo different types of the networks
(m=6, d=2)

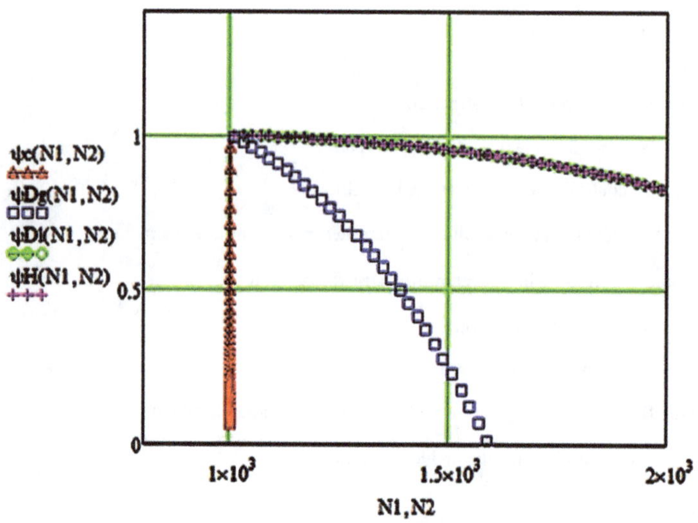

connected to the first switch (s1), the host h5 is connected to the second switch (s2), and the host h6, h7 is connected to the third switch (s3). The controller is connected to all switches, and the switches s2, s3 are connected to s1. The topology of the described network is presented in Fig. 16. To verify the network's performance, the connection between the nodes of the network was checked. After running the network, its performance was checked using the command "pingall", shown in Fig. 17.

Figure 16. Model of a centralized management level architecture

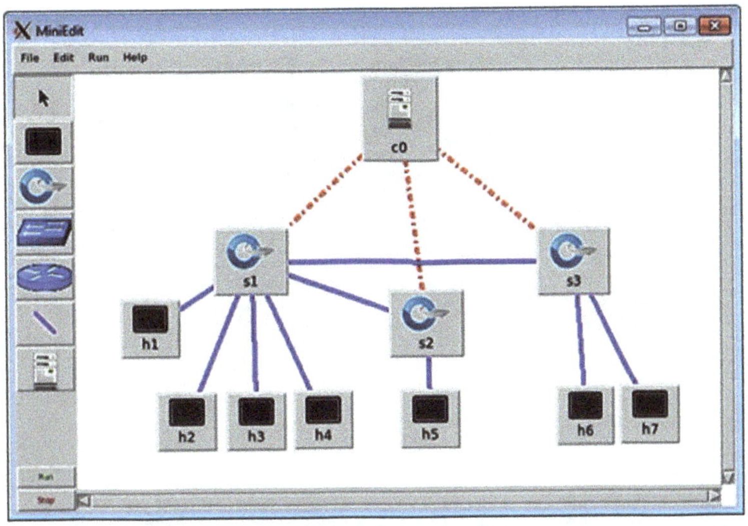

The pingall command was used to demonstrate the network's performance. The network works fine because all hosts are available and packets are not lost during transmission.

In the process of network operation, it is possible to stimulate the breakdown of the communication line, resulting in a loss of communication with the network node. An experiment was conducted when the switch s1 loses its connection with the host h4. As a result, the h4 host is not available to other hosts, ICMP packets do not arrive. The ping all command was used to demonstrate the breakdown.

The speed of data transmission on this network was also studied using the "ping" command. The first packet is transmitted the longest, because at first the switch does not have a record in the flow table. When the first package arrives, the switch sends it to the controller. The controller makes a decision and instructs all switches to add information about this path to the flowchart. After that, the switch will not ask the controller how to handle such packets, but will be forwarded in accordance with its flowchart. If a package arrives, about which the switch does not have information, then the switch will send the request again to the controller.

Let's check how the transfer rate will change with the increase in the number of hosts.

Figure 17. Ping all

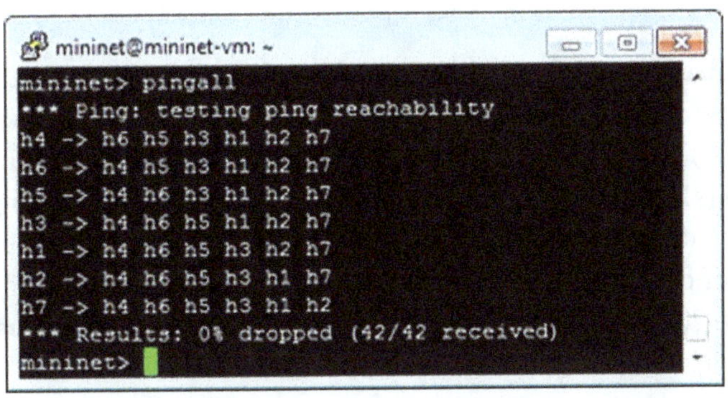

Figure 18. Ping all with breakdown

```
mininet@mininet-vm: ~
mininet> pingall
*** Ping: testing ping reachability
h4 -> X X X X X X
h6 -> X h5 h3 h1 h2 h7
h5 -> X h6 h3 h1 h2 h7
h3 -> X h6 h5 h1 h2 h7
h1 -> X h6 h5 h3 h2 h7
h2 -> X h6 h5 h3 h1 h7
h7 -> X h6 h5 h3 h1 h2
*** Results: 28% dropped (30/42 received)
mininet>
```

Figure 19. Ping host h7 from host h2

```
mininet@mininet-vm: ~
mininet> h2 ping h7
PING 10.0.0.7 (10.0.0.7) 56(84) bytes of data.
64 bytes from 10.0.0.7: icmp_seq=1 ttl=64 time=5.31 ms
64 bytes from 10.0.0.7: icmp_seq=2 ttl=64 time=0.241 ms
64 bytes from 10.0.0.7: icmp_seq=3 ttl=64 time=0.056 ms
64 bytes from 10.0.0.7: icmp_seq=4 ttl=64 time=0.049 ms
64 bytes from 10.0.0.7: icmp_seq=5 ttl=64 time=0.052 ms
64 bytes from 10.0.0.7: icmp_seq=6 ttl=64 time=0.045 ms
64 bytes from 10.0.0.7: icmp_seq=7 ttl=64 time=0.240 ms
64 bytes from 10.0.0.7: icmp_seq=8 ttl=64 time=0.054 ms
64 bytes from 10.0.0.7: icmp_seq=9 ttl=64 time=0.051 ms
64 bytes from 10.0.0.7: icmp_seq=10 ttl=64 time=0.056 ms
^C
--- 10.0.0.7 ping statistics ---
10 packets transmitted, 10 received, 0% packet loss, time 9001ms
rtt min/avg/max/mdev = 0.045/0.616/5.318/1.569 ms
mininet>
```

Leave the net as on rice, but we will increase the number of hosts to 34.

With the increase in the number of nodes, the scalability of the network is deteriorating. This is evident from the processing time of the request, the processing time of the query is larger than the network with 34 hosts than with seven hosts (Fig. 19). But with a small number of hosts, this is not so noticeable, because the scalability worsens within a small margin. With a large number of hosts, the network may stop functioning at all.

The data transfer rate is 66.9 Mb / s.

- **Global Decentralized Structure:** In the software environment MiniEdit has been developed simulation model of the network SDN. The simulation model represents two controllers (*c0, c1*), three switches (*s1, s2, s3*), seven hosts (*h1, h2, h3, h4, h5, h6, h7*). The first three hosts are connected to the first switch (s1), the hosts h4, h5, and the third switch (s3) is hosting h6, h7 to the second switch (s2). The controllers are connected with all switches by service channels, and switches s1, s2, s3 - channels of data transmission among themselves. The topology of the described network is shown in Figure 20. To verify the network's performance, the connection between the nodes of the network was checked.

Figure 20. Global decentralized structure

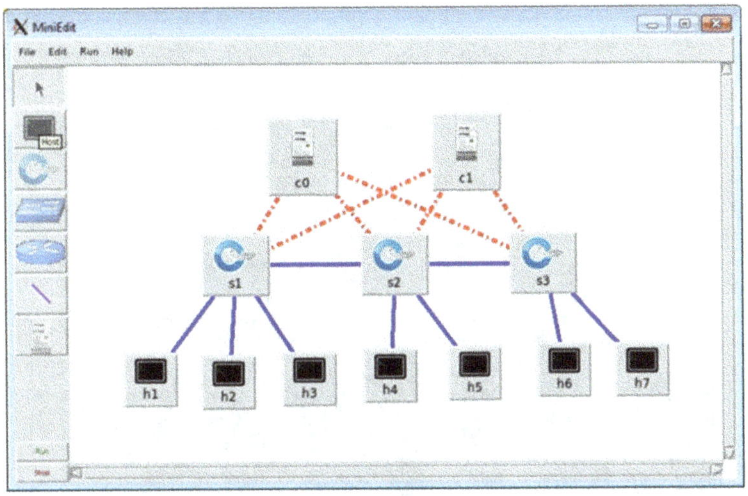

The speed of data transmission in this network was investigated using the command "ping". The first packet is transmitted the longest, because at first the switch does not have a record in the flow table. When the first packet arrives, the switch sends it to the controller. The controller makes a decision and instructs all switches to add information about this path to the flowchart. After that, the switch will not ask the controller how to handle such packets, but will be forwarded in accordance with its flowchart. If a package arrives, about which the switch does not have information, then the switch will send the request again to the controller.

We will conduct a study on the bandwidth of a global decentralized management level structure. To do this, we'll create a simple HTTP server based on the h5 host.

From the host h2, we will try to access it and download the data that is there using the h2 wget -O-h5 command. This is shown in Fig. 22.

After sending a request from the h2 server, the server responds to it. When the ACK response is received, the host h2 receives access to the required data on the server. The data transfer rate is 137 MB / s, which is higher than the centralized architecture. This shows that the global decentralized approach for building a network architecture is more advantageous compared to a centralized one.

- **Local Decentralized Management Level Structure:** In the software environment MiniEdit has been developed simulation model of the network SDN. The simulation model represents two controllers (c0, c1), three switches (s1, s2, s3), seven hosts (h1, h2, h3, h4, h5, h6, h7). The first three hosts are connected to the first switch (s1), the hosts h4, h5, and the third switch (s3) is hosting h6,

Figure 21. Creating a server

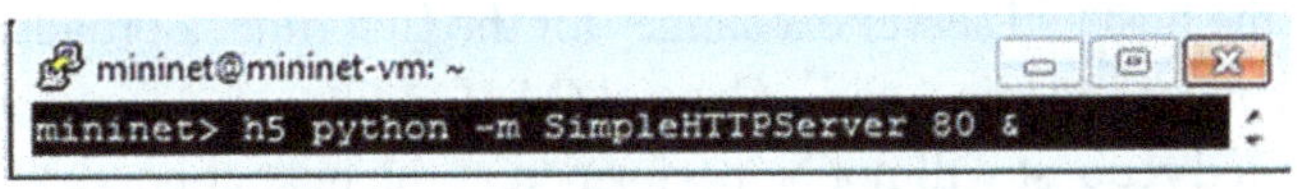

Figure 22. Server data

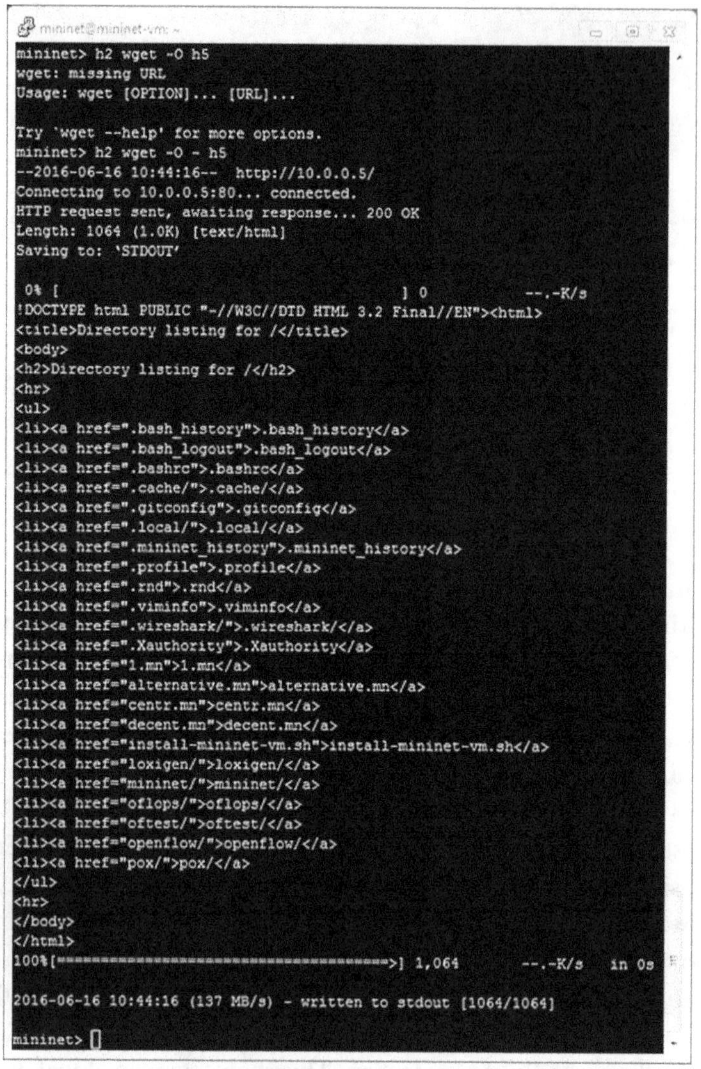

```
mininet> h2 wget -O h5
wget: missing URL
Usage: wget [OPTION]... [URL]...

Try `wget --help' for more options.
mininet> h2 wget -O - h5
--2016-06-16 10:44:16--  http://10.0.0.5/
Connecting to 10.0.0.5:80... connected.
HTTP request sent, awaiting response... 200 OK
Length: 1064 (1.0K) [text/html]
Saving to: 'STDOUT'

 0% [                                    ] 0           --.-K/s
!DOCTYPE html PUBLIC "-//W3C//DTD HTML 3.2 Final//EN"><html>
<title>Directory listing for /</title>
<body>
<h2>Directory listing for /</h2>
<hr>
<ul>
<li><a href=".bash_history">.bash_history</a>
<li><a href=".bash_logout">.bash_logout</a>
<li><a href=".bashrc">.bashrc</a>
<li><a href=".cache/">.cache/</a>
<li><a href=".gitconfig">.gitconfig</a>
<li><a href=".local/">.local/</a>
<li><a href=".mininet_history">.mininet_history</a>
<li><a href=".profile">.profile</a>
<li><a href=".rnd">.rnd</a>
<li><a href=".viminfo">.viminfo</a>
<li><a href=".wireshark/">.wireshark/</a>
<li><a href=".Xauthority">.Xauthority</a>
<li><a href="1.mn">1.mn</a>
<li><a href="alternative.mn">alternative.mn</a>
<li><a href="centr.mn">centr.mn</a>
<li><a href="decent.mn">decent.mn</a>
<li><a href="install-mininet-vm.sh">install-mininet-vm.sh</a>
<li><a href="loxigen/">loxigen/</a>
<li><a href="mininet/">mininet/</a>
<li><a href="oflops/">oflops/</a>
<li><a href="oftest/">oftest/</a>
<li><a href="openflow/">openflow/</a>
<li><a href="pox/">pox/</a>
</ul>
<hr>
</body>
</html>
100%[===================================>] 1,064       --.-K/s   in 0s

2016-06-16 10:44:16 (137 MB/s) - written to stdout [1064/1064]

mininet>
```

h7 to the second switch (s2). The controller c0 is connected to the switch s1, and the controller c1 is from s2, s3 by service channels. Switches s1, s2, s3 are connected by channels of data transmission among themselves.

Data was also intercepted with WireShark when the host h2 hits the server on the host h1. An analysis of the data transfer process between the server and the host is performed. First, the installation of a communication session is performed by triple-squeezing, then the host requests the server for data exchange, data transmission takes place. After the data has been transferred, the communication session ends with triple squeezing. When the host and server exchanges for the first time, a broadcast ARP request is sent to match the IP address and the Media Access Control (MAC address). After that, they are added to the ARP table and the repeated request will not be performed at the next connected.

Figure 23. Data transfer between host and a server

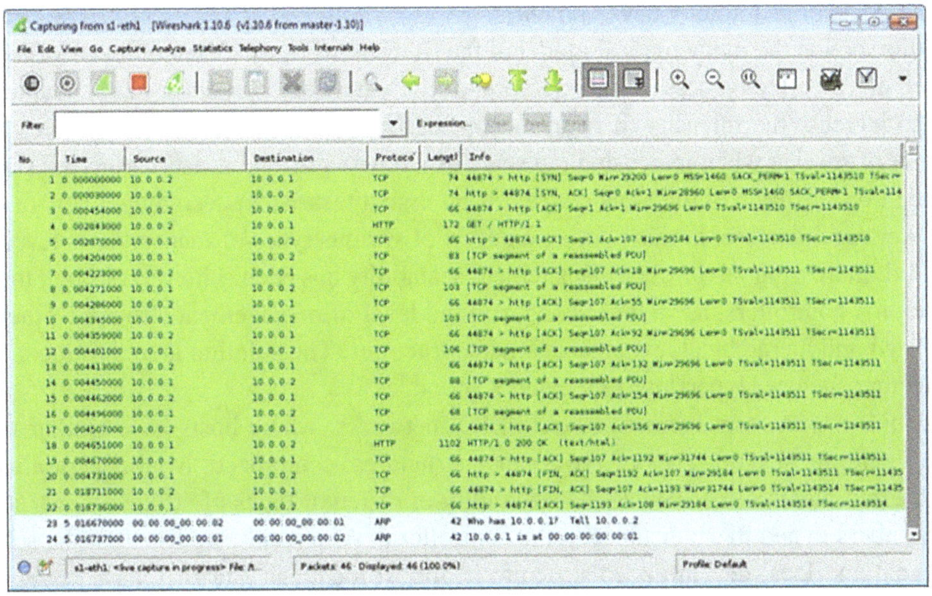

CONCLUSION

In this section, an analysis of the architecture and concept of SDN networks was conducted. There were differences from traditional networks, such as: separation of the control plane and the data transmission plane, the presence of a centralized control element (controller), the transfer of the functionality from the network equipment to the controller, the ability to use software applications to manage network resources.

The analysis of the possibilities provided by this approach to the implementation of computer networks is carried out. Consequently, software-configurable networks allow the use of advanced functionality for managing computer networks, namely: centralized management of all network resources (both communication channels and network equipment); the ability to use software applications to automatically manage the configuration of network equipment, provide QoS, prioritize traffic, provide additional network security features, etc.; automatic reconfiguration of the network in case of the equipment failure or the failure of the communication channel. In addition, SDNs simplify and manage network equipment by standardizing the controller interaction protocol and physical infrastructure. In this case, the controller-controlled equipment may perform functions that are not inherent to modern switches (for example, firewall functions). Also, the SDN network optimizes routing for data flows and provides tools for easier virtualization of networks.

We also investigated the disadvantages of SDN networks. It is revealed that at present, such networks are characterized by the following disadvantages: stopping the network at the failure of the controller, the possibility of error in the programming of applications, delay in obtaining information controller on the failures or load capacity of the channel, the complexity of building a network that has more than one owner through the inability to divide the functions of one controller. Were suggested methods to combat all of the above disadvantages. Due to the novelty of the technology, SDN networks also face shortages of skilled administrators of such networks, a smaller range of equipment for SDN networks, and a small number of programs for open source software controllers.

The prospects of introducing SDN networks were also substantiated, based on technological as well as financial aspects of technology development.

Also in this section the mathematical model of the SDN network for different levels of management was considered. The effectiveness of the operation of SDN networks depends on the adaptation of the controllers to increase the intensity of requests coming from the switches, and the ability to provide proper quality of service when increasing the scale of the network. Thus, estimating the scalability of a network built on the basis of the SDN concept is an urgent task when designing a new or expanding existing network architecture. In this part, an analysis of various types of management level structures was conducted. Increasing the level of management scalability has been achieved through the introduction of additional controllers, as well as changes in the level of management structure - the formation of decentralized and hierarchical level management structures. The solution is to develop a distributed SDN architecture such as Hyperflow, Onix, and Kandoo.

The scalability of the management level in SDN networks, where hosts have similar traffic, was considered. There are several types of governance structures: centralized, local decentralized, global decentralized, and hierarchical. To test the scalability of different types of structures, we set different controller numbers m and mean distance to the controller d. As a result, we find that the scalability of the hierarchical and local decentralized structure is almost identical. But they have better scalability compared to a centralized structure and a global decentralized structure.

The scalability of the hierarchical structure and the decentralized local structure becomes worse when the average distance to the controller increases, and it improves with an increase in the number of controllers.

This is evident from the experiments conducted, since the local decentralized network has the best characteristics: the speed of data transmission and the speed of processing requests. With an increase in the number of hosts, a local decentralized network has better performance than other structures with the same number of hosts.

We also configured the HTTP server on one of the hosts and connected to it from another host. An analysis of the data transfer process between the server and the host is performed.

It is proposed to use the architecture with a distributed controller, since it has better reliability. Though such a network is scaling worse, it has a big advantage: if one or more controllers fail, the network will remain operational and will transmit data until the work of the controllers or communication channels is restored.

FUTURE RESEARCH DIRECTIONS

For the future it is planned to use SDN equipment for the experimental studies to ensure the effectiveness of this technology solution. Also it is planned to investigate network slicing for mobile 5G and aviation networks using modern SDN infrastructure.

REFERENCES

Chernyak, L. (2012). *SDN - from concept to market*. Open Systems.

Haleplidis, E. (2014). *For CES Applicability to SDN-enhanced NFV*. Proc. Euro. Wksp. Software Defined Networks.

Haleplidis, E. (2015). *Software-Defined Networking (SDN)*. Layers and Architecture Terminology.

Hassas Yeganeh, S., & Ganjali, Y. (2012). Kandoo: a framework for efficient and scalable offloading of control applications. *Proceedings of the first workshop on Hot topics in software defined networks*, 19–24. 10.1145/2342441.2342446

Heller, S., & McKeown. (2012). The controller placement problem. *Proceedings of the first workshop on Hot topics in software defined networks, HotSDN.*

International Civil Aviation Organization (ICAO). (1999). *Manual of technical provisions for the aeronautical telecommunication network (ATN)*. Retrieved from https://www.icao.int/safety/acp/repository/_%20Doc9705_ed2_1999.pdf

Kim. (2011). The Evolution of Network Configuration: A Tale of Two Campuses. *Proc. 2011 ACM SIGCOMM Conf. Internet Measurement Conf.*, 499–514.

Koponen, Casado, Gude, Stribling, Poutievski, Zhu, … Hama. (2010). Onix: A distributed control platform for large-scale production networks. *OSDI, 10,* 1–6.

Korzhov, V. (2012). *Children's Diseases SDN*. Open Systems.

Matias, J. (2014). *FlowNAC: Flow-Based Network Access Control*. Proc. Euro. Wksp. Software Defined Networks.

mininet.org. (2014). *Mininet. An Instant Virtual Network on your Laptop (or other PC)*. Retrieved from http://mininet.org/

NagibinP.KrylosovD. (2014). *Ambiguity SDN*. Retrieved from http://nag.ru/articles/reviews/26333/neodnoznachnost-sdn.htm

Open Networks Foundation (ONF). (2012). *Software-Defined Networking: The New Norm for Networks, white paper.* Retrieved from https://www.opennetworking.org

Orlov. (2014). SDN. *Journal of Network Solutions*.

Porras, P. (2012). A Security Enforcement Kernel for OpenFlow Networks. Proc. 1st Wksp. *Hot Topics in Software Defined Networks*, *2012*, 121–126.

Qin, Dai, Huang, & Xu. (2017). Bandwidth-Aware Scheduling With SDN in Hadoop: A New Trend for Big Data. *IEEE Systems Journal, 11*(4).

Schmid, S., & Suomela, J. (2013). Exploiting locality in distributed sdn control. *Proceedings of the second ACM SIGCOMM workshop on Hot topics in software defined networking*, 121–126. 10.1145/2491185.2491198

Semenov, Y. A. (2014). *Network technology OpenFlow*. Moskow, Russia: SDN.

Skoldstrom, P. (2015). *Towards Unified Programmability of Cloud and Carrier Networks*. Proc. Euro. Wksp. Software Defined Networks.

Smelyansky, R. (2012). *Program-configurable networks*. Ruslan Smelyansky.

Tootoonchian, A., & Ganjali, Y. (2010). Hyperflow: a distributed control plane for OpenFlow. *Proceedings of the 2010 internet network management conference on Research on enterprise networking*, 3–3.

Tootoonchian, A., & Ganjali, Y. (2010). HyperFlow: A Distributed Control Plane for Open-Flow. *Proc. 2010 Internet Network Management Conf. Research on Enterprise Networking*.

Wanderer, J. (2014). *Case Study: The Google SDN WAN*. Retrieved from http://www.computing.co.uk/ctg/analysis/2235886/case-study-the-google-sdn-wan

Xie, H., Tsou, T., Lopez, D., Yin, H., & Gurbani, V. (2012). *Use cases for alto with software defined networks*. IETF Internet-Draft.

Yeganeh, H., Tootoonchian, A., & Ganjali, Y. (2013). On scalability of software-defined networking. *Communications Magazine, IEEE, 51*(2), 136–141. doi:10.1109/MCOM.2013.6461198

Zhu. (2016). *MCTCP: Congestion-aware and robust multicast TCP in Software-Defined networks*. IEEE.

KEY TERMS AND DEFINITIONS

API: Application programming interface.

BGP: Border gateway protocol.

Computer Network: A digital telecommunications network which allows nodes to share resources.

HTTP: Hypertext transfer protocol.

ICMP: Inernet control message protocol.

IDS/IPS: Intrusion detection system/intrusion prevention system.

Internet Protocol: The principal communications protocol in the internet protocol suite for relaying datagrams across network boundaries.

IP: Internet protocol.

IS-IS: Intermediate system to intermediate system.

ISP: Internet service provider.

MAC: Media access control.

Network Protocol: A set of rules that controls how data is sent between computers on the internet.

NVP: Networking virtualization platform.

ONE: Open network environment.

OSPF: Open shortest path first.

Packet Switching: A method of grouping data which is transmitted over a digital network into packets.

RIP: Routing information protocol.

Routing: The process of selecting a path for traffic in a network, or between or across multiple networks.

SDK: Software development kit.

SDN: Software-defined network.

Software-Defined Networking: An approach to cloud computing that facilitates network management and enables programmatically efficient network configuration in order to improve network performance and monitoring.

Telecommunications: The transmission of signs, signals, messages, words, writings, images, and sounds or information of any nature by wire, radio, optical, or electromagnetic systems.

VLAN: Virtual local area network.

Chapter 6
Data Science Tools Application for Business Processes Modelling in Aviation

Maryna Nehrey
National University of Life and Environmental Sciences of Ukraine, Ukraine

Taras Hnot
National University of Life and Environmental Science of Ukraine, Ukraine

EXECUTIVE SUMMARY

Successful business involves making decisions under uncertainty using a lot of information. Modern modeling approaches based on data science algorithms are a necessity for the effective management of business processes in aviation. Data science involves principles, processes, and techniques for understanding business processes through the analysis of data. The main goal of this chapter is to improve decision making using data science algorithms. There are sets of frequently used algorithms described in the chapter: linear, logistic regression models, decision trees as a classical example of supervised learning, and k-means and hierarchical clustering as unsupervised learning. Application of data science algorithms gives an opportunity for deep analyses and understanding of business processes in aviation, gives structuring of problems, provides systematization of business processes. Business processes modeling, based on the data science algorithms, enables us to substantiate solutions and even automate the processes of business decision making.

INTRODUCTION

This chapter will addresses challenges with Data Science algorithms in aviation. Successful business involves making decisions under uncertainty using a lot of information. Modern modeling approaches based on Data Science algorithms are a necessity for the effective management of business processes in aviation. Data science involves principles, processes, and techniques for understanding business processes through the analysis of data. The main goal of this chapter is to improve decision making using data science algorithms. There are a set of frequently used algorithms described in the paper. Linear,

DOI: 10.4018/978-1-5225-7588-7.ch006

logistic regression models, decision trees as a classical example of supervised learning and k-means and hierarchical clustering as – unsupervised learning. Application of Data Science algorithms gives an opportunity for deep analyses and understanding of business processes in aviation, gives structuring of problems, provides systematization of business processes. Business processes modeling, based on the Data Science algorithms, enables us to substantiate solutions and even automate the processes of business decision making.

Nowadays business environment of aviation is characterized by significant uncertainty, increasing competition, and globalization. For a successful business operation, it is necessary to make decisions, taking into account a large number of factors and a considerable volume of information. The effectiveness of business decisions depends on the ability to analyze existing information, to predict the development of business processes and the system vision of the whole business. Business processes modeling in aviation is the most difficult part of their analysis. Improving the business decision-making process is possible provided that current methods and models of business analysis, such as Data Science, are correctly applied.

DATA SCIENCE: ESSENCE, PRINCIPLES, AND TOOLS

Data science involves principles, processes, and techniques for understanding business processes through the analysis of data. The main goal of Data Science is the exploration of the complexities inherently trapped in data, business, and problem-solving systems.

Data Science is a science of learning from data. Data Science is a continuation of Data Mining and Predictive Analytics. This approach is multidisciplinary, it combines the methods and models of disciplines such as mathematics, statistics, probability theory, information technology, including signal processing, probabilistic models, machine learning, statistical training, data mining, databases, object recognition, visualization, uncertainty modeling, data warehousing, data compression, computer programming, and high-performance computing.

The essence of Data Science is the extraction of information based on the knowledge and skills from the various fields of activity necessary for gaining knowledge. The composition of such a set largely depends on the field of research. For specialists in this area of research - Data Scientist - generalized qualification requirements have been developed.

Data Science has a big list of tools: Linear Regression, Logistic Regression, Density Estimation, Confidence Interval, Test of Hypotheses, Pattern Recognition, Clustering, Supervised Learning, Time Series, Decision Trees, Monte-Carlo Simulation, Naive Bayes, Principal Component Analysis, Neural Networks, k-means, Recommendation Engine, Collaborative Filtering, Association Rules, Scoring Engine, Segmentation, Predictive Modeling, Graphs, Deep Learning, Game Theory, Arbitrage, Cross-Validation, Model Fitting, etc. Some of this tools were used in the next researches. Teaching data science, for example, were introduced (Brunner & Kim, 2016), Big data and Data Science methods presented in (Chen, Chiang & Storey, 2012), (George, Osinga, Lavie & Scott, 2016), (Kucherov, 2007), (Shoro, Soomro, 2015), (Xiong, Yu & Zhang, 2017), machine learning used (Parish & Duraisamy, 2016), Monte Carlo method presented (Patriarca, Di Gravio & Costantino, 2017), game theory and genetic algorithms combined (Periaux, Chen, Mantel, Sefrioui & Sui, 2001), Artificial Intelligence presented (Rizun & Shmelova, 2017). Data Science is fast developing. A large volume of information that grows with each passing year makes it possible to build high-precision models that simplify and partially automate the decision-making process. Models are being developed that implement the key data science algorithms for decision-making in business (Hnot, Nehrey, 2017).

The main approaches in Data Science are Supervised learning models and Unsupervised learning models.

Supervised Learning Models

Supervised learning is one of the methods of machine learning, in which the model learns on the basis of labeled data. Using Supervised learning is possible to decide on two types of tasks: regression and classification. The main difference between them is the type of variance that is predicted by the corresponding algorithm. In regression training, it is a continuous variable, in the classification, it is a categorical variable. To solve these problems, a large number of algorithms have been developed. One of the most common is a linear and logistic regression, a decision tree.

- **Linear Regression:** Regression analysis can be considered as the basis of statistical research. This approach involves a wide range of algorithms for forecasting a dependent variable using one or more factors (independent variables). The relationship between variables is expressed by a linear function:

$$y = b_0 + b_1x_1 + b_2x_2 + ... + b_nx_n,$$

x_i : factor *i*, based on which the forecast is based,
b_i : parameter of the model, the influence of the factor,
y : dependent variable for which the forecast is constructed.

The advantage of applying such an approach to modeling is the simplicity and clarity of the results, the speed of learning and the release of the forecast. The disadvantage is not always sufficiently high precision (since in business processes, the linear relationship between changes is rare).

As the example, linear regression in business process modeling is trends for time series when time values or index values are taken for an independent variable (for example, from 1 to n, where n is the number of elements in the time series). Trend allows you to predict the value for the next period. For example, research on real climate change has been conducted on the basis of the analysis of the average monthly temperature of soil and air in various areas for 1990-2011 (Figure 1) (Hnot, Nehrey, 2017). It is important to analyze this trend in order to model business processes in the aviation. Weather forecasting allows pilots, navigators, airline companies and businesses to ensure safe flights and save money by making a right decision.

Another example of applying linear regression is the optimizing prices. Consider an example of optimizing tickets prices per week in one of the flights. Having historical data on price and demand, this task can be solved in several stages:

1. To forecast the demand for a product by analyzing the time series for the next period, taking into account seasonal characteristics and growth of consumers.

Figure 1. Average monthly Ground and Air temperature in Ukraine

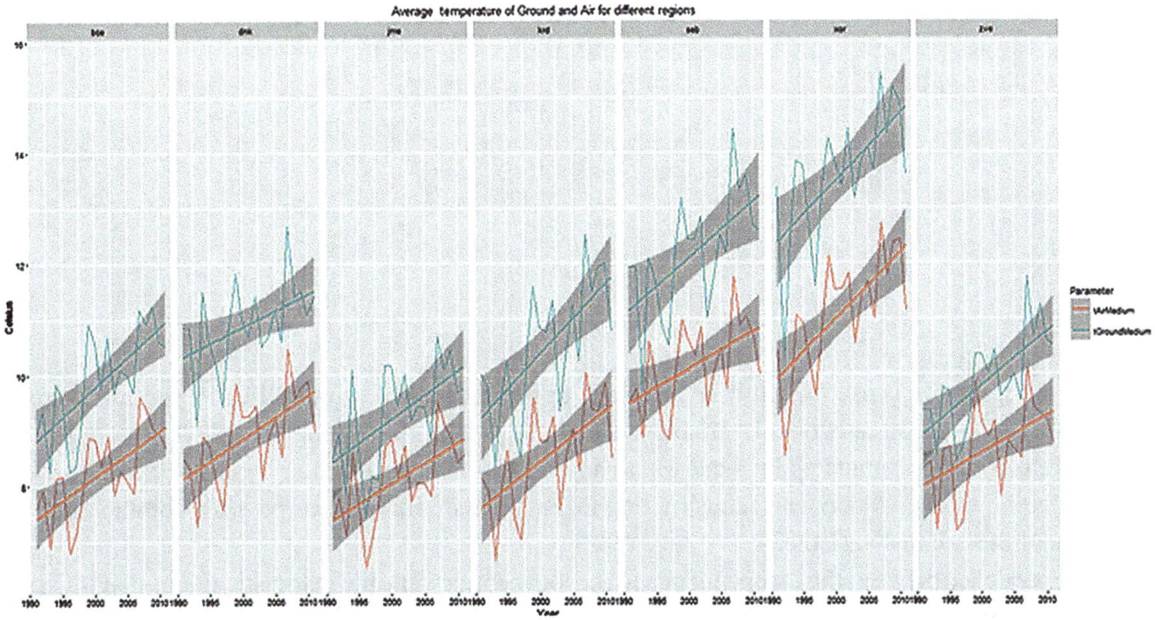

2. Estimate the linear function depending on demand from the price, first calculating the demand level for the next week based on the forecast. In addition, it is possible to add such dependent variables as the promotional product, the presence of ads in booklets/displays, on the Internet, etc. (Figure 2). In complex cases, such dependence can be expressed as a nonlinear function (for example, sigmoid).

Figure 2. Price - demand function of tickets (thousand hryvnyas)

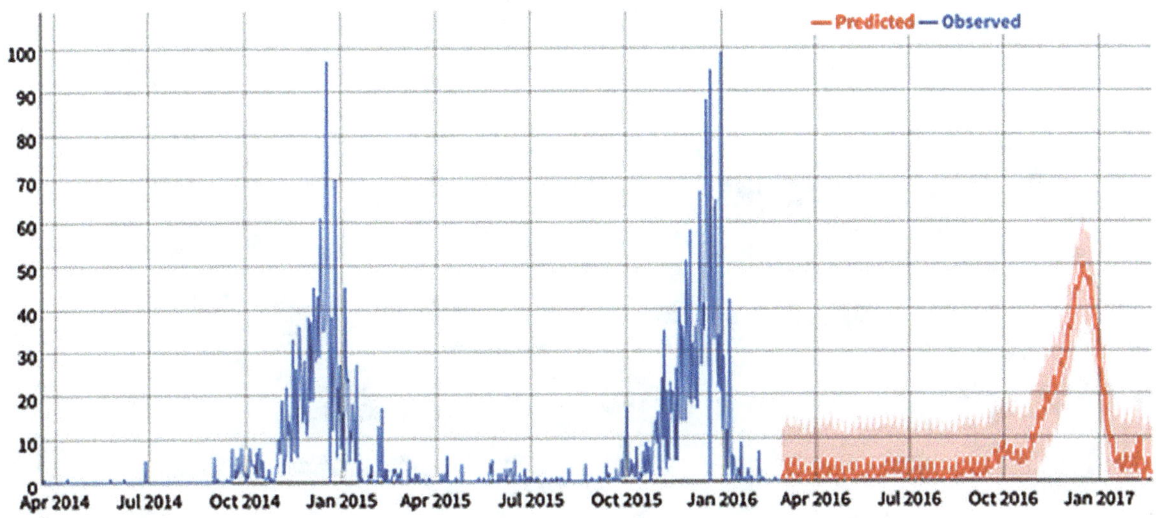

1. Taking into account the estimated function of demand, optimization of the quadratic function of revenue (Figure 3):

$$D = pu = p(a + bp) = bp^2 + ap,$$

where

D: profit,
p: price,
u: demand.

Logistic regression is used when it is necessary to predict the release of a binary variable using a dataset of continuous or categorical variables. Situations, where the parent variable has more than 2 possible values, can be simulated by a one-vs-all approach when constructing a logistic classifier for a possible output, or one-vs-one when constructing logistic classifiers for each possible combinations of categories of the original variable.

The dependence between the independent and the logarithmic variable in logistic regression is linear, the only difference with linear regression is sigmoidal functions, which converts a linear result in the probability of belonging to a class within [0; 1]:

$$p = \frac{1}{1 + e^{-(b_0 + b_1 x_1 + b_2 x_2 + \ldots + b_n x_n)}},$$

Figure 3. The function of profit of the tickets

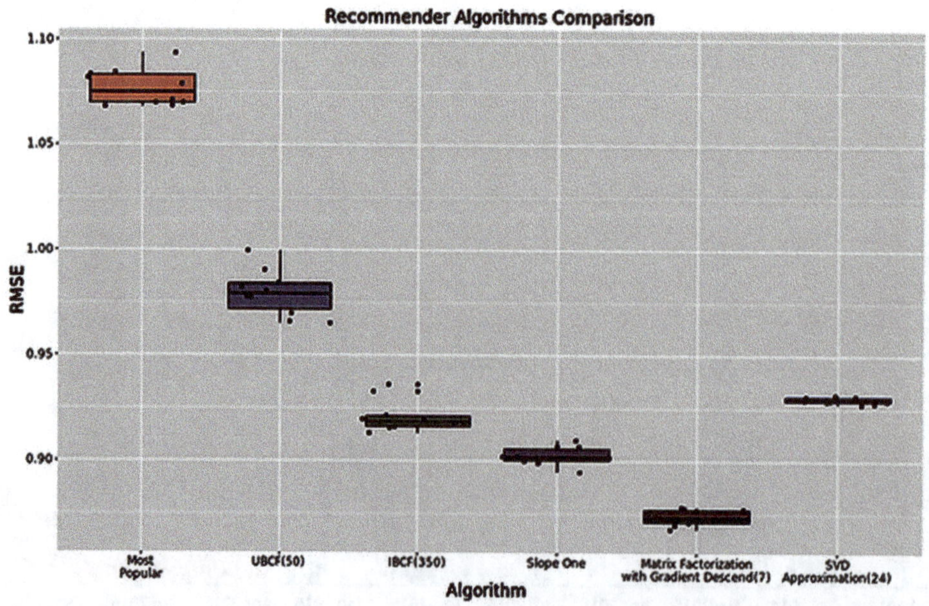

x_i : *i* factor, a base for the forecast,

b_i : *i* parameter of the model,

p : a probability of belonging to the class "1".

The advantages and disadvantages of logistic regression are due to the advantages and disadvantages of linear regression. This is the speed of the algorithm and the possible interpretation of the results, on the one hand, and a little accuracy - on the other.

Logistic regression is often used to construct vote counting models. An important factor in this is the interpretation of its results. The influence of each factor is clearly expressed by the magnitude of the coefficient b, which allows it to be clearly defined which of them positively and to what extent influence the decision. In Figure 4 shows a simple model of indicators, which predicts a loan client based on two factors: the age of clients and the term of the loan. This model is based on 1000 copies of the data set "German Credit Risk". As can be seen from the picture, the model assumes higher creditworthiness of clients with a term of lending up to 2 years and at the age of 30-40 years. The accuracy of such a model is ~ 60%, the construction of logistic regression across all 20 attributes, can achieve the accuracy of up to 80%. The black line on the graph reflects the boundary of the model's decision: it has a greater probability of a positive response> 50%.

A decision tree is an approach to both regression and classification. It is widely used in intelligent data analysis. The decision tree consists of "nodes" and "branches". The tree nodes have attributes that are used to make decisions. In order to make a decision, it is needed to go down to the bottom of the decision tree. The sequence of attributes in a tree, as well as the values that divide the leaves into branches, depends on such parameters as the amount of information or entropy that the attribute adds to the prediction variable.

The advantages of decision trees are the simplicity of interpretation, greater accuracy in decision-making simulation compared with regression models, the simplicity of visualization, natural modeling

Figure 4. Scoring model of the creditworthiness of clients

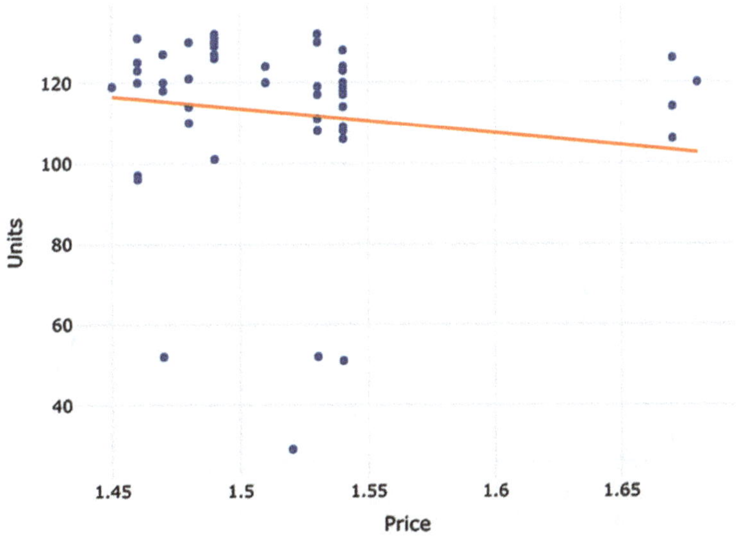

of categorical variables (in regression models it is needed to be coded by artificial variables). However, the decision trees have one significant drawback - fairly low predictive accuracy (James, Witten, Hastie & Tibshirani, 2013).

An example of applying a decision tree is the definition of the companies client classification algorithm - the construction of Loyalty Matrix. All clients are divided into 4 groups (TTruly Loyal, Accessible, Trapped, High risk) based on the answer to questions1 to 5 questions. In Figure 5 shows a tree that, based on three questions, allows us to predict the client's class with the accuracy of 98%.

Unsupervised Learning

Unsupervised learning describes a more complex situation in which, for each observation i = 1, ..., n, observation of the measurement vector x_i, but without any variables in the output y_i. In such data, the construction of linear or logistic regression models is impossible, since there are no predictive variables. In such a situation, a so-called "blind" analysis is conducted. Such a task belongs to the class of tasks of unsupervised learning, due to the absence of an output variable that guided the analysis. Unsupervised learning algorithms can be divided into algorithms for space reduction and clustering algorithms. The main task of clustering is to find patterns in the data that allow you to divide the data into groups and then in a certain way analyze them and give them an interpretation.

K-means is one of the most popular clustering algorithms, whose main task is to divide n-observations into k-clusters. The minimum sum of squares is the distance of each observation to the center of the corresponding cluster. This algorithm is iterative, at each step the cluster centers are re-indexed and redistributed observation between them until a stable result is achieved.

The benefits of such an algorithm of clustering are the simplicity, speed, and the ability to process large amounts of data. But the user must specify the number of clusters he wants to use for clustering before computing; the instability of the result (it depends on the initial separation of points between the clusters).

Figure 5. Decision Tree of the classification of companies clients

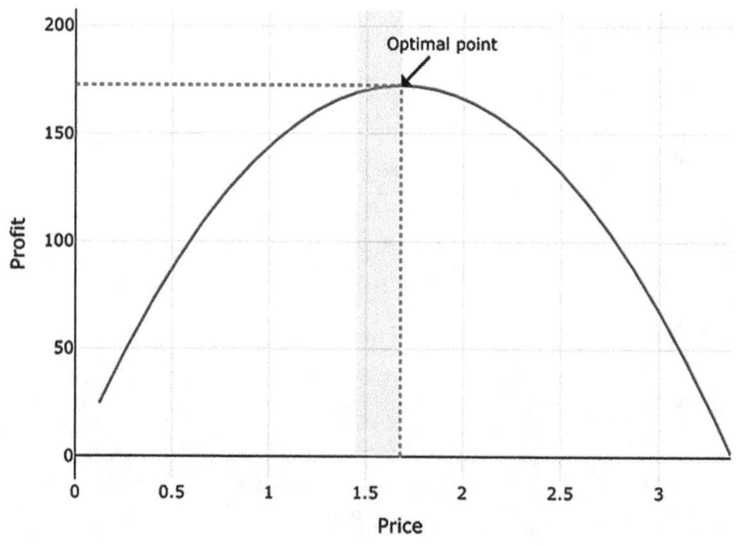

Figure 6 shows an example of using k-means for clustering users of the Internet service by coordinates. It allows you to split them into groups and form a delivery zone.

Hierarchical clustering is an alternative approach to clustering, which does not require a preliminary determination of the number of clusters. Moreover, the hierarchical clustering ensures the stability of the result and gives the output an attractive visualization based on the tree-like structure of observations/ clusters - dendrogram. This clustering algorithm uses different distance metrics and cluster agglomeration cluster criteria, which makes it very flexible to the data on which clustering is performed. However, the disadvantage of hierarchical clustering is the need to calculate the matrices of the distance between observations before agglomeration, which complicates the application of this algorithm for large data and data with many dimensions.

Figure 7 shows a dendrogram of customer segmentation based on features such as the number of weekends/weekdays transactions, the average number of purchases per week, and so on. Segmentation allows you to select groups of "similar" clients, for example, those who make purchases only on the weekend; those who buy mostly discounted goods, etc. This algorithm allows improving targeted marketing.

A time series is built by observations that have been collected with a fixed interval. It could be daily demand, or monthly profit growth rates, number of flights, etc. The time series analysis takes an important part in the analysis of data that covers the region, from the analysis of exchange rates to sales forecasting (Nehrey & Hnot T, 2017). One of the tasks of time series analysis is the allocation of trend and seasonal components and the construction of the forecast. There are a large number of algorithms have been developed, and we consider models such as ARIMA and Prophet.

ARIMA The ARIMA algorithm is one of the most common algorithms for forecasting time series. The basic idea is to use the previous time series values to predict the future. This can use any number of lags, which makes such an approach difficult in setting because it is necessary to select the parameter so as to minimize the error and not override the model. ARIMA is often used for short-term forecasting. A disadvantage is a complexity of learning a model in many seasonal conditions.

Figure 6. Clustering of Internet clients based on their coordinates

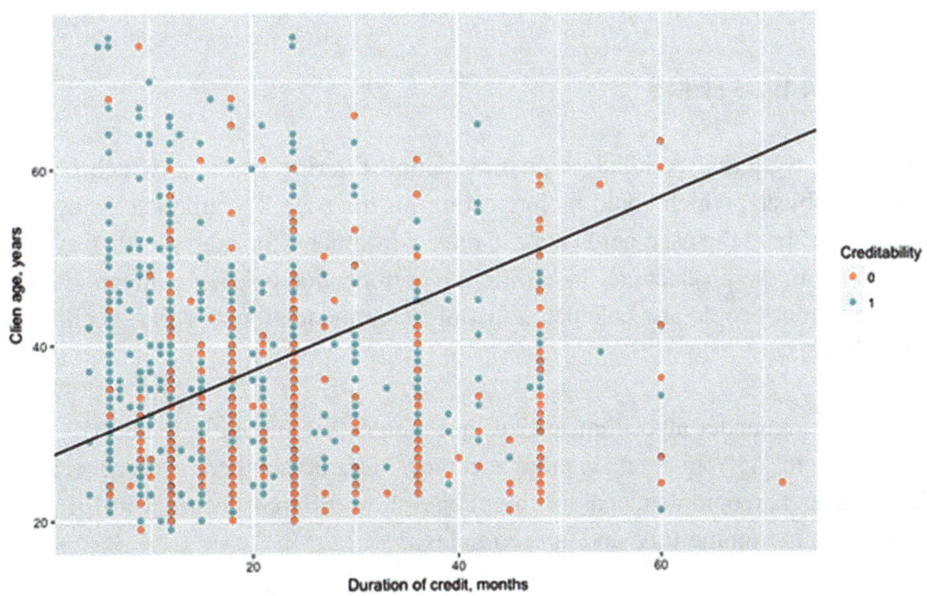

Figure 7. Dendrogram of customer segmentation

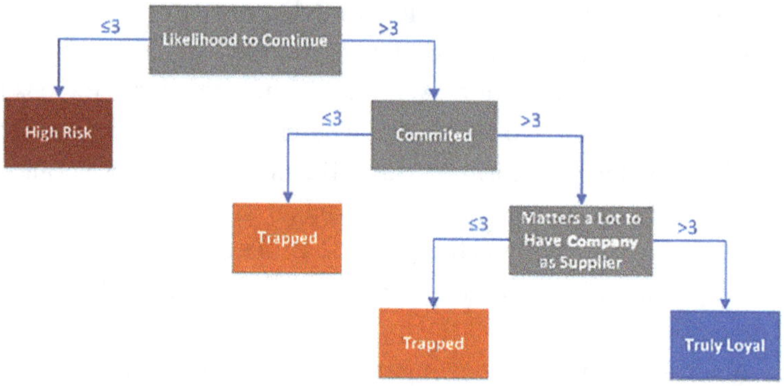

Figure 8 shows an example of forecasting for 1 week the number of orders in a restaurant (Hnot, Nehrey, 2017). One can clearly see seasonality in one day, which is inherent in the series of this kind.

Algorithm Prophet was developed by Facebook in the beginning of 2017 for forecasting based on time series (Nehrey & Hnot, 2017). It is based on an additive model in which nonlinear trends are of annual and weekly seasonality. This approach also allows to model holidays and weekends, thereby allowing to predict residuals in a time series. Also, the Prophet is insensitive to missed values, bias in the trend and significant residuals, which is an important advantage over ARIMA. Another advantage is the rather high speed of training, as well as the ability to use large-scale time series.

Figures 9, 10 shows an example of prediction with the Prophet. On the first of the charts - forecasting the entire category of goods, on the second - those products that are bought for Christmas. In the second case, only the seasonal components are taken into account and the “holiday” component is not modeled.

Although all of the above algorithms do not constitute a complete list of Data Science algorithms, however, in our view, they form the basis needed for business process modeling of aviation. The list of Data Science tools that can be used in business process modeling can be extended by such algorithms as ANOVA, neural networks, key component method, factor analysis, etc.

RECOMMENDER SYSTEMS

- **Recommender Systems:** A subclass of information filtering systems, which build sorted and ranked list of objects, which could be interesting for the user. To build such a system we could use information about the user, her history in the environment (for example, history of purchases), information about the product etc. In addition, recommender systems compare data of the same type from different people and calculate a list of recommendations for the specific user (Nehrey & Hnot T, 2017).

Such diversity in recommender algorithms evokes questions: “Which one is the best?”, “Which one is better for mine problem?”, “Which one is more accurate?” Of course, there is no general answer to all these questions. Different recommender algorithms could show different accuracy in different situations. They also are diverse in training time and tune complexity.

Figure 8. ARIMA for forecasting tickets demand

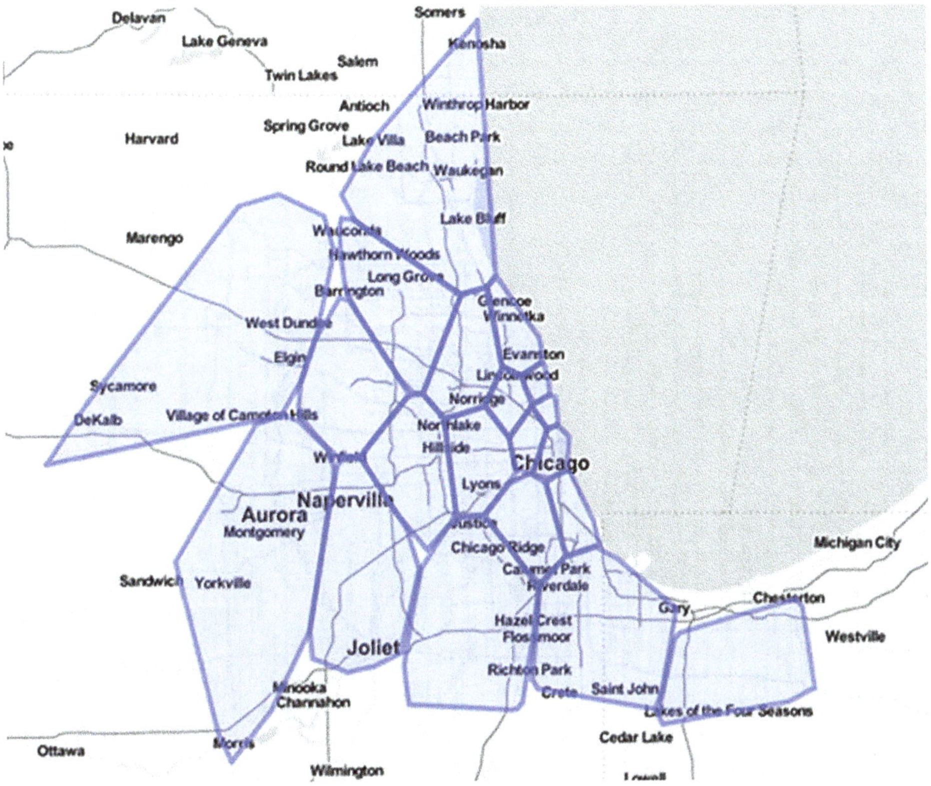

Figure 9. Forecasting the entire category of goods using the Prophet

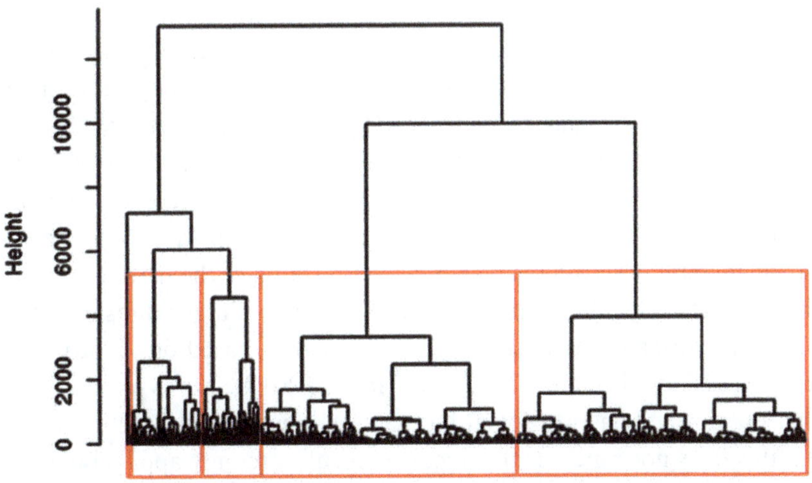

Figure 10. Time series analysis using the Prophet

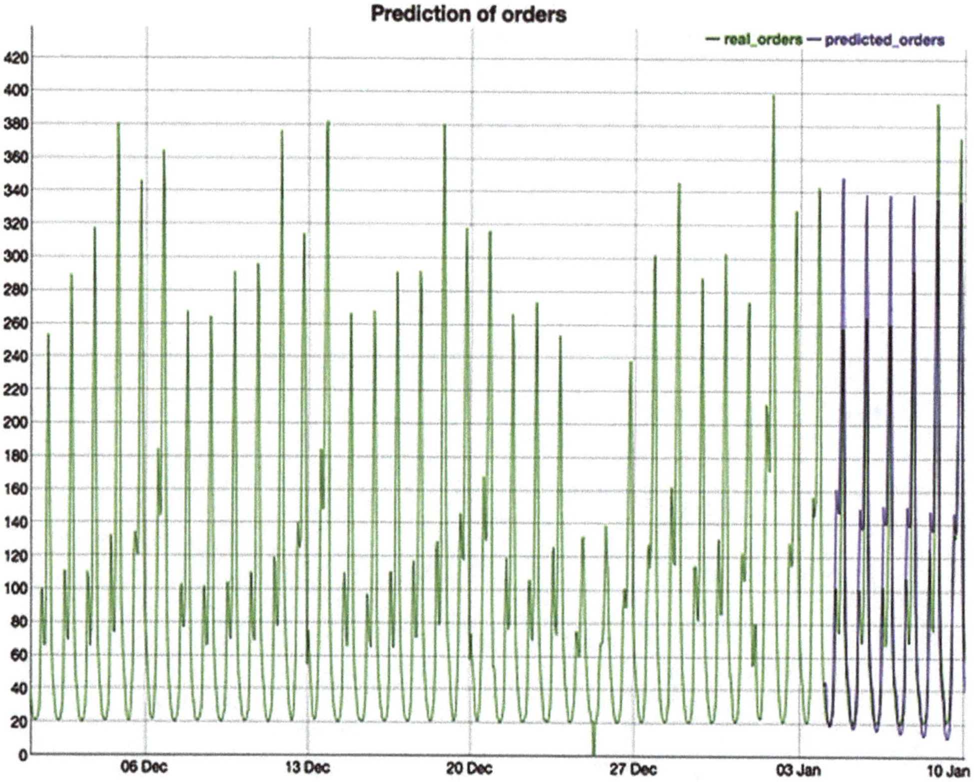

The research was focused on the comparison of a few different approaches to build recommender system. "Most Popular Product" is one of the most widely used techniques for basic recommendations extraction. User-based and item-based algorithms are also a powerful way to do recommendations. Despite the fact that these algorithms were developed more than 20 years ago, they show great interpretability and could be easily used for small datasets. Factorization-based techniques like "SVD Approximation" or "Matrix Factorization with Gradient Descend" are base for state-of-the-art approaches for mining recommendations in today's online world. There are a lot of different algorithms, which are related to factorization techniques, like non-negative/non-linear matrix factorization, weighted matrix factorization, etc. All of them are built based on the idea of decomposition of the matrix on two smaller one, the product of which should replicate the original matrix. This class of algorithms is well on small and big datasets, shows great performance and accuracy, as was shown based on "Matrix Factorization with Gradient Descend" example in the paper (Nehrey & Hnot T, 2017).

The analysis was performed on 1M MovieLens dataset, where we have rates in the range from 1 to 5 (Figure 11). But this does not conclude that received results are only applicable for data of the same nature. In online retail, there are two most popular data types: transactions and rating data. But transactions data could be transformed to a rating matrix as well, by counting product numbers bought by some customer and normalizing this score with a log or in some similar way. So, results, received in this paper could be applied to different situations, not only with explicit rating data. What is more important, research results applicable to situations where sparsity of rating matrix is approximately the same as in

Figure 11. Comparison RMSE of recommender algorithms

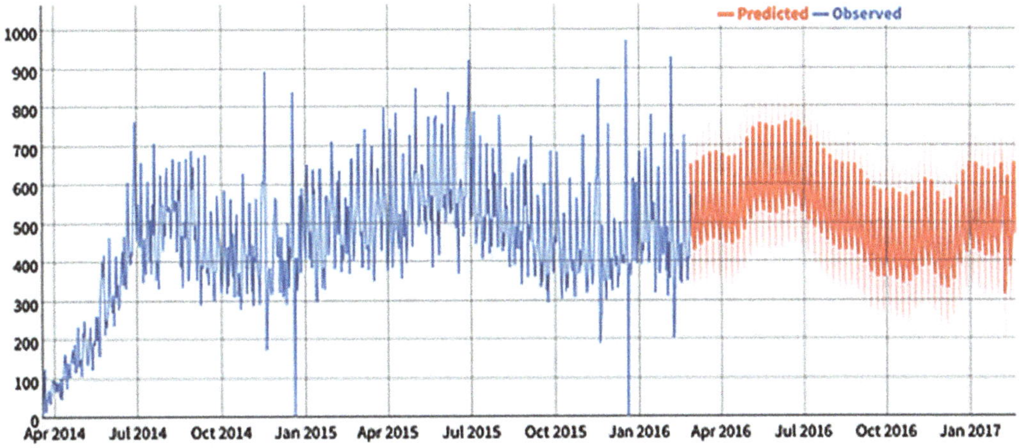

the test dataset (4.5%). Sparsity plays an important role and algorithms could treat themselves differently with smaller or larger level of sparsity.

FUTURE RESEARCH DIRECTIONS

Described approaches and algorithms are just some basic for business processes modeling, which could be applied to aviation problem. There are multiple examples of how all these methods could be used in aviation. With timeseries analysis we could predict future demand for tickets, using regression models we could determine the loyalty of the customers and so on.

Nowadays there are much more algorithms, which could be applied in this area. Like complicated non-liners algorithm for regression predictions. As an example, it could be a random forest, XGBoost, neural networks. With such method, we could build models for maintenance prediction, what is very crucial in aviation.

Another very important example is price oprimization. This problem is very hard in situation, when we want to encount time, customer information, fligh details.

CONCLUSION

To ensure successful business development, it is necessary to make a decision using modern approaches of business analytics - Data Science methods. Application of Data Science tools gives an opportunity to deeply analyze and understand business processes, promotes structuring of problems, provides systematization of business processes. The simulation of business processes, based on the Data Science tools, enables to substantiate solutions and even automate the processes of business decision-making.

REFERENCES

Aghabozorgi, S., Shirkhorshidi, A. S., & Wah, T. Y. (2015). Time-series clustering–A decade review. *Information Systems*, *53*, 16–38. doi:10.1016/j.is.2015.04.007

Alamdari, F., & Fagan, S. (2017). Impact of the adherence to the original low-cost model on the profitability of low-cost airlines. In *Low Cost Carriers* (pp. 73–88). Routledge.

Breese, J. S., Heckerman, D., & Kadie, C. (1998, July). Empirical analysis of predictive algorithms for collaborative filtering. In *Proceedings of the Fourteenth Conference on Uncertainty in artificial intelligence* (pp. 43-52). Morgan Kaufmann Publishers Inc.

Brunner, R. J., & Kim, E. J. (2016). Teaching data science. *Procedia Computer Science*, *80*, 1947–1956. doi:10.1016/j.procs.2016.05.513

Chen, H., Chiang, R. H., & Storey, V. C. (2012). Business intelligence and analytics: From big data to big impact. *Management Information Systems Quarterly*, *36*(4), 1165–1188. doi:10.2307/41703503

Davenport, T. H., & Patil, D. J. (2012). Data scientist: The sexiest job of the 21st century. *Harvard Business Review*. PMID:23074866

George, G., Osinga, E. C., Lavie, D., & Scott, B. A. (2016). *Big data and data science methods for management research*. Academic Press.

Gorakala, S. K., & Usuelli, M. (2015). *Building a Recommendation System with R*. Packt Publishing Ltd.

Hahsler, M. (2011). *Recommenderlab: A Framework for Developing and Testing Recommendation Algorithms*. Southern Methodist University.

Hauger, S., Tso, K. H., & Schmidt-Thieme, L. (2008). Comparison of recommender system algorithms focusing on the new-item and user-bias problem. In *Data Analysis, Machine Learning and Applications* (pp. 525–532). Berlin: Springer. doi:10.1007/978-3-540-78246-9_62

Hnot, T. V., & Nehrey, M. V. (2017). Data Science Algorithms in Modeling Business Processes. *Economics and Society.*, *12*, 187.

James, G., Witten, D., Hastie, T., & Tibshirani, R. (2013). *An introduction to statistical learning* (Vol. 112). New York: Springer. doi:10.1007/978-1-4614-7138-7

Kucherov, D. P. (2007). The synthesis of adaptive terminal control algorithm for inertial secondary order system with bounded noises. *Journal of Automation and Information Sciences*, *39*(9), 16–25. doi:10.1615/JAutomatInfScien.v39.i9.20

Matthews, B., Das, S., Bhaduri, K., Das, K., Martin, R., & Oza, N. (2013). Discovering anomalous aviation safety events using scalable data mining algorithms. *Journal of Aerospace Information Systems*, *10*(10), 467–475. doi:10.2514/1.I010080

Nehrey, M., & Hnot, T. (2017). Using recommendation approaches for ratings matrixes in online marketing. *Studia Ekonomiczne*, *342*, 115–130.

Parish, E. J., & Duraisamy, K. (2016). A paradigm for data-driven predictive modeling using field inversion and machine learning. *Journal of Computational Physics, 305*, 758–774. doi:10.1016/j.jcp.2015.11.012

Patriarca, R., Di Gravio, G., & Costantino, F. (2017). A Monte Carlo evolution of the Functional Resonance Analysis Method (FRAM) to assess performance variability in complex systems. *Safety Science, 91*, 49-60.

Pavlyshenko, B. M. (2016, August). Linear, machine learning and probabilistic approaches for time series analysis. In *Data Stream Mining & Processing (DSMP), IEEE First International Conference on* (pp. 377-381). IEEE.

Périaux, J., Chen, H. Q., Mantel, B., Sefrioui, M., & Sui, H. T. (2001). Combining game theory and genetic algorithms with application to DDM-nozzle optimization problems. *Finite Elements in Analysis and Design, 37*(5), 417–429. doi:10.1016/S0168-874X(00)00055-X

Ricci, F., Rokach, L., & Shapira, B. (2015). Recommender systems: introduction and challenges. In Recommender systems handbook (pp. 1-34). Springer US. doi:10.1007/978-1-4899-7637-6_1

Rizun, N., & Shmelova, T. (2017). Decision-Making Models of the Human-Operator as an Element of the Socio-Technical Systems. In *Strategic Imperatives and Core Competencies in the Era of Robotics and Artificial Intelligence* (pp. 167-204). IGI Global. Retrieved from http://rpubs.com/tarashnot/orders_full

Shoro, A. G., & Soomro, T. R. (2015). *Big data analysis: Apache spark perspective. Global Journal of Computer Science and Technology.*

Taylor, S. J., & Letham, B. (2018). Forecasting at scale. *The American Statistician, 72*(1), 37–45. doi: 10.1080/00031305.2017.1380080

Xiong, J., Yu, G., & Zhang, X. (2017). Research on Governance Structure of Big Data of Civil Aviation. *Journal of Computer and Communications, 5*(05), 112–118. doi:10.4236/jcc.2017.55009

Zubair, M., Khan, M. J., & Awais, M. M. (2012). Prediction and analysis of air incidents and accidents using case-based reasoning. In *Intelligent Systems (GCIS), 2012 Third Global Congress on* (pp. 315-318). IEEE.

KEY TERMS AND DEFINITIONS

Data Filtering: In IT can refer to a wide range of strategies or solutions for refining data sets. This means the data sets are refined into simply what a user (or set of users) needs, without including other data that can be repetitive, irrelevant or even sensitive. Different types of data filters can be used to amend reports, query results, or other kinds of information results.

Data Science: Is a broad field that refers to the collective processes, theories, concepts, tools and technologies that enable the review, analysis, and extraction of valuable knowledge and information from raw data. It is geared toward helping individuals and organizations make better decisions from stored, consumed and managed data.

Recommender Systems: Is a system that identifies and provides recommended content or digital items for users. As mobile apps and other advances in technology continue to change the way users choose and utilize information, the recommendation engine is becoming an integral part of applications and software products.

Supervised Learning: Is a method used to enable machines to classify objects, problems, or situations based on related data fed into the machines. Machines are fed with data such as characteristics, patterns, dimensions, color and height of objects, people or situations repetitively until the machines are able to perform accurate classifications. Supervised learning is a popular technology or concept that is applied to real-life scenarios. Supervised learning is used to provide product recommendations, segment customers based on customer data, diagnose disease based on previous symptoms and perform many other tasks.

Unsupervised Learning: Is a method used to enable machines to classify both tangible and intangible objects without providing the machines with any prior information about the objects. The things machines need to classify are varied, such as customer purchasing habits, behavioral patterns of bacteria and hacker attacks. The main idea behind unsupervised learning is to expose the machines to large volumes of varied data and allow it to learn and infer from the data. However, the machines must first be programmed to learn from data.

Chapter 7
Information Technology of the Aerial Photo Materials Spatial Overlay on the Raster Maps

Iryna Yurchuk
National Aviation University, Ukraine

Oleksiy Piskunov
National Aviation University, Ukraine

Pylyp Prystavka
National Aviation University, Ukraine

EXECUTIVE SUMMARY

The information technology that is researched in the chapter provides a spatial overlay of the images received by the camera of an unmanned aerial vehicle (UAV) and raster maps of open aerial photography services. Such software helps to solve issues of actualization of maps, observation of agricultural field yields, creation of terrain photo planes, monitoring, etc. "Frames and a Map Overlay Tools" is software developed in C# using .NET4.0. All algorithms that were used during the development of the complex are described in detail, as well as the flow diagrams of the data utilities from which the complex is composed. Despite the fact that the testing of this complex has shown poorly high speed in real time, the estimates will allow the possibility of its interactive use under conditions of further refinement.

INTRODUCTION

Every day processing of photographic materials practically has become necessary for the development of modern technical means for making pictures and monitoring with the help of UAVs. Some common tasks should be solved in different fields of usage. In particular, to overlay photos to the map of the area, the input data should be led to the same scale and transformed using angles of the camera's inclination. There are several methods to make such transformations.

DOI: 10.4018/978-1-5225-7588-7.ch007

There are many computer programs which allow users to overlay an image on a map (Google Earth; Digitals; MapTiler and etc). They are complex, multifunctional and some of them are expensive. For example, MapTiler which has many functions (one of them is an overlay of image over a map) costs $3500 and its support has to be paid in $200/hour. Integration of such big and complex software as a part of an end product development is time- and resource- consuming process. It is hard to remove some less usable functions from ready-made software. These costs are not reasonable in most cases.

In this section authors defined an information technology which is based on a simple algorithm, easy integrable and not overloaded by additional functionality. It helps to solve problems of the actualization of the maps, the observation of yields of agricultural fields, the aerial photography of pipelines, the monitoring of the status of waste heaps and the creation of photographic planes of the area, for example, see Karpov, 2012; Zheltov, Veremeenko, Kim, Kozorez & Krasilshchikov, 2009 and Nechausov & Zamirec, 2005.

REVIEW OF PREVIOUS LITERATURE

Modern technologies allow creating software for the photo spatial overlays on a map not for professional cartographers only, but also for the amateurs having even slite skills in programming. There are many blogs where the authors describe such own software or guides for professional software using any types of images (from digital cameras, phone cameras and etc.) and various types of maps. But most of them use GPS-based coordinates as input data to tie map with aggregated images (Monastyrskyi, 2017; Khramov, 2016; Google maps overlay; Cleveret, 2016; Adding a Google Earth overlay; Polymaps. Image overlay; Edwards & Titchenal, 2009; Overlay tiled images on a map).

In (Chyrkov & Prystavka, 2018) authors formulated the problem of the aerial photo materials spatial overlay on raster maps as an important component of the successful solution of suspicious objects location on the video stream from the UAV's camera.

It is necessary to add that many of such technologies have a special purpose (military, reconnaissance and etc.) and secret.

STATEMENT OF THE PROBLEM

Let consider the World Geodetic System 84 (WGS 84). The authors assume that UAV is defined by $(B_{BLA}, L_{BLA}, z_{BLA})$ on WGS 84, where B_{BLA} is the camera latitude, L_{BLA} - is the camera longitude and z_{BLA} - is the height of the camera's focus above the surface of the global ellipsoid and coordinates of the UAV's camera are the same as UAV coordinates.

There is also the $\xi\eta\zeta\ O'$ coordinate system, such that the point O' is a center of the vehicle mass and $\xi\eta\zeta$ are three principal axes, that describe a local position of UAV. The axis ξ is oriented from the vehicle tail to the vehicle nose according to the vehicle course, η is an axis directed from the left to right with regard to the pilot and it is parallel to wings and ζ is an axis directed from a top to a down and it's orthogonal to other axes. Let remark that $\xi\eta\zeta\ O'$ is the right-hand system.

The following angles are known: ω is the rotation angle of UAV around an axis ξ, φ is the rotation angle around η and κ is the rotation angle around ζ. These angles are called orientation angles of UAV.

Let assume that the authors obtain the following data (*photo*, B_{BLA}, L_{BLA}, $z_{BLA,}$ ω, φ, κ), where *photo* is a digital photo (also "frame" in further) and (ω, φ, κ) is the triple of the orientation angles at a moment of making a photo. What are the coordinates of photo points on the raster map of the area?

BACKGROUND

Let *O* be the original point on the Earth surface and there is a coordinate system *XYZO* such that the axis *OZ* is a normal to the surface at the point *O*, the axis *OX* is the tangent to the meridian at the point *O*, which positive direction coincides with the North and the axis *OY* is the tangent to parallel at the point *O*, which positive direction coincides with the East. This coordinate system is the left-hand system.

Let (x_{BLA}, y_{BLA}, z_{BLA}) be the coordinates of the unmanned aerial vehicle (UAV) defined in *XYZO*. The following orientation angles are known: ω - rotation of the UAV around ¾ (roll axis), φ- rotation of the UAV around η (pitch axis) and κ- rotation of the UAV around ζ (yaw axis). The UAV received data (*photo*, x_{BLA}, y_{BLA}, $z_{BLA,}$ ω, φ, κ), where the *photo* is a digital image frame, (x_{BLA}, y_{BLA}, z_{BLA}) - UAV coordinates and (ω, φ, κ) - the orientation angles of the UAV in space at the moment of fixing the image.

The authors remind the following mathematical formulas. If a point *A(x,y,z)* is rotated counterclockwise on an angle α as to the point *O*(0,0,0) around the *OX* axis of some right-hand Cartesian coordinate system then its new position has the following coordinates *A′ (x′, y′, z′)* and calculated as:

$$(x'y'z') = (xyz)\begin{pmatrix} 1 & 0 & 0 \\ 0 & \cos\alpha & \sin\alpha \\ 0 & -\sin\alpha & \cos\alpha \end{pmatrix} \quad (1)$$

The matrix at (1) is called a rotation matrix and denoted by $R_{ox}(\alpha)$.

The authors have to remark that $R_{ox}(\alpha)$ has different notations in cases of the clockwise rotation or the left-hand coordinate system.

Let assume that a point $P(x_{BLA}, y_{BLA}, 0)$ be the projection of the center of the camera's view field on the Earth surface in case of zero roll, pitch and yaw angles.

The algorithm for determining the camera's view field in the aerial photography at the coordinate system *XYZO* with given angles (ω, φ, κ):

1. Let's accord two coordinate systems ξηζ *O´* and *XYZO* by using the composition of the translation on ($-x_{BLA}$, $-y_{BLA}$, $-z_{BLA}$) and a transformation $Z = -z$;
2. Let's calculate $R = R_{\eta}(\varphi) \bullet R_{\zeta}(\omega)$, where $R_{ox}(\alpha)$ is a rotation matrix and • is an operation of right multiplication of the matrices. This operation is non-commutative, that means the order of multipliers affects on a result. In the other words, the rotation on the pitch angle should be done at first and then the rotation on the yaw angle;
3. Let's make transformations which are inversed to obtained at step 1. and get a point $P'(x', y', z')$;
4. Let's write an equation of the line *l* which contains points *O′* (its coordinates are given at a system *XYZO* as (x_{BLA}, y_{BLA}, z_{BLA}) and *P′*. Then the coordinates of a new point P'' (x'', y'', z'') which is common for the line *l* and the surface of the Earth (it means that $z'' = 0$) should be calculated.

5. Let P''' (x''', y''', z''') is obtained from the P'' by the translation on ($-x_{BLA}$, $-y_{BLA}$, 0), the rotation $R_\xi(\kappa)$ and the inverse translation (x_{BLA}, y_{BLA}, 0). The point P''' is the center of the camera's field of view on the Earth.

Similarly, the coordinates of points which belong to the boundary of the camera's field of the view and depend on the camera's characteristics are calculated (Piskunov, Yurchuk & Bilyanska, 2017).

Remark: For all calculations, linear formulas as at (Karpov, 2012) were used. It means that the neighborhood of a point at the Earth surface is approximated by the tangent plane. Such assumption simplifies the calculation and does not take into account the curvature of the Earth. So, it contributes to an increase in the errors.

MAIN PART

The information technology "Frames and a Map Overlay Tools" consists of following parts:

- Coor dinates conversion between the WGS 84 and two Mercator's projections. They are known as the Web Google Mercator projection (EPSG:3857) and the Real Mercator projection (EPSG:3395);
- Coordinates conversion between the WGS 84 and the Pulkovo 1942 on the Krasovsky ellipsoid, GOST R 51794-2008 (2009). In further, the Pulkovo 1942 geodetic datum will be denoted by SC 42 following the traditional notation in the post-Soviet space;
- Central projection of the aerial footage frame to the plane;

In addition, the "Frames and an Map Overlay Tools" also includes following additional components:

- u.PAP is the utility module for projection of the frame on the plane. While producing the projection of the frame it creates the geographic registration data (or the files of the world binding). With the help of this data, all calculations between the taken image pixels and the coordinates of that pixel on the ground are made. The u.PAP data stream schema is shown in Figure 1. World binding data files are shown as '.snapshot.jpgw' and they consist of coefficients of a polynomial transformation of the first order.

The output of u.PAP has the following structure:

1. **Set of '.FrameXXX.jpg' Files:** The sequence of frames which are projected onto the plane (the dot at the beginning of the file name is included) in one of the common formats;
2. **Set of '.FrameXXX.jpgw' Files:** The sequence of geographic registration files for the corresponding frame. For details of the file format (Gis-Lab). The format of the geographic registration file (world file);
3. **Set of '.track.mtr' Files (Projections of Fixed Points):** The CSV file whose string has the following format:
    a. **String:** The name of the file containing the image from the UAV at the current point;
    b. **String:** A unique point identifier;

Figure 1. The data stream of u.PAP: the ellipses are processes and the rectangles are data.

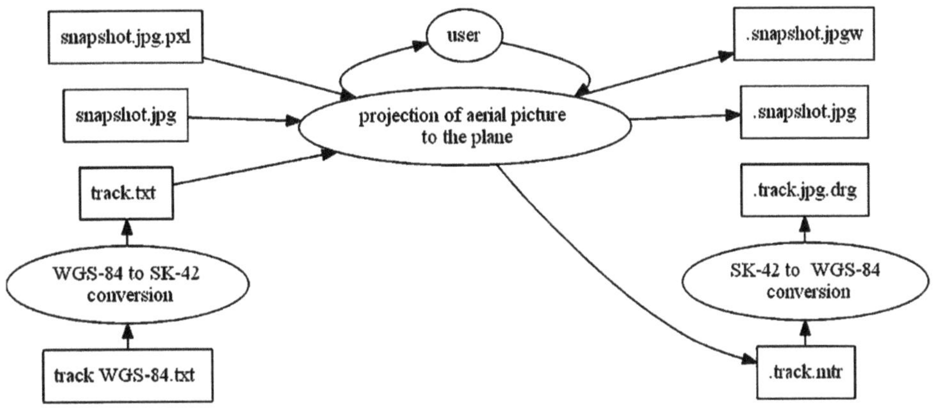

c. **Coordinate to the North:** The latitude in WGS 84;
d. **Coordinate to the East:** The longitude in WGS 84;
e. Optionally, the additional data to evaluate the accuracy of the transformation. These include: the distance from the coordinates of the focus, the overall angle of the camera, and so on;
f. Note to the point.

Since the projection of the frame requires the height of the camera's focus above the ground level, but the input data is the height of the UAV over the global ellipsoid, the utility has a special parameter for height correction. Yandex.Maps or OpenStreetMap (Evenden, 2003) can be selected as a source of raster maps providing service (see Open Street Map Project). The map can be displayed as one tile (256 ×256 pixels) or four tiles (512 ×512 pixels). The utility provides the ability to navigate through the sequence of frames, zoom in/out maps and also move around maps. The status string of the window displays the zoom level, camera focus coordinates for this snapshot, and pixel dimensions in meters.

- **u.imMap:** Utility for the overlay of the frame on raster maps (OpenStreetMap or Yandex.Maps).

For the successful action of u.imMap the algorithm for determining the field of view of a UAV's camera is essential. It's based on the formulas from *Background* subject to $x_{BLA} = B_{BLA}$, $y_{BLA} = L_{BLA}$ and $z_{BLA} = z_{BLA}$. The u.imMap data stream schema is shown in Figure 2.

The general u.imMap utility processing algorithm for each input track coordinate is the following:

- The input data is a track (coordinates in WGS 84) with converted frames by u.PAP and world binding files;
- u.inMap takes the next frame from the track. It converts the coordinates of camera focus (B_{BLA}, L_{BLA}, z_{BLA}) into SC 42 and downloads the raster map for (B_{BLA}, L_{BLA}). Using a frame and the world binding file the linear mapping of each frame pixel to the obtained map is built. The result is displayed to the operator.

The final output of u.imMap is the view of colored pixels of the aerial photo-aligned with the raster map. In Figure 3 an example of the result is presented.

Figure 2. The data stream of u.imMap: the ellipses are processes and the rectangles are data.

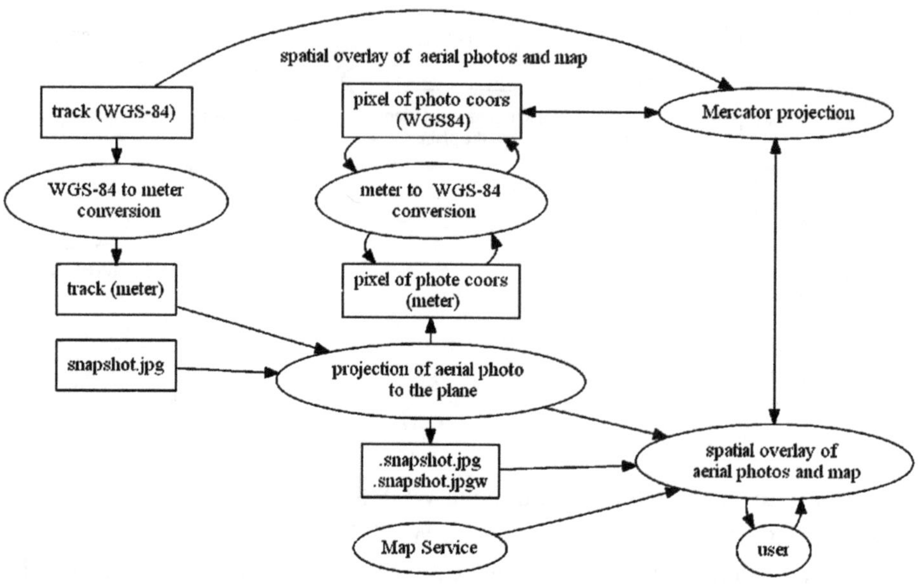

Using SC - 42(1942) and Krasovsky ellipsoid is conditioned by the presence of maps into such format as a primary source.

ILLUSTRATIVE EXAMPLE AND TESTING

The information technology of the aerial photo materials spatial overlay with the raster maps was tested on the real photos (6000×8000 pixels) that were obtained from the UAV aerial photo set. It has a satisfactory accuracy of the calculation in a near-real-time mode. Despite the fact that the testing of this complex has shown a not high enough speed for a real-time, there exist estimations which allow the possibility of its interactive use under conditions of further refinement (Sotnik, 2003).

Also, an experiment for a calculation of projection errors was conducted. A UAV footage frame was processed with photoluminescence filter (the scale on sample frame is 1:10000). 200 points were marked on the map. Then their coordinates were calculated for UAV frames in the pixels. A utility u.PAP issued their coordinates in meters. Both results were compared (see Figure 4). For making the conclusion circular error probability (CEP) parameter was used. It is a measure of the weapon system's precision. It is defined as the radius of a circle, centered on the mean, whose boundary is expected to include the landing points of 50% of the rounds. That is, if the given bomb design has a CEP of 100m (330 ft), when 100 are targeted at the same point, 50 will fall within the 100m circle around their average impact point. (The distance between the target point and the average impact point is referred to as bias.) In Figure 4 different types of projection errors as the different types of lines can be seen.

Also, the experimental data was processed by statistics methods and the following conclusion obtained: the error is increasing with increasing of the distance between the point (B_{BLA}, L_{BLA}, 0) and the projection of the center of the camera's field of the view on the Earth (see Figure 5)

Figure 3. An example of a spatial overlay.

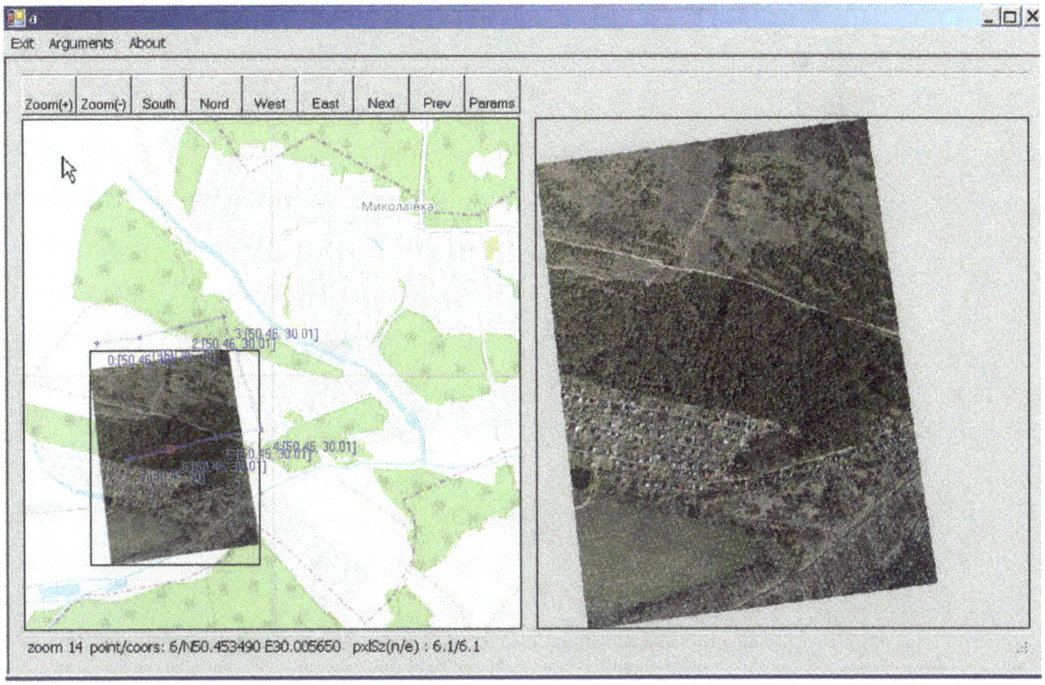

Figure 4. An example of the comparison: every marked point has a few offcuts. Each of them has one of four types of CEP (circular error probability): 1CEP, 2CEP, 3CEP, and 4CEP.

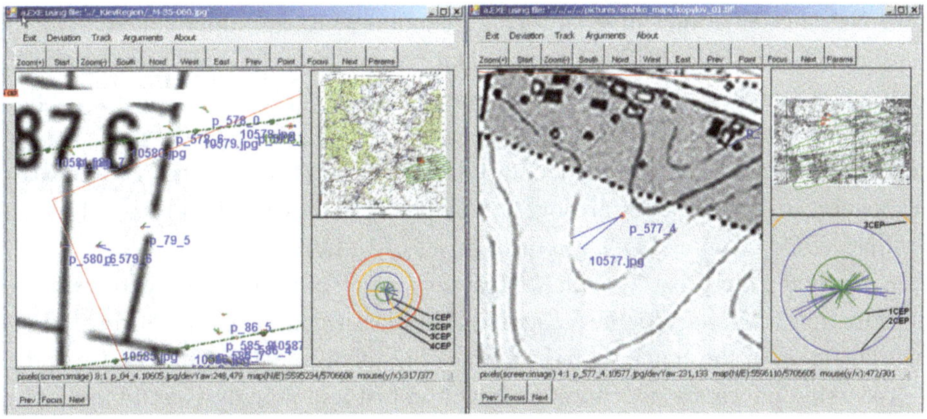

PRACTICAL RESULTS AND FUTURE RESEARCH

This information technology is a part of computer program installed at Raspberry Pi3 Model B microcomputer of the experimental model of the UAV (Chyrkov & Prystavka, 2018; Prystavka, Chyrkov & Sorokopud, 2017; Chyrkov, 2016), and implements the information technology of adaptive search, classification and recognition of objects with self-learning algorithm, and which is integrated into the

Figure 5. The graph of the dependency between the error value and the distance between the point (B_{BLA}, L_{BLA}, 0) and the projection of the center of the camera's field of the view.

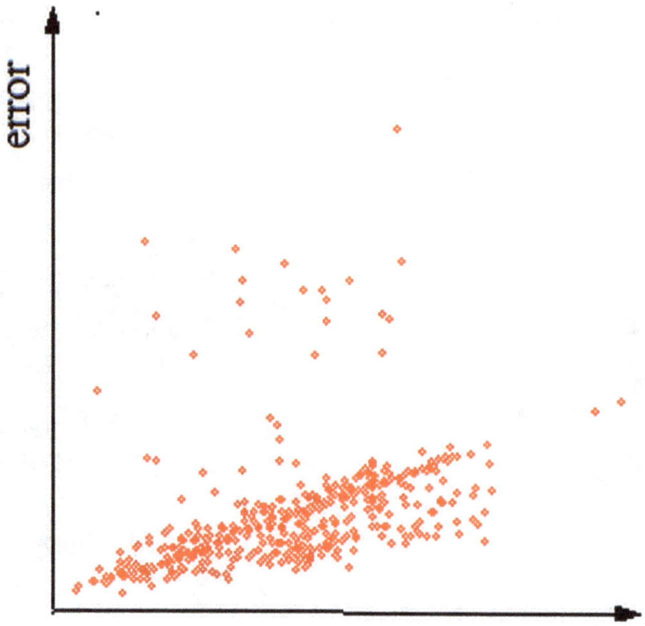

developed prototype of the software-hardware complex of search of suspicious objects (Chyrkov & Prystavka, 2018). It is obtained by the authors during the research work SSW N°1062-DB16 "Automation of recognition and classification of suspicious objects for video data from unmanned aircraft cameras" provided at the National Aviation University. This technology was also approbated at the International IT Championship "Golden Bite 2017" (Ukraine) and presented at the XIV International Specialized Exhibition "Weapons and Security - 2017" on October 10-13, 2017, Kyiv, the function of the automated search of suspicious objects for no conditions for the absence of reference samples. The subsystem of the determination of the coordinates of the identified objects gives less accuracy in comparison with the declared (but not certified) accuracy of analogs (development of the State Enterprise "Ukroboronprom", enterprises of the NGO "League of Defense Enterprises of Ukraine", etc.).

It is used by the State Enterprise "Orizon Navigation", which is an appropriate implementation act, in particular: information technology for the recognition and classification of objects from the UAV cameras and the synchronization of the frame with the data of the satellite navigation system in the automatic mode; Post-processing information technology based on histogram analysis and modified recognition method based on the analysis of special points based on the spline model of the image.

In further research, the possibility of this information technology will be expanded to the ability to operate by Universal Transverse Mercator (UTM). That makes it more competitive not only in the domestic market but also abroad.

CONCLUSION

The information technology of the aerial photo materials spatial overlay with the raster maps was tested as a part of UAV's experimental model and can be used for the generation of photomaps. It has a satisfactory accuracy of the calculation in a near-real-time mode which will be improved during further refinement.

REFERENCES

Adding a Google Earth Overlay. (n.d.). Retrieved from https://help.gpsinsight.com/docs/about-maps/using-3d-maps/adding-a-google-earth-overlay/

Chyrkov A. (2016). Poshuk pidozrilykh ob'jektiv na video z kamery bezpilotnogho litaljnogho aparatu na osnovi analizu ghistoghram [Finding suspicious objects on a video from an unmanned aircraft camera based on histogram analysis]. *Problems of creation, testing and operation of complex information systems: Collection of scientific work of Zhytomyr Military Institute named after S. P. Korolev*, *13*, 126–135.

Chyrkov, A., & Prystavka, P. (2018). *Algorithm for Suspicious Object Search on Video Stream from Aircraft Camera and an Example of its Application. In Advances in Computer Science for Engineering and Education. ICCSEEA 2018. Advances in Intelligent Systems and Computing* (Vol. 754, pp. 340–348). Cham, Switzerland: Springer.

Cleveret, T. (2016, December 5). *Google map overlayer.* Retrieved from http://cleveret.net/?page_id=161

Edwards, T., & Titchenal, S. (2009, November 19). *Creating image overlays for online mapping services.* Retrieved from http://railsandtrails.com/mapoverlays/imageoverlays.htm

Evenden, G. I. (2003). *Cartographic Projection Procedures for the UNIX Environment — A User's Manual.* Open-File Report 90-284. U.S. Geological Survey. Retrieved from http:// trac.osgeo.org/proj

Gis-Lab. (2010). *The format of the geographic registration file.* Retrieved from http://gis-lab.info/qa/tfw.html

Google Maps Overlay. (n.d.). Retrieved from https://www.w3schools.com/graphics/google_maps_overlays.asp

GOST R 51794-2008. (2009). *Globalnye navigatsionnye sputnikovye sistemy. Sistemy koordinat. Metody preobrazovaniya koordinat opredelyaemykh tochek* [Global navigation satellite systems. Coordinate systems. Methods for converting the coordinates of defined points]. Moskow: standardinform. Retrieved from http://ingeo-pro.ru/upload/SNIP/gost-r-51794-2008.pdf (in Russian)

Karpov, D. (2012). *Sshivka izobrazheniy, poluchennykh v rezultate aerofotosemki* [Cross-linking of images obtained as a result of aerial photography] (PhD thesis). St. Petersburg: St. Petersburg National Research University of Information Technologies, Mechanics and Optics. Retrieved from https://archive.li/o/fiaQi/is.ifmo.ru/projects/2012/karpov/description.pdf (in Russian)

Khramov, P. (2016, March 5). *"Maps" section: Photos and taxa linking to the maps markers*. Retrieved from http://insecta.pro/community/60496

Monastyrskyi, D. (2017, February 19). *Geotag photos. Part one.* Retrieved from https://monastyrskiy.ru/?p=3354

Nechausov, A., & Zamirec, O. (2005). Metodika sozdaniya mozaiki izobrazheniy na osnove dannykh bespilotnogo letatelnogo apparata [Technique of creation of a mosaic of images on the basis of data of an unmanned aerial vehicle] [in Russian]. *System Information Boxes, 133*, 51–56.

Open Street Map Project. (n.d.). Retrieved from http://wiki.openstreetmap.org/wiki/Main_Page

Overlay Tiled Images on a Map. (n.d.). Retrieved from https://docs.microsoft.com/en-us/windows/uwp/maps-and-location/overlay-tiled-images

Piskunov, O., Yurchuk, I., & Bilyanska, L. (2017). Vyznachennja oblasti bachennja kamery pry aerofotozjomci [Determination of camera vision field at aerophotography] [in Ukrainian]. *Science-Based Technologies, 35*(3), 204–208. doi:10.18372/2410-5431.35.11839

Polymaps. (n.d.). *Image overlay*. Retrieved from http://polymaps.org/ex/overlay.html

Prystavka, P., Chyrkov, A. & Sorokopud V. (2017) Eksperymentaljnyj zrazok avtomatyzovanoji systemy poshuku pidozrilykh ob'jektiv na video z bezpilotnogho povitrjanogho sudna [Experimental sample of automated search system for suspicious objects on unmanned aircraft video]. *Arms Systems and Military Equipment, 20*(2), 26–32.

Sotnik, M. (2003). Rabota s rastrom na nizkom urovne dlya nachinayushchikh [Working with a raster at a low level for beginners]. Retrieved from http:// habrahabr.ru/post/196578/ (in Russian)

Zheltov S., Veremeenko K., Kim N., Kozorez D. & Krasilshchikov N. (2009). *Sovremennye informatsionnye tekhnologii v zadachakh navigatsii i navedeniya bespilotnykh manevrennykh letatelnykh apparatov* [Modern information technologies in the tasks of navigation and guidance of unmanned maneuverable aircraft]. Moskow: House of Physical and Mathematical Literature Publishing. (in Russian)

KEY TERMS AND DEFINITIONS

Global Positioning System (GPS): A satellite-based radio navigation system owned by the United States government and operated by the United States Air Force. It is a global navigation satellite system that provides geolocation and time information to a GPS receiver anywhere on or near the Earth where there is an unobstructed line of sight to four or more GPS satellites.

Information Technology: The use of computers to store, retrieve, transmit, and manipulate data or information, often in the context of a business or other enterprise.

Map Overlay: The compiling onto a single map all the disqualified areas on the individual maps and then choosing among whatever qualified locations remain.

Pulkovo 1942 (SC-42): A coordinate system established in the Soviet Union in 1942 and provides parameters which are linked to the geocentric Cartesian coordinate system PZ-90. It was used in geodetic calculations, notably in military mapping and determining state borders.

Raster Map: A dot matrix data structure that represents a generally rectangular grid of pixels (points of color), viewable via a monitor, paper, or other display medium and encodes geographic data in the pixel values as well as the pixel locations.

Unmanned Aerial Vehicle (UAV): An aircraft without a human pilot aboard.

World Geodetic System (WGS): A standard for use in cartography, geodesy, and satellite navigation including GPS. It comprises a standard coordinate system for the Earth, a standard spheroidal reference surface (the datum or reference ellipsoid) for raw altitude data, and a gravitational equipotential surface (the geoid) that defines the nominal sea level.

Chapter 8
Guidance Algorithm for Unmanned Aerial Vehicles on a Basis System of Technical Viewing

Oleksii Pikenin
National Technical University of Ukraine "Igor Sikorsky Kyiv Polytechnic Institute", Ukraine

Oleksander Marynoshenko
National Technical University of Ukraine "Igor Sikorsky Kyiv Polytechnic Institute", Ukraine

EXECUTIVE SUMMARY

The chapter considers a description of developed control system for a group of unmanned aerial vehicles (UAV) that has a software capable to continue the flight in case of failures by using alternative control algorithms. Control system is developed on vision system by using methods of image recognition. Grouped coordinated flight of UAVs can significantly improve the performance of surveillance processes, such as reconnaissance, image recognition, aerial photography, industrial and environmental monitoring, etc. But to control a group of UAVs is a quite difficult task. In this chapter, the authors propose a model that corresponds to the principle of construction by the leading UAVs. In the case of using this model, the parameters of the system motion are determined by the direction of motion, the speed, and the acceleration of the UAVs' driving. The control system based on the methods of image recognition expands the possibilities of coordinating the group of UAVs.

BACKGROUND

Unmanned aircraft systems (UAS) are playing increasingly prominent roles in defense programs and defense strategy around the world. Technology advancements have enabled the development of both large unmanned aircraft (e.g., Global Hawk, Predator) and smaller, increasingly capable unmanned aircraft (e.g., Wasp, Nighthawk). As recent conflicts have demonstrated, there are numerous military applications for unmanned aircraft, including reconnaissance, surveillance, battle damage assessment, and communications relays.

DOI: 10.4018/978-1-5225-7588-7.ch008

Civil and commercial applications are not as well developed, although potential applications are extremely broad in scope, including environmental monitoring (e.g., pollution, weather, and scientific applications), forest fire monitoring, homeland security, border patrol, drug interdiction, aerial surveillance and mapping, traffic monitoring, precision agriculture, disaster relief, ad hoc communications networks, and rural search and rescue. For many of these applications to develop to maturity, the reliability of UAS needs to increase, their capabilities need to be extended further, their ease of use needs to be improved, and their cost must decrease. In addition to these technical and economic challenges, the regulatory challenge of integrating unmanned aircraft into national and international airspace needs to be overcome.

The terminology unmanned aircraft system refers not only to the aircraft, but also to all of the supporting equipment used in the system, including sensors, microcontrollers, software, ground station computers, user interfaces, and communications hardware. This text focuses on the aircraft and its guidance, navigation, and control subsystems (Beard & McLain, 2012).

One of the important and actual purposes of using UAVs is the application of it in mixed groups, including manned and unmanned aerial vehicles, or as a part of operating autonomously UAVs group. To solve the navigational task for UAVs group, we use a control system based on image recognition methods. The methods of image recognition are based on tracking, identification, and detection of mobile and stationary air and land objects. To create a system, it is necessary to analyze the problem of classifying terrestrial stationary objects for constructing the flight trajectory of a group.

Formation flight control of multiple UAVs is an active topic for numerous researches see (Das et al., 2002; Hammer et al., 2004;Soleymani & Saghafi, 2010; Gosiewski & Ambroziak, 2013), with the much practical application: reconnaissance, communication, search and rescue.

There are many research methods proposed for implementation of multiple UAVs control, especially for control of UAVs formation flying, such as leader following (Soleymani & Saghafi, 2010), behavior-based approach (Hammer et al., 2004), virtual leader (Gosiewski & Ambroziak, 2013) and artificial potential functions (Paul, et al., 2008). In these methods, most appropriative is our leaders' methods.

The big problem information control is a question of creation and full usage of the neighbor-to-neighbor communication and synchronization. The well-known today methods for communication and synchronization inside of formation of UAVs are methods of use of video information (Das et al., 2002), and methods of use of radio transmitting data (Paul, et al., 2008).

The current development of aviation sets the task of implementation of the formation flight of unmanned aerial vehicles.

The necessity of development of technology for control of formation flying UAVs now opens a very important area: the creation of between onboard unmanned navigation systems (BONUS) for UAVs with very limited weight and volume. This need is determined by the fact that the absence of equipment onboard the UAV BONUS can greatly limit their opportunities.

Development of systems like a BONUS goes by two approaches: the creation of autonomous systems that do not depend on ground guiding systems, and onboard systems using ground-based radio beacons (Paul et al., 2008). Each of these ways has its own advantages and disadvantages. The autonomous management system can solve the problem of formation flight without restrictions imposed by the channels for communication with ground control, as well as in radio jamming.

Development of a "guidance system" for an autonomous navigation system of the grouped co-ordinated flight of UAVs by using methods of image recognition. Implementation of control algorithms for recognition of mobile and stationary air and land navigation objects.

Model UAVs Formation Flying

Group of the moving system – UAVs is usually regarded as a system of connected rigid bodies with significant degrees of freedom. And the number of degrees of freedom increases significantly with the increase of the number of UAVs in the group that makes the model of spatial movement extremely difficult and unsuitable for solving the problem of synthesis of a coherent control of the whole set of aircraft. In this connection, widely used is model group UAV relative motion, according to which the group is allocated carrier body (vehicle) and the transportable body. As the carrier serves the leading UAV, guided aerial vehicles (UAVs) play a role of transportable bodies. At the same time, any type of order: column, front, diamond, bearing, a wedge or a mixed order - can be viewed as a set of pairs: a master-slave. It should be noted that the formation of such pairs can be created based on two principles. In one case, the binding is carried by driven UAV marching to the leading UAV. In the second case, all driven UAVs determine their movements relative to the general for all - the leading UAV.

Without over viewing of the details and the characteristics of each method for forming a UAV system, note that in this we consider a model that meets the principle of building by the leading UAV.

As already noted, the group traffic control problem is related to the need to study the motion of the aircraft that are in certain relationships. Therefore, in the foreground a study of their relative motion.

Two aircraft movement relative to each other is the difference between two absolute movements and has three degrees of freedom.

In considering the relative motion of the aircraft, you can use a variety of the coordinate systems. Each system has its advantages and disadvantages. The choice is determined by its specific task. However, there are general principles of selection that determine the desirability or necessity of a system of reference. Given the simplicity of the dynamics equations obtained for simplification of analysis tasks and integration of these equations for the joint analysis of UAV motion group as a reference trajectory leading UAV to choose a coordinate system.

The beginning of the coordinate system has to be advice to combine with a center of mass of the driven aircraft. At the driven UAVs measurement equipment for determining of the parameters of his relative motion relatively of leading UAV are placed. So the most appropriated coordinate system is a trajectory coordinate system.

The Structure of the "Guidance System" of the Grouped Co-ordinated Flight of UAVs

The "guidance system" creates additional navigation signals for the control system. The structure of the control system is constructed in such a way that it allows solving two subtasks of the UAVs guidance for mobile and stationary air and land navigation objects. The structural-functional diagram of such a "guidance system" is shown in Figure 1. Block of control system includes an optical sensor (OS) on a rotating unit (RU). OS transmits streaming video or photos of the observed scene (the Earth's surface and Space in front of UAV) on a computer system (CS), which recognize land and mobile air facilities on the basis of the generated descriptors of these objects in the navigation database. The onboard CS contains the algorithms for calculating the angular coordinates of the selected objects. After calculating the angular coordinates, CS transmits control signals to the autopilot.

Figure 1. The structure of the "guidance system".

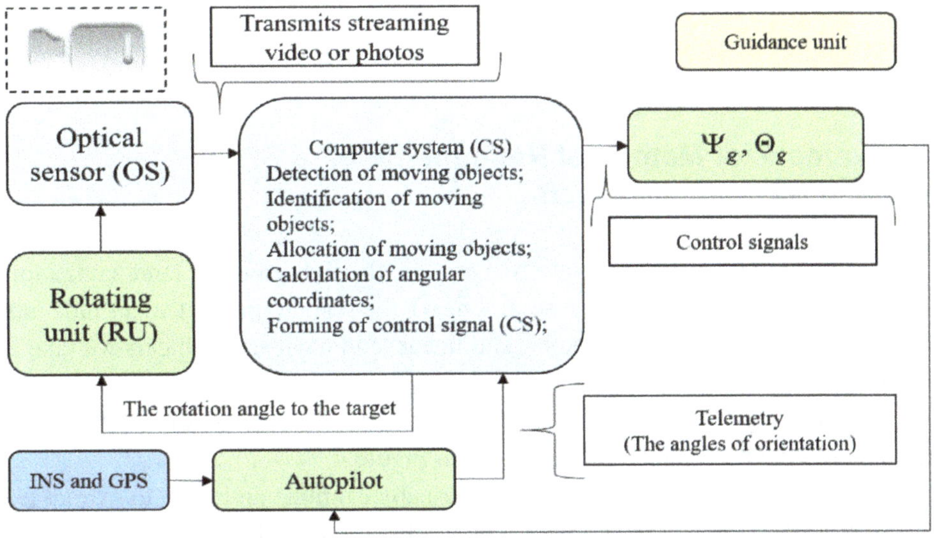

Control System for Flying of UAVs Formation.

The basis of the control group assigned flight system consists of the following modules-blocks: (1) - "guidance system" which includes a vision system (video camera) and pattern recognition algorithm (for forming control information about the position of the leading UAV); (2) - flight control system driven UAV which is based on control laws for driven UAV in accordance with the flight program for the group of UAVs; (3) - the navigation system to provide the necessary control, navigation and flight information.

The video system that installed on board each of the UAV, as well as pattern recognition algorithms, form a guidance system. Such a system, in accordance with the proposed algorithm discussed above, calculates the location data of the leading UAV. The data are converted into command signals characterizing the change in distance to the leading UAV, its vertical and horizontal displacements. An example of such output signals is shown in Figure 2. Further guidance system generates signals on the UAV control loops for the orientation angles of the pitch roll yaw, and to control a distance between the aircraft. In such control, commands are also included flight command for the order form (the order of flight) of UAVs in the group. After the control system, in other words, the autopilot commands supplied to UAV control

Figure 2. UAV flight control system.

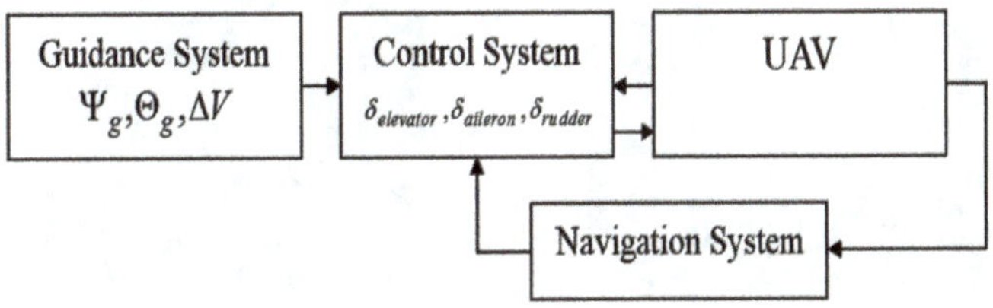

surfaces. At the same time, the autopilot includes a control and stabilization circuits not only on linear and angular displacements and for the velocities and accelerations. Achieving the required parameters for transients (overshoot, decay time, minimizing errors) is provided in our proposed system using the respective PI and PID controllers for longitudinal and lateral movement channels (Stevens & Lewis, 2003).

Illustrative Example for Method of Recognition of Stationary Land Navigation Objects

Stationary land navigation objects divide into two groups: the first group of land navigation objects at an altitude of fewer than 1.5 km (e.g., point, areal, linear), the second group at an altitude more than 1.5 km. At an altitude of fewer than 1.5 km, point, and linear land navigation objects we take as the basis for constructing the flight curve shown in Figure 3.

Buildings, lakes, railway stations, forests- especially important for the building of the flight trajectory as the surface area of land navigation objects shown in Figure.4.

Feature detection is the process where we automatically examine an image to extract features, that are unique to the objects in the image, in such a manner that we are able to detect an object based on its features in different images.

This detection should ideally be possible when the image shows the object with different transformations, mainly scale, and rotation, or when parts of the object are occluded. SIFT (Scale-Invariant Feature Transform) algorithm is a successful approach to feature detection introduced by Fazli et al., (2009). The SURF (Speeded-Up Robust Features) algorithm (Waldmann, 2002; Bay et al., 2009) is based on the same principles and steps, but it utilizes a different scheme and it should provide better results, faster.

To recognize such objects, we applied SURF algorithm (Lopez et al., 2003; Bay et al., 2009) which solves two problems - finding key points and creation of their image descriptors that are invariant to scale and rotation (orientation). This means that the description of the key points will be constant, even if the navigation object will change the size and position.

SURF algorithm looks for key points and compares them with the key point's descriptors from the database. Compiled navigation object will be a grouped descriptor of key points shown in Figure 5.

Figure 3. Detection of point and linear land navigation objects.

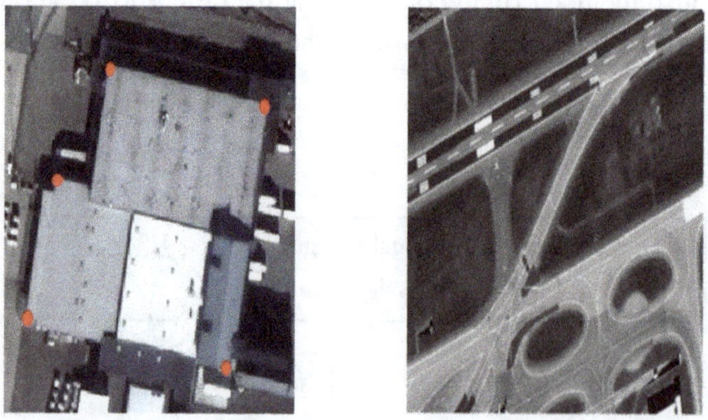

Figure 4. Detection of the area of land navigation objects.

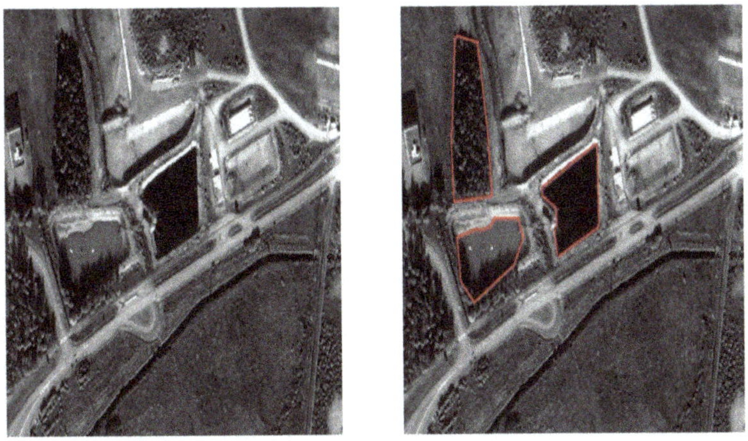

Figure 5. Visualization of descriptors of the linear and area land navigation objects.

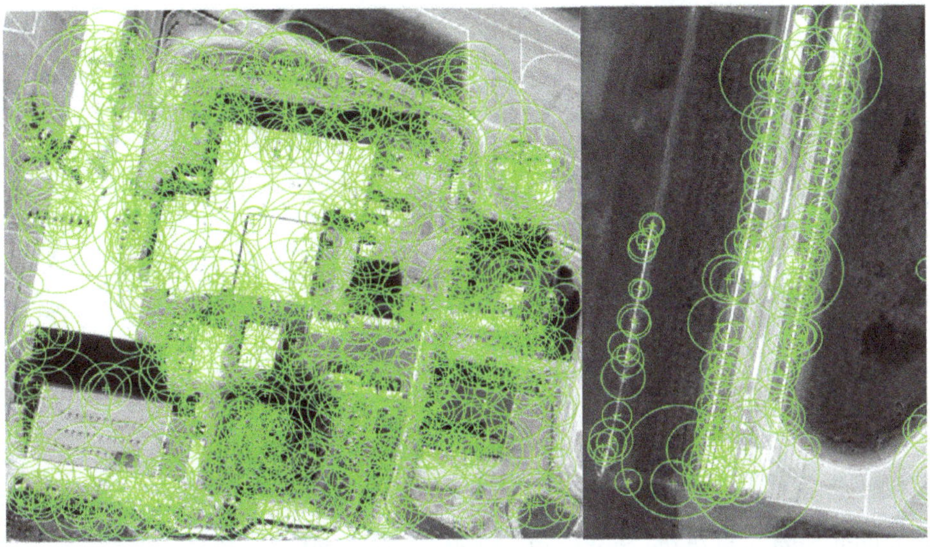

Theoretical Background and Illustrative Example for Proposed Method of Recognition of Mobile Land Navigation Objects

Technical means to implement the group flight of the aircraft are primarily those instruments and devices that allowing to define the parameters of the relative motion of the aircraft. They must receive the necessary distance measuring and angle measuring information. This measuring apparatus can be based on various physical principles for the radio, optical, quantum-mechanical see at (Stevens & Lewis, 2003; Beard & McLain, 2002).

In this section, we will assume that the origins of the gimbal and camera frames are located at the center of mass of the vehicle. The main reasons for the growing interest in small aircraft are that they provide an inexpensive platform for electro-optical (EO) and infrared (IR) cameras. Mobile land navigation objects will give us an opportunity to form the control system in a group, building a flight group and maintaining order in the ranks. To do this we have to match camera with a coordinate system. The camera's coordinate system and the coordinate system of the unmanned aerial vehicle are shown in Figures 6, 7.

Supposing that the optical axis of the camera coincides with a longitudinal axis of the UAV (OX^b axis of a body-fixed coordinate system $OX^bY^bZ^b$, $OX^CY^CZ^C$ - cameras local coordinate system, $OX^VY^VZ^V$ - airspeed coordinate system).

It is assumed that the origin of the reference frame and the camera located in the center of gravity of the aircraft.

Guidance for the goal (leader UAV) is obtained by two angles: Ψ_{az} azimuth to goal, θ_V elevation to goal. Where:

Figure 6. Matching camera with the coordinate system (top view)

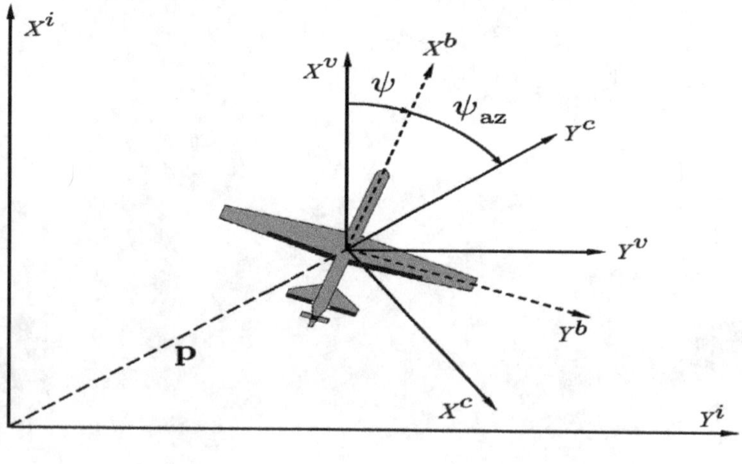

Figure 7. Matching camera with the coordinate system (side view).

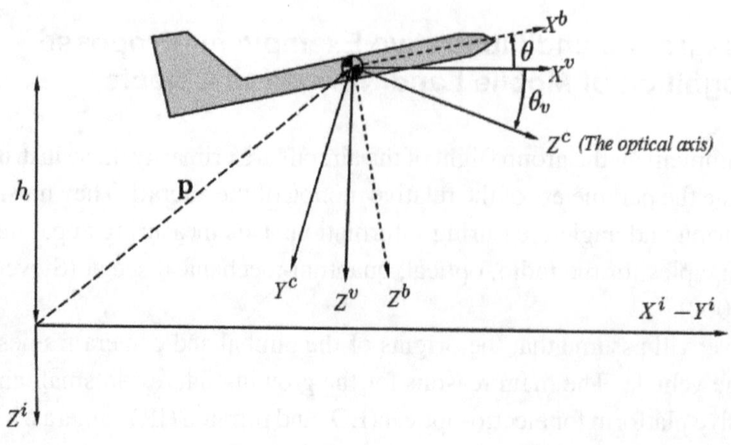

$$R_{\Psi_{az}} = \begin{bmatrix} \cos\Psi_{az} & 0 & -\sin\Psi_{az} \\ 0 & 1 & 0 \\ \sin\Psi_{az} & 0 & \cos\Psi_{az} \end{bmatrix}$$

is the matrix of rotation for azimuth;

$$R_{\theta_V} = \begin{bmatrix} \cos\theta_V & \sin\theta_V & 0 \\ -\sin\theta_V & \cos\theta_V & 0 \\ 0 & 0 & 1 \end{bmatrix}$$

is the matrix of rotation for elevation;

The resulting matrix of rotation for guidance to the goal is:

$$R_b^g = R_{\Psi_{az}} * R_{\theta_V} = \begin{bmatrix} \cos\theta_V \cos\Psi_{az} & \sin\theta_V \cos\Psi_{az} & -\sin\Psi_{az} \\ -\sin\theta_V & \cos\theta_V & 0 \\ \sin\Psi_{az} \cos\theta_V & \sin\theta_V \sin\Psi_{az} & \cos\Psi_{az} \end{bmatrix}.$$

If f - is a focal length in units of pixels, then converts the pixels P in meters. To simplify the discussion, we will assume that the pixels and the pixel array are square. To simplify the description, assume that the pixels and pixel array are squares. If the width of a square matrix of pixels in pixels is M and the v - cameras field of view is known, then the focal length f it can be written as:

$$f = \frac{M}{2\tan\left(\frac{v}{2}\right)}.$$

The position of the projection of the object expressed in the camera frame as $\left(P_{E_x}, P_{E_y}, f\right)$, where P_{E_x} and P_{E_y} define the position of the object in pixels. The distance from the origin of the camera system to a pixel position is $\left(P_{E_x}, P_{E_y}, f\right)$ $\sqrt{P_{Ex}^2 + P_{Ey}^2 + f^2}$ as shown in Figure 8.

Define the ort (single vector) for the direction of the goal:

$$\frac{l^c}{L} = \frac{1}{F}\begin{bmatrix} P_{E_x} \\ P_{E_y} \\ f \end{bmatrix} = \frac{1}{\sqrt{P_{Ex}^2 + P_{Ey}^2 + f^2}}\begin{bmatrix} P_{E_x} \\ P_{E_y} \\ f \end{bmatrix} = \vec{l} = \begin{bmatrix} \vec{l}_x \\ \vec{l}_y \\ \vec{l}_z \end{bmatrix};(1)$$

Figure 8. The camera coordinate system.

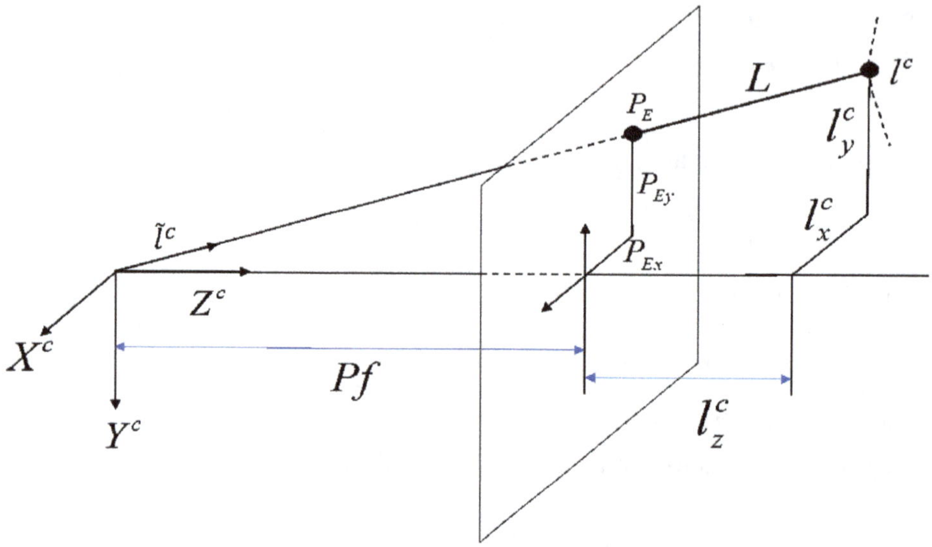

The task of formation of the unit vector pointing to the leading UAV reduced to the tracking of the image with leading UAV and definition of its geometric center.

Trajectory Equation of the Driven UAVs

The equations of the relative motion of the driven UAVs in the trajectory coordinate system can be represented as follows:

$$\Delta x_{gi} = x_{g1} - x_{gi}; \Delta y_{gi} = y_{g1} - y_{gi}; \Delta z_{gi} = z_{g1} - z_{gi};$$

$$\Delta Vx_{gi} = \frac{dx_{g1}}{dt} - \frac{dx_{gi}}{dt}; \Delta Vy_{gi} = \frac{dy_{g1}}{dt} - \frac{dy_{gi}}{dt}; \Delta Vz_{gi} = \frac{dz_{g1}}{dt} - \frac{dz_{gi}}{dt};$$

$$x_{gi}^{(1)} = \Delta x_{gi} \cos\theta_{vi} \cos\Psi_{azi} + \Delta y_{gi} \sin\theta_{vi} - \Delta z_{gi} \cos\theta_{vi} \sin\Psi_{azi};$$

$$y_{gi}^{(1)} = \Delta x_{gi} \sin\Psi_{azi} + \Delta y_{gi} \cos\Psi_{azi}; \tag{2}$$

$$z_{gi}^{(1)} = \Delta x_{gi} \sin\theta_{vi} \cos\Psi_{azi} - \Delta y_{gi} \sin\theta_{vi} \sin\Psi_{azi} - \Delta z_{gi} \cos\theta_{vi};$$

$$\Delta Vx_{gi}^{(1)} = \Delta Vx_{gi} \cos\theta_{vi} \cos\Psi_{azi} + \Delta Vy_{gi} \sin\theta_{vi} - \Delta Vz_{gi} \cos\theta_{vi} \sin\Psi_{azi};$$

$$\Delta Vy_{gi}^{(1)} = \Delta Vx_{gi} \sin\Psi_{azi} + \Delta Vy_{gi} \cos\Psi_{azi};$$

$$\Delta Vz_{gi}^{(1)} = \Delta Vx_{gi} \sin\theta_{vi} \cos\Psi_{azi} - \Delta Vy_{gi} \sin\theta_{vi} \sin\Psi_{azi} - \Delta Vz_{gi} \cos\theta_{vi}$$

where: $x_{gi}^{(1)}, y_{gi}^{(1)}, z_{gi}^{(1)}$ and $\Delta Vx_{gi}^{(1)}, \Delta Vy_{gi}^{(1)}, \Delta Vz_{gi}^{(1)}$ – the coordinates and the relative speeds of the UAVs to the leading UAV in the trajectory coordinate system of the leading UAV angles: elevation to goal θ_{vi} and azimuth to goal Ψ_{azi} for each UAV in formation.

The linearization of the equations (see Eq. (2)) provides a set of equations observation that defines the trajectory of the carried UAVs in relative to the trajectory of the leading (master) UAV.

$$\delta x_{gi}^{(1)} = c_{11}^{(i)}\delta\theta_1 + c_{12}^{(i)}\delta\Psi_1 + c_{13}^{(i)}\delta x_{g1} + c_{14}^{(i)}\delta y_{g1} + c_{15}^{(i)}\delta z_{g1} - c_{13}^{(i)}\delta x_{gi} - c_{14}^{(i)}\delta y_{gi} - c_{15}^{(i)}\delta z_{gi};$$

$$\delta y_{gi}^{(1)} = c_{22}^{(i)}\delta\Psi_1 + c_{23}^{(i)}\delta x_{g1} + c_{25}^{(i)}\delta z_{g1} - c_{23}^{(i)}\delta x_{gi} - c_{25}^{(i)}\delta z_{gi}; \quad (3)$$

$$\delta z_{gi}^{(1)} = c_{31}^{(i)}\delta\theta_1 + c_{32}^{(i)}\delta\Psi_1 + c_{33}^{(i)}\delta x_{g1} + c_{34}^{(i)}\delta y_{g1} + c_{35}^{(i)}\delta z_{g1} - c_{33}^{(i)}\delta x_{gi} - c_{34}^{(i)}\delta y_{gi} - c_{35}^{(i)}\delta z_{gi};$$

$\delta\alpha$ denotes (small) deviations from the normal steady-state condition.

The Mathematical Model of the Aircraft in the Trajectory Coordinate System

The vector form of the equation of motion of the aircraft as a solid body moving in space has six degrees of freedom is derived from the theorem: the theorem change of impulse, and the theorem of change of momentum of impulse.

The complete system of equations of motion of an arbitrary spatial individual UAV in trajectory coordinates are as follows (in the trajectory coordinate system are presented equation of the translational motion (force equations) of UAV). For the creation of a control system for formation of UAVs will be used linearized mathematic model – a linearized system of equations of the UAV motion.

$$\delta\dot{V}_i = a_{11}^{(i)}\delta V_i + a_{12}^{(i)}\delta\alpha_i + a_{13}^{(i)}\delta\beta_i + a_{14}^{(i)}\delta\gamma_i + a_{15}^{(i)}\delta\theta_i + a_{16}^{(i)}\delta\Psi_i + \ldots;$$

$$\delta\theta_i = a_{21}^{(i)}\delta V_i + a_{22}^{(i)}\delta\alpha_i + a_{23}^{(i)}\delta\beta_i + a_{24}^{(i)}\delta\gamma_i + a_{25}^{(i)}\delta\theta_i + a_{26}^{(i)}\delta\Psi_i + \ldots;$$

$$\delta\dot{\Psi}_i = a_{31}^{(i)}\delta V_i + a_{32}^{(i)}\delta\alpha_i + a_{33}^{(i)}\delta\beta_i + a_{34}^{(i)}\delta\gamma_i + a_{35}^{(i)}\delta\theta_i + a_{36}^{(i)}\delta\Psi_i + \ldots;$$

$$\delta\dot{\omega}_{x_i} = a_{41}^{(i)}\delta V_i + a_{42}^{(i)}\delta\alpha_i + a_{43}^{(i)}\delta\beta_i + a_{44}^{(i)}\delta\gamma_i + a_{45}^{(i)}\delta\theta_i + a_{46}^{(i)}\delta\Psi_i + \ldots; \quad (4)$$

$$\delta\dot{\omega}_{y_i} = a_{51}^{(i)}\delta V_i + a_{52}^{(i)}\delta\alpha_i + a_{53}^{(i)}\delta\beta_i + a_{54}^{(i)}\delta\gamma_i + a_{55}^{(i)}\delta\theta_i + a_{56}^{(i)}\delta\Psi_i + \ldots;$$

$$\delta\dot{\omega}_{z_i} = a_{61}^{(i)}\delta V_i + a_{62}^{(i)}\delta\alpha_i + a_{63}^{(i)}\delta\beta_i + a_{64}^{(i)}\delta\gamma_i + a_{65}^{(i)}\delta\theta_i + a_{66}^{(i)}\delta\Psi_i + \ldots;$$

$$\delta\dot{X}_{g_i} = a_{71}^{(i)}\delta V_i + a_{72}^{(i)}\delta\alpha_i + a_{73}^{(i)}\delta\beta_i + a_{74}^{(i)}\delta\gamma_i + a_{75}^{(i)}\delta\theta_i + \ldots;$$

$$\delta\dot{Y}_{g_i} = \delta\dot{H}_i = a_{81}^{(i)}\delta V_i + a_{82}^{(i)}\delta\alpha_i + \ldots + b_{84}^{(i)}\delta u_{iT} = a_{81}^{(i)}\delta V_i = -\delta V_{i0}\theta_{i0};$$

$$\delta\dot{Z}_{g_i} = a_{91}^{(i)}\delta V_i + a_{92}^{(i)}\delta\alpha_i + a_{93}^{(i)}\delta\beta_i + \ldots + b_{94}^{(i)}\delta u_{iT} = a_{91}^{(i)}\delta V_i = \delta V_{i0}\Psi_{i0};$$

where: V – module of air velocity; α – angle of attack; β – sideslip angle; $a_{ml}^{(i)}$ – linearized coefficients for aerodynamic forces, moments, thrust and gravity forces in trajectory coordinate system; θ – tilt angle of trajectory; Ψ – flight path angle; ω_{ij} – the angular velocities of rotation around the fixed body axis; γ, ϑ, ψ – roll, pitch, yaw angle; X_g, Y_g, Z_g – aircraft coordinates relative to normal coordinate system X_0, Y_0, Z_0; $H = -Z_g$ – altitude.

Method of Recognition of Small Mobile Air Navigation Objects

Preliminary analysis revealed that SIFT algorithm is a potentially useful algorithm for our task. The specified descriptor is different both in terms of stability and information capacity. It is important that the characteristics of the descriptors change when the type of region is changed.

The SIFT algorithm is an algorithm of image matching that is completely invariant to affine transformations and must cover all six affine parameters. The SIFT method (Seok-Wun & Yong-Ho, 2011) covers 6 parameters by normalizing rotations and displacements and modeling all scales from the current frame and complex descriptor in memory. For solving our tasks, we are interested in small-sized fast-moving objects, such as UAVs.

For testing SURF algorithm for mobile aerial moving objects we build descriptor model shown in Figure 9 and compare it with the descriptor model of SURF algorithm (Figure 10).

For the detection of the UAVs geometric center on the video frame, we combined two methods, SIFT and SURF. After the addition of two algorithms, we build a complex descriptor of UAVs object. The coordinates of the geometric center of the object in the frame can be calculated in accordance with the position of identified by the integrated UAVs model. This approach will provide trimming of uninformative zones by SIFT method and found false points by SURF method. Searching the geometric center of the UAVs model descriptor in the frame reduced to a simple approach: the arithmetic average of the specific points in the descriptor or using the moving average value approach. These significantly reduce the volume of calculations, which is important when we operate in real time.

The results of locating the position of the geometric center of the unmanned aerial vehicle in the frame (in different positions relative to the center of the frame) are shown in Figure 11.

Figure 9. Visualization of descriptor model of SIFT algorithm.

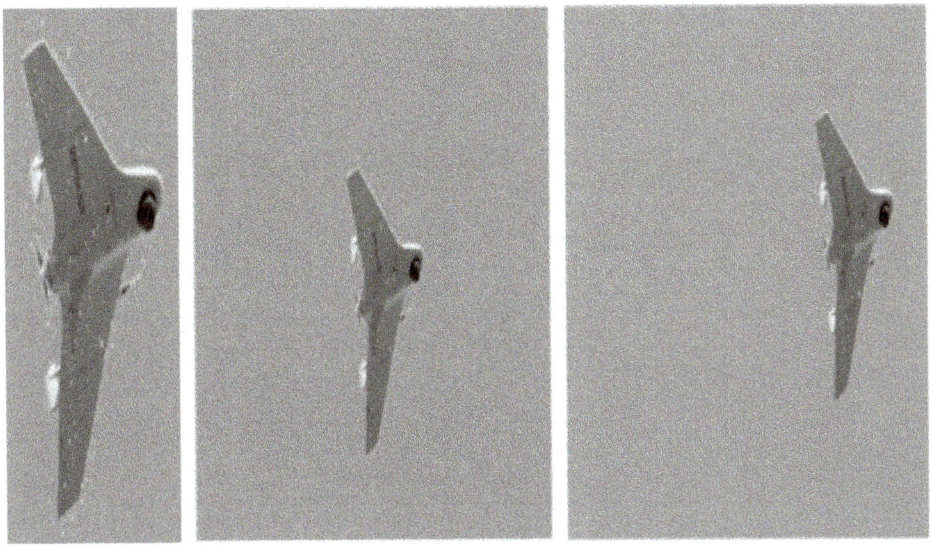

Figure 10. Visualization of descriptor model of SURF algorithm.

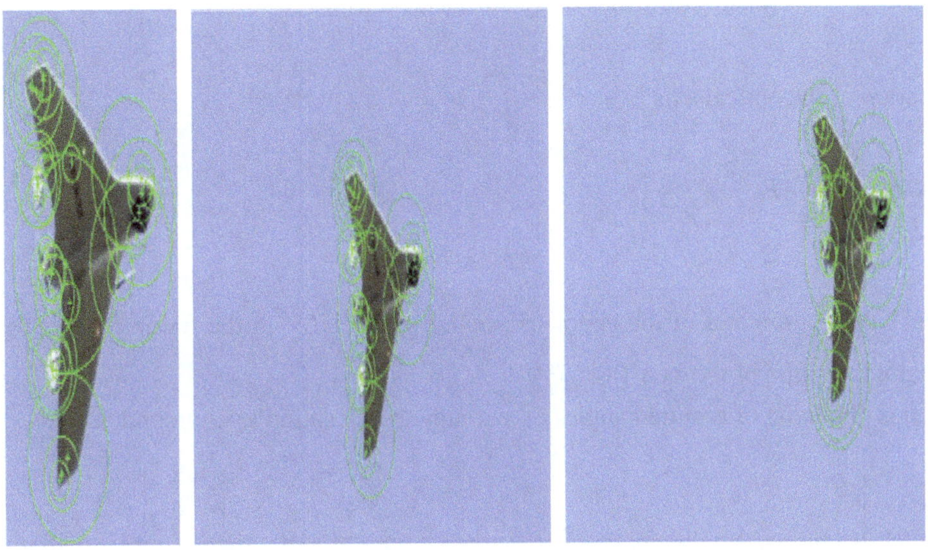

Describe a simple algorithm of guiding or targeting of the UAVs in formation flying. Assume the existence of the control tilts of the UAV frame that can be described in the form of control equations for azimuth and inclination angles:

$$\Psi_{az} = W^{\delta_{rc}}_{\Psi_{az}} \delta_{rc}, \Psi_{az} = W^{\delta_{ec}}_{\theta_V} \delta_{ec};$$

where δ_{rc}, δ_{ec} - commanded deflections of UAVs control surfaces, $W^{\delta_{rc}}_{\Psi_{az}}, W^{\delta_{ec}}_{\theta_V}$ - transfer functions of UAV.

Figure 11. Visualization of position the geometric center of the frame of leading UAV.

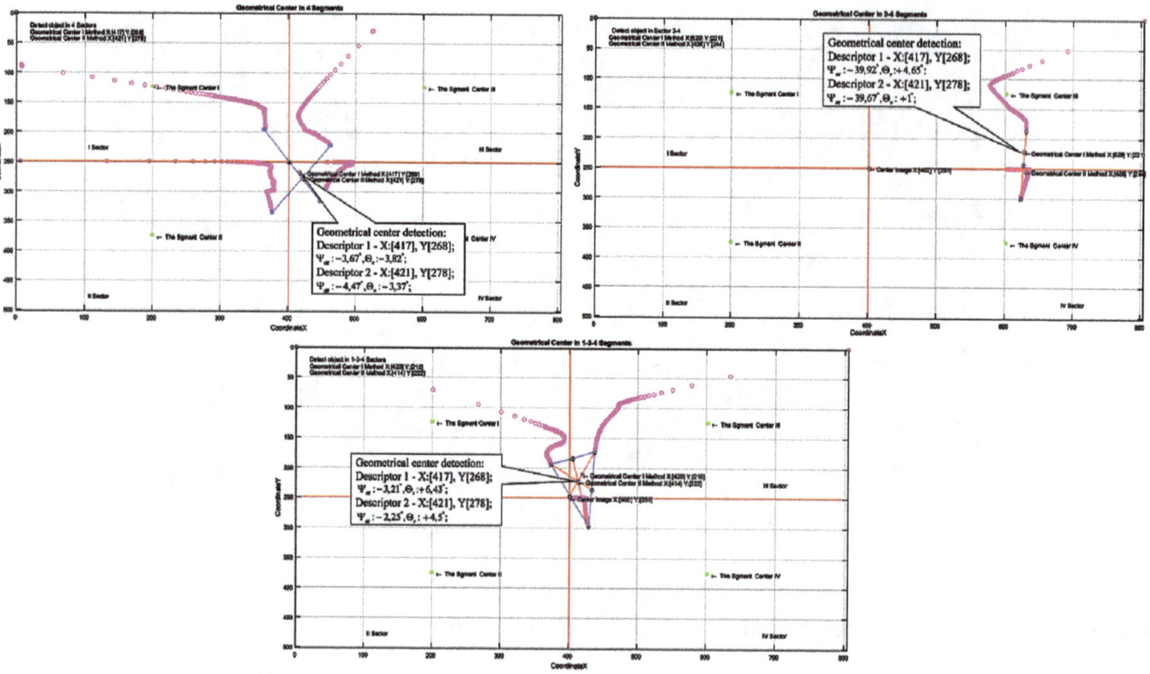

The equation of the optical axis is:

$$l = \frac{1}{\sqrt{P_{Ex}^{2} + P_{Ey}^{2} + f^{2}}} \begin{bmatrix} P_{E_x} & P_{E_y} & f \end{bmatrix}^{T}.$$

P_{E_x}, P_{E_y} - the coordinates of the geometric center of the UAV in the frame relative to the image center. Required position of the axis $l_r = R_b^{gT} l$.

Next step is obtaining of required angles of azimuth and elevation to align optical axes.

$$l_r = \begin{bmatrix} l_{xr} \\ l_{yr} \\ l_{zr} \end{bmatrix} = R_b^{gT} \begin{bmatrix} l_x \\ l_y \\ l_z \end{bmatrix};$$

$$l_r = \begin{bmatrix} l_{xr} \\ l_{yr} \\ l_{zr} \end{bmatrix} = \begin{bmatrix} \cos\theta_V \cos\Psi_{az} & \sin\theta_V \cos\Psi_{az} & -\sin\Psi_{az} \\ -\sin\theta_V & \cos\theta_V & 0 \\ \sin\Psi_{az} \cos\theta_V & \sin\theta_V \sin\Psi_{az} & \cos\Psi_{az} \end{bmatrix}^{T} \begin{bmatrix} l_x \\ l_y \\ l_z \end{bmatrix}.$$

To solve this equation of relative azimuth and elevation angles give as the following expressions for command angles:

$$\theta_V c = \tan^{-1}\left(\frac{l_y}{l_x}\right), \Psi_{az} = \sin^{-1}\left(-l_z\right). \tag{5}$$

So the servo commands are:

$$\delta_e = K_e\left(\theta_V c - \theta_V\right) = K_e\left[\tan^{-1}\left(\frac{l_y}{l_x}\right) - \theta_V\right];$$

$$\delta_r = K_r\left(\Psi_{az} c - \Psi_{az}\right) = K_r\left[\sin^{-1}\left(-l_z\right) - \Psi_{az}\right]. \tag{6}$$

According to the results obtained by determination of the object geometrical center, have the information about the focal length and the exact coordinates of the object in the frame where P_{E_x}, P_{E_y} using them in (1) and generate the control signals (5).

Modeling the System of Autonomous UAV Flight Control Group

For the modeling of the proposed control system was selected ZAGI UAV, geometrical and a complete list of aerodynamic characteristics of this UAV was used by Beard and McLain (2012). For the system shown in Figure 2, add the blocks with a mathematic model of the nonlinear behavior of UAV, (including the aerodynamic loads) instead of the "UAV" block. This module is based on a system of equations – the mathematical model of UAV.

Also for the unification of the proposed control system would be the necessary addition to the system structure module with parameter identification of the controlled object, for identification of aerodynamic derivatives, as individual parameters of each aircraft.

As the background for the creation, working algorithms of the guidance system are using equations (5) and (6) that describe the distance, angles of inclination and azimuth path to the leading UAV. The automatic control system (autopilot) includes control algorithms for the fulfillment of flight mission (speed, altitude, the direction of flight), and additional control commands concerning for creation the flight order of UAV formation.

Simulation of the proposed algorithm and the corresponding developed systems for providing UAV flight control group was conducted in MATLAB Simulink environment. As noted previously for the simulation model of control system example with two flying UAVs with flying order - line along the direction of flight of a leading UAV was described. The flight carried out horizontally at a given altitude. The procedure for the flight in the group is based on the equations (3), (4).

For the simulation of the guidance system and the formation of the next commands to control the flight of the UAV has been used a model of such a system. The simulation was carried out with a changing of the parameters of the horizontal flight of leading UAV. Horizontal and vertical displacement of the leading UAV and the corresponding change in the distance between a pair of UAV took place on the second and fourth seconds of the simulation. These signals with a distance of one hundred meters correspond to the horizontal and vertical displacement of the leading UAV 2 and 1 meter respectively.

Figure 12 shows the command outputs of the "guidance system" that characterize the vertical and horizontal displacement of the leading UAV.

The response of driven UAV in the case of maintaining the flight order in the UAVs group are shown in Figure 13.

As can be seen from Figure 13, the simulation results the system practically have no errors in the output parameters of the flight, as well as provided with a minimum overshoot and oscillation.

FUTURE RESEARCH DIRECTIONS

On the basis of the developed recognition algorithms and spatial detection of UAVs, the UAV group flight algorithms that execute the commands of the so-called "virtual" leader can be implemented as part of the group.

The implementation of such a modern autonomic approach requires the additional creation of a common system for exchanging data between the elements of the group. It also requires the development of a group (multi-channel control) system for UAV- members of the group. In this case, such a control system will be based on the commands and flight parameters of the virtual leader. The virtual leader is supposed to be a set of software algorithm with several variants of the flight task solution (for each of the group agents) and navigation parameters determined locally depending on the UAV group

Figure 12. Visualization of command output of "guidance system".

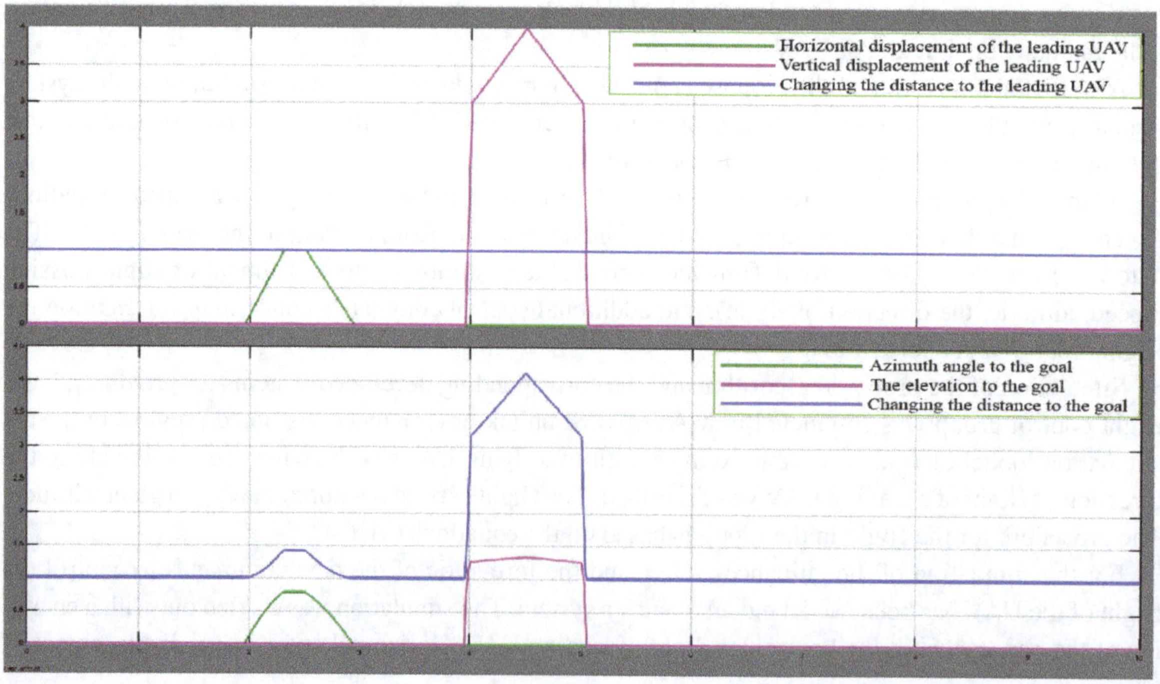

Figure 13. The reaction of the driven UAV as a response to the changing of the flight parameters of the leading UAV.

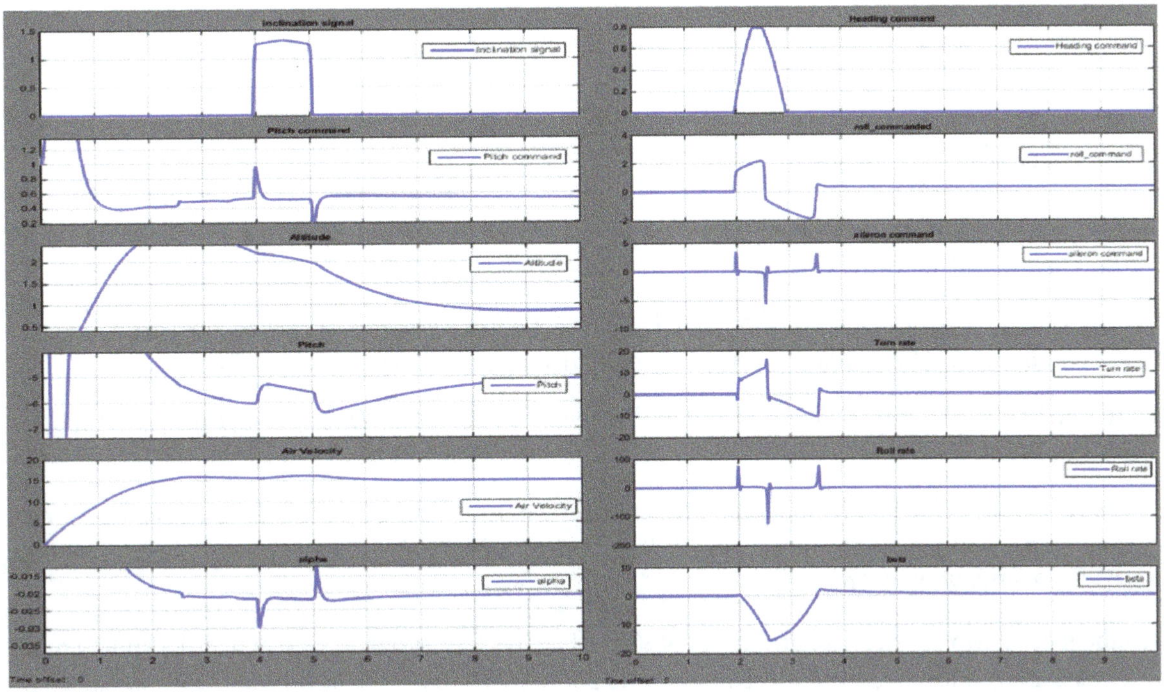

configuration and using additional subsystems and algorithms of the local spatial navigation detection of UAVs that are parts of the group.

CONCLUSION

In this work, an opportunity implementation of multiple UAVs control has been provided, especially for control of UAVs formation flying, using the leader following approach. Based on this obtained algorithms most appropriative methods of virtual leader behavior method will be researched and developed. For the creation neighbor-to-neighbor communication and synchronization, such algorithm with using electro-optical or infrared cameras and pattern recognition approaches have been proposed and realized. Also, their approach for implementation of such control system is given, as well as algorithms and principles of works, based on appropriate mathematical models. We developed a "guidance system" for an autonomous navigation system of the grouped coordinated flight of UAVs by using recognition methods of mobile and stationary air and land navigation objects. On the basis of the obtained results by finding the coordinates of the geometric center and the mathematical model of the camera we formed a mathematical model of the control signals for UAVs "guidance system".

ACKNOWLEDGMENT

This research was supported by Grant-in-Aid for Scientific Research of the National Technical University of Ukraine "Igor Sikorsky Kyiv Polytechnic Institute" [no. 0115U002524, UA].

REFERENCES

Bay, H., Ess, A., Tuytelaars, T., & Gool, L. V. (2009). Speeded-Up Robust Features. *Proceedings of the 9th European Conference on Computer Vision,* 404-417.

Bay, H., Ess, A., Tuytelaars, T., & Van Gool, L. (2008). SURF: Speeded Up Robust Features. *Computer Vision and Image Understanding, 110*(3), 346–359. doi:10.1016/j.cviu.2007.09.014

Beard, R. W., & McLain, T. W. (2012). *Small Unmanned Aircraft: Theory and Practice.* Princeton University Press.

Beard, R. W., McLain, T. W., Goodrich, M., & Anderson, E. P. (2002). Coordinated target assignment and intercept for unmanned air vehicles. *IEEE Transactions on Robotics and Automation, 18*(6), 911–922. doi:10.1109/TRA.2002.805653

Benallegue, A., Mokhtari, A., & Fridman, L. (2006). Feedback linearizationand high order sliding mode observer for a quadrotor UAV. *Proceedings of the 2006 International Workshop on Variable Structure Systems*, 365. 10.1109/VSS.2006.1644545

Castillo, P., Lozano, R., & Dzul, A. (2005). *Experimental implementation of linear and nonlinear control laws.* IEEE Control Systems Magazine.

Chao, X., Zeping, L., Guangquan, X., Zhiyong, F., Hongyan, T., & Haifeng, Z. (2015). *3D Reconstruction of Tree-Crown Based on the UAV Aerial Images* (Vol. 2015). Mathematical Problems in Engineering.

Choset, H., & Burdick, J. (2000). Sensor-based exploration: The hierarchical generalized Voronoi graph. *The International Journal of Robotics Research, 19*(2), 96–125. doi:10.1177/02783640022066770

Choset, H., Lynch, K. M., Hutchinson, S., Kantor, G., Burgard, W., Kavraki, L., & Thrun, S. (2005). *Principles of Robot Motion: Theory, Algorithms, and Implementation.* Cambridge, MA: MIT Press.

Das, A. K., Fierro, R., Kumar, V., Ostrowski, J. P., Spletzer, J., & Taylor, C. J. (2002). A Vision-Based Formation Control Framework. *IEEE Transactions on Robotics and Automation, 18*(5), 813–825. doi:10.1109/TRA.2002.803463

Fazli, S., Pour, H. M., & Bouzari, H. (2009). Particle Filter Based Object Tracking with Sift and Color Feature. *Second International Conference on Machine Vision*, 89–93. 10.1109/ICMV.2009.47

Felzenszwalb, P.F., Girshick, R.B., McAllester, D., & Ramanan, D. (2009). Object detection with discriminatively trained part based models. *IEEE PAMI, 99.*

Gosiewski, Z., & Ambroziak, L. (2013). UAV Autonomous Formation Flight Experiment with Virtual Leader Control Structure. *Diffusion and Defect Data, Solid State Data. Part B, Solid State Phenomena, 198*, 254–259. doi:10.4028/www.scientific.net/SSP.198.254

Hammer, J., Piper, G., Thorp, O., & Watkins, J. (2004). *Investigating Virtual Structure Based Control Strategies for Spacecraft Formation Maneuvers*. Providence, RI: AIAA Guidance, Navigation, and Control Conference and Exhibit. doi:10.2514/6.2004-4898

Lee, D., Lillywhite, K., Fowers, S., Nelson, B., & Archibald, J. (2006). An embedded vision system for an unmanned four-rotor helicopter. SPIE Optics East, Intelligent Robots and Computer Vision XXIV: Algorithms, Techniques, and Active Vision.

Lopez, J., Markel, M., Siddiqi, N., Gebert, G., & Evers, J. (2003). Performance of passive ranging from image flow. *Proceedings of the IEEE International Conference on Image Processing*, 1, 929–932. 10.1109/ICIP.2003.1247116

Luiz, G. B., Mirisola, & Dias, J. (2009). Exploiting Attitude Sensing in Vision-Based Navigation for an Airship. Journal of Robotics.

McLain, T., & Beard, R. (2000). Cooperative rendezvous of multiple unmanned air vehicles. *Proceedings of the AIAA Guidance, Navigation and Control Conference, (Denver, CO), AIAA 2000-4369.*

Panchal, P. M., Panchal, S. R., & Shah, S. K. (2013). A Comparison of SIFT and SURF. *International Journal of Innovative Research in Computer and Communication Engineering, 1*(2).

Paul, T., Krogstad, T. R., & Gravdahl, J. T. (2008). Modeling of UAV Formation Flight Using 3D Potential Field. *Simulation Modelling Practice and Theory, 16*(9), 1453–1462. doi:10.1016/j.simpat.2008.08.005

Rhudy, M., Yu, G., Haiyang, C., & Gross, J. (2015). Unmanned Aerial Vehicle Navigation Using Wide-Field Optical Flow and Inertial Sensors. Journal of Robotics.

Seok-Wun, H., & Yong-Ho, M. (2011). Multiple Object Tracking Using SIFT Features and Location Matching. *International Journal of Smart Home, 5*(4), 17–26.

Shin, J., & Kim, H. J. (2009). Nonlinear model predictive formation flight. *IEEE Transactions on Systems, Man, and Cybernetics. Part A, Systems and Humans, 39*(5), 1116–1125. doi:10.1109/TSMCA.2009.2021935

Soleymani, T., & Saghafi, F. (2010). Behavior-Based Acceleration Commanded Formation Flight Control. *International Conference on Control, Automation and Systems.*

Stevens, L. B., & Lewis, L. F. (2003). *Aircraft Control and Simulation* (2nd ed.). Hoboken, NJ: John Wiley & Sons, Inc.

Sumner, R. W., & Popovic, J. (2004). Deformation transfer for triangle meshes. *ACM Transactions on Graphics, 23*(3), 399–405. doi:10.1145/1015706.1015736

Waldmann, J. (2002). Line-of-sight rate estimation and linearizing control by an imaging seeker in a tactical missile guided by proportional navigation. *IEEE Transactions on Control Systems Technology, 10*(4), 556–567. doi:10.1109/TCST.2002.1014675

KEY TERMS AND DEFINITIONS

Descriptors: In computer vision, visual descriptors or image descriptors are descriptions of the visual features of the contents in images, videos, or algorithms or applications that produce such descriptions. They describe elementary characteristics such as the shape, the color, the texture or the motion, among others.

Machine Vision: MV is the technology and methods used to provide imaging-based automatic inspection and analysis for such applications as automatic inspection, process control, and robot guidance, usually in industry. Machine vision is a term encompassing a large number of technologies, software and hardware products, integrated systems, actions, methods, and expertise.

Pattern Recognition: Pattern recognition is the automated recognition of patterns and regularities in data. Pattern recognition is closely related to artificial intelligence and machine learning, together with applications such as data mining and knowledge discovery in databases, and is often used interchangeably with these terms.

Scale-Invariant Feature Transform: SIFT is an algorithm in computer vision to detect and describe local features in images.

Speeded Up Robust Features: In computer vision, speeded up robust features (SURF) is a patented local feature detector and descriptor. It can be used for tasks such as object recognition, image registration, classification, or 3D reconstruction.

Unmanned Aerial Vehicle: A UAV, commonly known as a drone, is an aircraft without a human pilot aboard. UAVs are a component of an unmanned aircraft system (UAS), which include a UAV, a ground-based controller, and a system of communications between the two. The flight of UAVs may operate with various degrees of autonomy: either under remote control by a human operator or autonomously by onboard computers.

Visual Navigation: Visual navigation is a technique often employed in light aircraft, which operate at relatively low speeds and heights when the weather is good and visual contact can be maintained with the ground for most of the flight.

Chapter 9
Information Technology for the Coordinated Control of Unmanned Aerial Vehicle Teams Based on the Scenario-Case Approach

Vladimir Sherstjuk
Kherson National Technical University, Ukraine

Maryna Zharikova
Kherson National Technical University, Ukraine

EXECUTIVE SUMMARY

The authors present a dynamic scenario-case approach to coordinated control of heterogeneous ensembles of unmanned aerial vehicles, which use coordination patterns of activity in similar situations described as spatial configurations affected by observed events. The method of obtaining deviations for approximate spatial configurations, which allows obtaining elements of the safe vehicle's trajectories. The method of qualitative safety assessment is presented. It uses a soft level topology to obtaining blurred boundaries of dynamic safety domains using fuzzy soft level sets and allows finding suitable compensations of vehicles' activity scenarios that can both keep the spatial configuration and satisfy all safety restrictions. The authors demonstrate that the proposed approach significantly reduces the computational complexity of problem solving and provides the acceptable performance.

BACKGROUND

Modern complex technical systems are becoming increasingly sophisticated, with a mixture of unmanned vehicles (UV) being used to solve various tasks that are dangerous to human life. In recent years UVs are widely used in various fields. The modern UVs are not remotely controlled by a human operator or a computer program, but they are autonomous and therefore are equipped with sophisticated sensors,

DOI: 10.4018/978-1-5225-7588-7.ch009

actuators, and on-board computer programs for intelligent control. Thus, autonomous vehicles must be able to make their own decisions in a highly dynamic, partially observable, and unpredictable environment. Significant advances in modern technologies ensure availability of various UVs of different sizes, equipment, and purposes working together in various environments as a team (Waslander, 2013).

One of the applications that motivates the use of multiple unmanned vehicles is forest fire fighting (Yuan, Zhang & Liu, 2015), where unmanned aerial vehicles (UAV) should provide surveillance and situation monitoring (Merino, Martínez de Dios & Ollero, 2015), and a variety of unmanned ground vehicles (UGV) such as bulldozers, excavators, fire hydrants, cisterns, trucks, etc., can be used as a team to combat forest fires. Obviously, each vehicle has different role and functions, but it executes a certain scenario jointly and simultaneously with the other UVs to achieve a mission objective (Sherstjuk, Zharikova & Sokol, 2018). Due to differences in vehicles features, destinations, their roles in a team, and various environments, such team is called heterogeneous ensemble of UVs. The ensembles may include objects moving in different environments. It is essential that such ensembles are very difficult to coordinate and control remotely . The most important role in such teams is played by UAVs. Due to lower cost of UAVs, it is possible now to build very large teams, which can perform search-and-rescue, first response, defense tasks etc (Sharifi, Mirzaei, Zhang & Gordon, 2015). Teams of vehicles are useful for complex, long-term, multi-task, multi-stage operations, and usually based on the swarming or flocking behavior (as birds or insects) (Toner & Yuhai, 1998).

However, in many cases the team of universal UAVs used as swarm is excessively expensive solution. This is particularly essential for forest fire-fighting and military applications where the cost of universal UAVs can be quite significant. The joint use of specialized UAVs of different classes aimed at solving very specific problems in complex missions, could be more appropriate solution. A team of vehicles can be organized in much more complicated way than a swarm; it may include a certain order and assign specific roles for vehicles in this order. At the same time, vehicles should select autonomously a relevant scenario of activity in dynamic environments within their role.

In general, the heterogeneous ensemble of UAV is a uniform ordered set of vehicles with different roles and functions, which jointly and simultaneously execute their scenarios of activity within a specified mission to achieve the objective. At the same time, the more complicated the ensemble structure and functions are , the more difficult the coordination of UAVs is. In the following, we will consider the complex structure of the UAVs' team as ensemble. One of the most important tasks of the ensemble's activity is a joint motion of its vehicles (Chen, Zhang, Xin & Fang, 2016). As a rule, the joint motion is limited by the space, by the given positional and functional restrictions, by the taken normative rules, and by the reaction of an environment, which gives rise to the different dynamic, navigation and situational disturbances into different points of space (Sherstjuk, Zharikova, Sokol & Tarasenko, 2017). Nevertheless, the most important limitation of the joint motion is a guaranteed safety of vehicles, so the problem of its maintenance is very essential. A growth of vehicles number and their size, a significant increase in their speed and density of movement within the confined space have lead to increase in the number of incidents and accidents, which in turn raises an important problem of ensuring a safe motion control. Beside that, the difference in the laws and features of UV motion in different environments leads to another significant impact. For example, UAV cannot stop on the fly as well as change its motion direction abruptly, while the most of UGVs can. Unfortunately, the problem of coordinated real time motion control for heterogeneous vehicles' ensembles still remains open, and has a great interest for research. Thus, the most topical issue for today is to develop the intelligent control and coordination system for heterogeneous UV ensemble under the confined navigation conditions, given a great amount of calculations with significant time limitations, which pose significant risks and threats to vehicles' safety.

When working in ensemble, vehicles should be able to communicate, coordinate, and cooperate with each other. There are many approaches for coordination control of UAVs, and there is extensive literature on this subject. The coordination and cooperation problems encountered in teams can widely range from collision avoidance to distributed planning (Tošić & Vilalta, 2010a).

The most common approach to address these problems is to use a bio-inspired techniques: it is believed that if vehicles can communicate, it should be possible to organize and coordinate them as a group based on the swarming or flocking behavior (Toner & Yuhai, 1998). Thus, most researchers devoted considerable attention to solving particular coordination tasks with swarms using a simple set of rules that mimic behaviors in natural systems (Mataric, 1995; Jadbabaie, Lin & Morse, 2003). This approach requires only limited information and is robust to individual failures; however, it relies on establishing consensus among vehicles that may take a very long time, so it is unacceptable for real-time applications with strict performance requirements in uncertain adversarial environment (Michael & Kumar, 2011). In a more complicated form UAVs' coordination and cooperation requires a more formal description of its rules and procedures. In behavior-based approaches, model-based or rule-based techniques are used for this purpose as well as algorithmization of interaction procedures (Lawton, Beard & Young, 2003). However, the complexity of UAV's interactions prevents the direct use of a model-based approach. A problem of a-priori building an exhaustive set of rules for coordination is not feasible and verification of this set is unrealizable in real-time systems, so the rule-based approach is unacceptable as well. Although these approaches allow detailed specifications and analysis of teams' behaviors, they are difficult to adapt to navigation and interaction in complex environments (Ben-Asher, Feldman, Gurl & Feldman, 2008). A graph theoretic framework (Jaidee, Munoz-Avila & Aha, 2013) was used for simplification of the set of interaction rules, so interactions can be represented as graphs or virtual structures where vertices are considered as UAVs and edges as the relationships among them. However, due to a high computational complexity it is difficult to guarantee a system performance using this approach (Lawton, Beard & Young, 2003). Other possible techniques of rules' set simplification such as hybridization, abstraction, statistical estimation, temporal logic also have this weaknesses (Ben-Asher, Feldman, Gurl & Feldman, 2008).

Some researchers have studied heterogeneous teams that include UAVs of various types (Waslander, 2013; Chen, Zhang, Xin & Fang, 2016), focusing on challenges of heterogeneous cooperation, including heterogeneous flocking and optimizing, providing formation control and stability, and so on. At the same time, they have considered UAVs having the same universal functionality, i.e. they have assumed that heterogeneity of the UV's team lies in difference of their environment of movement, but not in their functional roles. Nevertheless, UV ensemble is a class of teams, which include vehicles of different type as well as of different functionality and prescribed roles (Tošić & Vilalta, 2010b). Obviously, various challenges of UAV ensembles' joint motion coordination is not been fully reflected in the literature, and the proposed solutions are not always appropriate. Thus, the joint motion of the plurality of high-speed vehicles in confined space prevents the proper calculation of their correct trajectories in terms of safety using the classical approaches because the amount of computations increases exponentially. For example, in (De Sousa, Girard & Hedrick, 2004) the authors model distributed control problem in the framework of dynamic networks of hybrid automata, which has the complexity that considerably exceeds the reasonable limits.

Sharing UAVs' tasks presupposes a certain order for joint motion, which should be maintained stably in accordance with the mission execution plan. Potential fields method (Sharma, Vanualailai & Chand, 2009) allows to create distributed systems of UV team control, but does not allow the use of complex spatial structures, and restricts them to in-figure structures only. Method of iterative optimization of collective actions (Kaliaev, Gaiduk & Kapustian, 2009) works cyclically in discrete time. The complexity of this method lies in the formation of adequate assessments of the action effectiveness for each UAV, as well as in ensuring the convergence of the optimization algorithm under conditions of significant situation dynamics. Negotiation methods (Dias, Zlot, Kalra & Stentz, 2006) have a significant drawback as they need an intensive information exchange through a certain control center, which is a bottleneck.

Situational (behavioristic) methods (Nicolescu & Mataric, 2001) accept that each UV identifies the current situation and implements the corresponding action scenario independently. A case-based approach is proposed in (Sherstjuk, 2013) to solve coordination problems in large vehicle's groups based on the following assumptions: (a) activity of UAV's interactions is repeatable; (b) activity of UAV's interactions cannot be described as adequate models or well-defined rules and/or procedures. Thus, we can describe any activities of UAV's interactions as patterns. It means that there are coordination stereotypes for similar situations, and if there is an ability to collect patterns of group interaction for different classes of situations, we can use a case-based approach. However, situational approaches become possible in limited environments only, where an exhaustive description of all possible situations or their classes at least exists. Fuzzy logic methods (Tunstel, de Oliveira & Berman, 2002) based on fuzzy inference rules are widely used to develop UV's local control systems. Like the previous methods, this method has a significant drawback: it requires a-priori information, which is difficult to obtain in frequently changing and unpredictable environments. Another drawback is the need to process significant amount of information, which becomes unacceptable under condition of considerable situation dynamics. Coordination methods based on artificial neural networks (Ng & Trivedy, 2004) give rise to the similar problems.

According to the literature analysis, various challenges of UAV ensemble joint motion coordination has not been fully reflected in the literature, and the proposed solutions are not always appropriate. All above-mentioned approaches are suitable only to the teams of universal UAVs of the same type. The joint motion of the plurality of high-speed UAVs in confined space prevents the proper calculation of their correct trajectories in terms of safety and their order stability using the classical approaches because the amount of computations increases exponentially. Existing methods have the complexity that considerably exceeds the reasonable limits. This situation is totally exacerbated by imprecision and incompleteness of available information in highly dynamic, unpredictable, nondeterministic and partially observable environments. Thus, due to a high computational complexity, it is difficult to guarantee the UV ensemble safety and order stability using classical approaches (Lawton, Beard & Young, 2003). Therefore, the tasks of UVs' ensemble real time motion coordination have not been currently worked enough, and have a great interest for research. The problem addressed in this paper relates to coordination of joint motion of heterogeneous UV ensemble maintaining a stable order.

OUR APPROACH

We consider the UV team as a uniform ordered set of vehicles with different roles and functions, which jointly and simultaneously execute the scenarios of their activity according to the established roles within a specified mission in a certain area of interest (AOI). The most significant difference between the ensemble and the other forms of the UV team organization is the presence of the spatial configuration (order) given by the set of restrictions on the relative positions of vehicles, depending on the mission that is carried out. Most often, such restrictions are represented for each pair of vehicles as $\left(\beta, l\right)$, where β is a bearing and l is a distance. Since it is generally not possible to sustain precise values, the limit of the permissible deviations can be set, so that restrictions take the form of $\left(\beta \pm \Delta\beta, l \pm \Delta l\right)$.

Another essential difference is the presence of the scenario, or schema of the mission execution, which corresponds to vehicle's role. When moving, each vehicle should maintain safety conditions keeping spatial configuration and ensure ensemble's stability to achieve the mission goals. The scenario can be represented as a sequence of waypoints (*WP*), each of which is associated with a certain time point (*TP*) and specified motion parameters, e.g. speed, attitude, etc. Regardless of the whether the vehicle performs some function at the certain *WP* or not, it must be located in a given *WP* at a given *TP* with specific motion parameters. The ensemble scenario can also be represented as a sequence of scenes associated with the certain *TP*s, each of which is described as an aggregate of all vehicle's *WP*s and necessary motion parameters. It is crucial to ensure the necessary spatial position of the vehicles in each scene for the proper implementation of the scenario. However, it is difficult to provide even for homogeneous ensembles. Nevertheless, it is significantly complicated in heterogeneous ensembles due to different motion laws in different environments. For example, during the forest fire-fighting operation (Sherstjuk, Zharikova & Sokol, 2018) when a UGV (bulldozer) meets an insurmountable obstacle on its way, it is forced to change its route. Obviously, it changes the arrangement of *WP*s and *TP*s. At the same time, the UAV observing both the fire front and the actions of a bulldozer and other UGVs cannot stop immediately, it must continue moving adjusting its parameters, so that it is possible to lose sight of either the fire front, or the bulldozer, or the other UGVs. Beside that, there are various dynamic, navigation and situational disturbances changing the vehicle motion due to unpredictability of environmental effects. Therefore, the vehicle is forced to maneuver changing the motion parameters and consequently the trajectory. At the same time, other vehicles of ensemble can also start changing the motion parameters for safety reasons and for reasons of keeping the spatial configuration. Thus, this can lead to cyclic dependencies and cause disturbances of the predetermined spatial configuration. The appearance of various disturbing objects ahead of the ensemble is the most unpleasant, because vehicles begin to shying away from disturbing object, disrupting the spatial configuration, and in such a case existing vehicle control systems nothing can do about it.

We propose a scenario-case approach to coordinated control of heterogeneous ensembles of UAVs, which is closely related to (Sherstjuk, 2015) and uses activity scenarios as the main dynamic constituents of cases in contrast to the classical case-based approach where cases are static. Suppose that an ensemble is formed from a group of UAVs for a certain mission by applying restrictions on UAV's functions and roles. It is follows that vehicle's behavior in this ensemble obeys the general rules specific to UAV class, and is limited by (a) its role in the mission, (b) its technical characteristics that determine their capabilities, and (c) environmental factors, such as weather conditions and opponents' counteracting. Thus, the activity of each vehicle should follow the basic principles: (a) it needs to be aware of its role within a mission; (b) it needs to observe other members for collision avoidance and communicate with them in order to ensemble's stability; (c) it must necessarily achieve its mission goals. We assume that vehicle's activity in a certain mission can be described as a scenario of its actions, which corresponds to vehicle's role and aims at achieving its mission goal. The sequence of actions may change due to environmental effects and opponents' actions. It is obvious that we can collect scenarios for specific missions performed under certain conditions considered as situations. Since such collections of scenarios may constitute cases, we can use the case-based approach for coordinated control in similar situations.

In general, we can coordinate UAV ensemble using the scenario-case approach. Based on the dynamics of changes of the UAVs' spatial positions and disturbances, the approach allows finding solution as correct changes for the executed scenario. However, the obtained solution requires adaptation, and the search for a correct and safe path is one of the main elements of such adaptation. The most unpleasant thing is that security reasons often contradict the reasons of maintaining the spatial order. Moreover, both problems are solved in parallel, and they have an exponential complexity because of the number of considered objects, including disturbances. The aim of this work is to develop the scenario-case coordination method that allows UAV ensemble keeping spatial configuration satisfying safety conditions during the adaptation stage of the case-scenario decision cycle. We urgently need to reduce the computational complexity of the method because it should be acceptable to using in the vehicle onboard motion control system. Thus, we use the non-linear dynamics approach and the topological model. In this chapter, we build a spatial model, a model of situations, propose a bipolar safety/danger degree, and construct a soft level topology of the AOI, which enabled us to implement the proposed approach.

BUILDING THE SPATIAL MODEL

Let C be a three-dimensional Euclidean space that contains the AOI. Suppose e_1, e_2, e_3 is a basis in C such that the metric ξ_C remains uniform. Decomposition of each vector $v = \alpha_1 e_1 + \alpha_2 e_2 + \alpha_3 e_3$ gives coordinates $v(\alpha_1, \alpha_2, \alpha_3)$ of a certain point within the space C, which represent a position of this point in the space C at the time $t \in T$. We can build a spatial model of the AOI at the points' level, but we cannot use it in real-time because of its high computational complexity. Thus, we consider the spatial model of the AOI at two levels: the lower level uses the cells of equal size as the minimal spatial objects and the upper level uses the spatial regions having different sizes.

Low Level of the Spatial Model

Let's impose a metrical grid of coordinate lines with $\delta = \Delta\alpha_1 = \Delta\alpha_2 = \Delta\alpha_3$ within the subspace $X \subseteq C$ with the initial point $(\alpha_1 = 0,\ \alpha_2 = 0,\ \alpha_3 = 0)$ using the norm ξ_C and a linear map f such that coordinate lines form a set D of cubic cells of the size $\delta \times \delta \times \delta$, $f : C \to D$. Thus, the subspace X is discretized by a grid $D = \{d_{xyz}\}$ of isometric cubic cells d_{xyz}, where x, y, z correspond to e_1, e_2, e_3 respectively. Therefore, the location of each UAV is discrete and bounded by a specific cell.

Each cell $d \in D$ will fit the coordinates of its center O_d within the subspace X. Thus, the cell $d \in D$ is the spatial object of the minimal size. Each cell $d \in D$ is associated with a set of attribute values, which is called the cell state, via the values function $\{f(d,a) | a \in A\}$. The proposed discretization assigns equal values of the state attributes to each point belonging to a certain cell d, therefore each cell $d \in D$ can be reduced to a point of the subspace X. Correspondingly, a cell is a homogeneous area of the AOI in terms of attribute values A, therefore all points of the cell are A-indiscernible:

$$\left(\forall d_1, d_2 \in D\right)\left(\forall a \in A\right)\left[f\left(d_1, a\right) = f\left(d_2, a\right)\right].$$

Because a size of the cells can vary, the AOI's scale can also change.

Upper Level of the Spatial Model

In parallel with the discretization by cells, the subspace X can be divided into a finite set of disjoint objects presented as geometric shapes, which outline boundaries of the certain homogeneous areas of the AOI. Consider an attribute $A_S \in A$. Using A_S, we define the relation $\Re_D^{A_S}$ on a set of cells D:

$$\left(\exists A_S \in A\right)\Re_D^{A_S} = \left\{\left(d_m, d_n\right) \in D \times D \middle| \forall a \in A_S, f\left(d_m, a\right) = f\left(d_n, a\right)\right\}.$$

Denote $\Re_D^{A_S}\left(d\right)$ the equivalence class on the set of cells D generated by $\Re_D^{A_S}$. All the cells that belong to $\Re_D^{A_S}\left(d\right)$ have equal values of the attribute A_S:

$$\left(\exists A_S \in A\right)\left(\forall a \in A_S\right)\left(\forall d_m, d_n \in D\right)\left[\left(\Re_D^{A_S}\left(d_m\right) = \Re_D^{A_S}\left(d_n\right)\right) \Leftrightarrow f\left(d_m, a\right) = f\left(d_n, a\right)\right].$$

It means that all different points x, y of the cell $d_i \in \Re_D^{A_S}\left(d\right)$ have the same values of the attribute $A_S \in A$, as well as all different points y, z that belong to the different cells d_i, d_j of the equivalence class $\Re_D^{A_S}\left(d\right)$ also have the same values of the attribute A_S. Thus, we define an upper-level structural element of the spatial model as the homogeneous spatial area that is uniform in terms of attribute values and represented by the approximating set of cells. Such element is called a region and denoted by h. Thus, the AOI $\Xi = \left\langle D, H, \varsigma\right\rangle$ is determined by a set of cells D, set of regions H, and a linear isometric surjection $\varsigma : D \to H$. On this basis, each region $h_k \in H$ is approximated by an underlying set of cells $h_k = \cup_{j=1}^{n_k} d_j$, where n_k is a total number of cells in h_k. It should be noted that depending on requirements the AOI can be associated with various sets of regions simultaneously. Therefore, one cell may map to a plurality of regions.

VEHICLES ACTIVITY, MISSIONS AND GOALS

Suppose U is a set of UAVs, G is a set of targets, F is a set of UAV's parameters, and Φ is a set of their functions. Consider a given number of UAVs $u_1, \ldots u_n \in U$ and a given number of targets $g_1, \ldots g_m \in G$, which are dispersed over the AOI Ξ. Assume that features of each $u_i \in U$ are described by a set of parameters' values $F_{ui} = \left\{f_{i1}, \ldots f_{is}\right\}$. Therefore, each UAV u_i belongs to one class cl_k in a class set $Cl = \left\{cl_1, cl_2, \ldots cl_m\right\}$ depending on its technical features and capabilities. Furthermore, each UAV u_i

can perform at least one function φ_l in a function set $\Phi = \{\varphi_0,...\varphi_z\}$ depending on its class. We consider the targets $g_i \in G$ in the same way; in the sequel, both the goals and UAVs are called *objects*.

Suppose UAVs $u_1,...u_n \in U$ perform a given operation Op within a given AOI Ξ at a certain time.

Model of Vehicle's Activity

The activity of vehicle u_i can be represented in two aspects. First, vehicle u_i performs the movement function φ_0, and moves inside the AOI Ξ changing its position over the time avoiding collisions with other vehicles. Second, vehicle u_i can perform other functions, but to perform a certain function φ_j it should take a specific position at the moment $t_l \in T$.

Let $Pos(u_i)$ be a position of vehicle u_i such that $\langle Pos(u_i), t_l \rangle = (d_{xyz})$, $d_{xyz} \in D$. Suppose the tuple $\langle Pos(u_i), t_l, \varphi_j \rangle$ specifies performing the function φ_j in the position $Pos(u_i)$ at the time t_l. Then, a continuous sequence

$$\left[\langle Pos(u_i), t_l, \varphi_j \rangle ... \langle Pos(u_i), t_m, \varphi_k \rangle\right]$$

at the time interval $t \in [t_l,...t_m]$ is called *activity trajectory*, and is denoted by $Tr(u_i)$. It should be noted that $\langle Pos(u_i), t_l \rangle = \langle WP, TP \rangle_{(i,l)}$. Thus, a continuous sequence $\left[\langle WP, TP \rangle_{(i,l)} ... \langle WP, TP \rangle_{(i,m)}\right]$ is determined at the time interval $t \in [t_l,...t_m]$ for each u_i. Goals are always set as the final *WP* of the sequence. At the beginning, the activity trajectory $Tr(u_i)$ for each vehicle $u_i \in U$ is prescribed by a mission plan $Pl(u_i)$, which specifies important positions and performed functions. However, due to unpredictable environments the vehicle u_i is exposed to a number of dynamic and situational disturbances (overcast, weather conditions, and results of activity of other members or opponents), which require compensation such as changing, adding or deleting some pairs *WP/TP* in $Tr(u_i)$, i.e. maneuvering. The search for a suitable compensation is the task of the vehicle's scenario-case onboard control system. Of course, it should take into account all safety restrictions and keep the prescribed spatial configuration.

It is obvious that the compensation can be a series of *maneuvers* $m = [m_1,...m_k]$, where each maneuver is described as a single-step or multi-step change of values of one or more specific parameters $p_{li},...p_{vj}$ of one or more UAV's actuators $ac_l,...ac_v$. Each maneuver is associated with a specified point at the trajectory or in time. Changes of values of actuator's parameter p_{li} lead to change of values of some vehicle's activity parameters $f_{vi}, f_{yi} \in F$, and then affect the trajectory $Tr(u_i)$.

Operations and Missions

Consider a long-term operation Op in the given AOI Ξ. Operations are always performed in conditions of counteraction by the other side (opponent), whose objects are the targets in G. In general, the operation Op has a goal $\Gamma(Op)$ and consists of missions $M_1,...M_n$, which are performed sequentially or simultaneously and dedicated to a certain tasks. Each mission M_i also has its goal $\Gamma(M_i)$, but while the operation goal is given spatially (as a certain region) all over the time, the goal of the mission is usually associated with certain target(s). If the operation Op aims to establish control over a certain region h of the theatre Ξ, the appearance of one or several targets $g_i,...g_l \in G$ starts the mission M_j to eliminate them. Let R be a set of roles and $Sch(M_i)$ be a schema of the mission execution. Some vehicles $u_i,...u_m \in U$ allocated to perform a mission M_i constitute an ensemble En that has a tree-like structure $Str(En)$ in which every member $u_k \in U$ is assigned to a certain role $r_k \in R$, and pattern $Shp(En)$, which determines the spatial configuration of ensemble En. Thus, $Str(En)$ describes structural aspects as well as $Shp(En)$ describes geometric aspects of a mission's M_i order.

Suppose that the structure $Str(En)$ depends on mission's execution schema $Sch(M_i)$, and the latter depends on changing the spatial positions of targets as well as on their belonging to the certain classes in Cl. Thus, situations Sit can be described as a combination of a set of values of objects' parameters including their positions and spatial relations, and a set of values of environmental parameters. We assume that the mission execution schema $Sch(M_i)$ is the prototype, which corresponds to the stereotyped situation Sit and determines a set of scenarios for achieving the mission goal. In this schema, each role $r_k \in R$ is associated with a certain scenario Ω_k that contains prescribed activity trajectory $Tr(u_i)$. On the other hand, each mission is an implementation of a certain mission plan $Pl(M_i) = Pl(u_i) \circ ... \circ Pl(u_l)$, where $u_i,...u_l$ are performers of certain roles in ensemble, so the main components of mission's execution schema $Sch(M_i)$ are joint maneuvers of several performers, which are synchronized in time and space. Such components are called *tactics*. A choice of initial plan $Pl(M_i)$ based on the mutual spatial position of performers and observed goals can be disrupted by changing of any object's parameters and lead to change of the plan execution process. Any change of objects' parameters is called an *event*; it can or cannot change the mutual spatial position of objects. In the first case, it is necessary to change the mission plan, adjust the synchronization points in time and space, or add/delete tactics in some scenarios.

DESCRIPTION OF THE SITUATIONS

Assume that situation Sit within Ξ has a prototype for making a decision. In the scenario-case approach, this prototype is a case; it includes the descriptions of situation and solution. As the situation is described by the mutual spatial position of objects, we have to build the spatial configuration.

Definition of Spatial Areas

Let ξ_B be a metric in Ξ with the properties similar to ξ_C, and χ be anisometric surjection such that $\chi : \xi_C \to \xi_B$, which determines the subset of interacting vehicles, such that $\xi_B(d_1, d_2) = \|d_1 - d_2\| \to \mathrm{P}_j$. Suppose we have a set of limits $\rho_j = \{\rho_{j0}, \rho_{j1}, \ldots \rho_{jl}\}$ for each φ_j, then for each couple (u_i, u_k) with respect to φ_j we have

$$\xi_B(u_i, u_k) = \|Pos(u_i) - Pos(u_k)\| \to \rho_j.$$

Based on these limits we obtain some domain-dependent regions around u_i at each moment (Figure 1), each of which are generally presented as a sphere with a center $Pos(u_i)$ and radius ρ_{jl}.

We assume that u_i interacts with u_j, if and only if $\xi_B(u_i, u_j) < p_{1j}$. Accordingly, u_i interacts with u_j dangerously with danger degree k if $\xi_B(u_i, u_j) < p_{kj}$, and interacts with u_j critically if $\xi_B(u_i, u_j) \leq p_{mj}$. It is essential that the relationship χ is non-symmetric, i.e. $\chi_i(u_i, u_j) \neq \chi_j(u_i, u_j)$. An interaction set for vehicle u_j includes all vehicles, which interact based on χ_j with at least one u_i such that $u_i, u_j \in U$.

Given the ξ_B and P_j we can build a set of domains around each u_j starting from $Pos(u_j)$ at each moment t. For example, the set of domains can represent the domain of bounded activity H_j^1, the domain of hazardous activity H_j^2, and the domain of prohibited activity H_j^3, each of which generally is a sphere with a radius p_{1j}, p_{2j}, p_{3j} for H_j^1, H_j^2, H_j^3 respectively. Since these domains are connected

Figure 1. Spatial areas

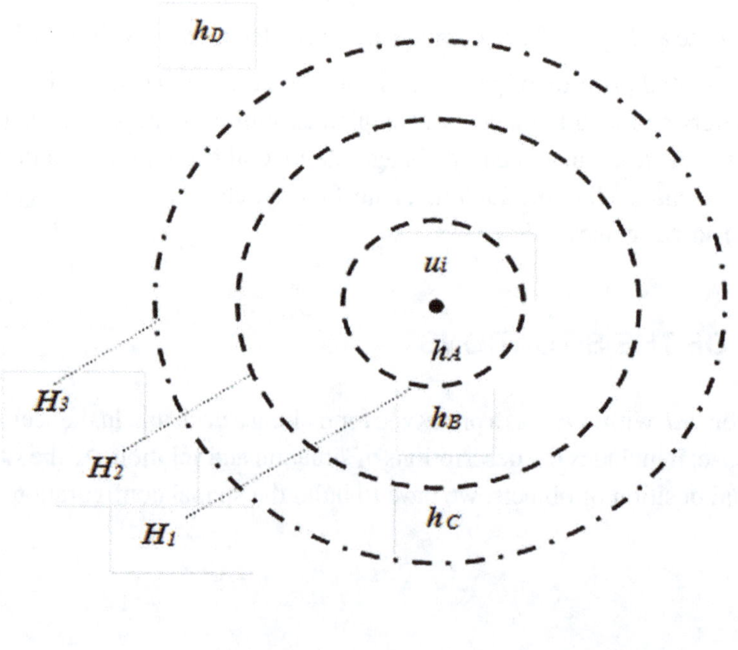

with a reference point $Pos(u_j)$ that moves along the activity trajectory $Tr(u_j)$, the domains H_j^1, H_j^2, H_j^3 also move inside Ξ together with the position of u_j. Using non-linear and/or non-uniform metrics as well as fuzzy or rough sets' methods, we can change the shape and blur the boundaries of built domains.

Thus, based on P_j we can obtain some domain-dependent regions around u_j at each moment, each of which are generally presented as a sphere with a center $Pos(u_j)$ and radius p_{ij}. Further, we can impose the space Ξ structure that divide this sphere into numbered or labeled sectors with a certain angular size, which are delimited by border lines with respect to $p_{1j},...p_{mj}$ as it's shown in Figure 2. For example, because of collision avoiding for φ_0 we determine safety areas of forbidden activity (h_A), dangerous activity (h_B), restricted activity (h_C), or unrestricted activity (h_D), which are the regions in Ξ delimited by border lines H^1, H^2, H^3. Thus, the object location is assigned to a concrete sector and specifies by its name, e.g. «C02».

If we assume that in different directions for different vehicles limiting safety norms can be established separately, taking into account the spatial configuration of the disturbances, we obtain the ability to define non-spherical safety domains as shown in Figure 3.

Determining the Vague Spatial Areas

The determining of the safety domains' vague boundaries can be performed using fuzzy or rough set approaches. The fuzzy sets (Zadeh, 1965) are weakly suitable for the safety conditions formalization

Figure 2. Spatial structure imposed for u_j

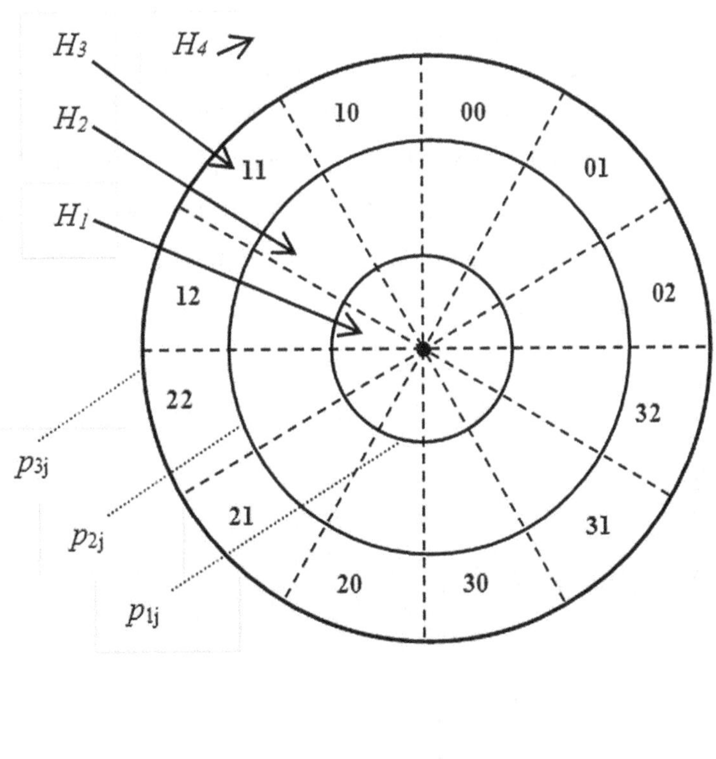

Figure 3. Determining non-spherical safety domains

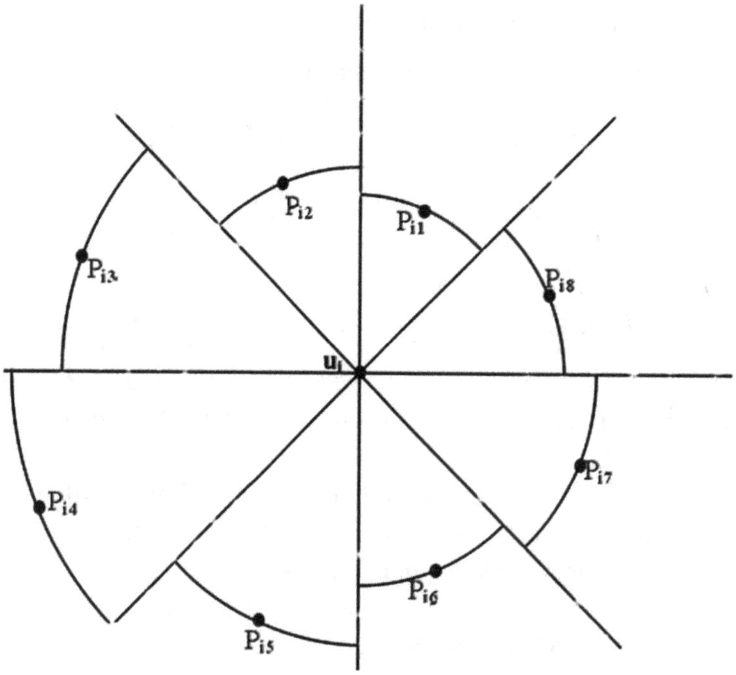

as their membership functions depend on many factors, which are poorly formalized, and the construction of these functions by expert method are impossible in real time. Let us use the rough set approach (Pawlak, Jerzy, Slowinski & Ziarko, 1995) to solve the problem. The uncertain safety conditions can be represented in a "rough" way based on the rough sets without any apriori information because the information about the area between their upper and lower approximations does not require assignment of a probability or possibility distribution or any membership functions.

For a consideration of some concept $X \in U$ and a given equivalence relation R a lower approximation $POS_R(X)$ consists of all objects that must belong to the concept, X and an upper approximation $NEG_R(X)$ consists of all objects that may belong to the concept X. The space between the lower and the upper approximations is the boundary area $BND_R(X)$ of the concept X, which consists of all objects that can not be unambiguously mapped to $POS_R(X)$ or $NEG_R(X)$ using knowledge available at the moment as it is shown in Figure 4:

$$\underline{R}X = \cup\{Y \in U / R : Y \subseteq X\},$$

$$\bar{R}X = \cup\{Y \in U / R : Y \cap X \neq \varnothing\},$$

$$POS_R(X) = \underline{R}X,\; NEG_R(X) = U - \bar{R}X,$$

$$BND_R(X) = \bar{R}X - \underline{R}X .$$

Thus, $POS_R(X)$ и $NEG_R(X)$ set the intervals, which contain the desired estimate values of the safe boundaries with some accuracy $\alpha_R(X) = card(\underline{R}X) / card(\bar{R}X)$. Getting aposteriori information about the values distribution can improve the accuracy of determining the concept X.

The boundaries of safety areas can be defined as approximate estimates of intervals specified by the boundary areas BND_{H1}, BND_{Hi}, ... BND_{Hm}, given that $POS_{H(i+1)} = NEG_{Hi} \cup BND_{Hi} \cup POS_{Hi}$ for each $H(i+1)$, for example as it is shown in Figure 5.

Determining the Vague Safety Areas

As we cannot estimate boundaries of obtained spatial areas precisely due to dynamic environments, we describe these boundaries approximately using intervals of maximum permissible values posed by boundary regions of associated rough sets shown in Figure 5. These intervals describe vague boundaries of spatial areas.

Figure 4. Determining the areas of rough set

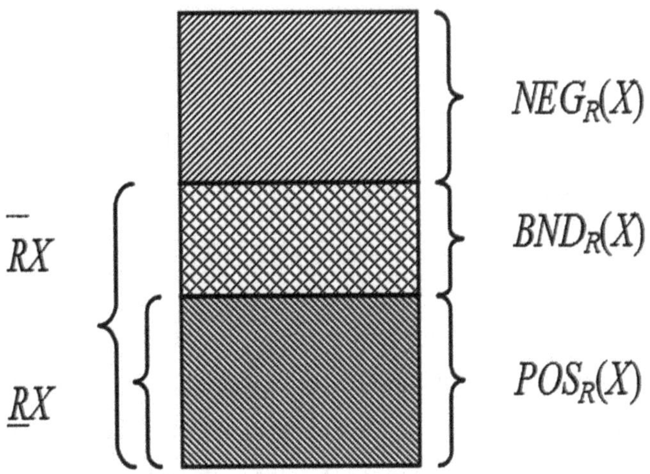

Figure 5. Determining boundaries of vague areas

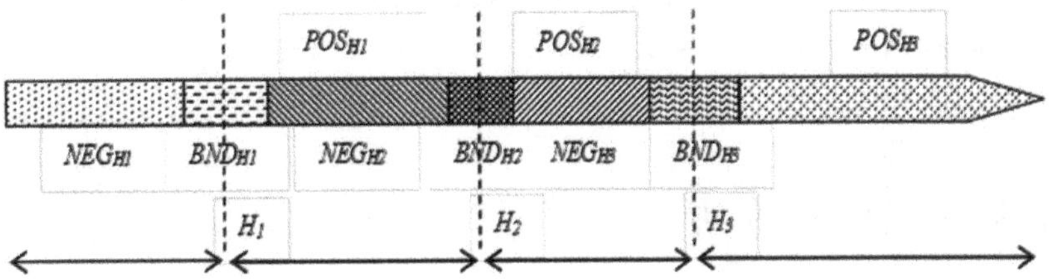

A joint activity of vehicles u_i and u_j is called mutually free if their trajectories $Tr(u_i)$ and $Tr(u_j)$ provide $H_i^1(t) \cap H_j^1(t) = \varnothing$ in all positions. The activity of u_j is called limited for u_i if $H_i^2(t) \cap H_j^2(t) \neq \varnothing$, k-dangerous for u_i if $H_i^k(t) \cap H_j^k(t) \neq \varnothing$, and critical for u_i if $H_i^m(t) \cap H_j^m(t) \neq \varnothing$. It is clear that every critical activity is dangerous and every dangerous activity is bounding. If u_j limits the activity of u_i, its domain $H_j^1(t)$ is a dynamic restriction area for $Tr(u_i)$.

The domain of possible activity $H_j^*(t)$ for u_j excludes the areas of static and dynamic restrictions:

$$H_j^*(t) = \Xi - \bigcup_{k=1}^{p} R_k^S - \bigcup_{i=1}^{n} H_{ij}^m(t).$$

Any interaction $\chi(u_i, u_j)$ such that $\xi_B(u_i, u_j) < p_{1j}$ is called situational disturbance of activity trajectory $Tr(u_j)$ with respect to u_i and is denoted by ω_{ji}.

Let's introduce a metric ξ_T on T such that $\forall t_i, t_j, t_k \in T$: *a)* $\xi_T(t_i, t_j) = 0 \Leftrightarrow t_i = t_j$; *b)* $\xi_T(t_i, t_j) = \xi_T(t_j, t_i)$; *c)* $\xi_T(t_i, t_l) \leq \xi_T(t_i, t_j) + \xi_T(t_j, t_l)$, and *d)* $\left\| t_i - t_j \right\|_T \to \mathrm{P}_j$, where $\mathrm{P}_j = \{p_{1j}, \dots p_{ij}, \dots p_{mj}\}$ is the time norm limit set.

Any interaction $\chi(u_i, u_j)$ such that $\xi_B(u_i, u_j) < p_{1j}$ and $\xi_T(u_i, u_j) \geq p_{1t}$ is called a threat to u_j that is a dangerous disturbance requiring immediate unconditional compensation, and is denoted by ω_{ji}^*.

Spatial Configuration

Thus, we can classify disturbances and build a spatial configuration as follows. We represent each disturbance ω_{ji} as $\omega_{ji} = \langle u_i, Pos(u_j), K_i \rangle$, where K_i is a certain class of u_i. At each *TP* we have a vector of parameters for each u_i such that:

$$\vec{u}_i = \langle t, Pos(u_i), \vec{m}_i, v_i, \vec{\phi}_i, \varphi_i, \psi_{ij}, l_{ij} \rangle,$$

where $\vec{m}_i$ is a velocity vector and v_i is its module, $\vec{\phi}$ is an angular velocity and φ_i is its module, ψ_{ji} is a bearing and l_{ij} is a distance from u_i to u_j. For each $u_i \in U$ classification depends on its observed motion parameters and is performed separately as K_{ij} for each known u_j. Thus, vehicles can be classified based on their motion parameters (altitude, velocity, and others) as "maneuvering / moving / stationary" at the moment, or as "moving closer / moving away / equidistant" at the time interval, or as "not dangerous / potentially dangerous / dangerous". The total classification is obtained by performing convolution operation as $K_i = K_{i1} \oplus \ldots \oplus K_{ik} \oplus \ldots \oplus K_{iz}$.

A tuple

$$V_j(t) = \langle (\omega_{j1}, \xi_{T1}, \xi_{B1}), \ldots, (\omega_{jn}, \xi_{Tn}, \xi_{Bn}) \rangle$$

defines a set of disturbances for vehicle u_j that are ordered with respect to ξ_T (as $\xi_{T1} \leqslant \xi_{T2} \leqslant ... \leqslant \xi_{Tn}$). Now, if we distribute the disturbances spatially across the sectors of the safety domain, $V_j(t)$ will be the spatial configuration for u_j at the moment $t \in T$ as shown in Figure 6. The spatial configuration for ensemble En is a composition of spatial configurations of its members, $V(t) = V_i(t) \circ ... \circ V_m(t)$ (Sherstjuk, 2015).

The Case Structure Based on Spatial Configuration

The case structure includes a description of the situation and corresponding vectors for an approximate estimation of safety domain boundaries. The search for a suitable case requires a given similarity function assessment for the observed situation with respect to the existing situations stored in the case base.

To build a similarity degree evaluation function we can use the well known nearest neighbor method based on measuring the coincidence degree for the case parameter values.

Consider a set of vehicles $\{u_i, u_j, u_k, u_m\}$ from the position of u_i as it is shown in Figure 7.

Suppose l is a distance function given on Ξ as

$$l_{ij} = l(u_i, u_j) = \xi_C(Pos(u_i), Pos(u_j)).$$

Figure 6. Description of the spatial configuration

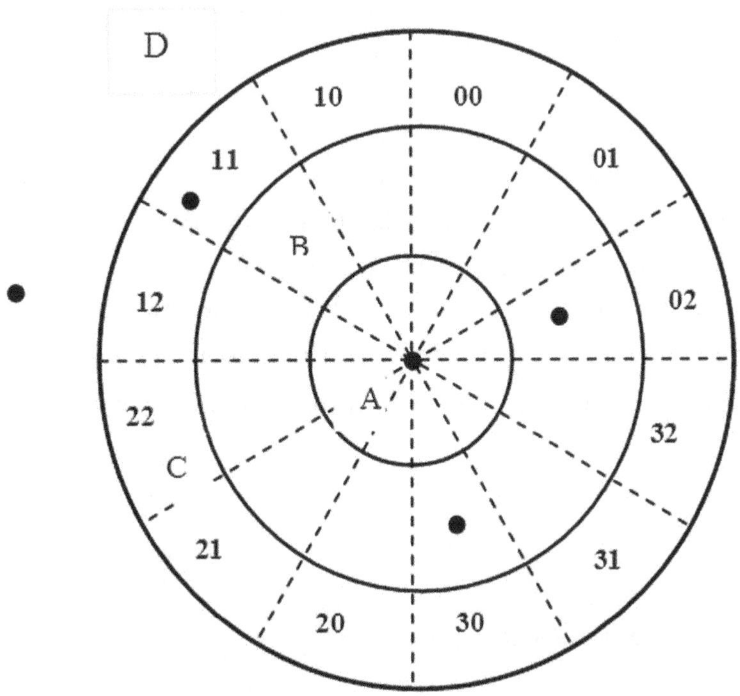

Figure 7. Determining the distances for u_i

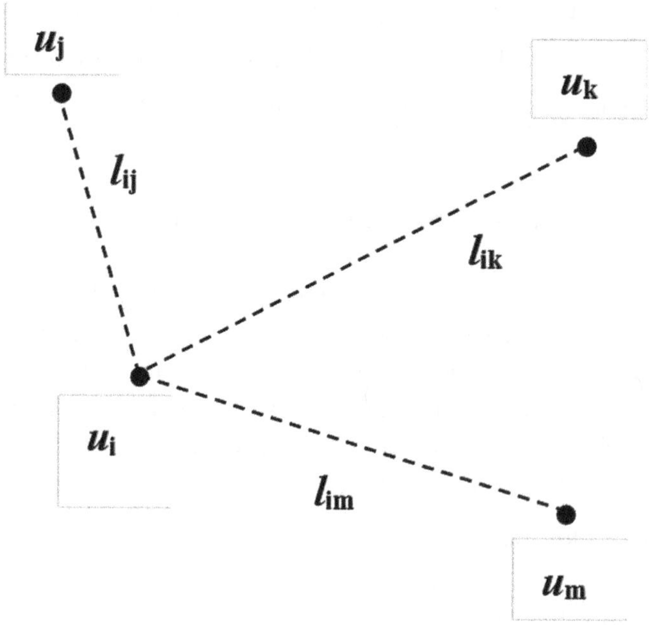

Let l has the following properties for each u_i, u_j, u_k: *a)* $l(u_i, u_j) = l(u_j, u_i)$; *b)* $l(u_i, u_j) > 0$; *c)* $l(u_i, u_i) = 0$; *d)* $l(u_i, u_k) \le l(u_i, u_j) + l(u_j, u_k)$. The last formula defines the triangular inequality, and provides a condition for evaluating distances among vehicles based on metric relationship.

However, determining the distance l_{ij} between vehicles does not give a complete description of their spatial arrangement. This information is not enough to find spatially similar situations, because being at a similar points in Ξ, vehicles can move with quite different speed and in different directions. The static component of the situation description should include bearings ψ_{ji} on the observed vehicles u_j, as well as the movement vectors $\vec{m}_j$ can describe the observed dynamics of the situation defining direction and speed of vehicles' movement, as it is shown in Figure 8.

The spatial description of the situation for u_i should enumerate other observable vehicles u_j with their positions $Pos(u_j)$, the calculated distances l_{ij} and bearings ψ_{ji}, as well as vectors of their movement $\vec{m}_j$, describing the direction ϕ_j and speed v_j. Thus, we have spatial description for each vehicle u_j as

$$Des_i(u_j) = \langle Pos(u_j), \phi_j, v_j, l_{ij}, \psi_{ji} \rangle$$

and spatial description of the situation for u_i as

$$Des_i = \langle Des_i(u_j), \ldots Des_i(u_k), \ldots Des_i(u_n) \rangle$$

Figure 8. Spatial description of the situation for u_i

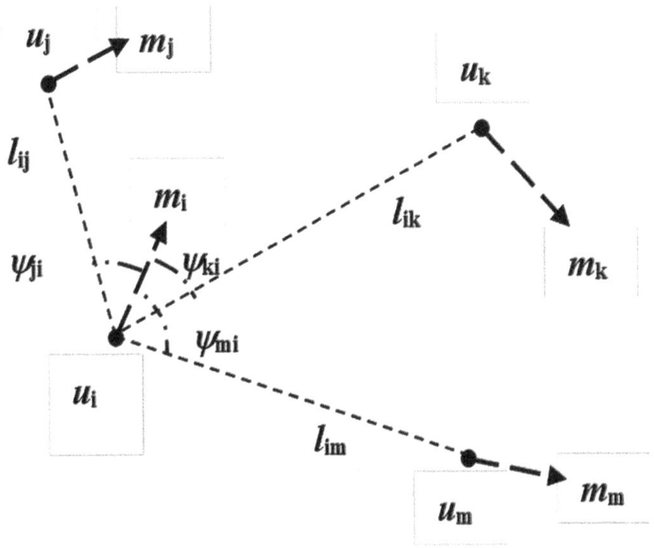

where n is a number of observed vehicles. Using a spatial description Des_i of the situation for u_i, we can determine the value of a spatial similarity function $SIM_S(u_i) = f(Des_i)$ based on the nearest neighbor algorithm *iDistance* (Jagadish, Ooi, Tan, Yu & Zhang, 2005). Since the boundaries of the safety domains strongly depend on the dynamic of the situation, we must also take into account the temporal component. When vehicles moving in Ξ, their distances and bearings are changing at different times as it is shown in Figure 9.

The spatio-temporal description of the situation for u_i should reflect the relative change of bearings and distances in time. If we have

$$Des_i(u_j) = \langle Pos(u_j), \phi_j, v_j, l_{ij}, \psi_{ji} \rangle$$

at the time $t \in T$ and

$$Des_i'(u_j) = \langle Pos'(u_j), \phi'_j, v'_j, l'_{ij}, \psi'_{ji} \rangle$$

at $t' \in T$ for some u_j, then we can obtain the relative changes of distances as $\Delta l = l'_{ij} - l_{ij}$ and bearings as $\Delta\psi = \psi'_{ij} - \psi_{ij}$. The spatio-temporal description for each vehicle u_j can be defined as $Dest_i(u_j) = \langle \Delta l_{ij}, \Delta\psi_{ji} \rangle$ and spatio-temporal description of the situation for u_i as

$$Dest_i = \langle Dest_i(u_j), \ldots Dest_i(u_k), \ldots Dest_i(u_n) \rangle$$

Figure 9. Spatio-temporal description of the situation for u_i

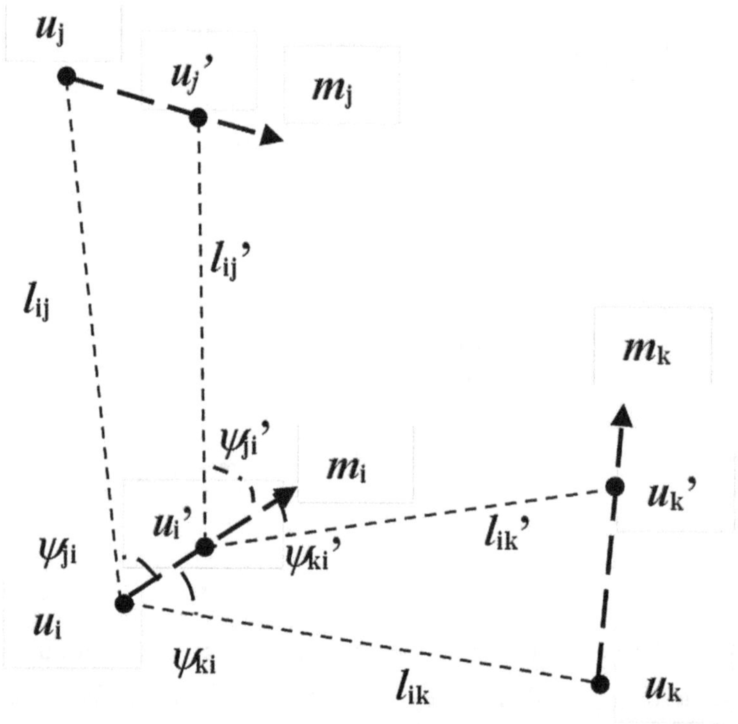

where n is a number of observed vehicles. Using a spatio-temporal description $Dest_i$ of the situation for u_i, we can determine the value of a spatio-temporal similarity function $SIM_{ST}\left(u_i\right)=f\left(Dest_i\right)$ as proposed in (Yoon & Shahabi, 2009).

Taking into account a huge amount of cases accumulated in the case base, we can split similarity function as

$$SIM\left(u_i\right)=SIM_S\left(u_i\right)\oplus SIM_{ST}\left(u_i\right)\oplus SIM_E,$$

where SIM_E is an environmental similarity. The similarity degree SIM_E is determined by comparing the wise situation's parameters to cases, whereby we can obtain the distance between the environmental parameters of the problem situation and the case situation as well as the maximal distance among them based on the parameters range (Haigh & Shewchuk, 1994). If we find a case describing an environmentally similar situation based on SIM_E, we can distinguish a subset of cases relevant for the problem situation, and there can be a lot of such cases.

On the next stage, we can apply $SIM_S\left(u_i\right)$ to distinguished subset of cases and obtain a restricted subset of spatially similar situations as cases for similar environmental conditions. Finally, we can find a subset of spatio-temporally similar situations stored as the cases using $SIM_{ST}\left(u_i\right)$. This subset will have a much smaller size, so that it is possible to identify the most similar situation effectively. Further, we can use the values contained in the founded case as a solution for obtaining the blurred boundaries of the safety domains and for danger / threat assessing.

SAFETY/DANGER MODEL

We can use an “attractor” and “repeller” concepts of the nonlinear dynamics to build the correct compensation. The “attractor” is an attracting manifold of the state space, as well as the “repeller” is a forbidding manifold of the state space, on the contrary. Thus, all *WP*s can be represented as attractors while all stationary and dynamic obstacles, including the positions of the other vehicles, can be represented as repellers. The trajectory corrections should be provided as searching the shortest path to the attractor avoiding the repellers as much as possible. Now we must define the influence of attractors and repellers on the safety of moving vehicles.

Safety/Danger Degree

Each cell of AOI can be evaluated either as a dangerous one or as a safe one (to a certain extent). We assume that for each cell d of the grid D its attribute A_S has a value of the safety/danger degree ω_d at each point of the given cell. Since none of the cell can be both safe and dangerous at the same time, the attribute A_S can be bipolar with domain $\left[-1,1\right]$ (Figure 10).

In this case, an estimate ranged $\left(0,1\right]$ is considered as the safety degree, as well as the estimate ranged $\left[-1,0\right)$ is considered as the danger degree. Zero means that the cell’s safety/danger status is not exactly known. Thus, we can assign a priori a value of 1 to each attractor, and value of -1 to each repeller. However, it will be difficult to operate with continuous real numerical estimates because of high computational complexity.

Safety/Danger Level

For each u_j we define a set of norm limits $\mathrm{P}_j = \left\{p_{1j},\ldots p_{ij},\ldots p_{mj}\right\}$ such that $p_{1j} > p_{ij} > p_{mj}$. However, calculating the exact values of such limits is a computationally complex task that is difficult to solve in real time. Beside that, norm limits can change dynamically due to the variability of the situation within the AOI because all vehicles continue to move.

We can make a safety assessment not for a complete set of disturbances but only for each particular cell of the AOI. This will prevent a full search among vehicles’ pairs, and reduce computations only for those cells that contain other vehicles or targets according to the spatial configuration. To reduce com-

Figure 10. Safety/Danger Bipolar Estimates

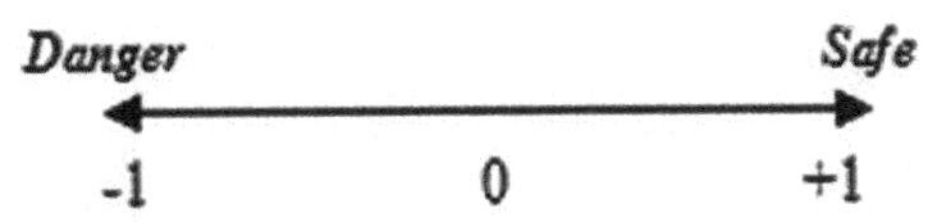

putational complexity, we can take advantage of a qualitative assessment of the safety/danger degree using level hierarchy. Imposing a partial order $\preccurlyeq_\rho$ onto the set of approximate limits $\{\rho_0, \rho_1, \dots \rho_l\}$ such that $\rho_0 \preccurlyeq_\rho \rho_1 \preccurlyeq_\rho \dots \preccurlyeq_\rho \rho_l$, we obtain an ordinal scale ϑ. Having the metric ξ_B and the ordinal scale ϑ, we can quantify values ω_d of safety/danger degree at each cell of AOI as it was proposed in (Zharikova & Sherstjuk, 2016). The number of scale's elements must be a compromise, since when it is greater then the model accuracy increases, but at the same time, the computational complexity also increases. An example of the 7-levels scale is shown in the Table 1.

Bipolar Level Soft Sets

Suppose the grid D is a universe and A is the cell's attribute set. A pair (F, A) is called a soft set of cells over D if and only if F is a mapping of A into the set of all subsets of the set D. In other words, the soft set is a parameterized family of subsets of the cell set D. The set (F, A_S), $A_S \in A$ of this family may be considered as the set of A_S-approximate elements of the soft set (Molodtsov, 1999). Consider the values of the attribute A_S with the domain $[-1,1]$ as degrees of membership of cells in the set of safe cells. Using the mapping $\tilde{F} : A_S \to [-1,1]$ we obtain the bipolar fuzzy soft set of cells $(\tilde{F}, A_S)$, which can substitute the crisp one. Let $\tau \in [-1,1]$. The τ-level soft set of a bipolar fuzzy soft set $(\tilde{F}, A_S)$ over D is a set (F_τ, A_S), defined by

$$F(A_S) = \{d \in D : \tilde{F}(A_S)(d) \geq \tau\}$$

Table 1. Safety/Danger levels

Limit	Safety/Danger Estimate	Safety/Danger Degree
ρ_0	Safe	1
ρ_1	Almost safe	0.8
ρ_2	Undangerous	0.4
ρ_3	Not known	0
ρ_4	Unsafe	-0.2
ρ_5	Critically dangerous	-0.6
ρ_6	Forbidden	-1

(Feng, Li & Leoreanu-Fotea, 2010). In this notion, τ serves as a given threshold on safety/danger level and can vary. Thus, the set $\left(F_\tau, A_S\right)$ contains all cells $d \in D$, for which the safety estimate is greater than τ. Moreover, a set with a smaller value of τ always includes elements of any other sets with higher values of τ.

Building the Level Soft Topological Model

Let us consider the AOI as an open connected subspace $X \subseteq C$ endowed with the topological properties (Allam, Bakeir & Abo-Tabl, 2008). We can use the indiscernibility relation $\Re_D^{A_S} \subseteq D \times D \times A_S$ on the set of cells D to build a topological space. $\Re_D^{A_S}$ should be constructed in such a way that all the cells, for which the value of the attribute A_S falls within the range between the certain low limit ρ_i and the certain high limit ρ_j, should not differ in the safety/danger level. Thus, all the cells such that $\rho_2 \le A_S \le \rho_1$ refer to the same (first) level of safety. The pair $apr_D = \left(D, \Re_D^{A_S}\right)$ is called the approximation space. The family of all open sets is denoted by $Def\left(apr_D\right)$, and the soft class containing a cell $d \in D$ is denoted by $\Re_D^{A_S}\left(d\right)$. The approximation space uniquely determines the topology $Def\left(apr_D\right)$ and the topological space $T = \left(D, Def\left(apr_D\right)\right)$. As mentioned above, all sets consisting the family $Def\left(apr_D\right)$ may be considered as bipolar level soft sets. In this case, $T = \left(D, Def\left(apr_D\right)\right)$ is the soft level topological space approximated by cells, and each $d \in D$ is the cell of the topological space. Therefore, all A_S-indiscernible cells belong to the same level τ_k of the soft topology as well as they constitute the same region h_{τ_k}. Because the attribute A_S is time-varying, the obtained soft topology is dynamic.

IMPLEMENTATION OF THE MULTI-LEVEL CASE-SCENARIO CONTROL OF UAV'S ENSEMBLE

The coordination control of heterogeneous UAV's ensembles is a complex multi-level process. It includes:

- **Operation (Higher) Level:** Which is dedicated to choose an adequate mission plan $Pl\left(M\right)$ (*case*) on the basis of spatial configuration $V\left(t\right)$. The mission execution schema $Sch\left(M\right)$ is adjusting to $Str\left(En\right)$ and $Shp\left(En\right)$ as a case adaptation process. Then the schema $Sch\left(M\right)$ splits into scenarios Ω_k for all roles $r_k \in R$ assigned to vehicles $u_i, u_j \in Str\left(En\right)$, and these vehicles start to execute their own scenarios.
- **Coordination (Middle) Level:** Which is aimed to execute scenarios. It transforms each scenario into a sequence of triads $\left\langle Pos\left(u_i\right), t_l, \varphi_j\right\rangle$ taking into account all available navigation, situational and time-critical constraints. In particular, this level is performed admissibility control for each position in triads, since each next position should not fall into the areas of forbidden or dangerous

activity. This level also receives events and checks for changes of the spatial configuration. In this case, operation level is notified and starts searching a new mission plan based on new spatial configuration. Otherwise, coordination level adjusts the positions and timings for coordinated implementation of joint maneuvers according to the scenario. High and middle levels together are implemented as hybrid event-oriented system Monsoon, which includes a scenario-case unit, a case base, and a model-based unit. The model-based unit is used for obtaining spatial areas. The scenarios and triggers for each class of events are written in *XML*-based SCDL (scenario definition language).

- **Individual Vehicle Control (Lower) Level:** Which implements the transformation of maneuvers to low-level control actions. It receives a sequence of triads $\langle Pos(u_i), t_l, \varphi_j \rangle$ at the input and generates values of parameters $p_{li},...p_{vj}$ for UAV's actuators $ac_l,...ac_v$ on output. This level is implemented as hybrid system Breeze, which includes a case-based and a model-based units. The model-based unit implements a kinematic model and is master, while the case-based unit is a slave. The case-based unit uses a constraint satisfaction algorithm in adaptation process to resolve given flying and time-critical constraints.

On the lower level, the scenario-case coordination control system obtains the safety assessment in a real-time as shown in Figure 11.

The task of finding suitable compensations can be solved by the following sequence of steps.

Firstly, we build the spatial model of AOI. On the second step, we place all the attractors, such as planned waypoints, in appropriate positions. As well, we also place all repellers such as obstacles, other vehicles, situation disturbances, etc., in appropriate positions.

Figure 11. Scenario-case coordination control system

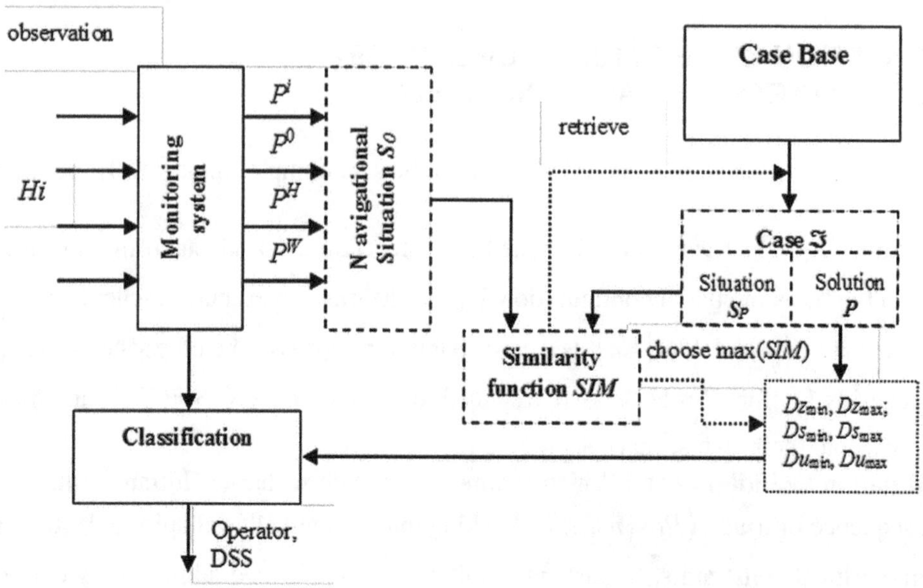

On the third step, we calculate the safety areas for all of repellers. On this stage, we should assign obtained values of the safety/danger degree ω_d to the attribute A_S for all cells of the grid D.

On the next step, in order to simplify the model, we transform three-dimensional space into two-dimensional space mapping all objects of the AOI on the horizontal plane $e_1, e_2, 0$, wherein perform a superposition of the values of the attribute A_S for all objects that are mapped to the same cell of the grid. Then we transform the safety/danger estimates into levels via the ordinal scale ϑ. At this moment, we already have built the soft topology model.

Now we can add an additional dimension with a scale $\left[-1,1\right]$, and build the corresponding surface of the permissible movement according to the safety/danger level values for each cell (Figure 12).

Repellers cut out dangerous areas from the surface. At the final step, we perform the search. We start from the threshold $\tau = 1$ and perform the τ-cut of the surface. If the path from the current position of vehicle and the planned WP exists, we regard this path as a planned trajectory. Otherwise, we lower the threshold either to the next (lower) safety level, or to the certain numerical value. For example, we may choose $\tau = 0.78$ and perform the new τ-cut of the surface (Figure 13).

If we find the desired path, the search process stops, and if not, we lower the threshold again and perform the next iteration. Such process can continue until the desired path is found or until the lowest safety permissible value is reached.

Figure 12. The surface of the permissible movement

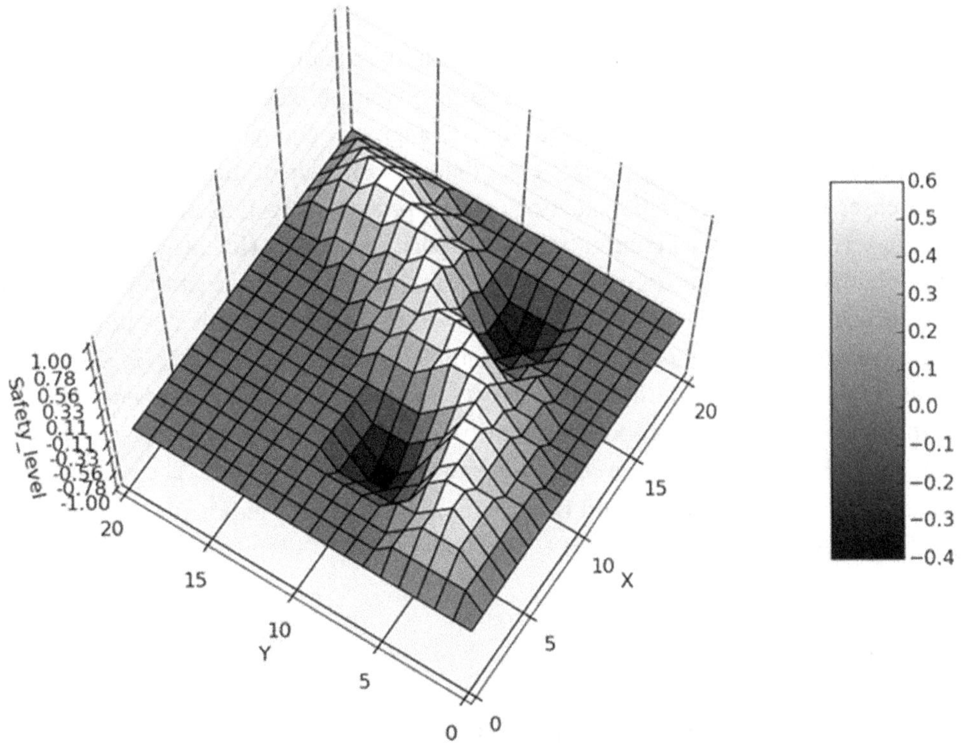

Figure 13. The plane of the 0.78-cut of the surface

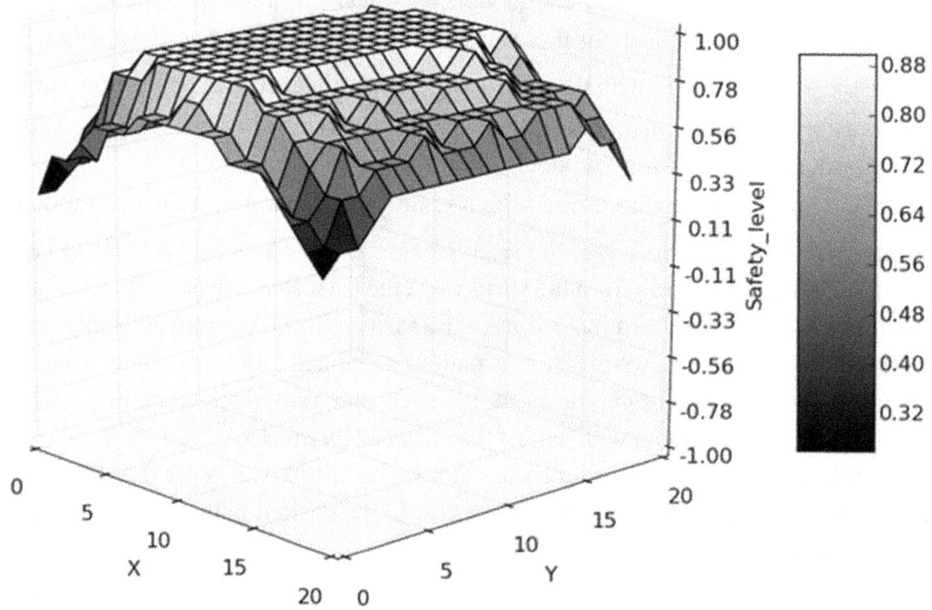

EXPERIMENT RESULTS

The proposed approach is implemented for the vehicle onboard control system prototype using embedded microcontroller STM32F429 and C++ programming language. The higher and middle layers of this system is built under a geoinformation system based on Django/GeoDjango framework with a lightweight PostgreSQL-compatible database engine. It is clear that performance of the control system depends on the number of cases involved in similar situations search process as well as on sampling of the spatial model, which describes the spatial configuration. We perform an experiment with the control system prototype that simulate the mission execution for the ensemble of 22 UAVs operating in 7 different roles.

The performance of the developed system depends on the number of vehicles in mission order, on the sampling of the spatial model (δ), and on the number of levels in the scale ϑ. The simulation is aimed at evaluation of influences of these parameters on the time of finding compensation. The results of the experiment showed that the proposed scenario-case approach provides acceptable performance for the given UV ensemble with a 12-leveled scale and a cell size up to 4 m. The optimal angular size for sectors is about 30°. Thus, the method is acceptable for solving the practical problems of motion coordination for heterogeneous ensembles of UAVs.

CONCLUSION

We proposed a scenario-case approach to coordinated control of heterogeneous ensembles of UAVs, which use patterns for activity in similar situations described as the spatial configurations that change by observed events. Due to using the rough sets for determine spatial areas and two-level sampling of space this approach is not sensitive to imprecise and incomplete observations. The main condition of proper implementation of this approach is a synchronization of each UAV's case bases in time and content. Another condition is an availability of enough competence to find a suitable case and choose an adequate set of scenarios, so our following work will focus on this issue. This work also presents the soft topology model and the method of finding the suitable compensation for vehicles' activity scenarios that can keep the prescribed spatial configuration and satisfy all safety restrictions for the vehicle onboard control system. The qualitative safety assessment is proposed, and the implementation of the method based on the soft level topology is presented. The developed system is apart of the complex vehicle onboard control system, which can pilot vehicles along the path prescribed in the scenario in accordance with the given spatial configuration satisfying the safety conditions. The proposed approach significantly reduces the computational complexity of problem solving. The developed onboard control system prototype provides the satisfactory performance.

FUTURE RESEARCH DIRECTIONS

The direction of further research is related to the use of soft fuzzy-rough level sets, which will allow improving the quality of safety/danger evaluating and the accuracy of the safety domain building significantly. Using the rough set, the concepts of dangerous and safe areas can be defined, and the boundary area of rough set can be blurred with the soft level set based on the α-cuts of fuzzy set of cells. Thus, a vague safety topology can be obtained efficiently.

REFERENCES

Allam, A., Bakeir, M., & Abo-Tabl, E. (2008). Some Methods for Generating Topologies by Relations. *Bulletin of the Malaysian Mathematical Sciences Society*, *2*(31), 35–45.

Ben-Asher, Y., Feldman, S., Gurl, P., & Feldman, M. (2008). Distributed Decision and Control for Cooperative UAVs using Ad-Hoc Communication. *IEEE Transactions on Control Systems Technology*, *16*(3), 511–516. doi:10.1109/TCST.2007.906314

Chen, J., Zhang, X., Xin, B., & Fang, H. (2016). Coordination Between Unmanned Aerial and Ground Vehicles: A Taxonomy and Optimization Perspective. *IEEE Transactions on Cybernetics*, *46*(4), 959–972. doi:10.1109/TCYB.2015.2418337 PMID:25898328

De Sousa, J., Girard, A., & Hedrick, J. (2004). Elemental maneuvers and coordination structures for unmanned air vehicles. *Proc. of the 43th IEEE Conf. on Decision and Control (CDC)*, *5*, 608–613. 10.1109/CDC.2004.1428711

Dias, M., Zlot, R., Kalra, N., & Stentz, A. (2006). Market-Based Multirobot Coordination: A Survey and Analisys. *Proceedings of the IEEE*, *94*(7), 1257–1270. doi:10.1109/JPROC.2006.876939

Feng, F., Li, Y., & Leoreanu-Fotea, V. (2010). Application of level soft sets in decision making based on interval-valued fuzzy soft sets. *Computers & Mathematics with Applications (Oxford, England)*, *60*(6), 1756–1767. doi:10.1016/j.camwa.2010.07.006

Goodwin, E. (1975). A statistical study of ship domain. *Journal of Navigation*, *28*(03), 328–344. doi:10.1017/S0373463300041230

Haigh, K., & Shewchuk, J. (1994). Geometric Similarity Metrics for Case-Based Reasoning. *Working Notes from the AAAI-94 Workshop on Case-Based Reasoning*, 182–187.

Jadbabaie, A., Lin, J., & Morse, A. (2003). Coordination of groups of mobile autonomous agents using nearest neighbor rules. *IEEE Transactions on Automatic Control*, *48*(6), 988–1001. doi:10.1109/TAC.2003.812781

Jagadish, H., Ooi, B.-Ch., Tan, K.-L., Yu, C., & Zhang, R. (2005). iDistance: An Adaptive B+-Tree Based Indexing Method for Nearest Neighbor Search. *ACM Transactions on Database Systems*, *30*(2), 364–397. doi:10.1145/1071610.1071612

Jaidee, U., Munoz-Avila, H., & Aha, D. W. (2013). Case-based goal-driven coordination of multiple learning agents. *Proc. of the 21st Int. Conf. on Case-Based Reasoning*, 164–178. 10.1007/978-3-642-39056-2_12

Kaliaev, I., Gaiduk, A., & Kapustian, S. (2009). *Models and Algorithms of Collective Control in Groups of Robots*. Moscow: Fizmatlit.

Lawton, J., Beard, R., & Young, B. (2003). A decentralized approach to formation maneuvers. *IEEE Transactions on Robotics and Automation*, *19*(6), 933–941. doi:10.1109/TRA.2003.819598

Lee, H., & Rhee, K. (2001). Developing Collision Avoidance System by Using Expert System and Search Algorithm. Int. *Shipbuilding Progress*, *48*, 197–212.

Likhachev, M., & Arkin, R. C. (2001). Spatio-Temporal Case-Based Reasoning for Behavioral Selection. *Proc. of Int. Conf. of Robot Applications (ICRA)*, 1627–1634.

Liu, Y., Yang, C., & Du, X. (2008). A CBR-based Approach for Ship Collision Avoidance. In Lecture Notes In. *Artificial Intelligence*, *5027*, 687–697.

Manathara, J., & Ghose, D. (2011). Reactive collision avoidance of multiple realistic UAVs. *Aircraft Engineering and Aerospace Technology*, *83*(6), 388–396. doi:10.1108/00022661111173261

Mataric, M. (1995). Designing and understanding adaptive group behaviors. *Adaptive Behavior*, *4*(1), 51–80. doi:10.1177/105971239500400104

Merino, L., Martínez de Dios, J., & Ollero, A. (2015). Cooperative Unmanned Aerial Systems for Fire Detection, Monitoring, and Extinguishing. In *Handbook of Unmanned Aerial Vehicles* (pp. 2693–2722). Springer Netherlands. doi:10.1007/978-90-481-9707-1_74

Michael, N., & Kumar, V. (2011). Control of ensembles of aerial robots. *Proceedings of the IEEE*, *99*(9), 1587–1602. doi:10.1109/JPROC.2011.2157275

Molodtsov, D. (1999). Soft Set Theory – first results. *Computers & Mathematics with Applications (Oxford, England)*, *37*(4-5), 19–31. doi:10.1016/S0898-1221(99)00056-5

Ng, K., & Trivedy, M. (1998). A Neuro-Fuzzy Controller for Mobile Robot Navigation and Multirobot Convoying. IEEE Trans. on Systems, Manufacture and Cybernetics, Part B. *Cybernetics*, *28*(6), 829–840. PMID:18256001

Nicolescu, M., & Mataric, M. (2001). Expirience-Based Representation Construction: Learning from Human and Robot Teachers. *Proc. of the IEEE Int. Conf. on Intelligent Robots and Sys.*, 2, 740–745.

Pawlak, Z., Jerzy, W., Slowinski, R., & Ziarko, W. (1995). Rough Sets. *Communications of the ACM*, *38*(11), 88–95. doi:10.1145/219717.219791

Pietrzykovski, Z., & Chomski, J. (2003). A Navigator Decision Support System in Planning a Safe Trajectory. In *Proc. of the 4th General Assembly Int* (pp. 4–10). Alexandria, VA: Ass. of Maritime Universities IAMU.

Pietrzykowski, Z., & Uriasz, J. (2009). The Ship Domain – A Criterion of Navigational Safety Assessment in an Open Sea Area. *Journal of Navigation*, *62*(01), 93–108. doi:10.1017/S0373463308005018

Qin, C., Wu, X., Jiang, J., & Zhu, A. (2016). Case-based knowledge formalization and reasoning method for digital terrain analysis – application to extracting drainage networks. *Hydrology and Earth System Sciences*, *20*(8), 3379–3392. doi:10.5194/hess-20-3379-2016

Ros, R., López De Màntaras, R., Sierra, C., & Arcos, J. L. (2005). A CBR system for autonomous robot navigation. In *Proc. of the 2005 Conf. on Artificial Intelligence Research and Development*, 131, 299–306.

Sharma, B., Vanualailai, J., & Chand, U. (2009). Flocking of Multi-agents in Constrained Environments. *Europen J. of Pure and Appl. Math.*, *2*(3), 401–425.

Sherstjuk, V. (2013). The case-scenario approach to control the dynamic objects. Pressing issues and priorities in development of the scientific and technological complex, 17, 97–103.

Sherstjuk, V. (2015). Scenario-Case Coordinated Control of Heterogeneous Ensembles of Unmanned Aerial Vehicles. *Proc. of 2015 IEEE 3rd Int. Conf. Actual Problems of Unmanned Aerial Vehicles Developments*, 275–279. 10.1109/APUAVD.2015.7346620

Sherstjuk, V., Zharikova, M., & Sokol, I. (2018). Forest Fire-Fighting Monitoring System Based on UAV team and Remote Sensing. *Proc. of 2018 IEEE 38th Int. Conf. on Electronics and Nanotechnology (ELNANO-2018)*, 99-106. 10.1109/ELNANO.2018.8477527

Sherstjuk, V., Zharikova, M., Sokol, I., & Tarasenko, E. (2017). Onboard Joint Motion Coordination System for Heterogeneous Ensembles of Unmanned Vehicles. *Proc. of 2017 IEEE 37th Int. Conf. on Electronics and Nanotechnology (ELNANO-2017)*, 480-484.

Toner, J., & Yuhai, T. (1998). Flocks, herds, and schools: A quantitative theory of flocking. *Physical Review. E*, *58*(4), 4828–4858. doi:10.1103/PhysRevE.58.4828

Tošić, P., & Vilalta, R. (2010a). A Unified Framework for Reinforcement Learning, Co-Learning and Meta-Learning How to Coordinate in Collaborative Multi-Agent Systems. *Procedia Computer Science*, *1*(1), 2217–2226. doi:10.1016/j.procs.2010.04.248

Tošić, P., & Vilalta, R. (2010b). Learning and Meta-Learning for Coordination of Autonomous Unmanned Vehicles: A Preliminary Analysis. *Proc. of 19th European Conf. on Artificial Intelligence ECAI-2010*, 163–168.

Tunstel, E., de Oliveira, M., & Berman, S. (2002). Fuzzy Behaviour Hierarchies for Multi-Robot Control. *International Journal of Intelligent Systems*, *17*(5), 449–470. doi:10.1002/int.10032

Waslander, S. (2013). Unmanned Aerial and Ground Vehicle Teams: Recent Work and Open Problems. Autonomous Control Systems and Vehicles, Intelligent Systems, Control and Automation: Science and Engineering, 65, 21–36.

Yoon, H., & Shahabi, C. (2009). Accurate Discovery of Valid Convoys from Moving Object Trajectories. *IEEE Int. Conf. on Data Mining (ICDMW'09)*, 636–643. 10.1109/ICDMW.2009.71

Yuan, C., Zhang, Y., & Liu, Z. (2015). A Survey on Technologies for Automatic Forest Fire Monitoring, Detection and Fighting Using UAVs and Remote Sensing Techniques. *Canadian Journal of Forest Research*, *45*(7), 783–792. doi:10.1139/cjfr-2014-0347

Zadeh, L. (1965). Fuzzy sets. *Information and Control*, *8*(3), 338–353. doi:10.1016/S0019-9958(65)90241-X

Zak, B. (2004). The problems of collision avoidance at sea in the formulation of complex motion principles. *International Journal of Applied Mathematics and Computer Science*, *14*, 503–514.

Zhao, J., Wu, Z., & Wang, F. (1993). Comments of ship domains. *Journal of Navigation*, *46*(03), 422–436. doi:10.1017/S0373463300011875

Zharikova, M., & Sherstjuk, V. (2016). Case-based approach to intelligent safety domains assessment for joint motion of vehicles ensembles. *Proc. of the 4th International Conference on Methods and Systems of Navigation and Motion Control (MSNMC)*, 245-250. 10.1109/MSNMC.2016.7783153

KEY TERMS AND DEFINITIONS

Attractor: An attracting manifold of the state space.

Maneuver: A single-step or multi-step change of values of one or more specific parameters of one or more unmanned aerial vehicles.

Repeller: A forbidding manifold of the state space.

Scenario: A sequence of scenes associated with the certain time points.

Unmanned Aerial Vehicle: An aircraft without a human pilot aboard.

Unmanned Ground Vehicle: A vehicle that operates while in contact with the ground and without an onboard human presence.

Chapter 10
Analysis and Optimization of Diagnostic Procedures for Aviation Radioelectronic Equipment

Maksym Zaliskyi
National Aviation University, Ukraine

Oleksandr Solomentsev
National Aviation University, Ukraine

Ivan Yashanov
National Aviation University, Ukraine

EXECUTIVE SUMMARY

In this chapter, the authors present the questions of aviation radioelectronic equipment operation. The structure of operation system is considered based on processes approach with adaptable control principles usage. Operation system contains processes of diagnostics and health monitoring. The authors consider the direct problem of efficiency estimation for diagnostics process, and main attention is paid to probability density function calculation for diagnostics duration. Simulation results were used for adequacy testing of these calculations. The authors also take into account the possibility of first and second kind errors presence. The inverse problem for diagnostics is defined and solved for mathematical expectation of repair time. In general case, the inverse problem can be solved for seven options of optimization.

INTRODUCTION

In September 1991, at the Tenth air navigation conference, ICAO members endorsed the concept of CNS/ATM, which allows civil aviation to overcome the known shortcomings of the existing system on a global scale and take advantage of the latest technology to ensure the predicted development of aviation in the 21st century. In 1996, a «Global Air Navigation Plan for CNS/ATM Systems» (Doc 9750) was developed, which is a strategic document for guidance in the implementation of CNS/ATM systems.

DOI: 10.4018/978-1-5225-7588-7.ch010

To provide the safety and regularity of flights, CNS/ATM contains three segments: 1) ground-based radioelectronic equipment (REE) of communication, navigation, surveillance; 2) onboard avionics; 3) satellite equipment. The interaction between these three segments is carried out using data links, which include equipment, protocols, and data processing technologies.

There are three stages of the life cycle of radioelectronic equipment: design, manufacture, and operation. The longest stage is the operation. Scientific, design and operational organizations strive to minimize the costs of expendable resources from the beginning of REE design to its recycling. Operation systems for REE can also be considered as objects of design and optimization (Solomentsev, Melkumyan, Zaliskyi & Asanov, 2015).

According to functioning terminology, the operation is a stage of device lifecycle at which its quality is realized, maintained and restored. Operation contains: equipment usage, transporting, storage, maintenance and repair (Goncharenko, 2015). Maintenance is a part of operation and contains the same processes as operation does, except for the process of the equipment using for the particular purpose (Goncharenko, 2017).

BACKGROUND

According to reference (Dhillon, 2006), it can be presumed that the main components of the operation system (OS) are as follows:

1. Structural elements of the operation system and their structure.
2. Technological processes.
3. Executors (engineering personnel maintaining radio flight support facilities).
4. Operational facilities.
5. Administrative regulatory documentation.
6. Consumables.
7. Informational resources.
8. Operational conditions (climatic conditions, electromagnetic compatibility).

Practice shows that important technological processes (TP) for the operation of aviation REE are commissioning, flight and ground monitoring for equipment technical condition, material, and technical supply, recycling, the extension of equipment life, etc.

The process of intended use is the main process of operation. All other processes are designed to maintain the efficiency of the main process.

During the REE operation, the degradation processes can occur. This is due to the influence of the environment, instability of power sources, electromagnetic compatibility, etc. As a result, there are changes in the trend of the diagnostic parameters towards the maximum permissible (pre-failure) levels. Therefore, the purpose of operation system is estimation of the current and future technical condition of REE, on the basis of which it is possible to generate and implement corrective and preventive actions (Solomentsev, Zaliskyi & Zuiev, 2013; Solomentsev, Zaliskyi, Nemyrovets & Asanov, 2015; Solomentsev, Zaliskyi, Kozhokhina & Herasymenko, 2017; Goncharenko, 2018).

The efficiency of the operation system functioning can be regarded as two aspects. The first aspect is related to the effect of the risk on radio flight support equipment and the stability of the air navigation service system functioning. Another aspect of efficiency is related to the minimization of the costs spent on radio flight support facilities.

Block diagram of operation system is shown in Figure 1.

Figure 1 contains OS components, which condition is under monitoring. The requirements establish the threshold values of the levels and conditions of the OS components in the corresponding phase spaces. The conformity assessment procedure (CAP) also includes a diagnostic procedure to determine the reasons for a possible failure. During the conformity assessment process, the errors of the first and second kinds are possible. The presence of errors reduces the efficiency and increases operating costs.

The efficiency assessment procedure is intended for determining the difference between the requirements for the levels and conditions of OS components and their real values (Solomentsev, Zaliskyi & Zuiev, 2016). As a result of such assessment, additional corrective actions can be generated.

All these procedures can be represented in the form of generalized operators. Each operator has an input, an output, a mathematical structure of data transformation. Operators can be complex, composite and include several algorithms.

ANALYSIS OF DIAGNOSTICS PROCEDURES

The main component of operation system is REE. During REE intended use, failures and damages are possible. For detection of failed aggregates, units and elements they use diagnostics procedures.

The procedure of REE technical condition diagnosing is realized as diagnostics program (DP) that determines the order of certain technological procedures fulfillment. If we fulfill DP, then we will find such an element of REE (or OS) that is the cause of failure. In this case, the technical condition of the REE element will be functional or non-functional (e.g. Hoyland & Rausand, 1994; Rausand, 2004; Smith, 2005).

After fulfillment of the diagnostics, the running repair procedures for REE are realized. These procedures provide restoring the serviceability of REE. At the same time, they initially implement the DP and then carry out procedures to restore the serviceability of REE.

Figure 1. Block diagram of operation system

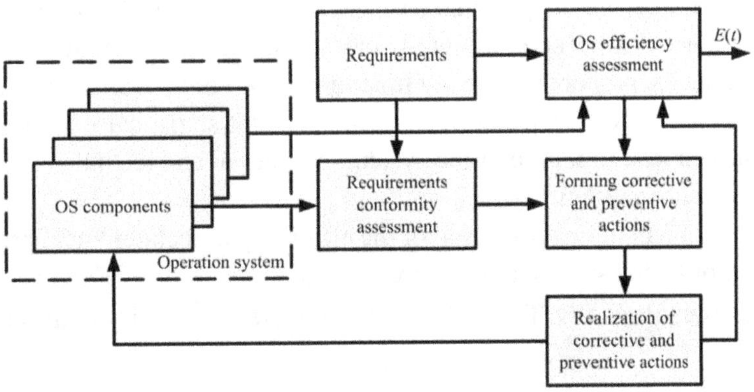

In the process of DP and running repair program development it is necessary to construct a diagnostic model (DM).

To construct the DM, the features of the structural, functional and electrical diagrams of REE are taken into account. DM can have several levels for diagnostics object (DO) decomposition (equipment, unit, circuit, etc.). In general, considering a specific REE, it is possible to construct several diagnostic models that correspond to different levels of decomposition.

In the process of diagnostic models construction it is necessary to take into consideration the following assumptions:

- For each element of the diagnostic model, the nominal values and allowable deviations of the input and output signals and the control points of the information monitoring for these signals are Known; in this case, we assume that we have corresponding measuring devices;
- the DM element is considered to be serviceable if in case of nominal input signals the value of the output signal is inside given tolerances;
- In case of exceeding the threshold values of at least one of the input signals, the output signal of the DM element also goes beyond the given tolerances;
- Any DM element can have only one output signal for any number of input signals;
- The DM should not have feedbacks between the elements since it creates uncertainty in the detection of the failures.

Thus, the performance of *i*-th element will be characterized by its output parameter X_i.

It is recommended to depict DP graphically in the following way: by rectangles – technological procedures of DP technical condition control at the DM elements outputs; by edges (arrows) – transition from one procedure to another; to complete every program link with a circle having an image of the detected failed DM element. Possible results of control procedures are shown on the edges (arrows), for example as "1", if DP value is within the tolerances limits, and as "0", if DP value is beyond the limits. When DP is finished, then running repair procedure is implemented. This procedure can consist of technological procedures of failed DM elements replacement and control of DO serviceable condition as the whole after its repair.

During DP design they develop alternative options, calculate the efficiency indicator and select the best according to a specific criterion.

In existing normative and regulatory documents and scientific literature, in case of assessing the efficiency of the TP of given type, as a rule, they calculate efficiency indicators in the form of mathematical expectations of resource costs. In the general case, such an approach isn't entirely correct as resource costs are random variables (RV). Therefore, in the case of theoretical calculations of TP efficiency, it is advisable to determine the probability density functions (PDF) or to evaluate these random variables variance and other moments. Uncompleted taking into account statistical parameters of fluctuations in the resources costs can lead to failure of the system of material and technical supply for TP and the inefficiency of the operation system.

Let's consider the problem of determination of the efficiency of technological procedures of searching failed elements in order to estimate the resources costs.

Let's consider the example of DO. This DO contains four elements series-connected – E_1, E_2, E_3, E_4 (Figure 2).

Figure 2. DO example

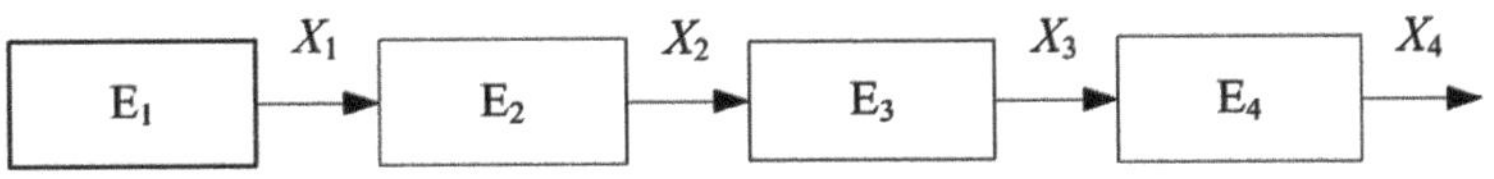

On Figure 2, X_i are informational signals at *i*-th element output. The value of X_i depends on element condition and the value of input signal X_{i-1}.

The example of DP is shown on Figure 3.

Program on Figure 3 was designed according to the method of half partition. In the diagram for DP there are three conformity assessment procedures for parameters X_2, X_3, X_1 and four decisions about non-compliance with the requirements of the elements E_4, E_3, E_2, E_1. Let's assume that after decision making about non-compliance with the requirements of *i*-th element it is necessary to implement an *i*-th set of technological operations STO_i for $i \in [1;4]$. The type of diagnostics diagram depends on the method of diagnostics, the model of DO and errors of first and second kind presence or absence. If there are no errors during decision making, then we implement only one STO, which allows us to correctly find the element in the DO with non-compliance presence.

In general, efficiency indicators are determined in regulatory documents. It is possible to use own system of efficiency indicators for diagnostics, which corresponds to the nature of the applied tasks. In this case, we can use the following indicators:

D is a probability of correct diagnostics. It is the complete probability of that diagnostics system determines that technical condition, in which DO is indeed;

$m_1(t_d)$ is a mean duration of diagnostics (mathematical expectation of one diagnostics duration);

$m_1(C_d)$ is an average cost of diagnostics (value estimating the mathematical expectation of one diagnostics duration);

$m_1(S_d)$ is an average labor intensity of diagnostics (value estimating the mathematical expectation of time taken by one diagnostics duration).

Let's consider the basic analytical correlations for diagnostics program. According to Figure 3, the duration of STO_1 is

Figure 3. DP example

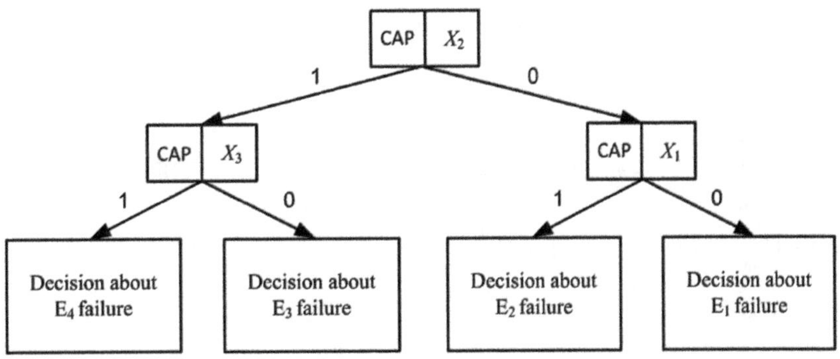

$$t_{\text{STO1}} = t_{\text{CA } x_1} + t_{\text{CA } x_2},$$

where $t_{\text{CA } x_1}$, $t_{\text{CA } x_2}$ are times of CAP implementation for X_1 and X_2. Parameters $t_{\text{CA } x_1}$ and $t_{\text{CA } x_2}$ in general case are RV, which properties are characterized by probability density function $f(t_{\text{CA } x_1})$, $f(t_{\text{CA } x_2})$. Taking in account these functions and using properties of functional transformations (Hahn & Shapiro, 1967) we can find conditional PDF $f(t_{\text{STO}i} \,/\, i\text{-th element failure (EF)})$ (for $i \in [1;\, n]$, where *n* is a quantity of DO elements). According to $f(t_{\text{STO}i} \,/\, i\text{-th EF})$ it is possible to determine unconditional PDF of DO diagnostics duration

$$f(t_d) = Q_1 f(t_{\text{STO1}} \,/\, \text{first EF}) + Q_2 f(t_{\text{STO2}} \,/\, \text{second EF}) + \ldots + Q_n f(t_{\text{STO}n} \,/\, n\text{-th EF});$$

$$\sum_{i=1}^{n} Q_i = 1.$$

where Q_i is a probability of *i*-th EF.

In formula (1) PDF $f(t_{\text{STO}i} \,/\, i\text{-th EF})$ must satisfy the normalization condition. If $\sum_{i=1}^{n} Q_i = 1$, then PDF $f(t_d)$ will also satisfy this condition

$$\begin{aligned}\int_0^\infty f(t_d)dt_d &= Q_1 \int_0^\infty f(t_{\text{STO1}} \,/\, \text{first EF})\, dt_{\text{STO1}} + \ldots + Q_n \int_0^\infty f(t_{\text{STO}n} \,/\, n\text{-th EF})\, dt_{\text{STO}n} \\ &= Q_1 + \ldots Q_n = 1.\end{aligned} \tag{1}$$

Formula (1) is valid for an arbitrary quantity of elements in DO. If $t_{\text{CA } x_i}$ isn't random, then $t_{\text{STO}i}$ will be non-random too and unconditional PDF $f(t_d)$ will be distribution Q_i weighted sum of delta-functions $\delta(t_{\text{STO}i})$ for $i \in [1;\, n]$.

Conditional moments $m_1(t_{\text{STO}i} \,/\, i\text{-th EF})$ and $\mu_2(t_{\text{STO}i} \,/\, i\text{-th EF})$ are equal to

$$m_1(t_{\text{STO}i} \,/\, i\text{-th EF}) = \int_0^\infty t_{\text{STO}i} f(t_{\text{STO}i} \,/\, \text{first EF})\, dt_{\text{STO}i},$$

$$\begin{aligned}\mu_2(t_{\text{STO}i} \,/\, i\text{-th EF}) &= \int_0^\infty \left(t_{\text{STO}i} - m_1(t_{\text{STO}i} \,/\, i\text{-th EF})\right)^2 f(t_{\text{STO}i} \,/\, \text{first EF})\, dt_{\text{STO}i} \\ &= m_2(t_{\text{STO}i} \,/\, i\text{-th EF}) - \left[m_1(t_{\text{STO}i} \,/\, i\text{-th EF})\right]^2.\end{aligned}$$

Unconditional mathematical expectation of the DO diagnosis duration $m_1(t_d)$ is determined by the formula

$$m_1(t_d) = \int_0^\infty t_d f(t_d) dt_d = Q_1 \int_0^\infty t_{STO1} f(t_{STO1} / \text{first EF})\, dt_{STO1}$$
$$+ \ldots + Q_n \int_0^\infty t_{STOn} f(t_{STOn} / n\text{-th EF})\, dt_{STOn}$$
$$= Q_1 m_1(t_{STO1} / \text{first EF}) + \ldots + Q_n m_1(t_{STOn} / n\text{-th EF}),$$

where $m_1(t_{STOi} / i\text{-th EF})$ is conditional mathematical expectation of *i*-th STO implementation duration in case of *i*-th element failure.

So mathematical expectation of random variable t_d is equal to the sum of conditional mathematical expectations of *i*-th STO implementation duration multiplied by Q_i values.

Equation for the unconditional variance of t_d is following

$$\mu_2(t_d) = \int_0^\infty \left(t_d - m_1(t_d)\right)^2 f(t_d) dt_d = m_2(t_d) - \left[m_1(t_d)\right]^2.$$

Second raw moment $m_2(t_d)$ is equal to

$$m_2(t_d) = \int_0^\infty t_d^2 f(t_d) dt_d = Q_1 \int_0^\infty t_{STO1}^2 f(t_{STO1} / \text{first EF})\, dt_{STO1} + \ldots +$$
$$+ Q_n \int_0^\infty t_{STOn}^2 f(t_{STOn} / n\text{-th EF})\, dt_{STOn} =$$
$$= Q_1 m_2(t_{STO1} / \text{first EF}) + \ldots + Q_n m_2(t_{STOn} / n\text{-th EF}).$$

In this equation conditional moment $m_2(t_{STOi} / i\text{-th EF})$ for i-th STO can be determined as

$$m_2(t_{STOi} / i\text{-th EF}) = \mu_2(t_{STOi} / i\text{-th EF}) + \left[m_1(t_{STOi} / i\text{-th EF})\right]^2.$$

To evaluate the veracity of these formulas, the statistical simulation was performed using the Monte-Carlo method for OD presented by the diagnostic model in Figure 2 and DP in Figure 3.

The simulation was performed under the conditions that PDF of the time resources for STO implementation is Gaussian with the parameters $m_1(t_{STOi} / i\text{-th EF})$, $\mu_2(t_{STOi} / i\text{-th EF})$ $(i \in [1; n])$. Table 1 shows data on the numerical values of the parameters of the general population that describes the diagnostic process of DO.

Table 1. Initial data for simulation of DO diagnostic program

Parameters of the General Population	Option N° 1	Option N° 2	Option N° 3	Option N° 4	Option N° 5
Q_1	0.4	0.25	0.1	0.2	0.1
Q_2	0.2	0.25	0.25	0.5	0.2
Q_3	0.1	0.25	0.4	0.15	0.3
Q_4	0.3	0.25	0.25	0.15	0.4
$m_1(t_{\text{STO1}})$	50	50	50	50	50
$m_1(t_{\text{STO2}})$	60	60	70	70	55
$m_1(t_{\text{STO3}})$	75	70	80	90	75
$m_1(t_{\text{STO4}})$	90	80	100	100	90
$\mu_2(t_{\text{STO1}})$	25	25	49	49	36
$\mu_2(t_{\text{STO2}})$	36	25	36	49	36
$\mu_2(t_{\text{STO3}})$	25	25	36	25	49
$\mu_2(t_{\text{STO4}})$	49	25	25	36	25

In the process of simulation, estimation of the mathematical expectation $m_1^*(t_d)$, variance $\mu_2^*(t_d)$, standard deviation $\sigma^*(t_d)$ and PDF $f^*(t_d)$ of one diagnostic procedure duration was performed.

Table 2 shows simulation results in the form of point estimates of parameters $m_1^*(t_d)$, $\mu_2^*(t_d)$, $\sigma^*(t_d)$, lower t_{d-} and upper t_{d+} thresholds of interval estimates of mathematical expectation $m_1^*(t_d)$, and also the results of theoretical calculations of parameters $m_1(t_d)$, $\mu_2(t_d)$, $\sigma(t_d)$. Interval estimates were calculated for the confidence probability $\gamma = 0.95$.

For a quantitative assessment of significant difference between the functions $f(t_d)$ and $f^*(t_d)$ we can calculate the chi-square criterion for the confidence probability $\gamma = 0.95$.

In the process of calculations, the following quantitative measures of chi-square criterion were calculated for five modeling options $\chi_1^2 = 10.8$, $\chi_2^2 = 11.7$, $\chi_3^2 = 11.3$, $\chi_4^2 = 9.5$, $\chi_5^2 = 9.1$ (for threshold level $\chi_{\text{thr}}^2 = 30.1$). The example of PDF for the fifth option of parameters set is shown in Figure 4.

Table 2. Results of simulation and theoretical calculations of diagnostic efficiency indicators

Option Number	Point Estimates			Interval Estimates	Calculation Results		
	$m_1^*(t_d)$	$\mu_2^*(t_d)$	$\sigma^*(t_d)$	t_{d-} / t_{d+}	$m_1(t_d)$	$\mu_2(t_d)$	$\sigma(t_d)$
1	66.2	318	17.8	64.90/68.00	66.5	324.6	18
2	65	143.7	12	63.90/66.07	65	150	12.2
3	79.6	233.7	15.3	78.10/80.90	79.5	249.3	15.8
4	73.3	283.3	16.8	71.90/75.03	73.5	306.2	17.5
5	74.8	256	16	73.06/75.93	74.5	267.7	16.4

Figure 4. Results of theoretical calculation and statistical simulation for determining the PDF of diagnostics duration (option of the initial data No. 5)

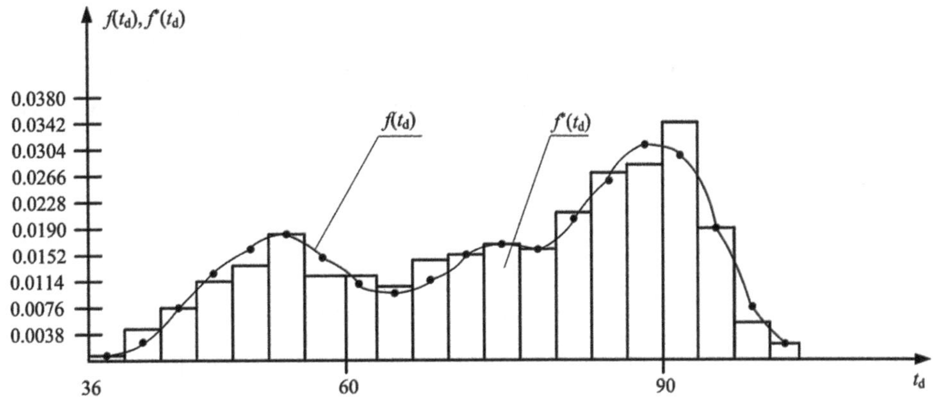

TAKING INTO ACCOUNT FIRST AND SECOND KIND ERRORS DURING DIAGNOSTICS PROCESS

In case of efficiency indicators calculating, it is appropriate to take into account the errors of first and second kinds for CAP, which are determined by conditional probabilities α and β. The neglect of the first and second kind errors during the compliance assessment (CA) can lead to increased resource costs (e.g. Barlow & Proschan, 1965; Dhillon, 2005).

Let's consider the processes of diagnostics and running repair (RR), which include the procedure for finding an inappropriate element, when $\alpha \neq 0$, $\beta \neq 0$. In this case, we also have four graphs, each of which will have four STO. Each graph will contain three STO associated with "false" element detection and only one – with "correct". The number of operations inside STO that are associated with the "false" detection of the element will be determined by the strategy accepted in these cases. For example, these strategies can include the repetition of the DP initially.

Let's consider the following strategy:

1. Replacing the failed element by objectively serviceable,

2. Control of the serviceability of the whole equipment (CSE),
3. If the result of the final control of the object doesn't satisfy the requirements, we will make the decision about the failure of the next element of equipment, starting from the first element. After that, the control of the serviceability of the whole equipment is performed again.

So it is clear that in the case of first and second kind errors presence resource costs for diagnostics increase.

Figure 5 shows the example of one of four graphs in case of first element failure for DO given in Figure 2. There are four STO in Figure 5. On the edges of DP we denote the first and second kind errors.

To analyze the DP efficiency, let's determine the PDF of resources cost (for this example resources cost is similar to time for DP implementation). Next, for engineering calculations, it is appropriate to determine mathematical expectations and variance of resources cost. If $\alpha \neq 0$, $\beta \neq 0$, we can calculate the mathematical expectation of correct diagnostics probability *D*.

More substantially, let's consider an example when the first element is objectively failed. On the edges of the graph in Figure 5, the error probability symbols are shown.

Due to the random nature of the results of control and measuring operations, all STO will also be random in terms of its number and duration. The probability of the implementation of certain STO – $P(\text{STO}i \,/\, \text{first EF})$ is conditional, and all these probabilities form a complete group of events:

$$\sum_{i=1}^{4} P(\text{STO}i \,/\, \text{first EF}) = 1, \tag{2}$$

where $P(\text{STO}i \,/\, \text{first EF})$ is a the probability of *i*-th STO performing in case of first element failure.

In addition, we assume that the probabilities of the first ($\alpha_i = \alpha$) and the second ($\beta_i = \beta$) kind errors for all elements are equal to each other.

Figure 5. The graph of DO conditions and running repair in case of first element E_1 failure and the presence of first and second kind errors

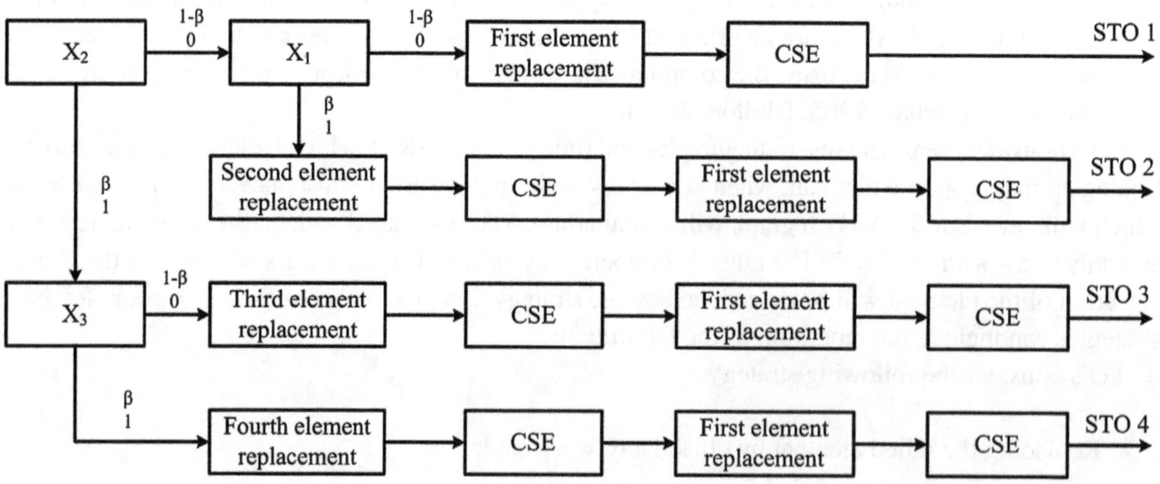

Then taking into account the Figure 5, we can calculate all probabilities $P(\mathrm{STO}i\ /\ \mathrm{first\ EF})$:

$$P(\mathrm{STO1}\ /\ \mathrm{first\ EF}) = \left(1-\beta\right)^2;$$

$$P(\mathrm{STO2}\ /\ \mathrm{first\ EF}) = \beta\left(1-\beta\right);$$

$$P(\mathrm{STO3}\ /\ \mathrm{first\ EF}) = \beta\left(1-\beta\right);$$

$$P(\mathrm{STO4}\ /\ \mathrm{first\ EF}) = \beta^2.$$

Let's assume that it is known that PDF of time for DP implementation for the case of first element failure is $f(t_\mathrm{d}\ /\ \mathrm{first\ EF})$. Then

$$\begin{aligned} f(t_\mathrm{d}\ /\ \mathrm{first\ EF}) &= P(\mathrm{STO1}\ /\ \mathrm{first\ EF})f(t_{\mathrm{STO1}}\ /\ \mathrm{first\ EF}) + \\ &+ P(\mathrm{STO2}\ /\ \mathrm{first\ EF})f(t_{\mathrm{STO2}}\ /\ \mathrm{first\ EF}) + P(\mathrm{STO3}\ /\ \mathrm{first\ EF})f(t_{\mathrm{STO3}}\ /\ \mathrm{first\ EF}) + \\ &+ P(\mathrm{STO4}\ /\ \mathrm{first\ EF})f(t_{\mathrm{STO4}}\ /\ \mathrm{first\ EF}). \end{aligned} \tag{3}$$

In accordance with the normalization condition we can write:

$$\begin{aligned} &\int_0^\infty f(t_\mathrm{d}\ /\ \mathrm{first\ EF})\,dt_\mathrm{d} = P(\mathrm{STO1}\ /\ \mathrm{first\ EF})\int_0^\infty f(t_{\mathrm{STO1}}\ /\ \mathrm{first\ EF})\,dt_{\mathrm{STO1}} + \ldots + \\ &+ P(\mathrm{STO4}\ /\ \mathrm{first\ EF})\int_0^\infty f(t_{\mathrm{STO4}}\ /\ \mathrm{first\ EF})\,dt_{\mathrm{STO4}} = P(\mathrm{STO1}\ /\ \mathrm{first\ EF}) + \ldots + \\ &+ P(\mathrm{STO4}\ /\ \mathrm{first\ EF}) = 1. \end{aligned}$$

After determining conditional PDF $f(t_\mathrm{d}\ /\ i\text{-th EF})$, it is possible to calculate the unconditional PDF of time for DP implementation:

$$\begin{aligned} f(t_\mathrm{d}) &= Q_1 f(t_\mathrm{d}\ /\ \mathrm{first\ EF}) + Q_2 f(t_\mathrm{d}\ /\ \mathrm{second\ EF}) + \\ &+ Q_3 f(t_\mathrm{d}\ /\ \mathrm{third\ EF}) + Q_4 f(t_\mathrm{d}\ /\ \mathrm{fourth\ EF}). \end{aligned} \tag{4}$$

Unconditional PDF $f(t_\mathrm{d})$ will satisfy the normalization condition, so:

$$\int_0^\infty f(t_d)dt_d = Q_1\int_0^\infty f(t_d / \text{first EF})dt_d + Q_2\int_0^\infty f(t_d / \text{second EF})dt_d +$$
$$+Q_3\int_0^\infty f(t_d / \text{third EF})dt_d + Q_4\int_0^\infty f(t_d / \text{fourth EF})dt_d = Q_1 + Q_2 + Q_3 + Q_4 = 1.$$

These formulas for conditional and unconditioned PDF can be generalized also in the case when there are n elements in the diagnostics object. In this case, the expressions for PDF of STO duration will depend on the number of elements in the object and DP type.

Let's consider the question of determining the moments for PDF, represented by the formula (4). Mathematical expectation $m_1(t_d)$:

$$\begin{aligned} m_1(t_d) &= \int_0^\infty t_d f(t_d)dt_d = Q_1\int_0^\infty t_d f(t_d / \text{first EF})dt_d + Q_2\int_0^\infty t_d f(t_d / \text{second EF})dt_d + \\ &+Q_3\int_0^\infty t_d f(t_d / \text{third EF})dt_d + Q_4\int_0^\infty t_d f(t_d / \text{fourth EF})dt_d = \\ &= Q_1 I_1 + Q_2 I_2 + Q_3 I_3 + Q_4 I_4, \end{aligned} \tag{5}$$

where

$$\begin{aligned} I_i &= P(\text{STO1} / i\text{-th EF}) m_1(t_{\text{STO1}} / i\text{-th EF}) + \\ &+P(\text{STO2} / i\text{-th EF}) m_1(t_{\text{STO2}} / i\text{-th EF}) + \\ &+P(\text{STO3} / i\text{-th EF}) m_1(t_{\text{STO3}} / i\text{-th EF}) + \\ &+P(\text{STO4} / i\text{-th EF}) m_1(t_{\text{STO4}} / i\text{-th EF}). \end{aligned} \tag{6}$$

The variance:

$$\mu_2(t_d) = \int_0^\infty \left(t_d - m_1(t_d)\right)^2 f(t_d)dt_d = m_2(t_d) - \left[m_1(t_d)\right]^2. \tag{7}$$

The second raw moment $m_2(t_d)$:

$$\begin{aligned} m_2(t_d) &= \int_0^\infty t_d^2 f(t_d)dt_d = Q_1\int_0^\infty t_d^2 f(t_d / \text{first EF})dt_d + Q_2\int_0^\infty t_d^2 f(t_d / \text{second EF})dt_d + \\ &+Q_3\int_0^\infty t_d^2 f(t_d / \text{third EF})dt_d + Q_4\int_0^\infty t_d^2 f(t_d / \text{fourth EF})dt_d = \\ &= Q_1 J_1 + Q_2 J_2 + Q_3 J_3 + Q_4 J_4, \end{aligned} \tag{8}$$

where

$$\begin{aligned} J_i = {} & P(\text{STO1} / i\text{-th EF}) m_2(t_{\text{STO1}} / i\text{-th EF}) + \\ & + P(\text{STO2} / i\text{-th EF}) m_2(t_{\text{STO2}} / i\text{-th EF}) + \\ & + P(\text{STO3} / i\text{-th EF}) m_2(t_{\text{STO3}} / i\text{-th EF}) + \\ & + P(\text{STO4} / i\text{-th EF}) m_2(t_{\text{STO4}} / i\text{-th EF}). \end{aligned} \tag{9}$$

These formulas can also be used to assess the efficiency of running repair procedures, but at the same time, each STO complements the operations of failed element replacement, and control of the serviceability of the whole equipment, taking into account repair strategy. To evaluate the veracity of these formulas, the statistical simulation was performed using the Monte-Carlo method for DO presented by the diagnostic model in Figure 2 and DP in Figure 5. The simulation was performed under the conditions that PDF of the time resources for STO implementation is Gaussian with the parameters $m_1(t_{\text{STO}i} / i\text{-th EF})$, $\mu_2(t_{\text{STO}i} / i\text{-th EF})$ ($i \in [1; n]$). Table 3 shows data on the numerical values of the parameters of the general population that describes the diagnostic process of DO. The table also contains probabilities of elements failure, α and β probabilities values, mathematical expectations of control actions $m_1(t_{\text{CA}i})$, CSE duration $m_1(t_{\text{CSE}})$, i-th element replacement duration $m_1(t_{\text{ER}i})$ and corresponding variances.

In the process of simulation, estimation of the mathematical expectation $m_1^*(t_{\text{d}})$, variance $\mu_2^*(t_{\text{d}})$, standard deviation $\sigma^*(t_{\text{d}})$ and PDF $f^*(t_{\text{d}})$ of one diagnostic procedure duration was performed.

Table 4 shows simulation results in the form of point estimates of parameters $m_1^*(t_{\text{d}})$, $\mu_2^*(t_{\text{d}})$, $\sigma^*(t_{\text{d}})$, lower $t_{\text{d}-}$ and upper $t_{\text{d}+}$ thresholds of interval estimates of mathematical expectation $m_1^*(t_{\text{d}})$, and also the results of theoretical calculations of parameters $m_1(t_{\text{d}})$, $\mu_2(t_{\text{d}})$, $\sigma(t_{\text{d}})$. Interval estimates were calculated for the confidence probability $\gamma = 0.95$. The example of PDF for the fifth option of parameters set is shown in figure 6.

Comparison of theoretical calculations of PDF of diagnostics duration in case of first and second kind errors absence ($f_1(t_d)$) and presence ($f_2(t_d)$) is shown in Figure 7.

APPROACHE TO THE INVERSE PROBLEMS SOLUTION IN CASE OF CONFORMITY ASSESSMENT

In order to substantiate and improve certain TP we need to solve two types of tasks – direct and inverse. Solving the direct problem, they form alternative options of design solutions and then calculate the numerical values of efficiency indicators for these alternative options and according to the given criterion choose the best option.

To solve an inverse problem for a given level of numerical value of a generalized efficiency indicator for DO or its structural components, we need to determine the numerical values of parameters this efficiency indicator depends on.

To solve an inverse problem it is necessary to know:

1. The DO structure,
2. DP type,

Table 3. Initial data for simulation of DO diagnostic program

Parameters of the General Population	Option Nº 1	Option Nº 2	Option Nº 3	Option Nº 4	Option Nº 5
Q_1	0.4	0.25	0.1	0.2	0.1
Q_2	0.2	0.25	0.25	0.5	0.2
Q_3	0.1	0.25	0.4	0.15	0.3
Q_4	0.3	0.25	0.25	0.15	0.4
$m_1(t_{CA1})$	10	20	30	50	60
$m_1(t_{CA2})$	5	6	7	8	10
$m_1(t_{CA3})$	50	40	30	20	10
$m_1(t_{CSE})$	2	6	8	10	5
$m_1(t_{ER1})$	1	2	3	4	5
$m_1(t_{ER2})$	30	25	20	15	10
$m_1(t_{ER3})$	20	35	30	10	5
$m_1(t_{ER4})$	10	15	20	25	30
$\mu_2(t_{CA1})$	3	3	3	3	3
$\mu_2(t_{CA2})$	5	4	3	2	1
$\mu_2(t_{CA3})$	1	4	6	8	11
$\mu_2(t_{CSE})$	1	2	1	1	2
$\mu_2(t_{ER1})$	1	1	4	2	3
$\mu_2(t_{ER2})$	1	2	3	3	4
$\mu_2(t_{ER3})$	1	1	3	4	2
$\mu_2(t_{ER4})$	1	2	4	5	2
α	0.02	0.04	0.05	0.05	0.05
β	0.03	0.02	0.03	0.04	0.05

Table 4. Results of simulation and theoretical calculations of diagnostic efficiency indicators

Option Number	Point Estimates			Interval Estimates	Calculation Results			Average Probability of Correct Diagnostics $m_1(D)$
	$m_1^*(t_d)$	$\mu_2^*(t_d)$	$\sigma^*(t_d)$	t_{d-} / t_{d+}	$m_1^*(t_d)$	$\mu_2^*(t_d)$	$\sigma^*(t_d)$	
1	42.915	165	12.8	42.595/43.235	42.743	154.8	12.4	0.94966
2	64.835	383.7	19.6	64.515/65.155	64.655	373.1	19.1	0.9409
3	88.14	829.3	28.8	79.82/88.46	88	811	28.5	0.91869
4	110.9	1375	37	110.58/111.22	110.7	1328	36.4	0.912495
5	64.57	772	27.78	133.58/134.22	64.12	770	27.75	0.9025

Figure 6. Results of theoretical calculation and statistical simulation for determining the PDF of diagnostics duration (option of the initial data No. 5)

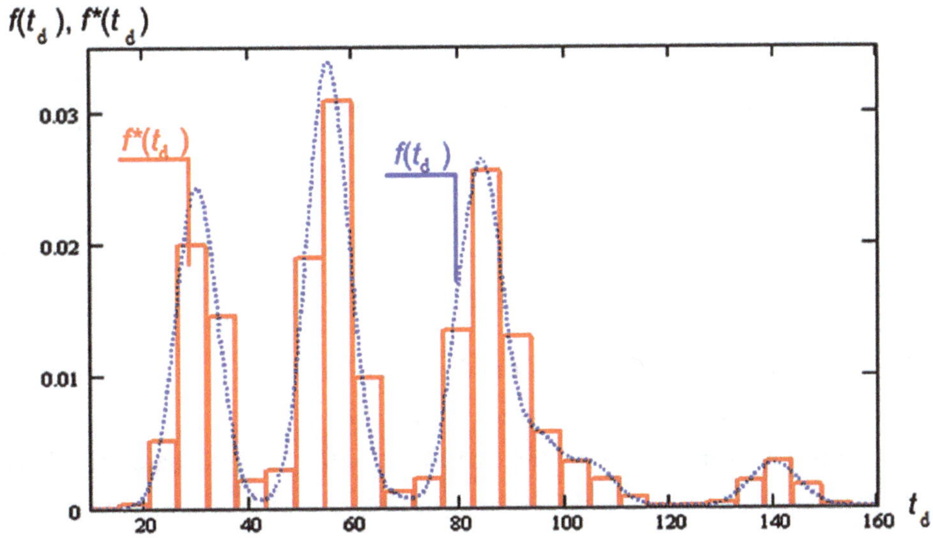

3. Probabilities of i-th element failure,
4. Values of resource costs for DP implementation, etc.

Let's consider the process of the running repair, which includes the procedure for finding an inappropriate element, when $\alpha \neq 0$, $\beta \neq 0$.

Let DO consists of n structural elements. For this object, diagnostic model and diagnostic program are known. In this case, the diagnostic model consists of m hierarchical levels of conformity assessment procedures, where $m \leq n-1$.

For inverse problem solution, a complete set of parameters is defined (the vectors of first and second kind errors, the vector of the duration of conformity assessment operations, the vector of the duration of elements replacement, etc.). These parameters affect one or more efficiency indicators. One or more parameters are defined as optimization parameters.

Figure 7. Comparison of PDF in case of first and second kind errors absence ($f_1(t_d)$) and presence ($f_2(t_d)$) (option of the initial data No. 5)

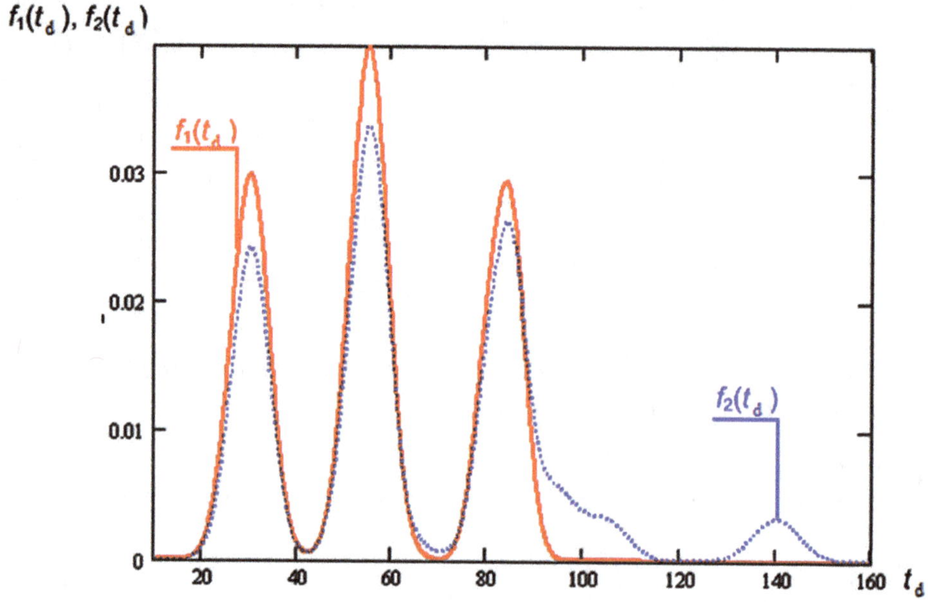

To simplify a problem let's assume $\alpha_i = \alpha$; $\beta_i = \beta$; $t_{CA_i} = t_{CA}$; $t_{ER_i} = t_{ER}$ for $i \in (0; m]$.

Efficiency indicators can be:

1. $m_1(t_{RR})$ is a mathematical expectation of running repair duration t_{RR};
2. $\mu_2(t_{RR})$ is variance of running repair duration t_{RR};
3. $m_1(t_{RR}) + K\sigma(t_{RR})$, where K is a constant coefficient; $\sigma(t_{RR})$ is a standard deviation of running repair duration t_{RR}.

Let's consider optimization case when efficiency indicator is mathematical expectation of running repair duration $m_1(t_{RR})$. We believe that optimization will performed for α, β and t_{CA}.

Since we know DM and DP, then we can find correlation

$$m_1(t_{rr}) = f(\alpha, \beta, t_{CA}).$$

There are seven options for inverse problems solution:

1. Optimization by α;
2. Optimization by β;
3. Optimization by t_{ca};
4. Optimization by α and β;
5. Optimization by α and t_{CA};

6. Optimization by β and t_{CA};
7. Optimization by α, β and t_{CA}.

Let's consider optimization by α. In this we need to find first derivative of function $f(\alpha, \beta, t_{\text{CA}})$ by α. As DP consists of m hierarchical levels of conformity assessment operations, function $f(\alpha, \beta, t_{\text{CA}})$ can be represented as polynomial of m degree relative to α with known coefficients (as we know β and t_{CA}), i.e.

$$f(\alpha, \beta, t_{\text{CA}}) = a_m(\beta, t_{\text{CA}})\alpha^m + a_{m-1}(\beta, t_{\text{CA}})\alpha^{m-1} + \ldots + a_1(\beta, t_{\text{CA}})\alpha + a_0(\beta, t_{\text{CA}}); \tag{10}$$

$$\begin{cases} \dfrac{\partial f(\alpha, \beta, t_{\text{CA}})}{\partial \alpha} = m a_m(\beta, t_{\text{CA}})\alpha^{m-1} + (m-1)a_{m-1}(\beta, t_{\text{CA}})\alpha^{m-2} + \ldots + a_1(\beta, t_{\text{CA}}); \\ 0 \le \alpha \le 1. \end{cases} \tag{11}$$

After equation (11) solution, we can find the optimum value of parameter α.

Optimization by β is performed similarly: equation (10) depends on parameter β; parameters α and t_{CA} are known.

In case of t_{CA} optimization function $f(\alpha, \beta, t_{\text{CA}})$ will be:

$$f(\alpha, \beta, t_{\text{CA}}) = a_1(\alpha, \beta)t_{\text{CA}} + a_0(\alpha, \beta).$$

Then

$$\frac{\partial f(\alpha, \beta, t_{\text{CA}})}{\partial t_{\text{CA}}} = a_1(\alpha, \beta) \ne 0.$$

So optimization by t_{CA} is impossible in case of $m_1(t_{\text{гг}})$ minimum calculation. Then we conclude that optimizations by (α, t_{CA}), (β, t_{CA}) and $(\alpha, \beta, t_{\text{CA}})$ are impossible too.

In case of optimization by α and β we need to solve the system of equations with conditions $0 \le \alpha \le 1$, $0 \le \beta \le 1$:

$$\begin{cases} \dfrac{\partial f(\alpha, \beta, t_{\text{CA}})}{\partial \alpha} = m a_m(\beta, t_{\text{CA}})\alpha^{m-1} + (m-1)a_{m-1}(\beta, t_{\text{CA}})\alpha^{m-2} + \ldots + a_1(\beta, t_{\text{CA}}) = 0; \\ \dfrac{\partial f(\alpha, \beta, t_{\text{CA}})}{\partial \beta} = m b_m(\alpha, t_{\text{CA}})\beta^{m-1} + (m-1)b_{m-1}(\beta, t_{\text{CA}})\beta^{m-2} + \ldots + b_1(\beta, t_{\text{CA}}) = 0. \end{cases} \tag{12}$$

To decide on the possibility of solving an optimization problem, it is necessary to calculate Hessian matrix.

Function has maximum (minimum) at given point (α_0, β_0), if:

$$G = \begin{vmatrix} \left.\dfrac{\partial^2 f(\alpha, \beta, t_{CA})}{\partial \alpha^2}\right|_{(\alpha_0, \beta_0)} & \left.\dfrac{\partial^2 f(\alpha, \beta, t_{CA})}{\partial \alpha \partial \beta}\right|_{(\alpha_0, \beta_0)} \\ \left.\dfrac{\partial^2 f(\alpha, \beta, t_{CA})}{\partial \beta \partial \alpha}\right|_{(\alpha_0, \beta_0)} & \left.\dfrac{\partial^2 f(\alpha, \beta, t_{CA})}{\partial \beta^2}\right|_{(\alpha_0, \beta_0)} \end{vmatrix} > 0$$

or

$$\begin{aligned} &(m(m-1)a_m(\beta, t_{CA})\alpha^{m-2} + (m-1)(m-2)a_{m-1}(\beta, t_{CA})\alpha^{m-3} + \ldots + 2a_2(\beta, t_{CA}))\times \\ &\times(m(m-1)b_m(\alpha, t_{CA})\beta^{m-2} + (m-1)(m-2)b_{m-1}(\beta, t_{CA})\beta^{m-3} + \ldots + 2b_2(\beta, t_{CA})) - \\ &-(m\frac{\partial a_m(\beta, t_{CA})}{\partial \beta}\alpha^{m-1} + (m-1)\frac{\partial a_{m-1}(\beta, t_{CA})}{\partial \beta}\alpha^{m-2} + \ldots + \frac{\partial a_1(\beta, t_{CA})}{\partial \beta})^2 > 0. \end{aligned}$$

If $\dfrac{\partial^2 f(\alpha, \beta, t_{CA})}{\partial \alpha^2} > 0$, then function has maximum at point (α_0, β_0), and minimum otherwise.

Let's consider mathematical equations for DO example (Figure 2).

We assume that the efficiency indicator is the mathematical expectation of running repair duration $m_1(t_{CA}, \alpha, \beta)$.

For this DO and DP we have

$$\begin{aligned} &m_1(t_{CA}, t_{ER}, t_{CSE}, \alpha, \beta) = Q_1((1-\beta)^2 m_1(t_{STO1} / \text{first EF}) + (\beta - \beta^2)m_1(t_{STO2} / \text{first EF}) \\ &+(\beta - \beta^2)m_1(t_{STO3} / \text{first EF}) + \beta^2 m_1(t_{STO4} / \text{first EF})) + Q_2(\alpha(1-\beta)m_1(t_{STO1} / \text{second EF}) \\ &+(1-\beta)(1-\alpha)m_1(t_{STO2} / \text{second EF}) + (\beta - \beta^2)m_1(t_{STO3} / \text{second EF}) \\ &+\beta^2 m_1(t_{STO4} / \text{second EF})) + Q_3(\alpha^2 m_1(t_{STO1} / \text{third EF}) + (\alpha - \alpha^2)m_1(t_{STO2} / \text{third EF}) \\ &+(1-\alpha)(1-\beta)m_1(t_{STO3} / \text{third EF}) + \beta(1-\alpha)m_1(t_{STO4} / \text{third EF})) \\ &+Q_4(\alpha^2 m_1(t_{STO1} / \text{fourth EF}) + (\alpha - \alpha^2)m_1(t_{STO2} / \text{fourth EF}) \\ &+(\alpha - \alpha^2)m_1(t_{STO3} / \text{fourth EF}) + (1 - 2\alpha + \alpha^2)m_1(t_{STO4} / \text{fourth EF})), \end{aligned}$$

In this case:

$$m_1(t_{STO1} / \text{first EF}) = t_{CA2} + t_{CA1} + (t_{ER} + t_{CSE}),$$

$$m_1(t_{STO2} / \text{first EF})) = t_{CA2} + t_{CA1} + 2(t_{ER} + t_{CSE}),$$

$$m_1(t_{\mathrm{STO3}}\ /\ \text{first EF}) = t_{\mathrm{CA2}} + t_{\mathrm{CA3}} + 2(t_{\mathrm{ER}} + t_{\mathrm{CSE}}),$$

$$m_1(t_{\mathrm{STO4}}\ /\ \text{first EF}) = t_{\mathrm{CA2}} + t_{\mathrm{CA3}} + 2(t_{\mathrm{ER}} + t_{\mathrm{CSE}}),$$

$$m_1(t_{\mathrm{STO1}}\ /\ \text{second EF}) = t_{\mathrm{CA2}} + t_{\mathrm{CA1}} + 2(t_{\mathrm{ER}} + t_{\mathrm{CSE}}),$$

$$m_1(t_{\mathrm{STO2}}\ /\ \text{second EF}) = t_{\mathrm{CA2}} + t_{\mathrm{CA1}} + (t_{\mathrm{ER}} + t_{\mathrm{CSE}}),$$

$$m_1(t_{\mathrm{STO3}}\ /\ \text{second EF}) = t_{\mathrm{CA2}} + t_{\mathrm{CA3}} + 3(t_{\mathrm{ER}} + t_{\mathrm{CSE}}),$$

$$m_1(t_{\mathrm{STO4}}\ /\ \text{second EF}) = t_{\mathrm{CA2}} + t_{\mathrm{CA3}} + 3(t_{\mathrm{ER}} + t_{\mathrm{CSE}}),$$

$$m_1(t_{\mathrm{STO1}}\ /\ \text{third EF}) = t_{\mathrm{CA2}} + t_{\mathrm{CA1}} + 3(t_{\mathrm{ER}} + t_{\mathrm{CSE}}),$$

$$m_1(t_{\mathrm{STO2}}\ /\ \text{third EF}) = t_{\mathrm{CA2}} + t_{\mathrm{CA1}} + 3(t_{\mathrm{ER}} + t_{\mathrm{CSE}}),$$

$$m_1(t_{\mathrm{STO3}}\ /\ \text{third EF}) = t_{\mathrm{CA2}} + t_{\mathrm{CA3}} + (t_{\mathrm{ER}} + t_{\mathrm{CSE}}),$$

$$m_1(t_{\mathrm{STO4}}\ /\ \text{third EF}) = t_{\mathrm{CA2}} + t_{\mathrm{CA3}} + 4(t_{\mathrm{ER}} + t_{\mathrm{CSE}}),$$

$$m_1(t_{\mathrm{STO1}}\ /\ \text{fourth EF}) = t_{\mathrm{CA2}} + t_{\mathrm{CA1}} + 4(t_{\mathrm{ER}} + t_{\mathrm{CSE}}),$$

$$m_1(t_{\mathrm{STO1}}\ /\ \text{fourth EF}) = t_{\mathrm{CA2}} + t_{\mathrm{CA1}} + 4(t_{\mathrm{ER}} + t_{\mathrm{CSE}}),$$

$$m_1(t_{\mathrm{STO1}}\ /\ \text{fourth EF}) = t_{\mathrm{CA2}} + t_{\mathrm{CA3}} + 4(t_{\mathrm{ER}} + t_{\mathrm{CSE}}),$$

$$m_1(t_{\mathrm{STO1}}\ /\ \text{fourth EF}) = t_{\mathrm{CA2}} + t_{\mathrm{CA3}} + (t_{\mathrm{ER}} + t_{\mathrm{CSE}}),$$

where t_{CSE} is a duration of CSE.

There are one-dimensional and two-dimensional optimizations. Let's consider examples.

1. One-dimensional optimization. According to equations (10) and (11) in the case of optimization by α we need to solve the equation

$$\frac{\partial m_1(t_{CA}, t_{ER}, t_{CSE}, \alpha, \beta)}{\partial \alpha} = Q_2(1-\beta)(m_1(t_{STO1} / \text{second EF}) - m_1(t_{STO2} / \text{second EF})) +$$
$$+Q_3(2\alpha m_1(t_{STO1} / \text{third EF}) + (1-2\alpha)m_1(t_{STO2} / \text{third EF}) - (1-\beta)m_1(t_{STO3} / \text{third EF}) - \tag{13}$$
$$-\beta m_1(t_{STO4} / \text{third EF})) + Q_4(2\alpha m_1(t_{STO1} / \text{fourth EF}) + (1-2\alpha)m_1(t_{STO2} / \text{fourth EF}) +$$
$$+(1-2\alpha)m_1(t_{STO3} / \text{fourth EF}) - (2-2\alpha)m_1(t_{STO4} / \text{fourth EF})) = 0.$$

After simplifications in equation (13), we will obtain

$$a_2(\beta, t_{CA}, t_{ER}, t_{CSE})\alpha_{min} = a_1(\beta, t_{CA}, t_{ER}, t_{CSE}),$$

$$\alpha_{min} = \frac{a_1(\beta, t_{CA}, t_{ER}, t_{CSE})}{a_2(\beta, t_{CA}, t_{ER}, t_{CSE})},$$

where

$$a_1(\beta, t_{CA}, t_{ER}, t_{CSE}) = Q_2(1-\beta)(m_1(t_{STO1} / \text{second EF}) - m_1(t_{STO2} / \text{second EF})) +$$
$$+Q_3(m_1(t_{STO2} / \text{third EF}) - (1-\beta)m_1(t_{STO3} / \text{third EF}) - \beta m_1(t_{STO4} / \text{third EF})) +$$
$$+Q_4(m_1(t_{STO2} / \text{fourth EF}) + m_1(t_{STO3} / \text{fourth EF}) + 2m_1(t_{STO4} / \text{fourth EF}));$$

$$a_2(\beta, t_{CA}, t_{ER}, t_{CSE}) = 2Q_3(m_1(t_{STO1} / \text{third EF})$$
$$-m_1(t_{STO2} / \text{third EF})) + 2Q_4(m_1(t_{STO1} / \text{fourth EF})$$
$$-m_1(t_{STO2} / \text{fourth EF}) - m_1(t_{STO3} / \text{fourth EF}) + m_1(t_{STO4} / \text{fourth EF})).$$

In case of optimization by β we need to solve the equation

$$\frac{\partial m_1(t_{CA}, t_{ER}, t_{CSE}, \alpha, \beta)}{\partial \beta} = Q_1(-2(1-\beta)m_1(t_{STO1} / \text{first EF}) + (1-2\beta)m_1(t_{STO2} / \text{first EF}) +$$
$$+(1-2\beta)m_1(t_{STO3} / \text{first EF}) + 2\beta m_1(t_{STO4} / \text{first EF})) + Q_2(-\alpha m_1(t_{STO1} / \text{second EF}) -$$
$$-(1-\alpha)m_1(t_{STO2} / \text{second EF}) + (1-2\beta)m_1(t_{STO3} / \text{second EF}) + 2\beta m_1(t_{STO4} / \text{second EF})) +$$
$$+Q_3(1-\alpha)(m_1(t_{STO3} / \text{third EF}) + m_1(t_{STO4} / \text{third EF})) = 0;$$

$$a_2(\alpha, t_{CA}, t_{ER}, t_{CSE})\beta_{min} = a_1(\alpha, t_{CA}, t_{ER}, t_{CSE});$$

$$\beta_{min} = \frac{a_1(\alpha, t_{CA}, t_{ER}, t_{CSE})}{a_2(\alpha, t_{CA}, t_{ER}, t_{CSE})},$$

where

$$a_1(\alpha,\ t_{CA}, t_{ER}, t_{CSE}) = Q_1(-2m_1(t_{STO1}\ /\ \text{first EF}) + m_1(t_{STO2}\ /\ \text{first EF}) + m_1(t_{STO3}\ /\ \text{first EF})) + \\ +Q_2(-\alpha m_1(t_{STO1}\ /\ \text{second EF}) - (1-\alpha)m_1(t_{STO2}\ /\ \text{second EF}) + m_1(t_{STO3}\ /\ \text{second EF})) + \\ +Q_3(1-\alpha)(m_1(t_{STO4}\ /\ \text{third EF}) - m_1(t_{STO3}\ /\ \text{third EF})).$$

$$a_2(\alpha, t_{CA}, t_{ER}, t_{CSE}) = 2Q_1(m_1(t_{STO1}\ /\ \text{first EF}) - m_1(t_{STO2}\ /\ \text{first EF}) - m_1(t_{STO3}\ /\ \text{first EF}) + \\ +m_1(t_{STO4}\ /\ \text{first EF})) + 2Q_2(m_1(t_{STO4}\ /\ \text{second EF}) - m_1(t_{STO3}\ /\ \text{second EF})).$$

In case of optimization by t_{CA} we assume

$$t_{CA1} = t_{CA2} = t_{CA3} = t_{CA}.$$

Then

$$\frac{\partial m_1(t_{CA}, t_{ER}, t_{CSE}, \alpha,\ \beta)}{\partial t_k} = Q_1((1-\beta)^2 + (\beta-\beta^2) + (\beta-\beta^2) + \beta^2) + Q_2(\alpha(1-\beta) \\ +(1-\beta)(1-\alpha) + (\beta-\beta^2) + \beta^2) + Q_3(\alpha^2 + (\alpha-\alpha^2) + (1-\alpha)(1-\beta) + \beta(1-\alpha)) \\ +Q_4(\alpha^2 + \alpha - \alpha^2 + \alpha - \alpha^2 + 1 - 2\alpha + \alpha^2) = Q_1(1 - 2\beta + \beta^2 + 2\beta - \beta^2) \\ +Q_2(\alpha - \alpha\beta + 1 - \alpha - \beta + \alpha\beta + \beta) + Q_3(\alpha + 1 - \alpha - \beta + \alpha\beta + \beta - \alpha\beta) + Q_4 \\ = Q_1 + Q_2 + Q_3 + Q_4 = 1 \neq 0.$$

So optimization by t_{CA} is impossible.

2. Two-dimensional optimization. In case of optimization by α and β first of all let's check the possibility of optimization problem solution. To do this we calculate Hessian

$$\frac{\partial^2 m_1(t_{CA}, t_{ER}, t_{CSE}, \alpha,\ \beta)}{\partial \alpha^2} = 2Q_3(m_1(t_{STO1}\ /\ \text{third EF}) - m_1(t_{STO2}\ /\ \text{third EF})) + \\ +2Q_4(m_1(t_{STO1}\ /\ \text{fourth EF}) - m_1(t_{STO2}\ /\ \text{fourth EF}) - m_1(t_{STO3}\ /\ \text{fourth EF}) + \\ +m_1(t_{STO4}\ /\ \text{fourth EF}));$$

$$\frac{\partial^2 m_1(t_{CA}, t_{ER}, t_{CSE}, \alpha,\ \beta)}{\partial \alpha \partial \beta} = Q_2(m_1(t_{STO2}\ /\ \text{second EF}) - m_1(t_{STO1}\ /\ \text{second EF}) + \\ +Q_3(m_1(t_{STO3}\ /\ \text{third EF}) - m_1(t_{STO4}\ /\ \text{third EF}));$$

$$\frac{\partial^2 m_1(t_{CA}, t_{ER}, t_{CSE}, \alpha, \beta)}{\partial\beta\partial\alpha} = Q_2(m_1(t_{STO2} / \text{second EF}) - m_1(t_{STO1} / \text{second EF}) + $$
$$+Q_3(m_1(t_{STO3} / \text{third EF}) - m_1(t_{STO4} / \text{third EF}));$$

$$\frac{\partial^2 m_1(t_{CA}, t_{ER}, t_{CSE}, \alpha, \beta)}{\partial\beta^2} = 2Q_1(m_1(t_{STO1} / \text{first EF}) - m_1(t_{STO2} / \text{first EF}) - m_1(t_{STO3} / \text{first EF})$$
$$+m_1(t_{STO4} / \text{first EF})) + 2Q_2(m_1(t_{STO4} / \text{second EF}) - m_1(t_{STO3} / \text{second EF})).$$

Taking into account the formula (12), we determine the first partial derivatives of the efficiency indicator and find the optimal numerical values $\left(\alpha_{opt}; \beta_{opt}\right)$ after system of equations solution

$$\begin{cases} a_2(\beta_{opt}, t_{CA}, t_{ER}, t_{CSE})\alpha_{opt} = a_1(\beta_{opt}, t_{CA}, t_{ER}, t_{CSE}), \\ a_2(\alpha_{opt}, t_{CA}, t_{ER}, t_{CSE})\beta_{opt} = a_1(\alpha_{opt}, t_{CA}, t_{ER}, t_{CSE}). \end{cases} \quad (14)$$

Finding α_{opt} from the first equation of the system (14) and substituting it in the second equation, we obtain the equation from which we define β_{opt}

$$\beta_{opt} = \frac{a_1\left(\dfrac{a_1(\beta_{opt}, t_{CA}, t_{ER}, t_{CSE})}{a_2(\beta_{opt}, t_{CA}, t_{ER}, t_{CSE})}, t_{CA}, t_{ER}, t_{CSE}\right)}{a_2\left(\dfrac{a_1(\beta_{opt}, t_{CA}, t_{ER}, t_{CSE})}{a_2(\beta_{opt}, t_{CA}, t_{ER}, t_{CSE})}, t_{CA}, t_{ER}, t_{CSE}\right)}.$$

FUTURE RESEARCH DIRECTIONS

Future research directions will be associated with the substantiation of the technological processes characteristics during the design and modernization of aviation radioelectronic equipment operation systems, including the processes of adaptive monitoring, control, and statistical data processing.

CONCLUSION

In this chapter, the authors present the questions of aviation radioelectronic equipment operation. The structure of operation system is considered based on processes approach with adaptable control principles usage. Operation system consists of REE, processes, personnel, normative and regulatory documents, consumable resources, informational recourses, means of operation, etc. The main element of operation system is radioelectronic equipment, and therefore the main process is intended use.

During operation, the equipment can fail, so diagnostics processes are considered in this chapter. Authors consider the direct problem of efficiency estimation for the diagnostics process, and main attention is paid to probability density function calculation for diagnostics duration. This approach allows describing operation costs more completely. Simulation results were used for adequacy testing of these calculations. Authors also take into account the possibility of first and second kind errors presence during the diagnostics process.

The inverse problem for diagnostics is defined and solved for the mathematical expectation of repair time. The inverse problem is considered for one-dimensional and two-dimensional cases. Results can be used for the design and improvement of REE operation systems.

REFERENCES

Barlow, R. E., & Proschan, F. (1965). *Mathematical Theory of Reliability*. New York: John Wiley and Sons.

Dhillon, B. S. (2005). *Reliability, Quality, and Safety for Engineers*. Boca Raton, FL: CRC PRESS.

Dhillon, B. S. (2006). *Maintainability, Maintenance, and Reliability for Engineers*. New York: Taylor & Francis Group. doi:10.1201/9781420006780

Goncharenko, A. V. (2015). Applicable aspects of alternative UAV operation. In *Proceedings of IEEE 3rd International Conference on Actual Problems of Unmanned Air Vehicles Developments (APUAVD 2015)*, (pp. 316 – 319). Kyiv, Ukraine: IEEE. 10.1109/APUAVD.2015.7346630

Goncharenko, A. V. (2017). Optimal UAV maintenance periodicity obtained on the multi-optional basis. In *Proceedings of IEEE 4th International Conference on Actual Problems of Unmanned Air Vehicles Developments (APUAVD 2017)*, (pp. 65 – 68). Kyiv, Ukraine: IEEE. 10.1109/APUAVD.2017.8308778

Goncharenko, A. V. (2018). Multi-optional hybrid effectiveness functions optimality doctrine for maintenance purposes. In *14th International Conference on Advanced Trends in Radioelectronics, Telecommunications and Computer Engineering (TCSET 2018)*, (pp. 771 – 775). Lviv-Slavske, Ukraine: Academic Press. 10.1109/TCSET.2018.8336313

Hahn, G. J., & Shapiro, S. S. (1967). *Statistical models in engineering*. New York: John Wiley & Sons, Inc.

Hoyland, A., & Rausand, M. (1994). *System reliability theory*. New York: John Wiley & Sons, Inc.

Rausand, M. (2004). *System reliability theory: models, statistical methods and applications*. New York: John Wiley & Sons, Inc.

Smith, D. J. (2005). *Reliability, Maintainability and Risk. Practical methods for engineers*. London: Elsevier.

Solomentsev, O., Zaliskyi, M., Kozhokhina, O., & Herasymenko, T. (2017). Reliability Parameters Estimation for Radioelectronic Equipment in Case of Change-point. In *Proceedings of Signal Processing Symposium 2017 (SPSympo 2017)*. (pp. 1 – 4). Jachranka Village, Poland: Academic Press.

Solomentsev, O., Zaliskyi, M., Nemyrovets, Yu., & Asanov, M. (2015). Signal processing in case of radio equipment technical state deterioration. In *Proceedings of Signal Processing Symposium 2015 (SPS 2015).* (pp. 1 – 5). Debe, Poland: Academic Press. 10.1109/SPS.2015.7168312

Solomentsev, O., Zaliskyi, M., & Zuiev, O. (2013). Radioelectronic equipment availability factor models. In *Proceedings of Signal Processing Symposium 2013 (SPS 2013)* (pp. 1 – 4). Jachranka Village, Poland: Academic Press.

Solomentsev, O., Zaliskyi, M., & Zuiev, O. (2016). Estimation of quality parameters in the radio flight support operational system. *Aviation.*, *20*(3), 123–128. doi:10.3846/16487788.2016.1227541

Solomentsev, O. V., Melkumyan, V. H., Zaliskyi, M. Yu., & Asanov, M. M. (2015). UAV operation system designing. In *Proceedings of IEEE 3rd International Conference on Actual Problems of Unmanned Air Vehicles Developments (APUAVD 2015)* (pp. 95 – 98). Kyiv, Ukraine: IEEE. 10.1109/APUAVD.2015.7346570

ADDITIONAL READING

Gertsbakh, I. (2000). *Reliability Theory with Applications to Preventive Maintenance*. New York: Springer.

Gnedenko, B. V., Belyayev, Y. K., & Solovyev, A. D. (1969). *Mathematical methods of reliability theory*. New York: Academic.

Kapur, K. C., & Lamberson, L. R. (1977). *Reliability in Engineering Design*. New York: Wiley.

Nakagawa, T. (2005). *Maintenance theory of reliability*. Springer.

Tartakovsky, A., Nikiforov, I., & Basseville, M. (2015). *Sequential analysis: hypothesis testing and change-point detection*. New York: Taylor & Francis Group.

KEY TERMS AND DEFINITIONS

Diagnostics: The process of defining the diagnostics object technical condition or searching a failed diagnostics object element.

Direct Problem for Diagnostics: A problem of probability density function calculation for efficiency indexes.

Efficiency of Diagnostics: Probability of correct diagnostics, mean duration of diagnostics, the average cost of diagnostics, average labor intensity of diagnostics.

Error of First Kind: The incorrect decision making about the serviceable condition of REE.

Error of Second Kind: The incorrect decision making about the non-serviceable condition of REE.

Inverse Problem for Diagnostics: Evaluation of diagnostics process parameters for which the efficiency index has optimum.

Operation System: A set of devices, means of operation, personnel, and documentation setting the rules of their cooperation for performing operation tasks. The operation is the stage of device lifecycle at which its quality is realized, maintained, and restored.

Probability Density Function (PDF): A function which represents the density of a continuous random variable and the likelihood that this random variable takes on some given value.

Chapter 11
Information-Measuring Technologies for UAV's Application:
Two Practical Examples

Vitalii Larin
National Aviation University, Ukraine

Nina Chichikalo
National Technical University of Ukraine, Ukraine

Georgii Rozorinov
National Technical University of Ukraine, Ukraine

Ekaterina Larina
National Technical University of Ukraine, Ukraine

EXECUTIVE SUMMARY

This chapter describes two practical examples of sensors application on an unmanned aerial vehicle. The first device is a proximity sensor allowing users to measure the rotating angle of UAV's elevator. The second example discovers a measuring unit established on the UAV and processed measuring information for landing the UAV. To perform exactness control of unmanned aerial vehicles actuating mechanisms, the control system must be supplied by devices providing precision definition of values of current operation factors of those mechanisms.

DOI: 10.4018/978-1-5225-7588-7.ch011

UAV'S SENSORS

An interest to create and scaling up unmanned aerial vehicles (UAV) application field is widening today. It is fair to say that low price and affordability of materials allowed creating radically new UAV designs, such as copters. Scaling up of UAV application field leads to R&D activities for outfitting them with new payload and adapting it for unpiloted or remote-controlled aerial vehicles. Apart from creating new and modernization of existing UAV designs, vigorous efforts are conducted for creating cost-efficient navigation and control means for various types of UAV. Modern hardware components allow implementing accurate navigation solution and complicated control algorithms, particularly for automatic UAV's.

A plenty of R&D and commercial aircraft builders in the world are concerned with creating UAV's of different specifications and design types. Remote controlled UAV's are predominant today. Such aerial vehicles are known to be controlled by a remote pilot that stays in a terrestrial control center to control a UAV flight by monitoring visual and navigational information received from UAV using a wireless information channel, makes a decision and transmits it to UAV using a control channel. The control chart described can vary in structure and certain implementation features, but a remote pilot or rather remote UAV operator is a necessary control loop element.

Another UAV type does not require an operator as a remote pilot. Such UAV's are controlled automatically using preset control algorithms. In this case, operator serves as a UAV maintenance engineer, i.e. adjusts navigation algorithms for a specific UAV task, verifies the functionality of UAV's onboard avionics and actuators and loads control system with algorithms adjusted for successful application task performance. No doubt, technical implementation of a really unmanned vehicle is much more difficult.

Thus the relevance of activities for a creation and technical implementation of robust algorithms for UAV flight control is undeniable.

The first example represents an original application of a barometric pressure sensor. The process of landing and touchdown of an unmanned aerial vehicle (UAV) with the set exactness requires the realization of new information technologies of control of parameters of space and operation process. So far the solution of mentioned questions of realization is a problem. A possibility of choice of well-known measuring facilities necessary for the operation of UAV is considered. The improved mathematical model adaptive to region and variability of parameters of real air environment is offered.

Pressure sensors include technical devices, output signals of which change depending on the pressure of the investigated environment behavior. At the modern market, the enormous amount of pressure sensors, intended for a wide circle of applications is presented. The optimal choice of a pressure sensor for this application can be executed stage-by-stage:

- Determination of purpose, type, and range of measurable pressure (absolute, extra or differential, working and maximal influencing pressure, static or dynamic pressures, service life);
- Analysis of metrology characteristics;
- Account of influence of surrounding conditions: temperature differentials, vibration, humidity, electromagnetic interference, electrostatic destruction, high-frequency interference on feed circuits;
- Presence of output chains protection, protection from short circuits;
- Analysis of requirements to an electric interface (analog or digital output signal);
- Account of requirements of mechanical characteristics for setting on an object;
- Economic efficiency.

Questions of increase of metrology characteristics of sensors, selection of their mechanical and electric interfaces, and also production on classes of protection and persistence to influences of the environment are not solved to the present time.

Classification of sensors for UAV on substantial technical characteristics is given in Table 1.

According to the type of sensing element, particular interest is in piezoresistive sensors of pressure using reverse piezo-effect. These sensors to a greater extent satisfy specified requirements.

SENSORS OF PRESSURE SELECTING

Without giving the full comparative evaluation of piezoelectric sensors of pressure, we will consider the serial unified converter Altitude – Pressure of the BMP085 type from the "Datasheet Preview for BMP085 by Bosch" (2011).

The piezoresistive effect is the basis of the creation of this converter, and the output signal after analog-digital transformation is available to the user. The received values of an output signal aren't values of atmospheric pressure and are connected with it by difficult dependence. For determination of this dependence, it is necessary to use 11 correcting coefficients. These coefficients are stored in the in-built EEPROM memory of the sensor and are strictly individual for each sensor. For temperature compensation sensor has the inbuilt analog converter of temperature from which signal needs also to be digitized, reproduced and redefined. At the same time, it should be borne in mind that when the pressure changes by 1 mm Hg. St., the height of the UAV varies by 10 ... 12 m.

The algorithm of operation of the BMP085 converter provides: activation; reading of the correcting coefficients; processing of a signal from temperature converter; expectation of the end of transformation; reading of result of transformation; determination of temperature; processing of a signal from pressure sensor; expectation of the end of transformation; reading of result of transformation; calculation of pressure. Control of the sensor is realized by means of the standard two-wire I2C interface. In addition, the sensor has the line of dumping (XCLR) and an output "the end of converting" (EOC).

For the measurement of temperature and humidity of the environment it is possible to use the low-cost digital sensor DHT11, which datasheet we can see in "Datasheet DHT11" (2010).

For contact with the microcontroller the single-wire bar connected to the output transistor with an open collector is used, therefore, it is necessary to use the connection of the load resistor R_H=5...10 kOm to the feed bar.

Table 1. Classification of sensors for UAV

On Nominal Range of Pressure, MPa	On Presence of Scheme of Signal Processing	On Type of Output Signal	On Type of Mechanical Joining
High and ultra-high ($P > 60$); – low, ultra-low, ($P < 0.1$); – average ($0.1 \le P \le 60$).	With unnormalized output signal With normalized output signal With passive temperature compensation With active temperature compensation With microprocessor signal processing	analog digital relay	field end flanged built-in submerged

For the development of measuring system, the hardware Arduino ATMega328 platform and the LabView 12 program and simulation environment was used. The hardware Arduino platform carries out an I2C converter role – UART and contact of the sensor with the computer by means of an UART converter chip – USBFT323. Processing and display of information are carried out by the program and simulation environment LabView 12.

The Applied Virtual Tool (AVT) developed in the environment of LabView12 displays the measured pressure in the form of the schedule on the WaveformChart indicator and also the measured pressure in a numerical look on the NumericIndicator and Gauge elements. Altitude, which is the function of pressure is calculated by the formula

$$Alt(P) = 44330\left(1 - \left(\frac{P}{P_0}\right)^{\frac{1}{5,255}}\right) \tag{1}$$

and it is displayed on the NumericIndicator and VerticalProgressBar elements.

EXPERIMENTAL RESEARCH IN LABVIEW

By means of elements of the package of the LabView 12 UARTVisa expansion, the communication with COM port is carried out. The appearance and the block chart AVT are shown in Figures 1, 2.

The schedule in Figure 3 shows the dependence of change of pressure during the time of day. Measurements were taken at an interval of one hour. For convenience of further use of results approximation and averaging of the received characteristic have been executed by means of modeling environment Mathcad 2000. The correlation coefficient between the taken and average dependence makes 0,946.

Figure 1. Front panel PVI

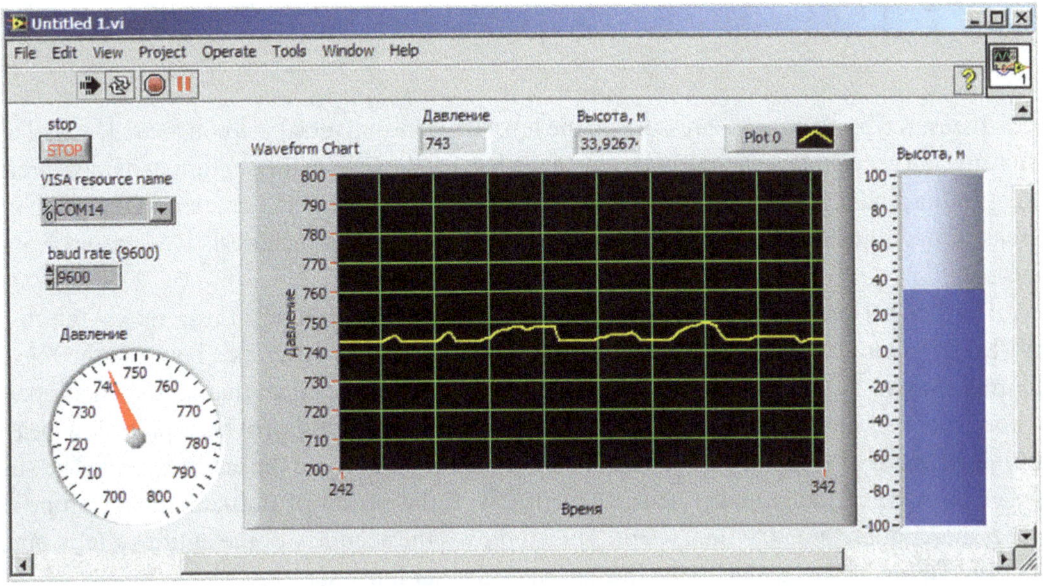

Figure 2. Block chart PVI

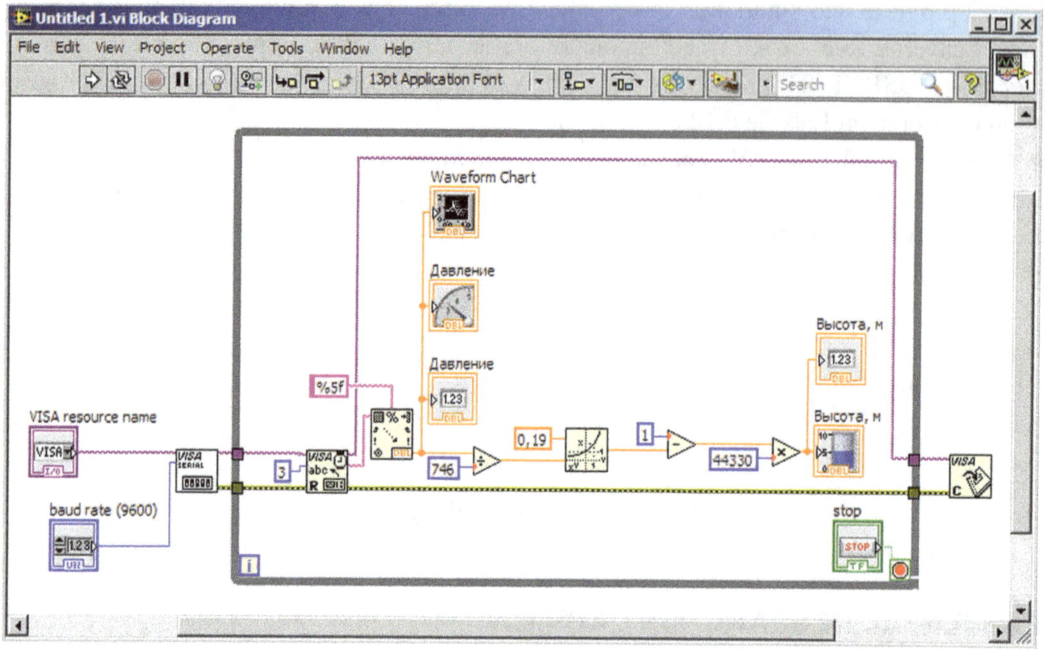

But pressure is a function of temperature and humidity of the environment. For definition of drift of indications of the altimeter on the basis of a barometer in addition, by means of the DHT11 sensor, temperature (Figure 4) and humidity (Figure 5) are taken. These data are also averaged. Correlation coefficients for the measured and average characteristics of temperature and humidity are equal 0,938 and 0,961, accordingly.

Pressure and Altitude are connected among themselves by barometric dependence (Stull, 2015).

$$h = \ln\left(\frac{P}{P_0}\right)\frac{RT}{-Mg} + h_0, \tag{2}$$

where P – measured pressure; P_0 – pressure on the level 0 m; R – universal gas constant; M – molar mass of gas; g – free fall acceleration; h_0 – altitude above sea level at the time of the beginning of measurements.

Taking into account the obtained data according to expression (2) the characteristic of change of indications of the altimeter has been received by the uncompensated piezoresistive sensor of pressure (figure 6).

The difference between the maximum and minimum values of Altitude, at rest, makes 9,5 m.

The BMP085 converter is the thermocompensated device and its difference between the maximum and minimum values of Altitude, at rest, makes 30 cm that is 30 times less, than in the uncompensated sensor. For use of the converter in control systems of aircraft or for the design of high-precision measuring systems it is necessary to enter the amendment on change of humidity of the environment. The solution of this task is usage of the humidity sensor together with the sensor of pressure and microprocessor data processing for obtaining the final result. This improves the accuracy of the entire system, but at the expense of reducing its speed, due to the low speed of the humidity sensors.

Figure 3. Measured and average characteristics of change of pressure

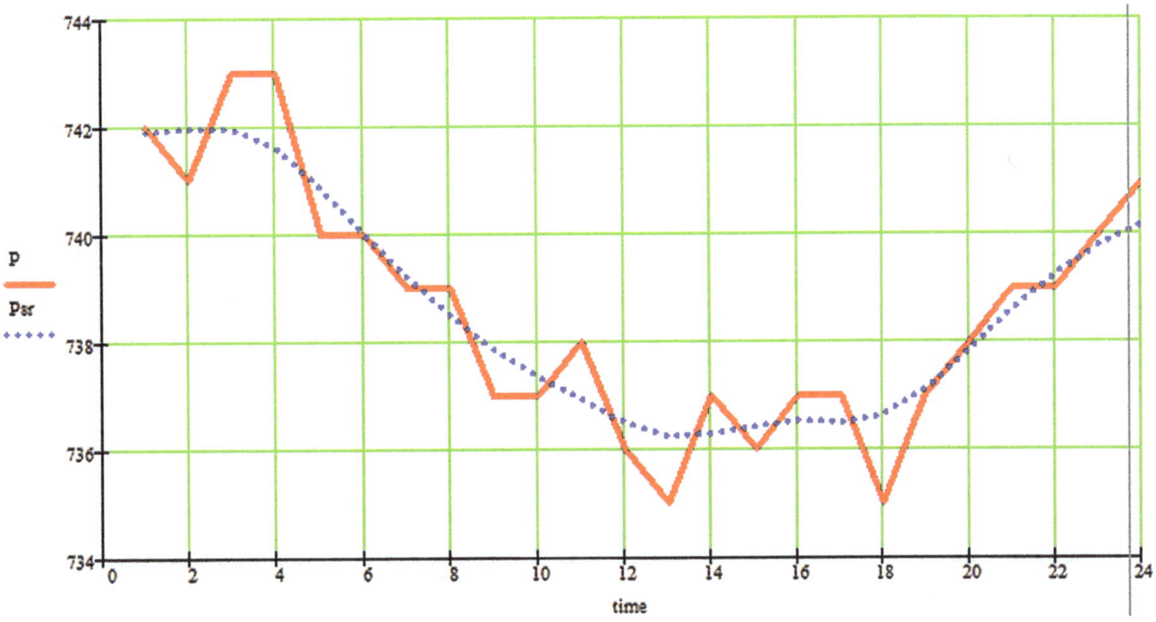

Figure 4. Measured and average characteristics of change of temperature

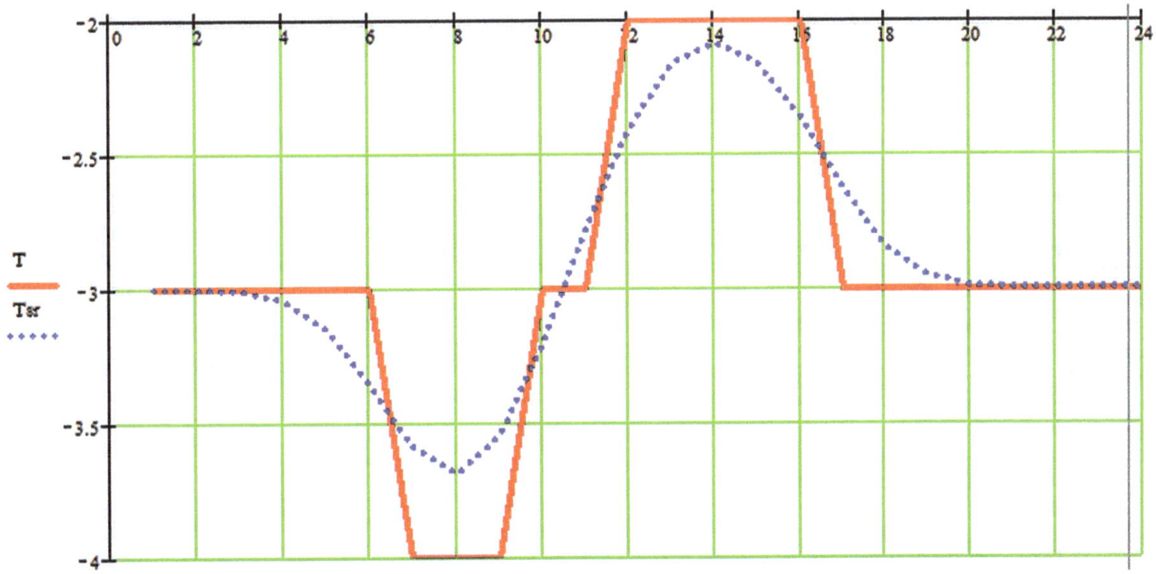

Figure 5. Measured and average characteristics of change of humidity

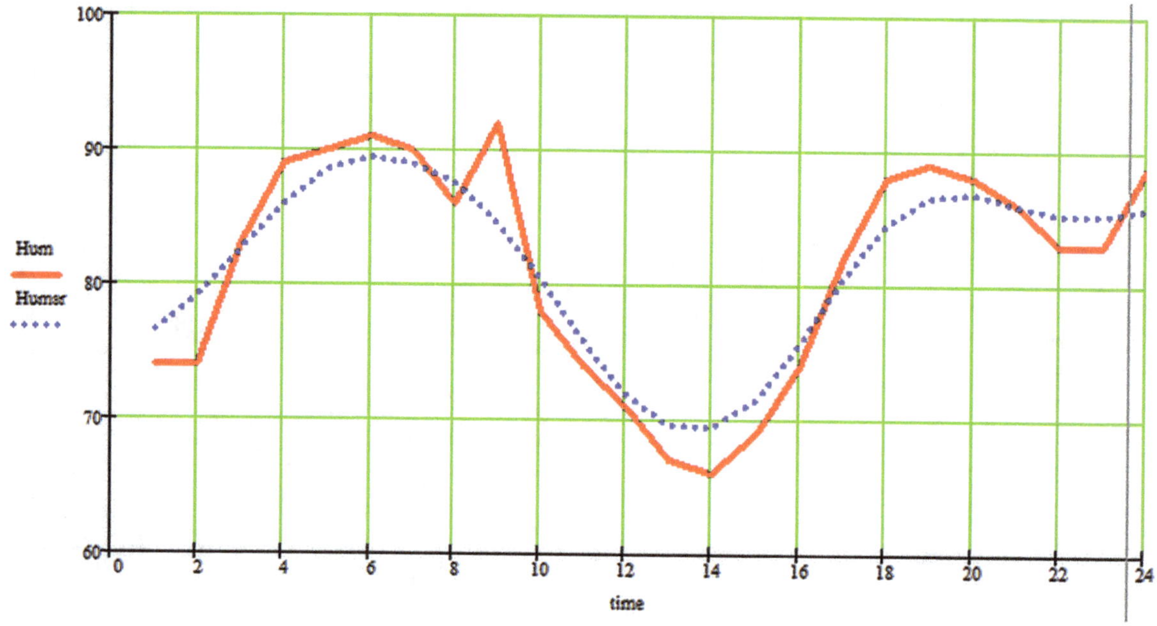

Figure 6. Measured and average characteristics of indications of the altimeter

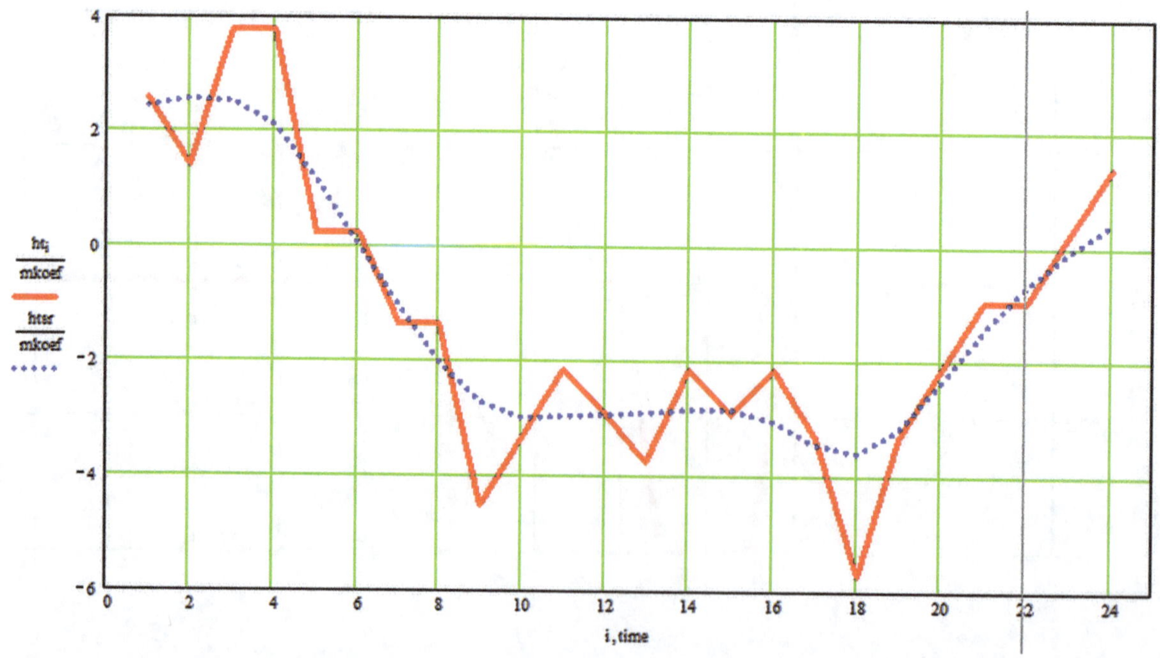

DESIGN OF FERRIMAGNETIC SENSOR FOR UAV

The second part of the chapter presents the description of the easy development of measured based feedback loop on the basis of proximity sensor having 3D- inductive coil as a sensitive element for measurement of rotation angle of an elevator. To perform exactness control of unmanned aerial vehicles actuating mechanisms the control system must be supplied by devices providing the precision definition of values of current operation factors of those mechanisms. As such devices can be developed or searched essential measurement transducers.

A servomotor is an electromechanical actuator that rotates UAV elevating rudders. A potentiometer with drive rod mechanically connected with servomotor's rotational gear is used as a feedback unit in low-price analog servo machines. Two signal wires from this potentiometer are connected to an electronic control circuit. The functional diagram of the servo machine implementing such a control method is shown in figure 7 (Woytcehovskii, 1977).

However in the presence of different negative factors, intrinsical to mechanical joints, for example – availability of micro-backlashes, affects on the exactness of potentiometer electrical signal. This feature, for one's turn, decreases of exactness of control of aircrafts elevator.

In the (Larin, 2012) is offered the modern non-contact manner of the measurement of the rotation angle of UAV's elevator. The sensitive element of a developed transducer is an inductive coil winded on the ferrimagnetic cup core, one side of which is opened on the purpose of magnetic flux propagation. The principle of operation of those measuring converter based on the inductive coils Q-factor variation when some metallic object intersects a zone of propagation of coils magnetic flux (Larin, 2009). On figure 7 the scheme of realization of developed manner is shown.

The sensitive element (1) is attached to the fuselage of an unmanned aircraft. The elevator rotation is realized by means of a mechanical rod (2) made from nonmagnetic substance. On this rod, the homing device (3) made as the metal plate is rigidly fastened. Since in the center of the coil a reach-through hole is placed, then the rod is put through this hole and in that way the rod will be moved free, and the homing device, in that way, will approach to or move away from the coil in the core. The variant of the

Figure 7. Diagram demonstrates realization of offered control manner

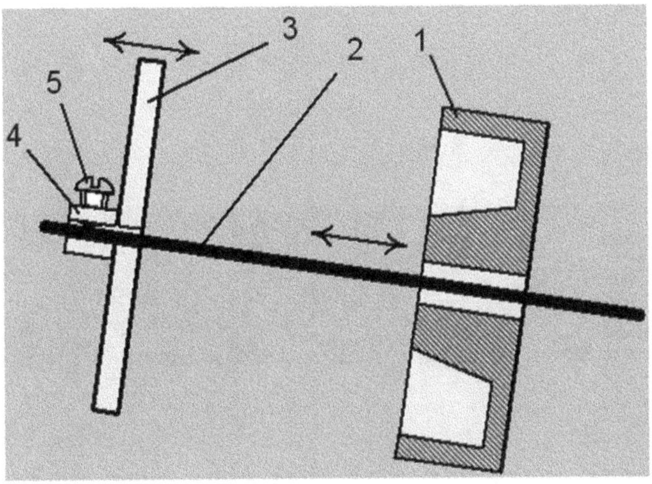

scheme of the capacitive oscillator is selected as a base scheme. as is well known (Larin, 2009; "Sensor products. Inductive sensing redefined", 2015), the full-scale output of those ferrimagnetic measurement transducers (FMT) amount no less than 25 mm. At that, the line part of FMT's converting performance occupies approximately 75-78%, and this fact greatly simplify development of measurement devises on the basis of such FMT.

A well-known electronics producer (Texas Instruments) introduced its LDCxxxx lineup in 2013 (where 'xxxx' – 4-digit identifier of a device) ("LDC sensor design application note", 2015). LDC devices use a PCB planar inductive element that, being combined with two capacitors (some models contain only one) forms an inductive sensor connected with LDC transducer IC. This sensor is embodied in a rectangular PCB. We can see planar inductor coil printed in the lower (soldering) PCB layer in figure 8. A transducer that performs inductance to code conversion is located in the center of the inductor coil. A standard USB port connector is located in the left part.

The sensor can be disconnected and located within measured process area. Such opportunity is provided with perforations that allow disconnecting the measurement circuit. Further connection of the measurement circuit to the PCB chipset is carried out by soldering flexible connection wires. Thus, a designer is given an opportunity to place the sensor in some distance if this is required in a specific measurement process. This option allows using such the device for rotation angle measurement using the proposed chart.

The basic lack of LDC's sensitive element is relatively small value of their full-scale output. In accordance with ("Sensor products. Inductive sensing redefined", 2015) it full-scale output equals 8 mm at external diameter value equaling 14 mm. (see Figure 9).

The other performances of PCB coil of LDC1000 sensor listed in the Table 2.

LDC1000 sensor measures the equivalent parallel resonance impedance Rp. The following equivalent circuit (figure 10) demonstrates a sensitive element connection. Rp can be calculated from formula where rs is the parasitic series resistance of the coil

$$R_p = \left(\frac{1}{r_s}\right) \times \left(\frac{L}{C}\right)$$

By using of 3D-coil we define mathematical dependence of output voltage of LC-loop from the value of equivalent tuned-circuit Q-factor,

Figure 8. One of LDCxxx's

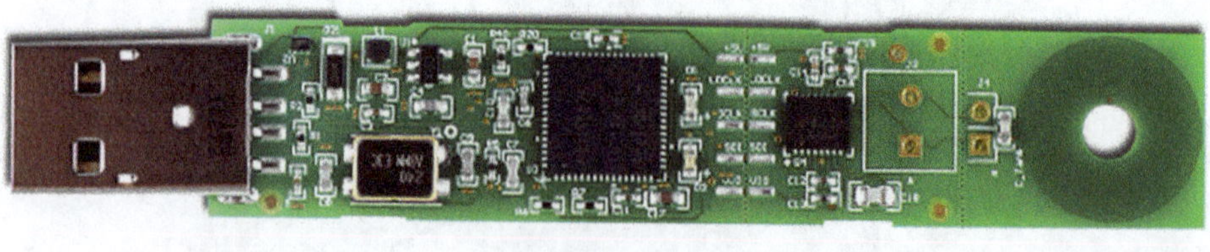

Figure 9. Typical RP versus Distance With a 14-mm PCB Coil

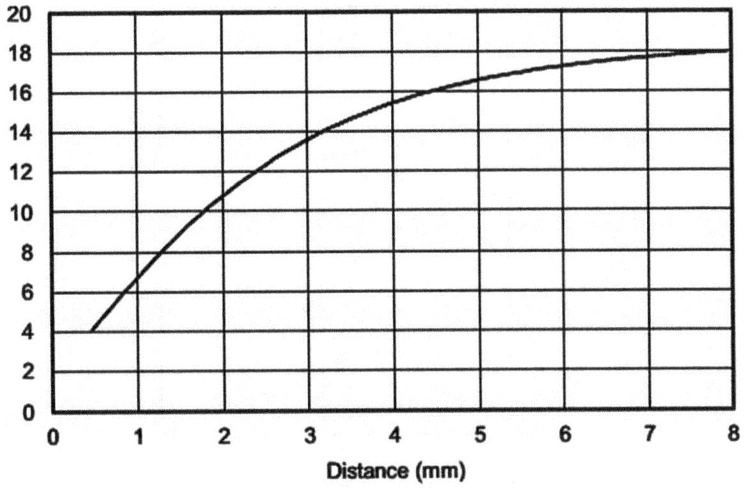

Table 2. PCB Coil characteristics

Parameter	Value
Layers	2
Thickness of cooper	1 oz
Coil shape	Circular
Number of turns	23
Trace thickness	4 mil
Trace spacing	4 mil
PCB core material	FR4
Rp at 1 mm	5 kΩ
Rp at 8 mm	12,5 kΩ
Nominal Inductance	18 μH

$$U_{out} = \frac{k \cdot \omega_0^2}{p^2 + \frac{\omega_0}{Q_e} \cdot p + \omega_0^2} \cdot U_{in} \tag{3}$$

where

U_{out} = voltage on the conditional output of LC-loop;
U_{in} = voltage on the conditional input of LC-loop;
ω_0 = resonant frequency of LC-loop;

Figure 10. Equivalent Resistance of Rp in parallel with LC loop

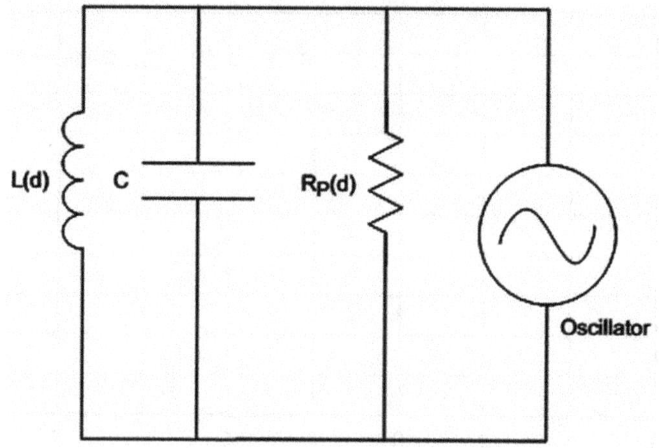

Q_e = equivalent Q-factor;
k = transfer constant on the resonant frequency.

Represented equation (3) demonstrates dependences of the LC-loop output voltage from the Q-factor. In this formula, the frequency value and value of output voltage (which is a power voltage for LC-loop) must be constant. These parameters stabilization are provided by circuit manner. In this way, only the value of the equivalent Q-factor will change at appearance in the area of coils electromagnetic flux propagation of metallic object.

By request of a user, the developer of device foresees an external LC oscillation loop connection. As well the user able to design required sensitive element of the transducer by the following guide:

- To calculate the required generation rate within the limits from 5 kHz to 5 MHz;
- To calculate required coil operation factors within the limits of tenth uH;
- To calculate the required operation factors of LC-oscillation loop;
- To calculate a value of interface filters capacitor.

SOLUTIONS AND RECOMMENDATIONS

All the calculations can be implemented by the two ways: as by using of well-known guides; as by using of existing automated design program. Especially, Epcos, the well-known manufacturer of different magnetic materials, including ferrites, offers proper design program for calculation of operation factors of core-based inductive coils. The benefit of that Epcos program is an operation factor and geometries of the core can be entered instantly just appropriate sort of ferrite article was selected. Texas Instruments offers own program solution for calculation of operation factors of the inductive coil. That program is named WEBENCH Inductive Sensing Coil Designer, and the program can be downloaded from the manufacturer site by an engineer or user.

Thus, a high precision 3D inductive angle sensor with digital output can be obtained by connecting a with LDCxxxx processing unit with a ferrimagnetic sensor that has three times as greater dynamic range than PCB coil. Its inductance lies within required limits. Its digital output can be used in aforesaid UAV elevating rudders angle control method. High precision linear 3D ferrimagnetic elements together with digital conversion unit featuring standard serial output are provided to form a high precision elevating rudders rotation angle control sensor. It defined the procedures for calculations of measurement inductive transducer performances. Using of simple computer automated design programs is offered with aim of user configuration of contactless proximity sensors for different kinds of structure of measurable objects.

REFERENCES

Datasheet Preview for BMP085 by Bosch. (2011). Retrieved April 4, 2018, from: https://www.datasheets.com/datasheet/bmp085-bosch-43859722.html

Datasheet DHT11. (2010). Retrieved April 11, 2018, from: https://ru.scribd.com/document/351339794/Datasheet-DHT11

Larin, V. J. (2009). Development of mathematical model of sensitive elements of ferrimagnetic transducer. *Eastern-European Journal of Enterprise Technologies*, *2/6*(38), 52–56. (in Russian)

Larin, V. (2009). *New test and development procedures for precision instruments and information-measurement systems design*. Donetsk, Ukraine: Weber. (in Russian)

Larin, V. (2012). Measurement of slewing angle of elevator of unmanned aircraft vehicle using ferri-magnetic transformer. [in Ukrainian]. *Radioelectronic and Computer Systems*, *4*, 29–34.

Sensor Design Application NoteL. D. C. (2015). Retrieved from: http://www.ti.com/lit/snoa930.pdf

Sensor products. Inductive sensing redefined. (2015). Texas Instruments. Retrieved from: http://www.ti.com/lsds/ti/sensors/inductive-sensing-multi-channel-ldc.page?DCMP=multichldcs&HQS=tlead-sensing-sva-psp-ssp-multichldcs-vanity-lp-en

Stull, R. (2015). *Practical Meteorology*. Retrieved April 19, 2018, from: https://www.eoas.ubc.ca/books/Practical_Meteorology/prmet/PracticalMet_WholeBook-v1_00b.pdf

Woytcehovskii, Ya. (1977). *Distance control of models. Manual for draftsmen and radio-amateur.* Moscow, USSR: Svyaz. (in Russian)

KEY TERMS AND DEFINITION

Applied Virtual Tool (AVT): Is an application programming interface developed as part of the LabVIEW software project, which allows designing "virtual instruments."

Electrically Erased Program Read Only Memory (EEPROM): User-modifiable read-only memory (ROM) that can be erased and reprogrammed (written to) repeatedly through the application of higher than normal electrical voltage.

End of Converting (EOC): A pin of a digital microcircuit that incorporates a conversion unit (usually an analog-to-digital converter), and on which after the end of the conversion a voltage level appears indicating the completion of the conversion.

Ferrimagnetic Measurement Transducer (FMT): A type of inductive sensor in which the conversion of the input non-electric quantity into the electric current of the inductor takes place.

Printed Circuit Board (PCB): Mechanically supports and electrically connects electronic components or electrical components using conductive tracks, pads, and other features etched from one or more sheet layers of copper laminated onto and/or between sheet layers of a non-conductive substrate.

Universal Asynchronous Receiver Transmitter (UART): Is a computer hardware device for asynchronous serial communication in which the data format and transmission speeds are configurable.

Unmanned Aerial Vehicles (UAV): Is a type of aircraft that operates without a human pilot onboard.

Chapter 12
Ensuring the Safety of UAV Flights by Means of Intellectualization of Control Systems

Konstantin Dergachov
National Aerospace University, Ukraine

Anatolii Kulik
National Aerospace University, Ukraine

EXECUTIVE SUMMARY

In this chapter, the authors present analysis of reasons for deficient safety of unmanned aerial vehicles (UAV) and further ground an approach to improve the safety by intellectualizing operation of the control system. Intellectualization results from the rational control owing to machine vision means used. A conception of building algorithms for visual evaluating position of the UAV that is equipped with a computer vision system is suggested. Algorithms are illustrated by related investigation of an adapted UAV. Both hardware and software means for realizing the visual estimation algorithms are presented.

BACKGROUND

International aviation salons manifest convincingly the trends present in aerospace technology. One of the most sustained of them is annual essential growth of UAVs. Another material trend is expansion of application areas and capability of drones.

Drones expansion is the result of their evident advantage comparing to piloted vehicles. The main benefit is unmanned aircraft functioning that brings to considerable simplification of the aircraft construction and reduction of avionics capacity, hence, the life cycle cost. In contrast, the functionality of the vehicle may rise through the possibility of management intellectualization.

DOI: 10.4018/978-1-5225-7588-7.ch012

Drones usage in various human activities is not wide enough, although there are a lot of spheres to potentially apply: cargo delivery, human habitat monitoring, forest fires monitoring, ensuring national security, mapping, traffic control, rational land-use, disaster assistance, help in case of man-made accidents, etc (Sładkowski, A. (Ed.). (2013).

To provide a demanded drones usage level in mentioned spheres of activity, it is required a higher safety of using pilotless aviation complexes (UAC) including the unmanned aerial vehicles (UAV).

The main reasons for the UAS deficient safety (Gulevich, SP, Veselov, Yu. G., Pryadkin, SP, & Tyrnov, SD (2012)) are the following:

1. Engine fault resulting in the uncontrolled fall of the fly vehicle.
2. Avionics failure that causes the flight mission failure, uncontrolled landing, objects collision, partial or complete UAV destruction.
3. Flight control room failure that results in UAV communication loss, hence, uncontrolled flight mission execution and the inability to receive monitoring information on the mission execution quality.
4. Exceeding the functionality related limits e.g. overstepping the design limits due to strong atmospheric turbulence, increased uncertainty of flight conditions, etc.
5. Breaking of the airframe and its structural elements (ailerons, rudder, wing partition) under degraded flight conditions that causes a UAV controllability deterioration and possible accident.
6. Mistakes occurrence while planning the flight mission, associated with the trajectory, distance to goal, duration of the flight, engine power resources.
7. Imperfect UAC maintenance leading to not specified emergency situation during the flight.

Thus, deficient UAV usage safety is explained by numerous contingency situations (sudden and unpredicted occasions) during the life cycle. Abnormal situations are the events, which are indefinite, first, regarding the time of appearance, and second because of the unknown peculiarity (specs) of the impact causing the abnormal situation. It is possible to automate the analysis of contingency events using theoretical approaches. It is possible to automate the contingency analysis procedure using for that different theoretical approaches. Techniques differ by the hypotheses base in use containing the mathematical models hypotheses adopted for the abnormal events, also by methods used to solve diagnosis inverse problems, as well as instrumentation means involved, and with expected implementation outcomes evaluation.

To study the UAC, which is a complex dynamic system assumed to present a lot of uncertain events to happen, one can apply the approach associated with using the principle of diagnosis and associated instruments of analysis allowing the detection of abnormalities. The approach utilizes dedicared means for identifying the causes of events with indeterminate characteristics occurrence (Isermann, R. (2006), Hajiyev, C., & Caliskan, F. (2013)). From a variety of known methods to diagnosing dynamic systems, the signal-parametric approach seems to be the most appropriate for the peculiar UACs (Kulik, A.S. (1991, September)). The approach is grounded on the hypothesis of uncertainty for fault's emerge time, its place, class, and specific physical reason. To eliminate the uncertainty, the principle of successive uncertainties evaluation, which is actually a set of interrelated and purposeful steps, is applied (actual diagnostic support) to allow the diagnosis of an abnormal situation promptly and with a given depth.

The full diagnosis of the off-nominal situation is just a premise to rise the UAC safe usage . After receiving complete information about causes of UFC (unmanned fly complex) operability loss it is necessary to solve a set of further tasks for parrying the issues that might happen due to the contingency, so as to reduce a risk of the crash progress; the latter implies the complex itself and the environment nearby. Recovery process automation associated with dynamic systems like the UFC has been a topic in many studies devoted to managing different reservings and corresponding redundancy (Hajiyev, C., & Caliskan, F. (2013)). The most near research on keeping the UAC efficiency provide for operability restoration through the flexible (complex) use of signal and parameters adjustments, as well as the software and hardware reconfiguration relyed on results of operational diagnosis (Kulik, A.S. (2014)).

Combination of approaches associated with the diagnosis and the restoration of control systems operability in condition of unexpected destabilizing influences, gave rise to a new theoretical direction - the use of operability rational management (control) for the aim of raising the autonomous vehicles safety (Kulik, A.S. (2014)).

UAV Rational Control Based on Operations Intellectualization

Rational control of objects operational integrity is based on a number of provisions. The objective to determine a reason of vehicle malfunction is a complicated problem for UAV designers and maintenance people. This is commonly known. A topical provision of this paper is intellectualization of such a difficult function as the search of the source of malfunctions is and parrying it in real time.

The need of using tools of artificial intelligence (Russell, S.J., & Norvig, P. (2016)) to provide a diagnosis and restoration of the UAV operation integrity, hence providing the required level of the flight safety in common, is explained by the following reasons below.

1. Conventional hand-on technology of the diagnosis and restoration is grounded on heuristic knowledge acquired by specialists during maintenance operations. Formalization of such knowledge may be done by using artificial intelligence tools only.
2. Mechanisms for specialists' heuristic judgements implementation needed to form the base of abnormal situations diagnosis can be downloaded onto the management computer only with help of the machine means of inference.
3. Practical procedures for carrying out a repair work to restore the UAC integrity that would be grounded on the diagnosis outcomes can be effectively formalized and realized with production models.

The artificial intelligence instruments aimed to formalize the diagnosing and restoring efficiency procedures can be supplemented with a means of solving the inverse problems and tasks of resources allocation. These tools are based on the models of UAC functioning wthat has been used while designing. The models represent the knowledge on processes occurring in both regular and off-nominal situations. These knowledges usage will allow to supply a more complete knowledge base to the diagnosing and efficiency restoration procedures. In general, this will make it possible to ensure the UAV mission fulfillment.

Rational Control of an UAV Equipped by a Computer Vision System

As a rule, autonomous robots are to work in partially or completely unknown conditions (Shim, D., Chung, H., Kim, H. J., & Sastry, S. (2005, August)). In order to provide an adaptation of the path to the situational uncertainty, one can use the appropriate processing of visual information obtained from the video cameras. The adaptation efficiency depends on the environment diagnosis quality. The development of the algorithmic support for the environment diagnosis process constitutes a topical problem, one of possible solutions of which is presented in the chapter.

A progressive trend of autonomous robots (AR) development is to extend their applications area by improving the functionality of the motion automatic control systems. (Achtelik, M., Bachrach, A., He, R., Prentice, S., & Roy, N. (2009, April) The environment where the movement occurs is characterized by situational uncertainty due to a number of factors.

The autonomous robots operate typically in conditions of uncertain factors (causes) which destabilize the navigation tasks fulfillment.

For the fly robots, to which the UAVs belong, the destabilizing influences could be:

- Atmosphere condition (windless weather or the wind presence, sunlight brightness, rain, fog, etc.);
- State of terrestrial and celestial visual landmarks (their displacements, changes of the reflecting surface, etc.);
- Appearing of obstacles (interference) on the flight path (other aircrafts, disturbance impacts) (Kucherov, D., Kozub, A., & Kostyna, O. (2016, October)).

Destabilizing impacts since being a possible malfunction reason need to be identified and evaluated in course of carrying out the related missions. In other words, destabilizing factors produce the situational uncertainty of the fly vehicle environment.

Such situational indeterminacy can be reduced through procedures for external environment operational diagnosis during the autonomous robot movement along the route (Bonin-Font, F., Ortiz, A., & Oliver, G. (2008)). Diagnostic procedures should allow to detect obstacles, localize and identify them, i.e., to ensure a complete diagnosis. The information is demanded in real time in order to promptly create on autonomous robot board such an obstacle avoidance procedure that will allow rationally solve navigation problems (Blösch, M., Weiss, S., Scaramuzza, D., & Siegwart, R. (2010, May)) in the conditions of traditional time and energy resource constraints. Similar diagnostic procedure functions should be performed when other destabilizing factors appear.

Due to a big variety of cheap and perfect video cameras presented on the market nowadays, it is appropriate to implement computer vision complexes directly onboard the fly vehicle for solving tasks of diagnosing the environment. For autonomous machine vision systems laid on mobile objects, a single-board platform Raspberry Pi is usually used, along with Raspbian operating system that has been recommended by the manufacturer. In this environment, the basic programming language is Python

When programming in Python, the major tasks of machine vision as regard the image and video data processing are conventionally solved using the Pillow library (Python Imaging Library), that provides for a fairly complete set of functions and techniques for processing images and video. However, sharing resources of the Pillow library and the Open CVV library (Open Source Computer Vision Library) is assumed to be more efficient.

The OpenCV usage in conjunction with the Python is convenient for creating the simple and clear applications, performing experiments and synthesizing various prototypes. On the Python, practically all functionality of the OpenCV library is available. When using the vector-matrix ideology similar to that adopted in Matlab, it is possible to obtain sufficiently fast algorithms and provide for the task solution in real time.

When solving problems of the type considered, such combination of the hardware and software seems to be optimal in terms of the price and performance.

The Concept of Designing Algorithms for Estimating UAV With the Computer Vision System

When designing the computer vision system for a present-day UAV or other topical applications, a developer is being in the frame of constraints always. A few may be like following.

1. It is mostly wished for a simple, reliable and reasonably cheap hardware platform.
2. The hardware needs using a standard programming means.
3. It is necessary to provide for a given measurement accuracy of information parameters,
4. The need to ensure that the system operates with adequate performance (usually, a vision system runs in real time or so). And others.

Many contemporary systems deal with processing and using the video information that that has been obtained by the airborne image recorder (AIR) mounted on the UAV. In these cases one needs solving the vehicle control tasks within constraints mentioned. Such projects are aiming to use the Raspberry Pi single-board computers and Python programming language in conjunction with the OpenCV image processing library. The joint usage of Python and OpenCV allows real-time video data processing with up-to-date algorithms implemented.

An invariably urgent problem of the computer vision technology is to obtain an effective solution of the problem on the plant detection and following motion parameters evaluation with a necessary accuracy, in real time, and using simple hardware and software.

The following tasks associated with the fly vehicle motion and related estimating need solving:

1. Reliable object-of-observation detection under different scene lighting and against the background of possible interference.
2. Spatial localization within the frame to within a pixel performance.
3. When used the video sequence, one needs computing the trajectory parameters (distance to origin, the angular attitude and velocity parameters).

Conventionally, these problems are being solved using algorithms of the optical flows analysis (e.g. by Lucas-Canada methods), objects motion estimating by the method of comparing blocks with the use of correlation-extreme algorithms (Krasnov, L. A., Liamtsev, S. E. (2016)). The methods have insufficient accuracy, demand vast computational capacity, and hence are of low speed. In machine vision systems of mobile applications, the mentioned drawbacks limit usage of those conventional algorithms.

Lower in the section, the designing of effective algorithms for objects detection and parameters evaluation based on video observation data will be in scope. The algorithms are designed for real-time running on a Raspberry Pi computer using the Python programming language and the OpenCV library. The proposed solution to the task of detecting and positioning the viewable objects in a video sequence frame is based on the knowing and using of color characteristics of the object observed.

In order to meet the stated objectives and implement actual algorithms, the most essential attribute of the recorded sequence was chosen. Namely, such a color component of the color model of frames, which best corresponds to the observed object color and reliably distinguishes the latter one from the scene background under different lighting conditions.

The procedure of converting the selected color component of the frame onto a halftone picture together with the following image binarization fitted to the selected brightness threshold is next carried out to obtain the exact spatial localization of the observation object. After, a set of zero order (m00) and first order (m01, m10) invariant moments of the binary image is calculated to determine the visible object center coordinates.

At the final stage of video data processing, the image coordinate system is converted to the right central rectangular coordinate system, and computation of trajectory parameters of the observed object is carried out.

A particular attention were given to constructing the interactive program interface and graphical displaying the results of data processing.

One of features of writing codes on the Python is the need to produce the required programming capabilities at the expense of connecting external libraries.

To solve this problem, we can use the following OpenCV library resources:

```
import sys,
import cv2,
import numpy as np,
import  math.
```

The analysis of facilities specification allows constitute that such a programming software configuration does not exert a significant load on the Raspberry Pi processor and gives rise a good prerequisites for processing data in real time. If the video processing performance must be high, all the basic program procedures should be using standard OpenCV functions, which have been optimized by developers by the moment of program release.

Let's consider more carefully some details included in the technical solutions development. Let's take a look now at more details of the technical solutions adopted.

To detect and reliably recognize a visible object in the frame, the technique of color features identification has been used. The color space is a data representation model based on the use of certain color coordinates. Conventional Web-camera video signals are images sequences presented in RGB format, the use of which is inefficient since the search of the regions of demanded color supposes analysis of all three color components, items R, G, and B. If you want implementing the image objects search by the color and brightness feature then the color model HSV is needed, where H, S, and V stand for hue, saturation, and value and define corresponding image characteristics. These are shown on the diagram (Figure 1) and normalized as follows:

- **Hue:** 0 – 360°;
- **Saturation:** (0 – 100) %;
- **Value:** (0 – 100)%.

Hence, after receiving the recorded video data one needs converting the RGB color space onto the HSV space by using the OpenCV function.

```
hsv=cv2.cvtColor(frame,cv2.COLOR_BGR2HSV.
```

A definite choice of components corresponding to the visible object color that can be displayed at different scene lighting conditions, has much effect on the detection quality. Any point taken on H scale determines a two-dimensional region with different meanings S and V. However for a predetermined H value, the target area in that two-dimensional space is more reliably determined by the condition:

```
V > Vmin и S > Smin,
```

where Vmin, Smin are certain constants.

Therefore, it is customary to select the target range of H scale by specifying two values – Hmin and Hmax [4 - 8]. For example, if one needed detecting blue color then HSV parameters value would be in the range, respectively:

```
low_range = np.array([100,100,100])
up_range = np.array([140,255,255]),
```

and the red color range:

```
lower_range = np.array([0,50,100])
up_range = np.array([10,255,255]).
```

Figure 1. Color model HSV

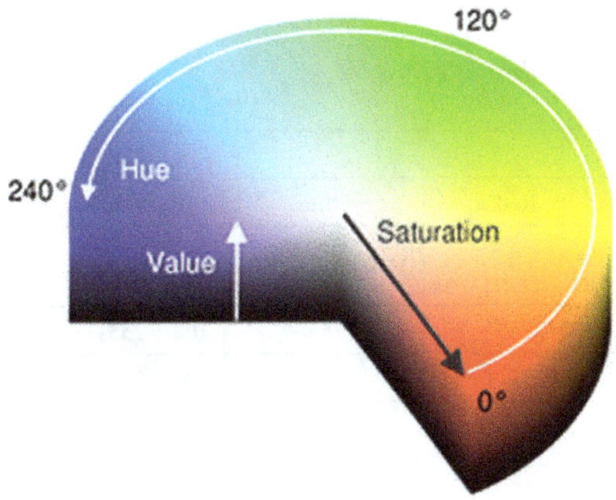

It should be noted that this approach has a significant drawback – only the bright color components are reliably recognized due to limits based on Vmin and Smin values. Therefore, it is almost impossible to allocate areas for dark colors (for example, brown), also for shades of gray. That is why, for successful objects search and their tracking on the base of color characteristics, it is desirable that the observed object would comprise bright, deep color markers (red, blue, or green).

Approximate ranges for the HSV model parameters chang regarding the simple colors are shown in the Table 1.

Note that the procedure for selecting the working range of HSV space parameters for the observable object is better to be performed interactively depending on the current lighting conditions and the object color. Such an approach implemented onto the algorithm considered below, can be represented by the following steps sequence.

1. Color components range selection. For this, a software interface is usually developed to allow convenient setting of the color components values range.
2. Performing filtering with the use of the customized color mask having a threshold function

```
mask_blue = cv2.inRange(hsv, low_range,up_range).
```

3. Frame image representation in shades of gray with pixels brightness range 0 to 255 values. This procedure allows reliable detecting the object by its color feature but does not completely remove the background noise containing resembling color components.
4. Selection of the brightness clipping threshold that depends on the noise magnitude. This function should also be interactive to perform the effective background noise cleaning.
5. This stage requires the image binarization that consists of, first, removing the color components, and then image clipped by brightness threshold binarization:

```
mask_blue = cv2.threshold(mask_blue, p, 255, cv2.THRESH_BINARY)[1].
```

6. Determination of the observed object's coordinates in the current frame. For solving the problem when dealing with the binary images received in the result of color picture transformation, one can compute the moment invariants of the black-and-white image and obtain the solution for object's coordinates with acceptable accuracy. Image moments are calculated using the function

Table 1. Ranges of HSV model for simple colors

Color Space HSV	Color				
	Red	Blue	Yellow	Black	White
H	324 ÷ 360 и 0 ÷ 30	190 ÷ 236	40 ÷ 60	0 ÷ 360	0 ÷ 360
S	130 ÷ 255	130 ÷ 255	100 ÷ 255	0 ÷ 255	0 ÷ 60
V	50 ÷ 255	50 ÷ 255	100 ÷ 255	0 ÷ 90	150 ÷ 255

```
moments = cv2.moments(thresh, 1).
```

7. Computing coordinates of the observed object's center relative to the video frame center. Since the origin of the frame coordinates is assumed to locate in the upper left corner of the shot, we need moving to the central rectangular coordinate system with coordinates of the origin being half the frame size. Shifting is performed by the parallel transposition technique. Such a transition makes it easy to calculate and display the trajectory parameters (distance to target, angular coordinates, velocity); actual implementation is not difficult.

For carrying out step 6, the <*moments*> function is used. This function is capable to compute moment's array up to the third order. However, for computing the object center coordinates, the moments of the first order m01 and m10 along with the zero order moment m00 are required only. They are defined as follows:

```
dM01 = moments['m01']
dM10 = moments['m10']
dArea = moments['m00'].
```

Note that the moment m00 is the number of all unit pixels which belong to the selected object, and the moments m01, m10 are the X and Y coordinates sums of these pixels. It is clear that to determine the coordinates of the observed object's center it is necessary to normalize these moments with respect to the zero order moment.

In carrying out this procedure it is advisable to perform an additional operation that requires the threshold correction and assists filtering out the false objects the appearing of which is probable when the color filter is in work. However, if there exist a priori information on the visible object dimensions then by the condition 'if dArea > N' where N is the number of unit pixels of the moment m00, it becomes possible to eliminate the false objects. In the following example, the program will only respond to moments containing more than 100 pixels:

```
if dArea > 100:
    x = int(dM10 / dArea)
    y = int(dM01 / dArea).
```

In this example, only the random blue color flashes are eliminated that occupied pretty small area in the whole frame exposure (smaller than the visible object area).

It should be noted that the moment function is not able to determine the quantity of objects in a frame. Therefore, if for example two independent objects are present in the frame then they are interpreted as a one entire object. Hence, the center coordinates will be found placed between the two. This results in rough anomalistic errors in the trajectory parameters calculation. In such cases, it is necessary to employ more accurate processing techniques, for example, assessing the object shape and/or seeking other morphological features.

Within the adopted methods framework, a synthesis of a few algorithms which running utilized a color selection procedure has been performed for detecting and estimating visible objects motion parameters. The algorithms can be used in stationary and mobile video surveillance systems for UAV detection, UAV trajectory parameters evaluation, UAV flight parameters estimation. All the algorithms are developed on Python using the library OpenCV on Windows. This allows carrying out software design and testing with the use of a conventional computer. With this respect, a lot attention was given to the interactive interface that had provided for preliminary tuning of all information processing regimes and visual control convenience.

When applying UAVs, an important research direction concerns a problem of determining the UAV flight parameters by using video images data since the video signal is the most helpful information carrier. Therefore one of progressive design approaches is the advanced vision sensors and image processing algorithms development; hereafter they would be able to replace the entire navigation sensors complex that is in use on aircrafts presently. This will allow degrading the level of situational uncertainty. In the following sections, we present results of the experimental research on the machine vision systems which are used to estimate the UAV motion parameters, namely the vehicle angular attitude and flight altitude.

EXPERIMENTAL RESEARCH OF THE ANGULAR ATTITUDE ESTIMATING FOR THE UAV EQUIPPED BY A VISION SYSTEM

To solve the problem of determining the angular position of a UAV, it is necessary first to provide the solving the UAV positioning problem that would use a video image. Let's consider methods and algorithms required to solve the task of positioning a UAV in space relative to surrounding items.

The solving of the proposed problem can be performed in two options that follow.

1. Onboard the UAV, one fixes a video camera that shoots the scene in the surrounding space. Following the shots received, it is necessary to examine and evaluate the outcomes of the space attitude changes of video camera carrier.
2. A stationary external video camera looks through a space area in which a moving object is present and evaluate the parameters of change of the object location.

First Option

To solve the problem of finding the coordinates of a UAV we need evaluation of how much the UAV has moved in a serial video frames during the session. To avoid excessive calculations, it will be enough to track on the current image certain characteristic places of the UAV framework (Saripalli, S., Sukhatme, G. S., Mejias, L. O., & Cervera, P. C. (2005, April)), the change in position of which in the coming frame unambiguously specifies the position of the UAV at the current instant. We call such places as reference points. Since the reference points on the shoot have two coordinates (the image is two-dimentional), whilst the reference point in space has 3 coordinates, then it is necessary to get a transformation algorithm. The following are the three basic implementation stages.

1. First you need to determine the camera domestic parameters, which means a carrying out of the internal calibration at the required accuracy. Note: the procedure may be performed only once.

2. The second stage is the solution of the task to specify reference points on the images of the recorded scene. This problem can be solved in various ways, but ensuring the reliance and necessary accuracy of the references recognition is of particular importance.
3. The final stage of the procedure is the evaluation of the object's spatial position change by calculating the unknown coefficients of the rotation matrix and displacement vector in a given coordinate system.

Thus, the task of determining the change in space attitude of a mobile vehicle is divided into three parts:

- Preparative determination of domestic parameters for the video camera;
- Finding the appropriate reference points on images;
- Calculation of the rotation matrix and camera shift vector in a given coordinate system.

For experimental study, the unmanned aerial vehicle T-10, which is a dynamically scaled 1: 5 model of SU-27fighter aircraft was chosen. The aircraft is built on a normal twin-tail aerodynamic scheme with a fixed wing. It is designed to perfom a purposeful controlled movement within the atmosphere at different including the small altitudes; a flight grounds the principle of utilizing the aerodynamic flight and producing the aerodynamic control moments.

To carry out the experiment, the T-10 model was fixed in a controlled three-dimensional suspension.

To simulate the motion of the UAV, the laboratory stand that intends a changing of the UAV angular attitude is used. The lab stand is actually a suspension assembly involving a metal frame on which the UAV itself, balancing springs, drives, and a potentiometer sensor unit are mounted. The neutral position of the control object is provided by the balancing springs. Changing the model pitch, roll, and yaw angles is performed by drives. The potentiometer sensors unit provides for feedback of all three control channels related to three-dimensional variation of the UAV angular attitude.

The onboard inertial measurement unit (IMU) that is placed in the point of center of mass of the model contains three gyroscopic sensors for yaw, roll, and pitch angles measurements, also a set of angular velocity sensors with two related sensors for each of axis of the base coordinate system.

The analog-to-digital conversion module serves to acquire and appropriately adapt analog sensor signals. The basic controller exchanges data between the model and the personal computer-server and is supposed to control both the suspension actuators and on-board servo drives. It acts as the stand computational means and provides a logic of the UAV airframe movement.

Figure 2. Kinematic scheme of the stand with suspension

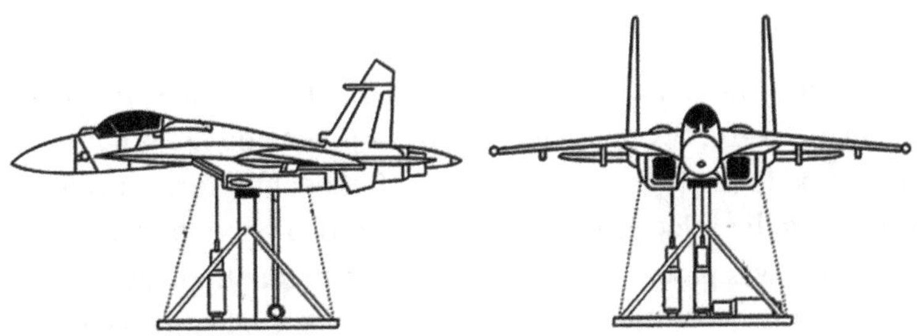

Figure 3. Visibility of the reference points to determine the UAV angular movement

The facility provides for a two-way communication between the personal computer-server and client computers interconnected via the wireless communication channel with the help of related system. The clients are able to receive data from sensor unit in real time.

The model position is controlled by a joystick connected to the server computer.

A few material constraints were accepted while carrying the experiments. They concern, first, the fact that the video camera constitutes the solely sensor for perceiving the surrounding space, and second, the UAV computer vision was monocular.

The video camera is fixed in the center of mass of the experimental model and in such a way that special details (control points) will be displayed in the frame, as shown in the figure (Figure 3).

According to the plan for inestigations, UAV was making evolutions thus changing its angular position. The experiment objective was to determine these changes by evaluating a shift of reference points in the video camera frames, i.e., the UAV angular attitude was assumed to be estimated using on-board video measurements.

The results of the task solution that display the calculated values of the roll angle estimates corresponding to the item of the video frame sequence are shown in Fig. 4.

The value of yaw and pitch angles have been estimated similarly.

Option 2

Now, the tracking video camera is placed so that the model T-10 becomes seen in the frame for following recording. On the model T-10, two reference points of green and red color that imitate the landing lights of the MBPLA can be found (Fig. 5). Next the T-10 model performs evolutions about the center of mass. MBPLA model angular relocations have been recorded on video during the experiment.

After shooting, the video was processed (using the program that implements the algorithm that will be described later) to output results of the UAV angular attitude evaluation (see Fig. 6).

Figure 4. The roll angle estimation outcomes obtained from recorded video frames

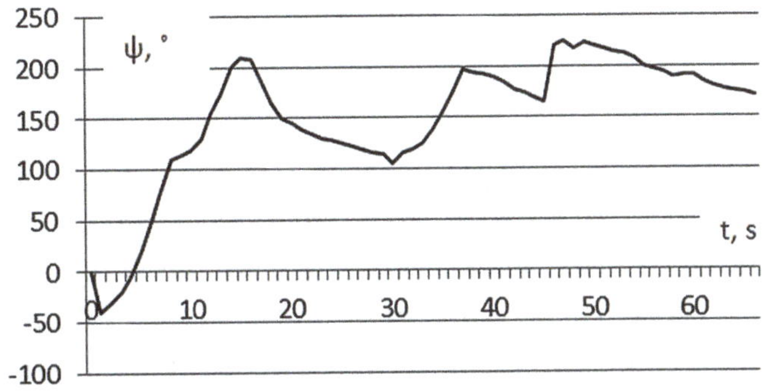

Figure 5. Presentation of the experiment with the tracking camera

Figure 6. The roll angle evolution of the UAV model: shooted, recorded, and processed

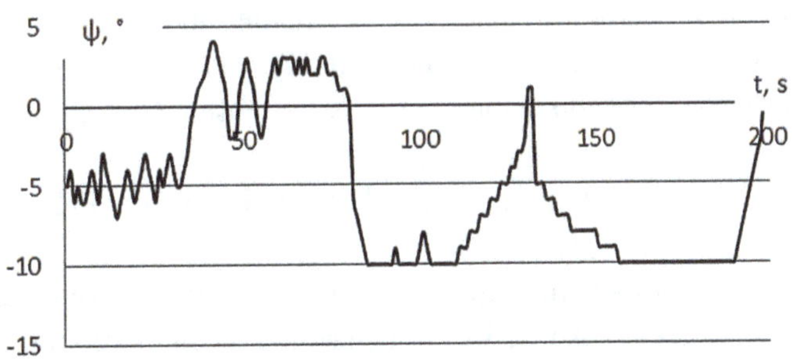

Section Summary

The section demonstrates two ways of solving a practice problem dedicated to experimental evaluation of the space attitude of the fly vehicle by using a video information. Qualitative analysis of the research results on determining the vehicle angular position shows the potential of using similar algorithms in computer vision systems of mobile robots and the systems for estimating the parameters of their motion. The first option, when the tracking video camera has been placed onboard the aircraft, can be successfully used in vision systems of mobile robots and UAVs, whereas the option with the external video camera that is shooting the scenes containing the object of interest from a constant position, serves for cases where the objective control for moving targets motion (mobile robots, UAV) is needed.

Experimental Research of the Multirotor UAV Altitude Estimating

The results of experimental research of the UAV flight height over the local relief are presented in this section. This task is of broad practical importance. For instance, the UAVs usage in fine agriculture, for performing works related to fertilizer distribution, etc.

Existing sensor systems for determining flight altitude do not allow to obtain accurate estimates. This is due to the following reasons.

1. The GPS-sensors allow you obtaining the altitude above sea level only, what significantly narrows such systems application area.
2. Ultrasonic measurement systems do not allow letting the required accuracy.
3. The laser measurement instruments may not be used for designing accurate systems associated with functioning over different underlying surfaces, for example, grassy terrain condition. This fact can lead to off-specified errors in altitude determination.

One of advanced ways for solving the problem is using video images data shoot by the onboard video camera and generating the effective processing algorithms allowing to accurately determine the flight altitude.

The work bench model that has been used to carry out related experiments is shown in Fig. 8. A multi-rotor quad-copter drone equipped by the computer vision system was chosen for the case study.

For mounting the system on board the drone, two hardware options were suggested and examined.

The first option is based on using a Raspberry Pi computer and a low-resolution USB camera. A board of the processing unit of such a system is shown in Fig. 8-9.

The second option utilizes the *Action* camera as a vision sensor. The option 2 hardware representation can be seen in Fig. 10.

For appropriate hardware implementation, we need certain Raspberry Pi 2 (model B) specifications. They are given in Table 2.

The primitives of different colors with known a priori geometric size were placed beneath the drone when it was flying. The experiment purpose was to determine the autonomous robot altitude above the arrangement of reference primitives when using different systems of computer vision. The computer system algorithm was fitted to detect and extract a specified primitive (of known color, shape and size) obtaining supposingly zero height reference level. The video images sequence obtained by the robot's

Figure 7. Photo of the case-study experiment with drone's flight height examination

Figure 8. Exterior of Raspberry Pi 2 board (model B)

Table 2. Technical characteristics of Raberry Pi 2 (model B) platform

	Parameter	Value
1	Processor (CPU)	Broadcom BCM2836 ARM Cortex-A7 Quad Core 900 MHz
2	Number of processor cores	4
3	RAM	1 GB
4	Memory	MicroSD
5	Networking	Ethernet 10/100
6	Video Output	HDMI
7	USB Ports	4
8	Audio output	3,5 Jack
9	GPIO	40 pin
10	Power supply	5V 2A
11	Additional features	2 diodes
12	Dimensions, weight	85,6 x 53,98 x 17mm, 45 grams

computer vision system was used as raw data to finally calculate the UAV altitude. After processing the outcomes found used to be compared with the reference data of the UAV flight controller (e.g. GPS).

Fig.11-12 illustrate the examples of video frames obtained by the two option hardwares associated with the on-board computer vision system.

The algorithm of altitude evaluation by the use of video record processing includes several operations . They are presented below by appropriate steps sequence.

1. Video capture device initialization and frame grabbing.
2. Frame scaling.

Figure 9. Prototype of the hardware based on Raspberry Pi computer and USB camera

Figure 10. Prototype of the hardware with the Action Camera

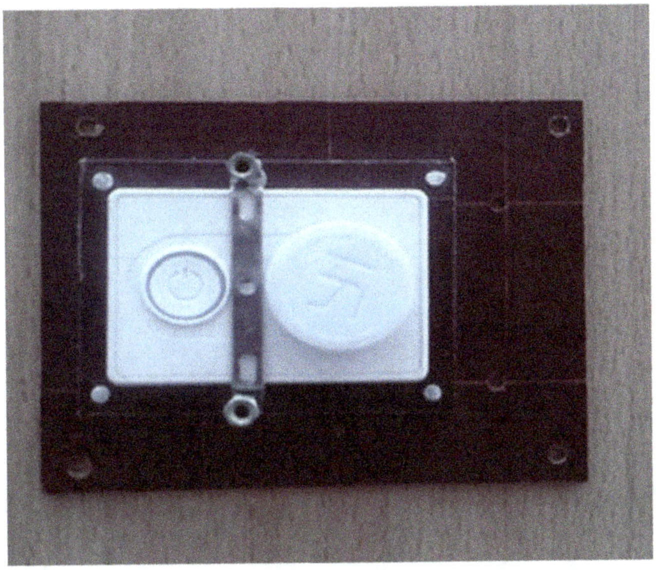

Figure 11. Image shot made by the on-board USB camera

3. Setting digital filter parameters.
4. Frame transformation from RGB to HSV color space.
5. Digital filter actuation.
6. Running the medians smoothing (blurring), which is similar to other averaging methods (in fact performed by the code MedianBlur from OpenCV library); the operation is used to processes image edges when filtering the noise.
7. Frame binarization by the brightness threshold.
8. Running the image-edge sensor.

Figure 12. Image shot of the on-board Action Camera

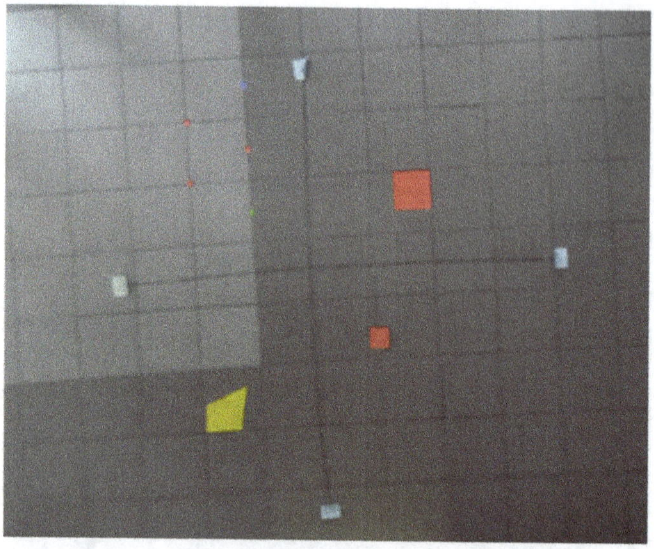

9. Geometric primitive detection (extraction from the frame, detected item's surface area calculation).
10. Altitude value calculation (distance to the geometric primitive computed via a known proportionality coefficient).

Fig. 12 represents the chart of the code execution.

The results of determining the UAV altitude obtained by the computer vision system for two arrangement options are shown in Fig. 13-19.

Analysis of the results obtained allows summarizing as follows:

1. The proposed algorithm for determining the flight altitude can be used in UAV's computer vision system.
2. To determine the UAV flight parameters, the two computer vision hardware options can be used: the Raspberry Pi microcomputer-based board and a low-resolution USB camera; or the system with using the *Action* camera.
3. Absolute accuracy while measuring the flight altitude was ± 0.5 m for the vision system based on Raspberry Pi and USB camera. For the computer vision system based on Action camera, the absolute error constitutes ± 1 m.
4. The comparison of outcomes let stating that a low resolution USB video camera may be used in the UAV computer vision system aimed to measure the fly vehicle altitude.

FUTURE RESEARCH DIRECTIONS

In future, it is planned to develop an intelligent control system for a UAV with the computer vision system to provide the automatic flight in condition of numerous obstacles and contingency. Such system allows

Figure 13. Chart of the algorithm for computing the flight height of the UAV with a computer vision system

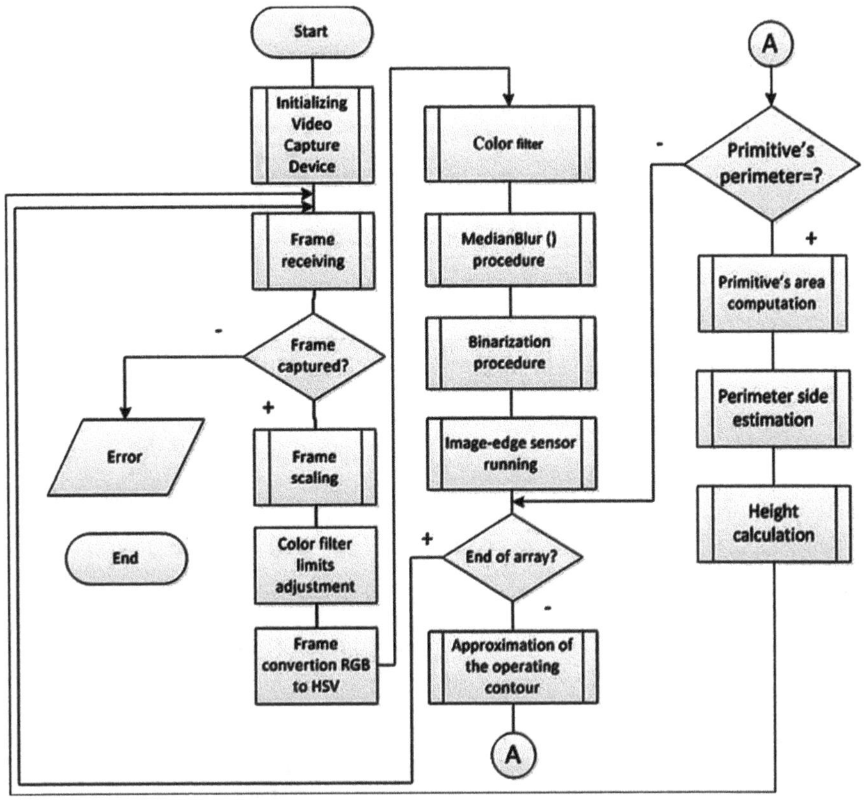

Figure 14. Flight altitude data from the UAV flight controller

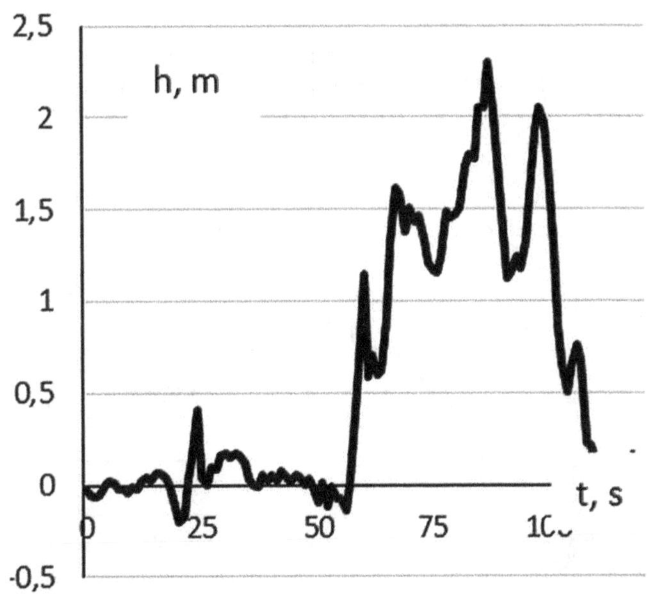

Figure 15. Flight altitude data from the UAV flight controller

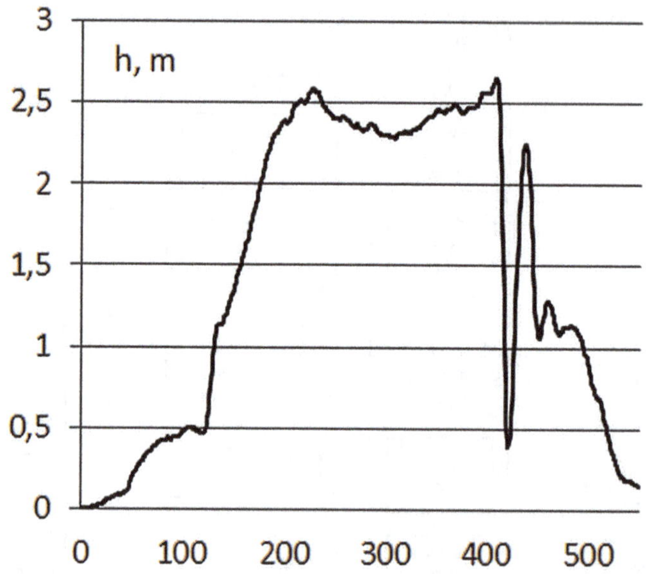

Figure 16. UAV flight altitude estimation by the computer vision system (option 1)

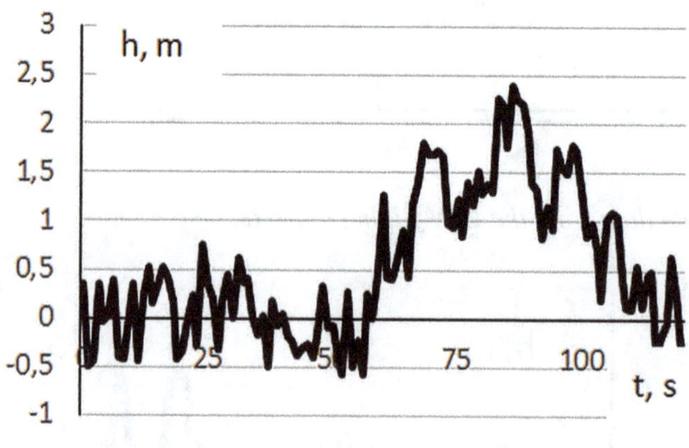

Figure 17. AV flight altitude estimation by the computer vision system (option 2)

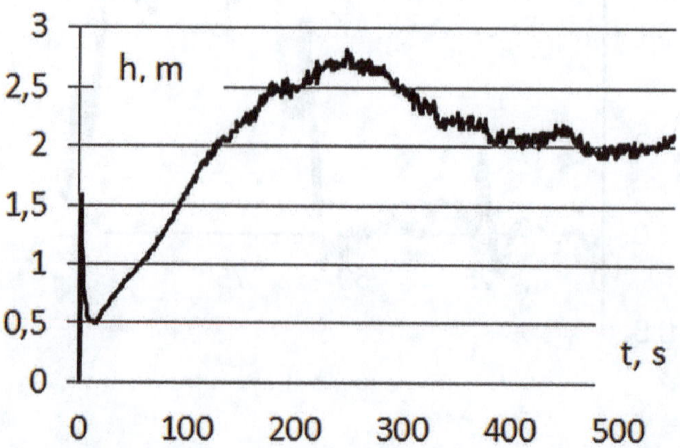

Figure 18. Contastive analysis of altitude estimation: flight controller data – blue; computer vision system (option 1) – red

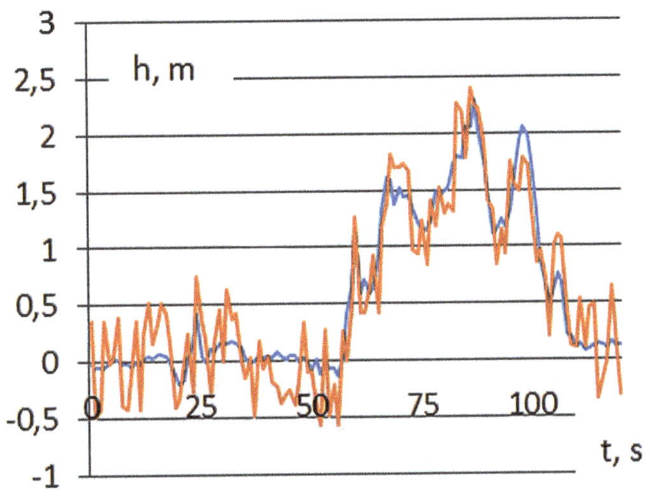

Figure 19. Contastive analysis of altitude estimation: flight controller data – blue; computer vision system (option 2) – red

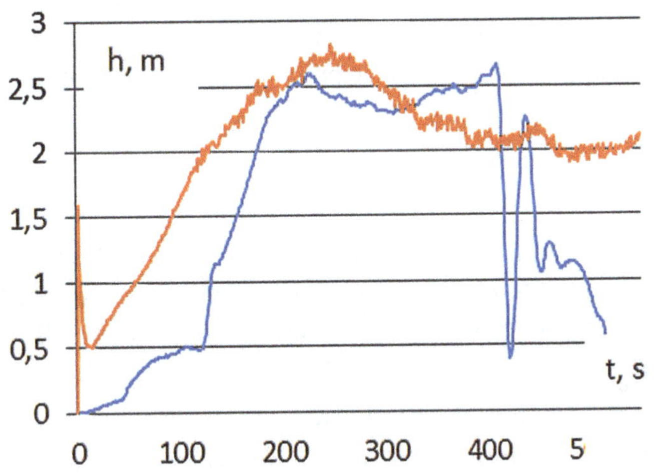

implementing the described principles of rational control that imply to increase the UAV performance capabilities, expand the practical application areas, and progress in the solving UAV safe use problem.

To create such systems, further improvement of image processing algorithms is topical since the latter allow providing the UAV by better navigational information, and giving the possibility to automatically synthesize the safe flight path in the environment with various obstacles. The development and applying of newest reliable parallel algorithms for images reference point detection in real time will be of broad demand. Such algorithms should employ the effective procedures on images brightness equalization and may be constructed with implementing artificial neural networks technique, e.g. Haar Cascade method.

CONCLUSION

The UAV, as autonomous robots, operate under conditions of situational uncertainty. They can adapt themselves to that uncertainty owing to intelligent functions of the control system which are based on using video information provided by the on-board camera. In turn, the adaptation efficiency depends on the quality of outdoor environment diagnosis. A development of algorithmic support for the process of diagnosing the environment is a topical problem.

From the practical UAV application point of view, the important effect is the video data usage for determining of the UAV flight parameters, because the video signal is the most informative source of information. That is why a topical issue is designing of contemporary video sensors together with appropriate image processing algorithms to replace the present complexes of UAV navigation sensors. The goal is decreasing the level of situational uncertainty.

Constructing the proposed algorithms is based on parallel placing of the color segments of the image onto the HSV color model with the image preprocessing. It is intended to be implemented on the Raspberry Pi single-board microcontroller platform together with the software recommended by Raspbian operating system manufacturer and using of Python programming language along with Open Source Computer Vision Library.

The experimental study of the proposed algorithm realizations allows constituting that such the algorithm may be used as a part of the machine vision system software of the autonomous UAVs. When solving the problem of determining the angular attitude and/or the height of the UAV it is found that a high resolution of the video sensor is not so important. In contrary, simple and cheap sensors can be used instead for a qualitative solution of the navigation problem, e.g., a low resolution USB video camera.

REFERENCES

Achtelik, M., Bachrach, A., He, R., Prentice, S., & Roy, N. (2009, April). Stereo vision and laser odometry for autonomous helicopters in GPS-denied indoor environments. In *Unmanned Systems Technology XI* (Vol. 7332, p. 733219). International Society for Optics and Photonics. doi:10.1117/12.819082

Barba-Guaman, L., Calderon-Cordova, C., & Quezada-Sarmiento, P. A. (2017). Detection of moving objects through color thresholding [Detección de objetos en movimiento a través de la umbralización del color]. *Iberian Conference on Information Systems and Technologies*.

Blösch, M., Weiss, S., Scaramuzza, D., & Siegwart, R. (2010, May). Vision based MAV navigation in unknown and unstructured environments. In Robotics and automation (ICRA), 2010 IEEE international conference on (pp. 21-28). IEEE.

Bonin-Font, F., Ortiz, A., & Oliver, G. (2008). Visual navigation for mobile robots: A survey. *Journal of Intelligent & Robotic Systems*, *53*(3), 263–296. doi:10.100710846-008-9235-4

Gulevitch, S. P., Veselov, Yu. G., Pryadkin, S. P., & Tyrnov, S. D. (2012). Analysis of factors affecting safety of flight of unmanned aerial vehicles (UAV). Causes of accidents of UAVs and methods of preventing them. Science and education Russia. Bauman Moscow State Technical University.

Hajiyev, C., & Caliskan, F. (2013). *Fault diagnosis and reconfiguration in flight control systems* (Vol. 2). Springer Science & Business Media.

Isermann, R. (2006). *Fault-diagnosis systems: an introduction from fault detection to fault tolerance*. Springer Science & Business Media. doi:10.1007/3-540-30368-5

Krasnov, L. A., Liamtsev, S. E. (2016). Research of antijammingness of algorithms of the trajectory measuring on viewdatas [Issledovaniye pomekhoustoychivosti algoritmov traektornykh izmereniy po videodannym]. *Radioelectronic and Computer Systems, 4*(78), 113-117.

Kucherov, D., Kozub, A., & Kostyna, O. (2016, October). Group behavior of UAVs in obstacles presence. In *Methods and Systems of Navigation and Motion Control (MSNMC), 2016 4th International Conference on* (pp. 51-54). IEEE.

Kulik, A. S. (1991, September). Fault diagnosis in dynamic Systems via signal-parametric approach. In *IFAC/IMACS Symp. Baden-Baden* (*Vol. 1*, pp. 157-162). Academic Press.

Kulik, A. S. (2014). Rational control of aerospace objects operability under the influence of the destabilizing actions. *Aerospace Engineering and Technology*, (1), 31-38.

Lucas, B., & Kanade, T. (1981). An Iterative Image Registration Technique with an Application to Stereo Vision. *7th International Joint Conference on Artificial Intelligence (IJCAI): materials of the international seminar*, 123-129.

Mojsilovic, A. (2005). A computational model for color naming and describing color composition of images. *IEEE Transactions on Image Processing, 14*(5), 690–699. doi:10.1109/TIP.2004.841201 PMID:15887562

Russell, S. J., & Norvig, P. (2016). *Artificial intelligence: a modern approach*. Pearson Education Limited.

Saripalli, S., Sukhatme, G. S., Mejias, L. O., & Cervera, P. C. (2005, April). Detection and tracking of external features in an urban environment using an autonomous helicopter. In *Proceedings of the 2005 IEEE International Conference on Robotics and Automation* (pp. 3972-3977). IEEE. 10.1109/ROBOT.2005.1570728

Shim, D., Chung, H., Kim, H. J., & Sastry, S. (2005, August). Autonomous exploration in unknown urban environments for unmanned aerial vehicles. In *AIAA Guidance* (p. 6478). Navigation, and Control Conference and Exhibit. doi:10.2514/6.2005-6478

Sładkowski, A. (Ed.). (2013). *Some Actual Issues of Traffic and Vehicle Safety. Faculty of Transport*. Silesian University of Technology.

ADDITIONAL READING

Dergachov, K. Y, Krasnov, L. A., Piavka, E. V. (2017) Algorithms of finding out objects and estimation of parameters of their motion in the systems of technical sight [Algoritmy obnarugeniye obyektov i otsenki i parametrov ikh dvigeniya v sistemakh tekhnicheskogo zreniya]. Radioelectronic and computer systems, 4 (84), pp. 28-39.

Gao, X., Yeh, H.-G., & Marayong, P. A. (2017). A high-speed color-based object detection algorithm for quayside crane operator assistance system. 11th Annual IEEE International Systems Conference, SysCon, proceedings.

Martynova, L. A., Koryakin, A. V., Martynova, L. A., Lantsov, K. V., & Lantsov, V. V. (2012). Determination of coordinates and parameters of movement of an object on the basis of image processing [Opredelenie koordinat i parametrov dvijeniya obiekta na osnove obrabotki izobrajeniy]. *Computer Optics*, *36*(2), 266–273.

Sabelnikov, P. Y. (2012). A comparison of objects contour with partially misresented shap. Journal of Qafqaz University. *Mathematics and Computer Science*, *34*, 47–58.

Severgnini, F. M. Q., M. L. Oliveira M. L., Mendes V. M. S., Peixoto Z. M. A. (2016). Object Tracking by Color and Active Contour Models Segmentation. IEEE Latin America Transactions, 14 (3), pp. 1488-1493.

KEY TERMS AND DEFINITIONS

Color Space HSV: Is a color model in which the color coordinates will be hue, saturation, and value.

Computer Vision: Implements a complex process of extracting, identifying, and converting video information, which includes six basic steps: 1) obtaining (perceiving) information, or sensing; 2) preconditioning, or preprocessing; 3) segmentation; 4) description; 5) recognition; and 6) interpretation.

Computer Vision Systems: Sensory assemblies that provide shooting of the work scenes and objects, and images converting, processing, and interpretation using an UAV on-board computer and further the transfer of results to the management device.

Destabilizing Factors: Are such objectively existing disturbing impacts that are need to be identified and evaluated in the course of carrying out the respective missions.

Off-Nominal Situations (Contingency Events): Are uncertain events with respect to both a moment of their occurrence and to an obscure feature of the reason causing the situation.

Chapter 13
Computer Modelling of Autonomous Satellite Navigation Characteristics on Geostationary Orbit

Fedir Shyshkov
National Aviation University, Ukraine

Valeriy Konin
National Aviation University, Ukraine

EXECUTIVE SUMMARY

Satellite systems are a fast-developing and broad field of study. The use of global navigation satellite systems for relatively autonomous spacecraft navigation holds a lot of interest for researchers. It is extremely expensive to research space applications as live experiments. Therefore, computer modelling comes in handy when there is a need to analyze important factors in space environment. The chapter describes the radionavigation field model that uses the off-nadir satellites. This model allows estimation of the availability and accuracy characteristics of autonomous satellite navigation in space up to the geostationary orbit in order to provide the necessary research data.

INTRODUCTION

Global navigation satellite system (GNSS) is an infrastructure of different satellite constellations and augmentations that is used to provide accurate position and time information worldwide. This is a simplest and cheapest global method of navigation available to a wide range of users. There are currently only two fully deployed satellite constellations. They are Global Positioning System (GPS) and Global Navigation Satellite System (GLONASS) that have the capability to provide the service worldwide. They have found many different uses in the modern world.

DOI: 10.4018/978-1-5225-7588-7.ch013

The market of GNSS is a constantly increasing sphere that encompasses a variety of different applications. According to the GNSS market report (European GNSS Agency[GSA], 2017), GNSS is widely used in smartphones and other small devices intended for personal use; calculation the location of vehicles, including unmanned, on the ground, in the sea or in the air; such accurate fields as agriculture, surveying and timing. In 2017 there were 5.8 bln GNSS devices in use, expecting to increase this number to 8 bln before 2020. Even though this market is mostly dominated by smartphones, billions of people in the professional market segments still benefit from it. This broad market requires many specialists in the sphere and it is necessary to attract the attention of workers, students, and researchers to it.

There still appear new possible uses for GNSS as the market is in the phase of development. One of such uses was somewhat unexpected for most of the researchers and deals with the satellite navigation in space. The wide use of unmanned spacecraft requires a cheap and reliable way of navigation as the radio link communication with the ground is somewhat lacking. The chapter proposes a model to study the autonomous satellite navigation in space and geostationary orbit specifically using computer modeling.

BACKGROUND

The idea to extend the range of satellite navigation from the surface up to the near-Earth space has appeared somewhere in mid-20th. This mostly were attempts to formalize the service volume in space and to produce a viable solution for the lack of satellite signals in space above Earth. The researchers produced different ideas like the use of back-lobe signals of the navigation satellites' antennas radiation patterns, signals passing through main-lobes and side-lobes of the antennas radiation patterns of the off-nadir satellites and surface-based "pseudo"-satellites. After the 1997 and AMSAT-OSCAR-40 (Moreau et al., 2002) launch, the opinions shifted towards the signals that are transmitted through main-lobes and side-lobes of the navigation satellites' antennas from behind Earth, as it was proven that a sensitive receiver can receive the weak GPS signals.

According to (US Department of Defense [DOD], 2008a, 2008b, 2008c), the near-Earth space is divided into terrestrial service volume (TSV) and space service volume (SSV). The TSV covers the near-Earth space from the surface and up to 3000 km for GPS, while the SSV covers the volume from TSV up to geosynchronous orbit altitude that is approximately 36000 km. For visual information see Figure 1.

The SSV is divided into medium and High Earth orbit altitudes including the geosynchronous orbits. The navigation satellites are orbiting at about 20000 km, so how can they be available at 36000 km? The traditional navigation is still partially possible in the medium altitudes, but, with the increase of altitude, it becomes evident that the number of available satellites is not enough to find the user's location. Therefore, the need for a new source of signal arises. The signals from the off-nadir satellites have become such a source, they are shown in Figure 2.

Figure 2 shows the antenna radiation pattern of an off-nadir satellite. The Earth is plotted in the center of the picture as a circle, the satellite is plotted on the left. The satellite's antenna radiation pattern is divided into the main and side lobes. Even though side-lobe signals are usually harmful, they have become a way to increase the number of available satellite signals in space. Signals passing through the main petals are partially shaded by the Earth, but some of the signals passing through the part of the main-lobe diagram can still be received. The signals that are passing through the side-lobes are much weaker than the signals passing through the main lobe of the radiation pattern, but they can still be received in the geostationary orbit by utilizing a sensitive receiver. Moreover, the spacecraft(SC) requires

Figure 1. The near-Earth space as defined by (DOD, 2008)

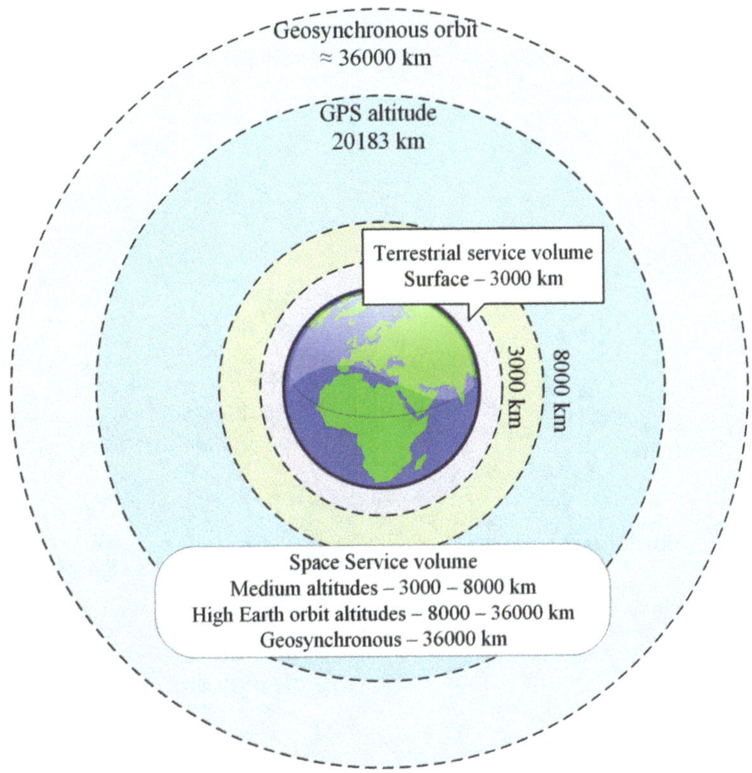

an antenna turned in the direction of the off-nadir satellites to receive the signals. In this chapter, the SC is considered to be equipped with an omnidirectional receiving antenna that allows receiving weak signals.

Many details about SSV and side-lobes are covered in (Bauer, 2015, Parker, 2017). The peculiarities of SSV for other satellite navigation constellations can be found in (European Space Agency [ESA], 2012, Kogure & Kishimoto, 2012, Zhan & Jing, 2014).

COMPUTER MODELLING OF AUTONOMOUS SATELLITE NAVIGATION

Forming Radio Navigation Field Model

The determination of user navigation characteristics using the satellite constellations in near-Earth space is carried out in the so-called radio navigation field (RNF) that is formed by radio signals of navigation satellites. If only one satellite constellation is used, it is necessary that the user simultaneously receives signals from at least four satellites to determine the coordinates in a three-dimensional space as there are four variable parameters. The fourth required parameter is the user clock error, as satellite clocks are much more accurate than the ones usually installed in the receiving equipment. In the case of multiple constellations, each one requires a variable for clock corrections, which in turn also increases the minimum number of satellites required (MNSR) to use several constellations in one solution. The MNSR can be calculated by a simple formula:

Figure 2. The main-lobe and side-lobe signals of the off-nadir satellites

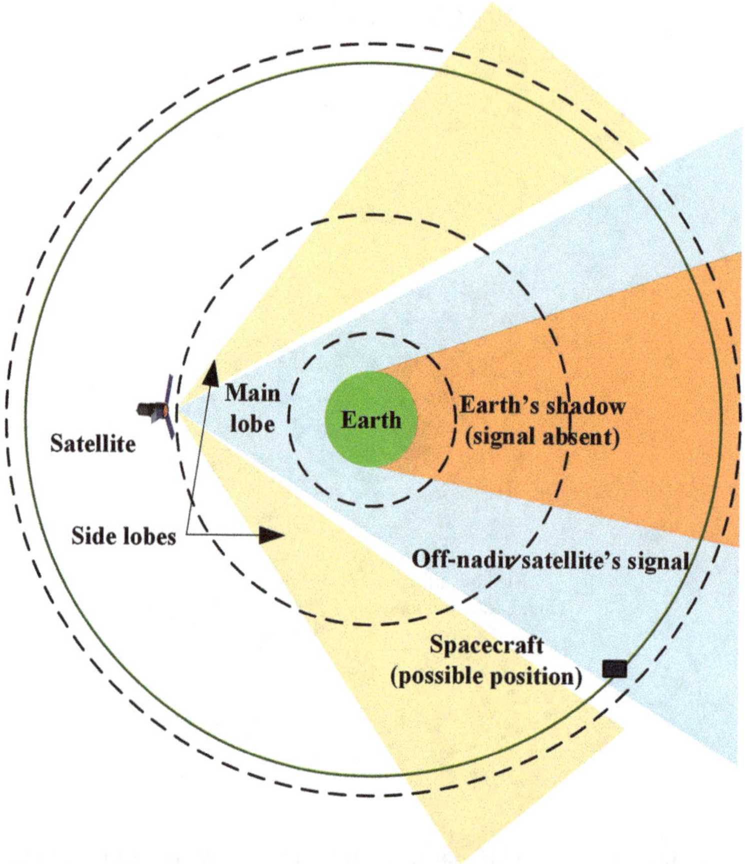

MNSR = 3 + (number of satellite constellations used)

For example, at least five satellites are required if two satellite constellations are used or at least seven satellites in the case of four constellations.

The availability of the required number of satellites is a necessary, but not sufficient condition. It is especially evident in the SSV. Their geometrical position also has a great impact on the resulting solution. This condition can be estimated with the use of dilution of precision (DOP) factors.

The RNF can be considered stable or unstable depending on the ability to receive the minimum required number of signals from the navigation satellites in any given moment of time. When it is theoretically possible to receive the minimum number of signals required (four in case of a single constellation) without any external factor influence (obstacles, jamming, etc. in case of Earth's surface) the RNF is considered stable. The (DOD, 2008) defines the next limits for a stable RNF: up to 3000 km above Earth's surface for GPS, essentially these are the limits of the TSV. In the case of SSV, the sufficient number of satellites is less probable the higher the user ascends, therefore the RNF in this volume is considered unstable.

While the radio navigation field model works pretty well for TSV it requires several considerations for SSV. In order to estimate the accuracy and availability of the navigation solution, several important components should be present.

Estimating Navigation Satellite Orbit

The motion of navigational satellites in Earth's orbit follows the laws of celestial mechanics, as they do not have much difference from other artificial objects in the near-Earth space. The force of Earth's gravity, as the main force, and the gravitational potentials of other objects of the solar system, form the satellites' orbits. However, the total magnitude of these potentials is much smaller than Earth's influence. Therefore, for certain applications, orbit calculation is simplified to the problem of undisturbed motion, which only takes into account the effect of Earth's gravity, and other factors are discarded. Such orbits are called Kepler orbits. This is an idealized orbit variant used to calculate the trajectory of the satellite, since the real trajectory over time, under the influence of disturbing factors, deviates from the pre-calculated line and needs to be corrected. The Kepler model can be used for insignificant time intervals, after which it is necessary to correct the orbital parameters.

The following Kepler parameters have to be known to define the orbit of the satellite:

1. **Semimajor Axis (a):** This parameter is a half of the major axis, and thus runs from the center to the perimeter through the focus of the ellipse that represents orbit.
2. **Eccentricity (e):** This parameter determines the amount by which the orbit of the navigation satellite deviates from a perfect circle
3. **Inclination (i):** This parameter measures the tilt of the satellite's orbit around the Earth, in simpler words it defines the angle between the orbital plane and the reference plane.
4. **Longitude of the Ascending Node (Ω):** This parameter is the angle from a reference direction, to the direction of the ascending node. For geocentric orbits, Earth's equatorial plane is taken as the reference plane, and the First Point of Aries is taken as the origin of longitude.
5. **Argument of Periapsis (ω):** This parameter defines the angle between the ascending node and the periapsis. It is measured in the direction of motion.
6. **True Anomaly (ν):** This parameter defines the position of the orbiting body along the trajectory, measured from periapsis.

The model uses an almanac (Kelso, 2014) to estimate the satellites' orbit and ephemerids. The algorithm to decode the almanac and determine the ephemerids of the signal is commonly known and is basically given in the technical documentation (GPS Directorate, 2013). This element is valid for both TSV and SSV. Before the calculation of exact ephemerids, it is vital to know the exact moment of time when the signal from the GNSS satellite was sent.

Determining Time

GPS Time (GPST) is determined by the GPS control segment on the basis of atomic clocks on board of the satellites and in the monitoring stations. The GPST was synchronized to coordinated universal time (UTC) at midnight of January 5th to 6th in 1980. The time scale is continuous and does not have leap seconds. The largest element of the countdown is a week that is 604 800 seconds. Since the navigation system has a continuous time scale, the time of the GPST may differ from the UTC. In 2015 this difference was 15 seconds, now it has increased even more. The actual difference between scales should not exceed a millisecond but is usually within microseconds. Such precision is controlled by the GPS control segment itself and is given in the form of navigation data to properly correlate GPST and UTC.

Knowing GPST is necessary to determine the actual position of the navigation satellite in orbit and its ephemerids.

Defining the Antenna Radiation Pattern Angle

The antenna radiation pattern (ARP) angle is necessary to determine if the satellite signal can or cannot be received in orbit. This is an angle formed by two straight lines, the first line goes between the center of mass of the satellite and Earth's center of mass, the second line goes between the satellite center of mass and the spacecraft's center of mass. Figure 3 illustrates this angle.

Antenna Radiation Pattern

Originally GNSS was designed exclusively for the terrestrial use, only partially covering the near-Earth territory, therefore their main-lobe signals are focused on Earth and attempt to cover it. In order to assess the possibilities of receiving signals from navigational satellites in space, it is first necessary to know their ARP. The satellite's antennas array contains 12 radiating elements, which are located on two concentric circles and form two sub arrays (Konin & Harchenko, 2010). Eight radiating elements are located on the outer circle with a radius of 438.82 mm (outer sub array). The other four are located on the inner circle of radius 162.4 mm (inner sub array). The radiating element is a spiral antenna. The sub-arrays are excited by phase-wise electromagnetic waves that are formed by systems of power distribution. An internal four-element sub array receives approximately 90% of the power supplied to the antenna as a whole.

Simplified calculation of the antenna array pattern can be accomplished using the following relationships:

$$\left.\begin{aligned} U_{\theta}(\theta,\varphi) &= U_{M}(\theta,\varphi)U_{\theta u}(\theta,\varphi) \\ U_{\varphi}(\theta,\varphi) &= U_{M}(\theta,\varphi)U_{\varphi u}(\theta,\varphi) \end{aligned}\right\},$$

Figure 3. ARP angle of the satellite

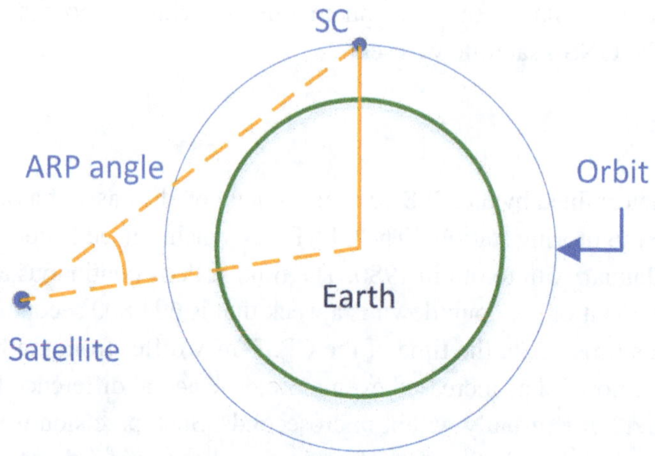

where $U_{\varphi}(\theta,\varphi)$, $U_{\theta}(\theta,\varphi)$, $U_{\varphi u}(\theta,\varphi)$, $U_{\theta u}(\theta,\varphi)$ are the antenna and emitter radiation patterns in the planes θ and φ, respectively; $U_M(\theta,\varphi)$ is the multiplier of the antenna array; θ and φ are the angles of the spherical coordinate system: the antenna array aperture lies in the plane X, axis Y and the angle φ are deduced from the X-axis, the Z-axis is directed to the Earth's side from the normal to the aperture, and the angle θ is deduced from the Z-axis, the center of the coordinate system coincides with the center concentric circles forming an antenna.

The antenna array multiplier is calculated as follows:

$$U_M(\theta,\varphi) = \sum_{i=1}^{12} U_i \exp\left[jk\left(x_i sin\theta \cdot cos\varphi + y_i sin\theta \cdot sin\varphi + z_i cos\theta\right)\right],$$

where: U_i is the amplitude of the electromagnetic wave that excites the i-th emitter; x_i, y_i, z_i are the coordinates of the phase center of the radiator.

Figure 4 shows a typical antenna radiation pattern for a GPS satellite for both L1 and L2 frequencies.

Angle Θ is plotted on the horizontal axis, the vertical axis contains the directivity of the antenna in decibels. The Earth's part of the satellite service is located in the range from – 13.8° to +13.8° of the ARP, this range also includes the atmosphere. Satellite signals in the range of ± 13.8 ° to ± 23.5 ° and from ± 13.8 ° to ± 26 ° are the main-lobe parts of the ARP at frequencies L1 and L2 / L5 respectively that are intended for space use. The signals from ± 30 ° to ± 60 ° are considered to be side-lobe while using both frequencies. The published data in (Marquis, 2014) was used to build the figure.

Figure 4. Typical ARP of a GPS navigation satellite for L1 and L2 frequencies for $\varphi = 0$

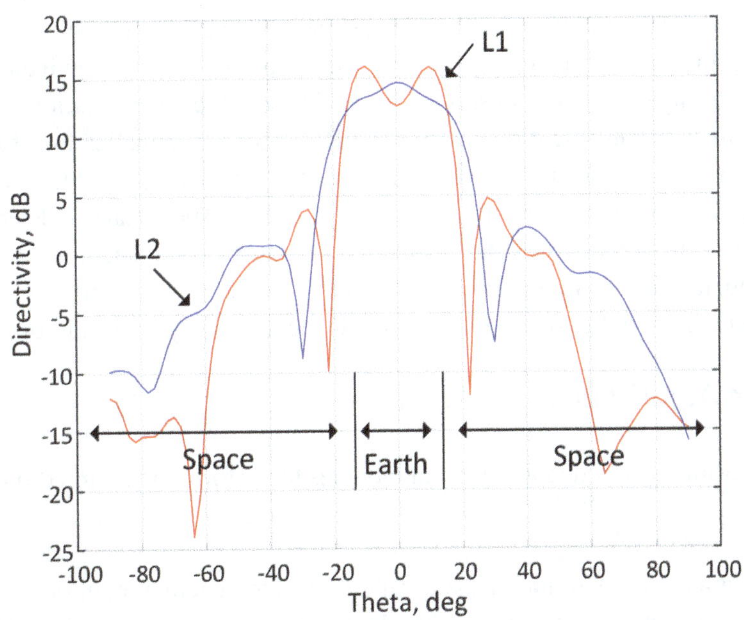

The signals on different frequencies have different radiation pattern and therefore the SSV can benefit from their separate use (Konin, Shyshkov & Pogurelskiy, 2017). With an angle Θ of about 40° in Figure 4, the side lobe directivity is high for the frequency L2 and low for the frequency L1. The difference is about 8 dB. In this case, it is most likely that the navigation receiver will not be able to receive the signal at the frequency L1, but will be able to accept it at a different frequency. That is, in many cases, even in the absence of a signal at the frequency L1, it can still be received at the frequency L2, and vice versa. To receive the signals the Navigator receiver can be used (Wintemitz et al., 2004, Keesey, 2010, National Aeronautics and Space Administration [NASA], 2016).

Estimating the Radio Navigation Field

Assume that the coordinates of the area, where the SC is located, are known. The coordinates of the SC in general form are calculated using the pseudorange values of the visible navigational satellites. Visible navigational satellites may belong to different navigation constellations.

Under the geometric visibility of navigational satellites, it is understood the theoretical possibility to receive a signal from a navigational satellite using a receiver on a spacecraft, which is due to the features of the ARP of the satellite and the shading from the Earth in the case of the off-nadir satellites.

Similar to the standard TSV, the SSV requires at least four satellites for real-time estimation of SC coordinates for a single satellite navigation constellation.

BUILDING THE ALGORITHM

How the Algorithm Works

The algorithm designed based on the given model takes a certain number of input parameters. These parameters include: the date and time in user-recognizable format; the GNSS almanac; user-defined coordinates in geographical coordinate system; the modelling time interval that includes the number of steps and a value of a single step; the constraints for visible satellite determination; statistical values of the distribution of the user equivalent range error (UERE); number of realizations for each moment of time; chosen satellite constellations; different constants. The next step is to extract data from almanac and to find the range information for the navigation satellites that is necessary to determine which satellites are visible. The next step involves the calculation of statistical information and DOP factors. Essentially, in this step, the algorithm recalculates the coordinates of the SC while including the UERE, a pseudo-receiver work. The next step is to finally provide the text and graphical information for analysis.

Structure of the Algorithm

Structurally the algorithm is divided into several modules to perform the required functions. The modules are:

1. **Input Data Generation Module:** This module is used to generate input data, including both the configurations and values of constants, the file generated during the work of this module will be used in further modules.

2. **Almanac Analyze Module:** This module is used to read, decrypt and store the data of the almanac for further use, the target almanac file is defined during the work of the previous module.
3. **General Calculations Module:** This module is used to calculate the satellite ephemerids and to define different angle information, including the ARP angle. Finds and stores the visible satellites and their relative data.
    a. **Calculation of Time:** Converts time from the human-understandable date to the exact GPST second.
    b. **Converting User Coordinates**: Converts the defined geographical user coordinates to Earth-centered Earth-fixed system (ECEF) and East North Up system (ENU) for further use.
    c. **Calculation of Satellite Range Data**: Includes the calculation of satellite ephemerids, elevation, pseudoranges, ARP angle, and other related data.
    d. **Choosing Visible Satellites:** Parses range data to determine whether the satellite is visible from the user's position or not.
    e. **Store Visible Satellites Data:** Stores the data of the visible satellites for further use, including their source, meaning if they are normally seen satellite signals, off-nadir main-lobe signals or off-nadir side-lobe signals.
4. **Statistical Information Calculations Module:** This module is used to calculate all the statistical data and DOP factors.
    a. **Generation of Errors:** Generates the errors of pseudoranges for visible satellites for the defined number of realizations for each moment of time. The errors are assumed to have the normal distribution with the defined sigma value of UERE.
    b. **Recalculation of Pseudoranges:** Recalculates the 'ideal' pseudoranges to include the generated range errors to find the user's position.
    c. **Calculation of User Coordinates:** Uses the least squares method to find the coordinates of the user.
    d. **Coordinate Transformation**: Transforms the calculated coordinates to ENU for easier human recognition of errors.
    e. **Calculation of Errors:** Calculates the difference between the coordinates obtained in the previous step and the ideal coordinates that were user-defined before the program start.
    f. **Calculation of DOP Factors:** Calculates all available DOP factors, including horizontal, vertical, position, time and geometric.
    g. **Store the Data:** Stores the obtained data for further use.
5. **Graphical Output Generation Module:** This module calculates any missing data, especially availability, required for figures and provides the graphical representations for improved human understanding.

Performance Characteristics

Several performance characteristics are applicable to satellite navigation. According to the (International Civil Aviation Organization (ICAO), 2005), there are four main ones:

1. **Accuracy:** This parameter is defined as the difference between the estimated and actual aircraft position.
2. **Integrity:** This parameter is a measure of trust in the correctness of the system.

3. **Continuity:** The ability to work uninterrupted during the whole period of operation.
4. **Availability:** This parameter shows the portion of time when the other three characteristics simultaneously fit the specified operational requirements.

The algorithm calculates the values of accuracy and availability. Even though it is possible to estimate continuity with the model it was somewhat disregarded. Integrity is not applicable as there is no real equipment or receiver-based algorithms to ensure the correct work of the GNSS.

User Equivalent Range Error

In order to calculate the accuracy, the algorithm injects certain errors in the pseudorange. These errors are called user equivalent range errors and are assumed to have a normal distribution. According to (DOD, 2008a) these errors can be calculated in the following way:

$$\text{UERE} = \sqrt{\Delta_{sys}^2 + \Delta_{user}^2},$$

where Δ_{sys} is the system component of the error, Δ_{user} is the user component of the error.

The system component is a square root of the sum of squares of the errors of satellite constellation's system segment. It can be calculated in the following way:

$$\Delta_{sys} = \sqrt{\sum_{k=0}^{n} \Delta_k^2}.$$

The errors of system segment are different depending on how long ago the almanac data was issued.

The user component is assumed to be 1m, but it might be better as SSV receivers tend to have better accuracy than their cheap TSV counterparts.

Three different values of UERE were defined in such a way depending on how fresh is the almanac data:

1. **Fresh UERE:** Has the value of 3.24 m for fresh almanac data.
2. **Standard UERE:** Has the value of 4.12 m, issued for in-between time.
3. **Old UERE:** Has the value of 6.63 m for old but usable almanac data.

These values are for non-augmented GPS-receiver. It is possible to improve the UERE by, for example, sending the corrections to satellite ephemerids via radio-link from the ground.

Forming a System of Equations

As it was mentioned in the part about radio navigation field, the spacecraft is required to receive the signals from at least four satellites in the case of a single navigation constellation. The satellites have known ephemerids and pseudorange towards the user. Pseudoranges are measured between the phase centers of the transmitting satellite antenna of the navigation satellite and the receiving antenna of the spacecraft. Figure 5 depicts a situation where four off-nadir satellites are used for the purpose of navigation.

Figure 5. Visible satellites from the user point of view

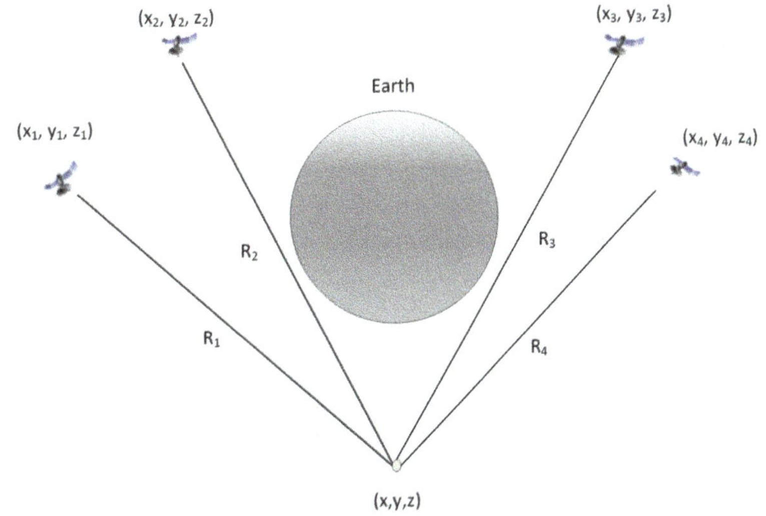

In the figure, R_i is the pseudorange to the i-th satellite, (x; y; z) are the coordinates of the user, (x_i; y_i; z_i) are the ephemerids of the i-th satellite.

In the real system, the pseudorange is an estimated value and is measured as the product of the speed of electromagnetic waves propagation and the time during which the satellite signal reaches the spacecraft that is the user of GNSS in this chapter. This time is measured in the user's equipment. The next equation shows this relationship:

$$R_i = c * \Delta t_i,$$

where R_i is the pseudorange to the i-th satellite, Δt_i is the time of the signal passing between the phase centers of the antennas of the i-th navigation satellite and the spacecraft, c = 299 792 458 m / s, the velocity of the electromagnetic waves propagation in space.

The value of the pseudorange can also be calculated using the coordinates of the i-th satellite (x_i; y_i; z_i) and the user (x; y; z):

$$R_i = \sqrt{\left(x - x_i\right)^2 + \left(y - y_i\right)^2 + \left(z - z_i\right)^2}.$$

It is clear that the above equation contains three unknowns, and therefore, it is necessary to have at least three equations in the system to determine the coordinates of the user. Since the time scales of the user and time scales of navigation satellite constellation are not synchronized, the error resulting from the difference must be taken into account when determining the pseudorange. Given that pseudoranges should be calculated exactly at the same moment of time as well as synchronization of satellite scales between the satellites in a separate satellite constellation, the difference between user scales and navigation satellite can be considered constant, but unknown.

Therefore, the equation above transforms in the following form when the time scale difference $\Delta\tau$ is taken into account:

$$R_i = \sqrt{(x - x_i)^2 + (y - y_i)^2 + (z - z_i)^2} + c\Delta\tau .$$

However, the difference in time scales between the navigation satellite and the user is not the only source of errors, so in general, the above equation can be written as follows:

$$R_i = \sqrt{(x - x_i)^2 + (y - y_i)^2 + (z - z_i)^2} + c\Delta\tau + \Delta_i ,$$

where Δ_i is the error of the i-th satellite pseudorange estimation due to the errors of determining the ephemeris, frequency-time support, the velocity of radio waves propagation in the troposphere and the ionosphere in the case of TSV, noise in the reception area, noise of the reception channel of the user equipment, and also solar radiation, etc. Essentially this is EURE.

Therefore, if only four satellites are present the equation above transforms in a following system of equations:

$$\begin{cases} R_1 = \sqrt{(x - x_1)^2 + (y - y_1)^2 + (z - z_1)^2} + c\Delta\tau + \Delta_1 \\ R_2 = \sqrt{(x - x_2)^2 + (y - y_2)^2 + (z - z_2)^2} + c\Delta\tau + \Delta_2 \\ R_3 = \sqrt{(x - x_3)^2 + (y - y_3)^2 + (z - z_3)^2} + c\Delta\tau + \Delta_3 \\ R_4 = \sqrt{(x - x_4)^2 + (y - y_4)^2 + (z - z_4)^2} + c\Delta\tau + \Delta_4 \end{cases}$$

In the case of additional satellite constellations, a different $\Delta\tau$ should be used for the given system and the minimal number of satellites required should be increased.

For the modeling purpose, Δ_i is generated based on the normal distribution, the expectation is considered to be zero, the standard deviation is chosen depending on how fresh are the almanac data. The three values for fresh, standard or old UERE are given above.

$$R_{measured} = R_{ideal} + \Delta = R_{ideal} + normrnd(\mu, \sigma)$$

Then a system of equations is formed with the new values of $R_{measured}$. There are usually about 400 realizations for each satellite for a given moment of time. The system of equations is solved using the least squares method.

Solving the System of Equations

The calculation of the user's coordinates with the use of multiple GNSS constellations in the ECEF system is performed using the next formula:

$$[\delta\ X] = \left\{[G]^T \cdot [G]\right\}^{-1} \cdot [G]^T \cdot [\delta\ PR]$$

where

$$[\delta\ X] = [X^{(m+1)}] - [X^{(m)}] = \begin{bmatrix}\delta x & \delta y & \delta z & \delta\ h_{s1} & \delta\ h_{s2} & \dots & \delta\ h_{sn}\end{bmatrix}^T$$

is the increment of the user coordinates on the latest iteration of the calculation,

$$[\delta\,\mathrm{Pr}_i] = [R^{(m+1)}] - [R^{(m)}]$$

is the increment of pseudorange on the latest iteration, $i = 1, \dots, n$ are the numbers of satellites used in the solution.

While using multiple GNSS systems the following matrix of observations can be formed:

$$[G] = \begin{pmatrix} \frac{-(x_{s1}-x)}{R_{s1}^{meas.}} & \frac{-(y_{s1}-y)}{R_{s1}^{meas.}} & \frac{-(z_{s1}-z)}{R_{s1}^{meas.}} & 1_{s1} & 0_{s1} & \dots & 0_{s1} \\ (.)_{s1} & (..)_{s1} & (\dots)_{s1} & 1_{s1} & 0_{s1} & \dots & 0_{s1} \\ \frac{-(x_{s2}-x)}{R_{s2}^{meas.}} & \frac{-(y_{s2}-y)}{R_{s2}^{meas.}} & \frac{-(z_{s2}-z)}{R_{s2}^{meas.}} & 0_{s2} & 1_{s2} & \dots & 0_{s2} \\ (.)_{s2} & (..)_{s2} & (\dots)_{s2} & 0_{s2} & 1_{s2} & \dots & 0_{s2} \\ \dots & \dots & \dots & \dots & \dots & \dots & \dots \\ \frac{-(x_{sn}-x)}{R_{sn}^{meas.}} & \frac{-(y_{sn}-y)}{R_{sn}^{meas.}} & \frac{-(z_{sn}-z)}{R_{sn}^{meas.}} & 0_{sn} & 0_{sn} & \dots & 1_{sn} \end{pmatrix},$$

where s1, s2, …, sn show the satellite constellation of the given satellite.

To solve the navigation problem in the ENU coordinate system, you must switch from the ECEF to the ENU coordinate system. Both of them are Cartesian and the conversion procedure is written in the following way:

$$\begin{bmatrix}\Delta x_{ENU} \\ \Delta y_{ENU} \\ \Delta z_{ENU}\end{bmatrix} = \begin{bmatrix} -\sin\lambda & \cos\lambda & 0 \\ -\sin\varphi\cos\lambda & -\sin\varphi\sin\lambda & \cos\varphi \\ \cos\varphi\cos\lambda & \cos\varphi\sin\lambda & \sin\varphi \end{bmatrix} \cdot \begin{bmatrix}\delta x \\ \delta y \\ \delta z\end{bmatrix}_{ECEF}$$

However, the ECEF coordinates contain components that are related to the divergence of the time scales of different systems:

$$[\delta T]_{ECEF} = \begin{bmatrix} \delta h_{s1} \\ \delta h_{s2} \\ \dots \\ \delta h_{sn} \end{bmatrix}$$

In contrast to the transformation of coordinates, time components do not change when converted from ECEF to ENU and the relationship between them is established by the relationship:

$$[\delta T_{ENU}] = [E] \cdot [\delta T]_{ECEF}$$

or:

$$\begin{bmatrix} \delta h_{s1} \\ \delta h_{s2} \\ \dots \\ \delta h_{sn} \end{bmatrix}_{ENU} = \begin{bmatrix} 1 & 0 & 0 & 0 \\ 0 & 1 & 0 & 0 \\ 0 & 0 & \dots & 0 \\ 0 & 0 & 0 & 1 \end{bmatrix} \cdot \begin{bmatrix} \delta h_{s1} \\ \delta h_{s2} \\ \dots \\ \delta h_{sn} \end{bmatrix},$$

where *[E]* is the matrix of ones.

When combining the matrices the next expression is obtained:

$$\begin{bmatrix} \Delta x_{ENU} \\ \Delta y_{ENU} \\ \Delta z_{ENU} \\ \delta h_{s1} \\ \delta h_{s2} \\ \dots \\ \delta h_{sn} \end{bmatrix} = [S] \cdot \begin{bmatrix} \delta x \\ \delta y \\ \delta z \\ \delta h_{s1} \\ \delta h_{s2} \\ \dots \\ \delta h_{sn} \end{bmatrix},$$

where

$$[S]=\begin{bmatrix} -\sin\lambda & \cos\lambda & 0 & 0 & 0 & 0 & 0 \\ -\sin\varphi\cos\lambda & -\sin\varphi\sin\lambda & \cos\varphi & 0 & 0 & 0 & 0 \\ \cos\varphi\cos\lambda & \cos\varphi\sin\lambda & \sin\varphi & 0 & 0 & 0 & 0 \\ 0 & 0 & 0 & 1 & 0 & 0 & 0 \\ 0 & 0 & 0 & 0 & 1 & 0 & 0 \\ \dots & \dots & \dots & \dots & \dots & \dots & \dots \\ 0 & 0 & 0 & 0 & 0 & 0 & 1 \end{bmatrix}$$

The matrix [S] is unitary and has the next expression valid for it:

$$[S]\cdot[S]^T=[S]^T\cdot[S]=[E],$$
$$[S]^T=[S]^{-1}.$$

In view of the above, transform the coordinates from the ECEF system into the ENU:

$$[S]\cdot[' X]=[S]\cdot\{[G]^T\cdot[G]\}^{-1}\cdot[G]^T\cdot[' R].$$

When transforming the right part:

$$\begin{aligned}[S]\cdot[\delta X]&=\left\{[G]^T\cdot[G]\cdot[S]^{-1}\right\}^{-1}\cdot[S]^{-1}\cdot[S]\cdot[G]^T\cdot[\delta PR] \\ &=\left\{[S]\cdot[G]^T\cdot[G]\cdot[S]^T\right\}^{-1}\cdot[S]\cdot[G]^T\cdot[\delta PR]\end{aligned}.$$

When defining $[S]\cdot[G]^T=[H]^T$, $[G]\cdot[S]^T=[H]$, the next expression is obtained for coordinates in ENU:

$$\begin{bmatrix}\Delta x_{ENU} \\ \Delta y_{ENU} \\ \Delta z_{ENU} \\ \delta h_{s1} \\ \delta h_{s2} \\ \dots \\ \delta h_{sn}\end{bmatrix}=\{[H]^T\cdot[H]\}^{-1}\cdot[H]^T\cdot[\delta PR].$$

Dilution of Precision

Using the diagonal elements of the matrix:

$$[G^T G]^{-1} = \begin{bmatrix} g_{11} & \cdots & \cdots & \cdots \\ \cdots & g_{22} & \cdots & \cdots \\ \cdots & \cdots & g_{33} & \cdots \\ \cdots & \cdots & \cdots & g_{44} \end{bmatrix},$$

where T defines the transposed matrix.

Using the elements from the matrix above, the dilution of precision factors can be calculated:

$$GDOP = \sqrt{g_{11}^2 + g_{22}^2 + g_{33}^2 + g_{44}^2},$$
$$PDOP = \sqrt{g_{11}^2 + g_{22}^2 + g_{33}^2},$$
$$HDOP = \sqrt{g_{11}^2 + g_{22}^2}$$
$$VDOP = \sqrt{g_{33}^2},$$
$$TDOP = \sqrt{g_{44}^2}.$$

The GDOP stands for geometric dilution of precision and is the most general among the factors mentioned, it includes both the position and time dilution of precision. The PDOP stands for position dilution of precision and deals only with position components in 3-dimensional space. HDOP and VDOP are the horizontal and vertical dilution of precision, respectively. It is quite hard to distinguish the horizontal and vertical planes in space for the actual spacecraft so this is more common in the TSV.

The interesting thing about the DOP components is that they are only dependent on satellite geometry, the mutual position of the visible satellite and the user. The effect of DOP is constant for the given geometrical setup and does not depend on the UERE. According to (DOD, 2000a, 2000b, 2008a), the following formula can be used for approximate estimation of accuracy:

$$Accuracy = UERE \times DOP.$$

In this case, accuracy is measured as RMS. This clearly illustrates the value of DOP in determining accuracy. Some of the errors may require the specific DOP, for example, HDOP is required to estimate the User Horizontal Navigation Error (UHNE).

In the TSV the given values of DOP are usually somewhat small. The DOPs are considered good if they are below 1, average when they are up to 5 and somewhat adequate if they do not exceed 20.

Estimation of Availability and Accuracy

In order to evaluate the characteristics of autonomous navigation in the geostationary orbit, it is necessary to consider such characteristics as availability and accuracy. According to (DOD, 2017), the autonomous navigation satellite navigation for a spacecraft in the SSV has the accuracy requirements of 100 m for a single sigma in three-dimensional space.

Accuracy is determined in the ENU coordinate system (Diggelen, 2007). Assume that ΔE_i, ΔN_i, ΔU_i are the estimated errors of the East, North and Up components for the i-th realization, and n is the total number of realizations. In the case when user navigation error (UNE) is required. Essentially this is a root-mean-square error. It can be calculated as a square root of the average of the squared error for all three coordinates. It is calculated in the following way:

$$UNE = \sqrt{\frac{1}{n} * \sum_{i=1}^{n} \left(\Delta E_i^2 + \Delta N_i^2 + \Delta U_i^2 \right)}.$$

Even though the horizontal and vertical planes in space are quite ambiguous, the model still provides the related accuracy data. The root-mean-square horizontal error (RMSH) is calculated in a next way:

$$RMSH = \sqrt{\frac{1}{n} * \sum_{i=1}^{n} \left(\Delta E_i^2 + \Delta N_i^2 \right)}.$$

The root-mean-square vertical error (RMSV) is calculated in a next way:

$$RMSV = \sqrt{\frac{1}{n} * \sum_{i=1}^{n} \left(\Delta U_i^2 \right)}.$$

The algorithm also calculates the percentile 95% error. It means that x% of the positions calculated have an error lower or equal to the accuracy value. For example, if there is an accuracy of 5m (95%), then it means that in 95% of the time the positioning error will not exceed 5m. Calculated separately for each coordinate.

To estimate the availability several types of operation are defined:

1. **Movement in Space Typical Operation (TO1):** The general movement in space which is limited by UNE of 556 m.
2. **Operation with Autonomous Navigation (TO2):** The value of UNE for this type of operation should be less than 100 m
3. **Approaching to Target (TO3):** The value of the UNE in East and North components should not exceed 40 m and 20 m for the Up component.

To calculate the availability for the given operation the next equation is used:

$$Av = \frac{\sum_{i=1}^{n} \begin{cases} 1, & UNE \le AL \\ 0, & UNE > AL \end{cases}}{n},$$

where AL is the alert limit value for the accuracy in accordance with the requirements for operations given above.

The Output Data

According to the structure of the algorithm, the output data are in the form of several data arrays. These are the results from the work of the algorithm, specifically the general calculations module and statistical data calculations module. The next data arrays are present in the algorithm:

1. **Visible Satellites Array:** The total number and actual satellites' pseudo random noise (PRN) numbers. The general calculations module generates it.
2. **Visible Satellites Range Data Array:** This data array provides the range information of the visible navigation satellites. The general calculations module generates it. It contains:
   a. **PRN Number:** The actual number of the visible satellite in order to identify it.
   b. **ARP Angle:** The stored information about the satellite ARP angle.
   c. **Satellite Coordinates:** Stores the satellite coordinates in the ECEF coordinate system.
   d. **Satellite Azimuth:** The azimuth angle of the navigation satellite.
   e. **Satellite Elevation:** The elevation of the satellite from the given point. Negative for off-nadir Satellites.
   f. **Satellite Source:** Defines whether the satellite signal is transmitted from the navigation satellite in direct view, an off-nadir main-lobe or side-lobe signal.
3. **Statistical User Data Array:** Statistical data array for assessing the accuracy of the provided navigation solution of the user. Statistical information calculations module generates this data. The array contains:
   a. **Calculated User Coordinates**: Contains the user coordinate data for each moment of time and for each realization that corresponds to it.
   b. **Coordinate Difference:** The difference between user-defined coordinates and the calculated ones.
   c. **ACC 95%:** The calculated error with the percentile 95% accuracy. It is calculated separately for each coordinate.
   d. **Root-Mean-Square Horizontal Error (RMSH):** The two-dimensional RMS error in the horizontal plane.
   e. **Root-Mean-Square Vertical Error (RMSV):** The RMS error in the vertical plane.
   f. **User Navigation Error (UNE):** The three-dimensional RMS error.
   g. **Dilution of Precision (DOP):** The multiple calculated DOP factors. Includes GDOP, PDOP, HDOP, VDOP and TDOP data arrays.
4. **Availability Data Array:** The array containing the calculated availability that is required to build the graphical representations.

The Graphical Representations

Graphical Output Generation Module is used to generate the graphical data for the better representation of research results. The actual graphs are generated from data in the MATLAB software (Konin & Shyshkov, 2016, Konin, Shyshkov & Pogurelskiy, 2016).

The produced output data allows plotting a wide range of graphs. The module is able to plot the satellites seen over certain duration (Figure 6), the satellite appearance frequency (Figure 7), the satellite sky plot to define the position of satellites with respect to their elevation from user's position (Figure 8), satellite position in space (Figure 9), error of position determination (Figure 10), a complex graphical model to provide the availability information (Figure 11, Figure 12) and other graphical data representations. Figures from 6 to 10 use only the signals passing through the main lobes of the satellites' antennas of the two satellite constellations, namely GPS and GLONASS. The data were generated for a time period of a day with a one-minute step with 400 realizations for a single point. Figures 11 and 12 are chosen to represent the actual difference in the availability that comes when using both the signals passing through the main-lobe and side-lobes of the antenna of a navigation satellite for a number of points on the geostationary orbit.

Figure 6 shows the exactly which satellites are seen in the given moment of time. The horizontal axis shows time in hours since the simulation start, the vertical axis shows the PRN numbers of the satellites. The strips show when the satellite was visible over a duration of time. It is possible to see if the multiple satellites are present in order to perform navigation. Satellites with PRN numbers up to 37 belong to the GPS, satellites with PRN numbers from 38 to 64 belong to GLONASS as per almanac format.

Figure 7 shows the frequency of when the certain number of satellites appears. As per the definition of MNSR, at least five satellites are required to obtain navigation solution with the use of two satellite constellations, but it is possible to have only a single constellation present with four satellites. Therefore, only about a third part of the observations is truly unavailable due to having fewer satellites than required.

Figure 6. Visible satellites over a period of a day

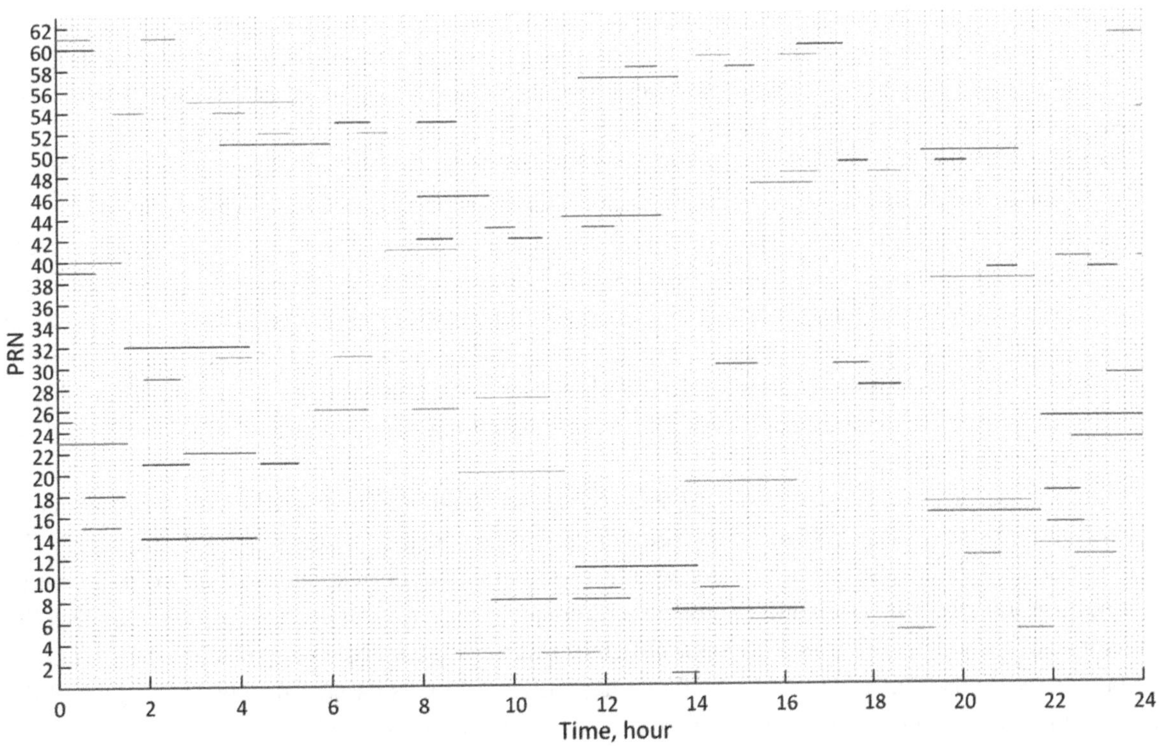

Figure 7. Satellite appearance frequency

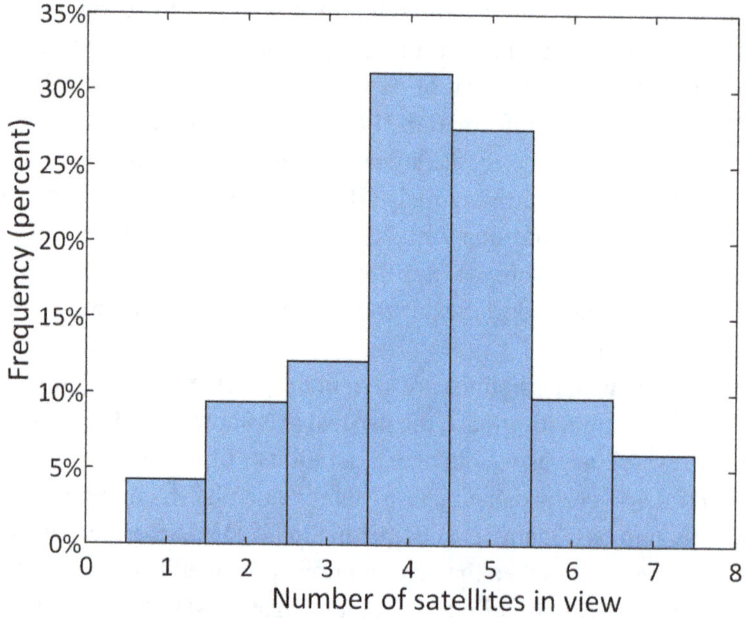

But even the excessive number of satellites cannot guarantee that a position solution can be obtained as it is heavily influenced by satellite geometry. The setups with the maximum number of satellites, namely 7, are available in 6% of total time.

Figure 8 describes the position of satellites relative to the user's position. As all of the satellites are off-nadir, their elevation is negative. The interesting fact, which is valid for the user on the geostationary orbit even when side-lobe signals are enabled, is that the navigation satellites are "collecting" in the center of the figure and all of their elevations are less than -55° and never achieve -90° as this part is overshadowed by Earth.

Figure 9 describes the satellite geometry for the best DOP factor achieved. In given case, the PDOP factor is equal to 39.6722 at the 11 hours 32 minutes since the simulation start. The circles in the left part of the figure show the positions of the navigation satellites with their respective PRN numbers, the circle in the right part with letter P above it shows the position of the user. The lines connect the satellites' centers of mass with Earth's center of mass and with the user's center of mass. There are seven navigation satellites present, among them, satellites with PRN numbers 3, 11, 8 and 9 belong to GPS, satellites 44, 43, 57 belong to GLONASS. In this case, user's position could have been determined with the GPS exclusively as there are enough satellites present, but not with the help of GLONASS, as there are only three satellites.

Figure 10 describes the distribution of UNE with respect to time. The horizontal axis holds time in hours from the start of the simulation, the vertical axis shows the user navigation error in meters. The minimum meaning the best, accuracy value was 133.8783 m, which corresponds to the time point with a minimum PDOP, the total availability of the measurement in space typical operation (TO1) was 23.47%, which means that only 23.47% had the values of UNE less than 556 m. No value matches the conditions for other operations. This is a typical situation when only the signals passing through main-lobe of the antenna radiation pattern are used, making such navigation extremely unreliable.

Figure 8. Sky plot of satellite configuration for best DOP

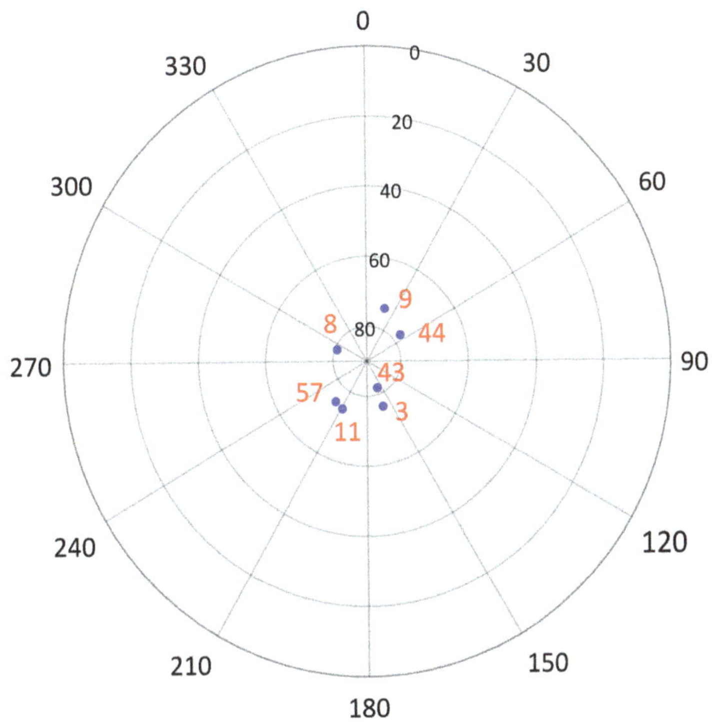

Figure 9. Satellite geometrical configuration for best DOP

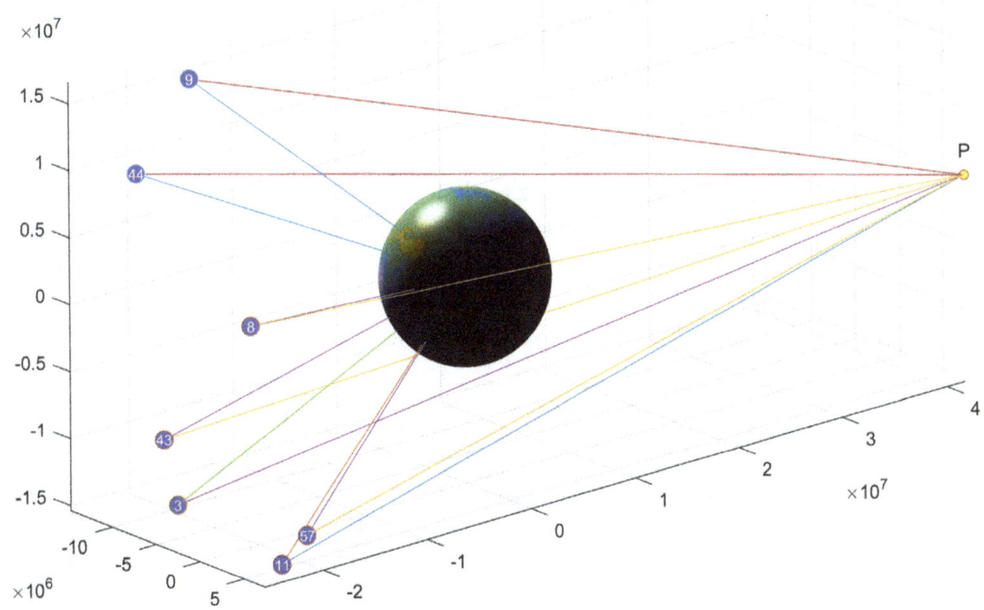

Figure 10. The UNE of user's position determination with the standard UERE

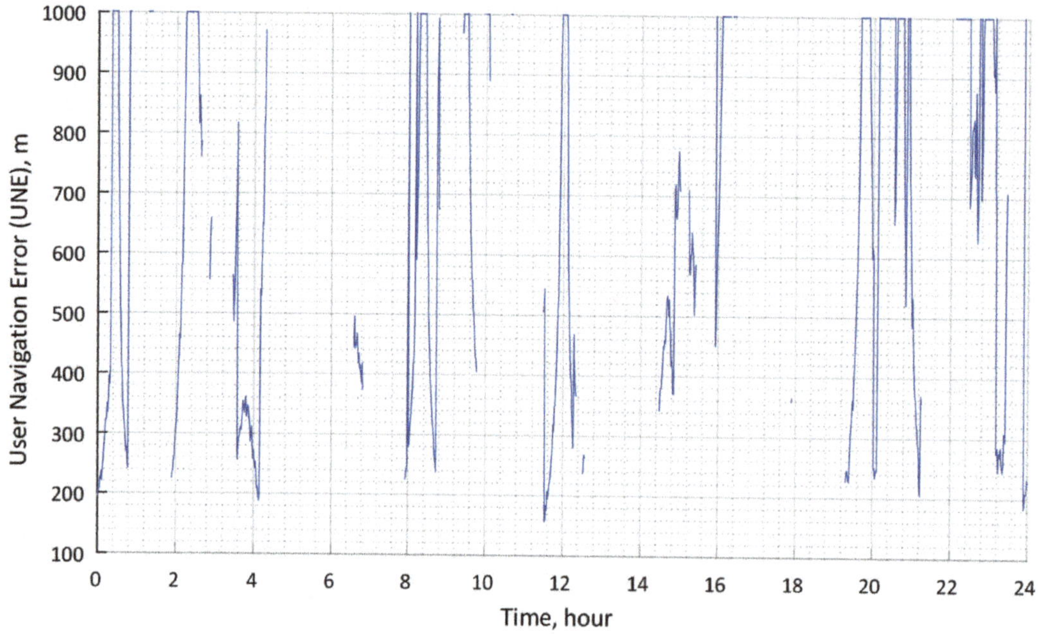

Figure 11. Complex graphical availability model for GPS and GLONASS with main-lobe signals only

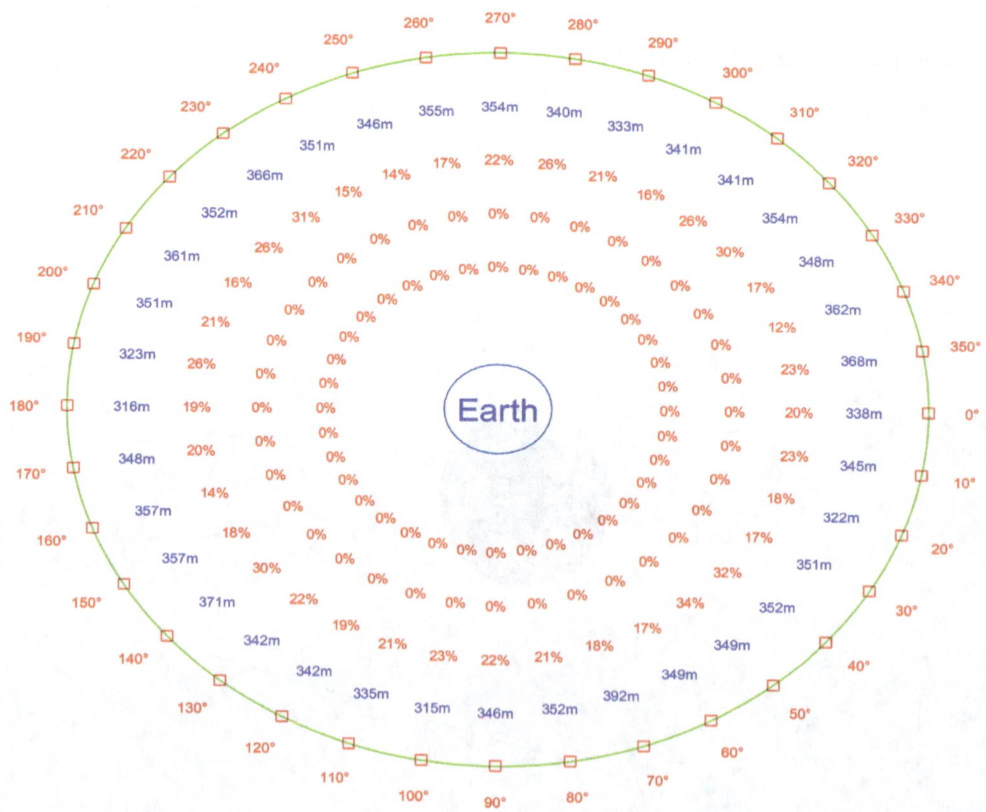

Figure 11 describes the availability of satellite signals for the typical operations mentioned in the chapter. This figure was generated for a 24-hour simulation with a 5-minute step for multiple points across the geostationary orbit. Only the signals passing through the main lobes of the antennas of GPS and GLONASS constellations were used. The values plotted outside of the circle defining the longitude of the user's possible position on the geostationary orbit. The circle with square points represents the geostationary orbit and respective user's position. The next circle shows the mean values of UNE for TO1 in meters. The three inner circles represent the availability for TO1, TO2, and TO3 respectively when going from outer edge to inner part of the figure. The values of availability of TO1 in different points range from 14% to 34%. The total availability for all of the points combined is 21.29%, which means only 2207 time moments are considered available among the total number of 10368. No values satisfy the TO2 or TO3 requirements. The figure clearly shows that it is impossible to satisfy the conditions of autonomous navigation with the use of navigation satellites when only signals passing through main-lobe are available.

Figure 12 describes the availability of satellite signals for the typical operations for GPS and GLONASS signals passing through main-lobes and side-lobes of the satellites' antennas radiation patterns. The figure's layout is identical to Figure 11, therefore it is not explained again. Almost 100% of the TO1 and TO2 time moments are available, 91% moments are available for TO3. The value of UNE ranges from 16.85 m to 44.92 m. Compared to the use of main-lobe signals only, this allows speaking

Figure 12. Complex graphical availability model for GPS and GLONASS with main-lobe and side-lobe signals

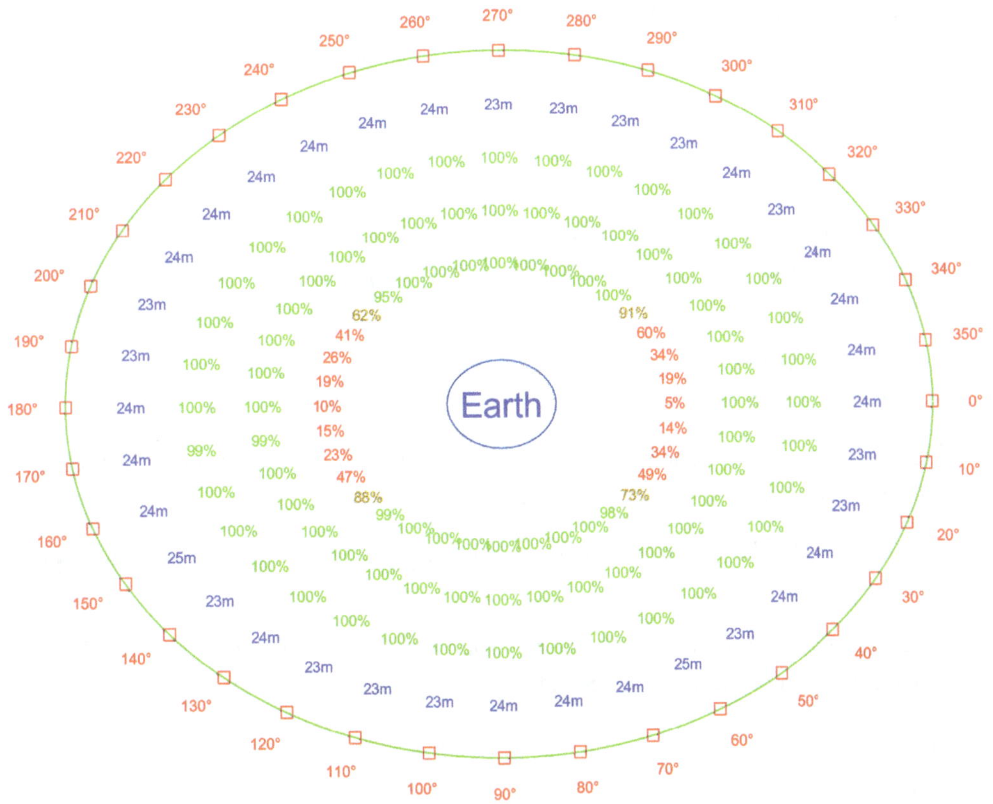

about stable autonomous satellite navigation. Therefore, the side-lobe signals are necessary to keep the radio navigation field stable.

FUTURE RESEARCH DIRECTIONS

The integration of a receiver autonomous integrity monitoring (RAIM) algorithm during the statistical part and estimation of integrity information. Another direction is the additional analysis of continuity characteristic. The next step of the research might involve the use of the GNSS-emulators in order to provide GNSS-signals, essentially moving from the use of almanac and calculated pseudoranges to working with actual signals and measured pseudoranges.

CONCLUSION

The radio navigation field model, which is described in the chapter, is used to generate the availability and accuracy of performance data. It is possible to use a variable number of satellite constellations for given position solution. The model uses the off-nadir satellites to provide the necessary amount of the visible satellites to determine the user's position. The model uses the navigation signals that are transmitted through the main-lobe and side-lobes of the satellites' antennas radiation patterns. The signals transmitted through the side-lobes hold the key spot in the current satellite navigation in space.

The model can be used both for the research and education purposes. It is a cheap solution to research, analyze and learn the important factors connected with the process of satellite navigation in the near-Earth space and an asset for striving future navigation specialists.

REFERENCES

Bauer, F. H. (2015). *GPS Space Service Volume (SSV) Ensuring Consistent Utility Across GPS Design Builds for Space Users*. Retrieved from https://www.gps.gov/governance/advisory/meetings/2015-06/bauer.pdf

Diggelen, F. (2007). *Update: GNSS Accuracy: Lies, Damn Lies, and Statistics*. Retrieved from: http://gpsworld.com/gpsgnss-accuracy-lies-damn-lies-and-statistics-1134/

European GNSS Agency. (2017, June 9). *GNSS Market Report Issue 5*. Retrieved from https://www.gsa.europa.eu/system/files/reports/gnss_mr_2017.pdf

European Space Agency. (2012). *Galileo and Space Service Volume*. Retrieved from: http://www.unoosa.org/pdf/icg/2012/icg-7/wg/wgb2-2.pdf

GPS Directorate. (2013, September 24). *Interface Specification IS-GPS-200 (rev. H)*. Retrieved from https://www.gps.gov/technical/icwg/IS-GPS-200H.pdf

International Civil Aviation Organization. (2005). *Doc 9849 AN/457 Global Navigation Satellite System (GNSS) Manual.* Retrieved from https://www.icao.int/Meetings/PBN-Symposium/Documents/9849_cons_en[1].pdf

Keesey, L. (2010). *Navigator Technology Takes GPS to a New High.* Retrieved from https://www.nasa.gov/topics/technology/features/navigator-gps.html

Kelso, T. S. (2014). *Definition of a Yuma Almanac* Retrieved from https://celestrak.com/GPS/almanac/Yuma/definition.asp

Kogure, S., & Kishimoto, M. (2012). *MICHIBIKI (QZS-1) and SSV.* Retrieved from: http://www.unoosa.org/pdf/icg/2012/icg-7/wg/wgb2-3.pdf

Konin, V., & Harchenko, V. (2010). *Sistemy sputnikovoj radionavigacii* [Satellite radionavigation systems]. Kyiv: Holtech.

Konin, V., & Shyshkov, F. (2016). Autonomous navigation of service spacecrafts on geostationary orbit using GNSS signals. *Radioelectronics and Communications Systems, 59*(12), 562–566. doi:10.3103/S0735272716120049

Konin, V., Shyshkov, F., & Pogurelskiy, O. (2016). Estimation of coordinates on geostationary orbit by using GNSS signals. 2016 IEEE Radar Methods and Systems Workshop (RMSW), 32-35. doi:10.1109/RMSW.2016.7778544

Konin, V., Shyshkov, F., & Pogurelskiy, O. (2017). Differences in measurements with separate use of frequencies L1 and L2 for the application of satellite navigation in near-earth space. *2017 IEEE Microwaves, Radar and Remote Sensing Symposium (MRRS)*, 67-70. doi: 10.1109/MRRS.2017.8075028

Marquis, W. (2014). *The GPS Block IIR/IIR-M Antenna Panel Pattern Appendix B – SV-Specific Patterns, Data.* LMCO.

Moreau, M. C., Davis, E., Carpenter, J. R., Kelbel, D., Davis, G., & Axelrad, P. (2002). *Results from the GPS Flight Experiment on the High Earth Orbit AMSAT OSCAR-40 Spacecraft.* Paper presented at the ION GPS 2002 Conference, Portland, OR.

National Aeronautics and Space Administration. (2016). *NASA Goddard Space Flight Center Navigator GPS receiver.* Retrieved from: https://technology.nasa.gov/patent/Navigator_GPS_Receiver

Parker, J. J. K. (2017). *Refining the GPS Space Service Volume (SSV) and Building a Multi-GNSS SSV.* Presented at the SCaN Workshop on Emerging Technologies for Autonomous Space Navigation. Retrieved from https://www.nasa.gov/sites/default/files/atoms/files/session_1_-_3_refining_the_gps_space_service_volume_and_building_a_multi-gnss_ssv_joel_parker_1_0.pdf

US Department of Defense. (2000a). *GPS IIF Operational Requirements Document.* ORD.

US Department of Defense. (2000b). *GPS III Capability Development Document.* CDD.

US Department of Defense. (2008a). *Global Positioning System Standard Positioning Service Performance Standard (Ed.4).* Retrieved from https://www.gps.gov/technical/ps/2008-SPS-performance-standard.pdf

US Department of Defense. (2008b). *SS-SYS-800 GPS III System Specification.* Author.

US Department of Defense. (2008c). *2008 Federal Radionavigation Plan.* Retrieved from https://www.navcen.uscg.gov/pdf/2008_Federal_Radionavigation_Plan.pdf

US Department of Defense. (2017). *2017 Federal Radionavigation Plan.* Retrieved from: https://www.navcen.uscg.gov/pdf/FederalRadioNavigationPlan2017.pdf

Wintemitz, L., Moreau, M., Boegner, G., & Sirotzky, S. (2004). *Navigator GPS Receiver for Fast Acquisition and Weak Signal Space Applications.* Retrieved from: https://ntrs.nasa.gov/archive/nasa/casi.ntrs.nasa.gov/20040171175.pdf

Zhan, X., & Jing, S. (2014). *Space Service Volume (SSV). Characteristics of BDS.* Retrieved from: http://www.unoosa.org/pdf/icg/2014/wg/wgb03.pdf

ADDITIONAL READING

Akim, E., Zaslavsky, G., Stepanyants, V., Tuchin, A., Yaroshevsky, V., & Tuchin, D. … Dmitreeva G. (2006). *Matematicheskiye modeli, metody, algoritmy i programmy v sovremennykh zadachakh avtonomnoy navigatsii iskusstvennykh sputnikov Zemli. Otchet.* [Mathematical models, methods, algorithms and programs in modern tasks of autonomous navigation of artificial earth satellites. Report]. [PDF document]. Retrieved from: ftp://ftp.kiam1.rssi.ru/pub/gps/lib/report/rep2006.pdf

European Space Agency. (2017) *About space debris.* Retrieved from http://www.esa.int/Our_Activities/Operations/Space_Debris/About_space_debris

Kessler, D. J., & Burton, C.-P. G. (1978). Collision Frequency of Artificial Satellites: The Creation of a Debris Belt. *Journal of Geophysical Research, 83*(A6), 2637–2646. doi:10.1029/JA083iA06p02637

Moreau, M., Winternitz, L., & Boegner, G. J. Jr. (2014) Navigator GPS Receiver for Fast Acquisition and Weak Signal Space Applications. Presented at the *ION GNSS Meeting*, Long Beach:CA.

NASA Goddard Space Flight Center. (2010) *Project report. On-orbit satellite servicing study.* Retrieved from https://sspd.gsfc.nasa.gov/images/nasa_satellite%20servicing_project_report_0511.pdf

Ziemnicki, P. (2017). Growing Importance of Satellite Navigation. On Earth and in the Outer Space [Analysis]. Retrieved from: https://www.defence24.com/growing-importance-of-satellite-navigation-on-earth-and-in-the-outer-space-analysis

KEY TERMS AND DEFINITIONS

Accuracy: The difference between the real and calculated user position.

Availability: The percent of the time when the navigation characteristics meet the requirements of given operation.

Dilution of Precision Factors: The measure of how the satellite geometry influences the accuracy of the navigation solution.

Navigation Satellite Constellation: A group of navigation satellites that is synchronized in time and can be used to obtain the position solution.

Off-Nadir Satellites: The satellites that are located behind with respect to the receiver's position.

Radio Navigation Field (RNF): This is a field formed by radio signals of navigation satellites.

Space Service Volume (SSV): The special volume in the near-Earth space where navigation satellites can be used to determine the user's position.

APPENDIX: ABBREVIATIONS

- Antenna Radiation Pattern (ARP)
- Coordinated Universal Time (UTC)
- Department of Defense (DOD)
- Dilution of Precision (DOP)
- European GNSS Agency (GSA)
- European Space Agency (ESA)
- Global Navigation Satellite System (GNSS)
- Global Navigation Satellite System (GLONASS)
- Global Positioning System (GPS)
- Global Positioning System Time (GPST)
- Minimum Number of Satellites Required (MNSR)
- Radio Navigation Field (RNF)
- Receiver Autonomous Integrity Monitoring (RAIM)
- Root-mean-square horizontal error (RMSH)
- Root-mean-square vertical error (RMSV)
- Spacecraft (SC)
- Space Service Volume (SSV)
- Terrestrial Service Volume (TSV)
- User Horizontal Navigation Error (UHNE)
- User Navigation Error (UNE)

Chapter 14
Application of Computer Modelling in Adaptive Compensation of Interferences on Global Navigation Satellite Systems

Valerian Shvets
National Aviation University, Ukraine

Svitlana Ilnytska
National Aviation University, Ukraine

Oleksandr Kutsenko
National Aviation University, Ukraine

EXECUTIVE SUMMARY

Modern society is characterized by the increased use of global navigation satellite systems (GNSS), which is inseparably linked with the interference immunity ensurance. The most effective way to protect against interferences is an introduction into the receiver structure of adaptive interference compensators. However, the most of proposed methods have been designed for radiolocation and communication and use a priori information about the transmitted signal. Since as structure of GNSS signal differs from the radar and communication systems, GNSS does not know the time-frequency structure of the useful signal in advance, which excludes the possibility of using a number of widely known methods. In this chapter, the authors propose a method, which does not use a priori information about a useful signal, and a new direct method for calculating the inverse correlation matrix of interference in adaptive antennas of interferences compensators.

DOI: 10.4018/978-1-5225-7588-7.ch014

ANALYSIS OF GLOBAL NAVIGATION SATELLITE SYSTEMS VULNERABILITIES

Modern society is characterized by the increased use of Positioning, Navigation, and Timing (PNT) services, which provide the basis for the effective functioning of many industries. In particular, PNT is an essential part of modern transportation systems, digital systems, telecommunications systems, command and control of precision weapons. The main suppliers (providers) of PNT services are Global Navigation Satellite Systems (GNSS), which are presented now by Global Positioning System (GPS, USA) (Federal Aviation Administration [FAA], 2013 a, 2013 b), GLObal NAvigation Satellite System (GLONASS, Russia) (Russian Institute of Space Device Engineering, 2008). The European Community sets up its own GALILEO system and China sets up BeiDow system for these purposes. GNSS provides with the data, using which one can determine the position of any user in space with an accuracy of one meter and by time with the accuracy of dozens and units of nanoseconds in any point of the globe and near-Earth space at any given time and in any weather.

After the first years of the intensive development and implementation of satellite navigation and time technologies, the more thorough analysis of the use of GNSS as the sole source of coordinate-time information, and more sober approach to the prospects of using GNSS begins. First, this is due to the GNSS vulnerability under the influence of unintentional and intentional interferences. The vulnerability of civilian GNSS receivers had been known for a long time (Littlepage, 1998; Pinker, Walker and Smith, 1998; Lyusin et.al., 1998; Ward, 1994; Gilmore, 1998; Key, 1995; Bond, 1998), but receiver manufacturers and their users rarely consider it. Only when the US Department of Defense has intensified its activities related to the use of GPS in the military environment (NAVWAR), it became apparent that deliberate interferences to the civilian receivers should be taken into account as an important factor. Military testings conducted in the New York (United States) (Forssel and Olsen, 2003) area have shown, that a number of receivers installed on board of civil aviation aircrafts have lost the ability to track GPS signals (due to severe drops in the carrier-to-noise ratio of several GPS satellites' L1 C/A code) at the approach phase at the International Newark Liberty Airport (NJ, USA). After an investigation by the FAA, it was discovered that a truck driver had installed a low-cost PPD on his vehicle (Colby et al., 1997).

The analysis of transport systems based on the use of GPS signals was carried out by (Winer, et al., 1996), (Wallis, 1999; Colby, 1997), (Report of the Commission to Assess United States National Security Space Management and Organization, 2001), (Corrigan, 1999). One of the most important and relevant reports on the research in this area was the Volpe Center Report on GPS vulnerability (John, 2001), which concluded that the GPS system, like other radio navigation systems, was vulnerable to unintended and intentional interferences and that such interference was a threat to security and can have serious consequences for the economy and the environment. The report concludes that the growing use of GPS in civilian infrastructure makes it an increasingly attractive target for hostile actions by individuals and groups. At the same time, the commercial availability of equipment for interference was detected (Forssel and Olsen, 2003; Rodgers, 1991).

Thus, the GNSS vulnerability under the influence of unintentional and intentional interference is now a generally known fact. This vulnerability equally applies to all systems: GPS, GLONASS, GALILEO and BeiDow, since as the principles of their construction and frequency ranges are quite close. Currently, the radio navigation community is actively discussing the vulnerability of GNSS and the search for back-up systems. In this regard, it is necessary to analyze the main sources of unintentional interferences, possible ways of setting up the intentional interference for consumer's equipment of GNSS and the prospects for improving the reliability of coordinate and time information under interference influence.

Based on vulnerability analysis, the aviation community has defined requirements for the operational characteristics of GNSS in respect of civil aviation. The ICAO recommendations are given in Table 1 (ICAO, 2014).

The data given in Table 1, illustrate high requirements to GNSS, which makes possible the use of satellite navigation systems as the primary means for flying over the ocean and as additional systems for flights in the airport, including non-precise and precise approach (using GNSS augmentation systems). However, as an auxiliary means, GNSS can't be the only navigation system on board, considering the possibility of GNSS signals degradation or their loss under the influence of interferences.

Global Navigation Satellite Systems Threats

A fundamental requirement to GNSS by the transport consumers is to ensure the integrity, availability, and continuity of navigation service, and in the first place, ensuring the continuity of service in the conditions of intentional interferences (jamming). Other factors for ensuring continuity (work under unintentional radio interferences) have been initially taken into account when developing and deploying GNSS systems.

Receivers of GNSS signals are potentially exposed to deliberate interferences because the signal power from a satellite is only $2.5 \bullet 10^{-16}$ W or -156 dB / W near the Earth's surface (Konin & Harchenko, 2010). Interferences can lead to the inability to detect a signal, by capturing incorrect signals and to errors in measurements of navigation parameters. Ultimately, the effects of interferences lead to a breakdown of the navigation task solution, or to the occurrence of navigation errors that exceed the requirements of consumers.

Based on the conditions of a high radio-technical vulnerability of GNSS, the threat of intentional interference is divided into two groups:

Table 1. Signal-in-space performance requirements (ICAO, 2014)

Typical operation	Accuracy horizontal 95%	Accuracy vertical 95%	Integrity	Time-to-alert	Continuity	Availability
En-route	3.7 km (2.0 NM)	N/A	$(1\text{-}10^{-7})$ /h	5 min	$1\text{-}1\times10^{-4}$/h to $1\text{-}1\times10^{-8}$/h	0,99 to 0,99999
En-route, Terminal	0.74 km (0.4 NM)	N/A	$(1\text{-}10^{-7})$ /h	15 s	$1\text{-}1\times10^{-4}$/h to $1\text{-}1\times10^{-8}$/h	0,999 to 0,99999
Initial approach, Intermediate approach, Non-precision approach (NPA), Departure	220 m (720 ft)	N/A	$(1\text{-}10^{-7})$ /h	10 s	$1\text{-}1\times10^{-4}$/h to $1\text{-}1\times10^{-8}$/h	0,99 to 0,99999
Approach operations with vertical guidance (APV-I)	16.0 m (52 ft)	20 m (66 ft)	$(1\text{-}2{*}10^{-7})$ in any approach	10 s	$1\text{-}8\times10^{-6}$ per 15 s	0,99 to 0,99999
Approach operations with vertical guidance (APV-II)	16.0 m (52 ft)	8.0 m (26 ft)	$(1\text{-}2{*}10^{-7})$ in any approach	6 s	$1\text{-}8\times10^{-6}$ per 15 s	0,99 to 0,99999
Category I precision approach	16.0 m (52 ft)	6.0 m to 4.0 m (20 ft to -13 ft)	$(1\text{-}2{*}10^{-7})$ in any approach	6 s	$1\text{-}8\times10^{-6}$ per 15 s	0,99 to 0,99999

1. The threat from terrorist influence, the parameters of which are limited by the dimensional and energy capabilities of the interferences setting equipment. This suppression is characterized typically by low power (up to 100 W) due to the available weight and dimensions of the portable equipment, a small time of suppression due to the limited power of portable power supplies, a small possibility of choosing the type of signals, usually eliminating the possibility of using imitation and retransmission interference. For such type of interferences, it is possible and necessary to develop measures to reduce the threat level.
2. The threat of combat (military) power, which can be conducted in the context of conflicts and hostilities in the area of transport operations and in neighboring zones, has practically no restrictions on energy and possible types of interference. This type of suppression can provide suppression of the receiver GNSS on the side lobes at a distance of direct visibility, reaching dozens and hundreds of kilometers. Therefore, the counteraction to such interference in the civilian use of GNSS can be realized only by methods of system duplication, the transition to other navigation systems or procedural security methods.

Thus, the following conclusions can be done:

1. The threat of terrorist suppression of GNSS can be partly reflected by a number of hardware measures that increase the noise immunity of the consumer receiver, and system measures inside the onboard consumer equipment.
2. The threat of combat (military) suppression can't be turned off at the level of onboard consumer equipment and means of GNSS service providers and therefore requires the transition to backup navigation systems and/or procedural security methods.

Possibilities to Create Interferences to Global Navigation Satellite Systems Threats

Unintentional interferences to GNSS. The peculiarity of the organization of radio-counteraction to the radionavigation fields of GNSS is the use of the physical fact that their energy levels near the Earth's surface have extremely low potential. The frequency bands, within which GNSS is operating, were previously allocated on a primary basis to the ground-based mobile services, satellite fixed communication "Inmarsat" and some aerodrome services. Harmonics of decimeter (very high frequency, VHF) transmitters of local television stations can also cause unintentional interferences to GNSS. Thus, even in non-combat situations, there are conditions for a so-called frequency electromagnetic conflict between the GNSS and all the above-mentioned civilian services (Table 2) (Radio Technical Commission for Aeronautics [RTCA], 2008).

In the technically developed countries, the main unintentional source of interferences to the satellite radio navigation system is a well-developed cellular network of base stations of mobile phones and individual personal mobile devices, Digital Video Broadcasting (DVB-T) (Wildemeersch, et.al., 2010). A mobile phone with the output power of 1W can already be considered as an unintentional interference for GNSS and, as a consequence for non-precision approach system (category 2 and 3) (The Commission, 2001). Therefore, with the new location of radiating radio equipment, the analysis of so-called electromagnetic compatibility (EMC) by specially organized services is a mandatory procedure in all advanced countries.

Table 2. Possible sources of unintentional artificial interferences

The interfering signals frequency range, MHz (channel number)	Source of interfering signals	GPS frequencies L1:1563-1587 MHz L2:1215-1237 MHz; L5:1164-1192 MHz	GLONASS frequencies: 1246-1256.5; 1602-1615.5; After the 2005 year: 1242.94-1247.75; 1598-1604.25 MHz	GALILEO frequencies E1:1559-1591 MHz; E5:1164-1215 MHz; E6:1260-1300 MHz
1533	Radioline	–	–	–
~ 500	3 harmonic	+	+	+
66 and 67 TV channels	2 harmonic	+	+	+
22 and 23 TV channels	3 harmonic	+	+	+
157 VHF (very high frequency)	10 harmonic	+	–	+
131 i 121 VHF	12 and 13 harmonics	+	+	–
525 frequency of DME cristal	3 harmonic	+	–	–
1575	Unmodulated carrier	+	–	+
> 1610	GLOBAL STAR	–	+	–
1240 – 1243.25	Digital transmission (packet radio)	–	+	–
1242 –1242.7	Amateur radio relay stations	–	+	–
1243 – 1260	Amateur TV transmitters	–	+	+
1250 – 1259	Air traffic control radar	–	+	+
108 – 118	Aviation communication and data transmission channel	+	+	+

There is the following technical explanation concerning the issue of the civilian use of GNSS. Due to the low level of GNSS signals on the Earth's surface the presence of unintentional interference from the above-mentioned services, exceeding the level of GNSS signals, leads to the so-called "multi-user effect", actually, the suppression of weak signals by the strong ones even if their types of modulation and access are different. Thus, the use of code division of signals of the CDMA type in the GPS in the same frequency band under the presence of a powerful radio frequency interference leads to the technical difficulties in the coordinated filtering of useful radio-navigation signals and even to the failure of the navigation task.

Deliberate GNSS interferences. When organizing the deliberate radio jamming on the enemy's radionavigation means, we distinguish the so-called *power suppression* (by the "brute force" method) and *intelligent or "smart" suppression.*

One of the *power suppression* types could be the pulsed energy. With the help of a powerful energy storage device discharge, a powerful over broadband rectangular pulse is formed, which ruins irreversibly the input circuits of GNSS receivers, even up to the ruining the solid-state semiconductor components – input low noise transistors and mixing diodes. When using for a radio-counteraction regular ultrahigh-frequency generator of noise like quasi-continuous or pulsed radiation of up to tens of watts, then the conditions appear for overloading the input and other analog amplifying channels of GNSS receivers due to their limited dynamic range. This leads to the failure of normal operation of the digital compatible code tracking filters, and, as a result, to the failure of navigation task.

The *power suppression* or the *brute-force radio-counteraction* is called typically *jamming*. It is a deliberate in-band emission of electromagnetic radiations (Huang, 2016). It can always be detected and in most cases be compensated by different means, in particular, by the so-called smart antenna or adaptive antenna array (Corrigan, 1999). A smart antenna of GNSS receiver with the help of special adaptive algorithms automatically forms nulls in beam pattern on the jammers directions.

When talking about *"smart"* or *intelligent GNSS interferences* the one differs the *spoofing* and *meaconing*.

GNSS spoofing is a technique of fooling a victim receiver into providing false Position, Velocity and Time (PVT) information by deliberately broadcasting GNSS-like signals (John, 2001), (Ruegamer and Kowalewski, 2015), (Psiaki and Humphreys, 2016), (Huang, 2016). GNSS spoofing is more difficult to detect than jamming, and therefore it may result in some serious situations due to unawareness of user about it (Liu, Y. et al., 2018). Due to its open structure and low power of satellite signals, civil GNSS service is vulnerable to spoofing (John, 2001). Spoofing of classified signals using cryptographic signal protection like the military GPS P(Y) or the Galileo PRS is practically impossible (Ruegamer and Kowalewski, 2015).

However, even classified signals are not safe from *meaconing attacks.* The aim of meaconing is the same as spoofing, but meaconing means recording and rebroadcasting of authentic GNSS signals. When the receiver tracks the signals generated by meaconing hardware without noticing it, the receiver will get instead of its own correct position the one from the meaconing hardware (Ruegamer and Kowalewski, 2015).

Transmitters of GNSS Interferences

The journalists have published the little known for the general public facts that at one time the British Airways had problems with GPS navigation in the sky of France because, in that area, the interference transmitter with a power of 4 W was secretly tested. A similar situation appeared with the DC-10 aircraft over New York, when the Air Force Research Lab (Rome, NY) tested the transmitter without informing the civil aviation of this event. In this regard, it is interesting to note that the editorial board of the well-known journal TSI – Telesatellite International for the editor's money has collected a transmitter of GPS interferences and successfully tested it on a straight section of the highway. As a result, the GPS receiver mounted at the car fixed a change in the speed vector from the South to the East direction and deviation of the location to 6 km.

Recently, the GNSS users concern more on intentional interferences, since at the modern level of radio-electronic components development it becomes quite easy to create own jamming device. In addition, there is an increasing number of intentional interferers readily available on the Component-off-the-self (COTS) market, mostly sold over the internet, even if their use is illegal in most countries (Ruegamer and Kowalewski, 2015), (Kuusniemi, 2012), (Boynton, 2014).

From the above information, we can say that interferences might be *unintentional* and *intentional (deliberate).* The deliberate ones could be of *power suppression* type, i.e. *jamming,* or the *intelligent* ones, i.e., *spoofing* and *meaconing attacks.*

The intelligent type of deliberate interferences are not the scope of this chapter and are mentioned here just to have a complete picture of a situation with GNSS vulnerabilities.

The GNSS equipment equally exposed to all types of interferences, the efficiency of which strongly depends on its power. Table 3 shows the extremely dangerous distances of the source of the interference (jamming), depending on the mode of operation.

From the Table 3, it is evident that the simple suppression system with insignificant power (noise like, harmonic) can lead to the creation of a zone of suppression in hundreds of kilometers.

PROBLEM STATEMENT

The navigation equipment of the consumer often works in the conditions of interference coming from directions different from the arrival of satellite signals. The following types of interference occur in GNSS receivers:

- Noise barrier;
- Noise is aimed at the frequency and coordinated with the spectrum from the signal;
- Noise impulse noise;
- Narrowband with a spectrum of less than 500 kHz (up to the emission of a monochromatic signal at the receiver frequency);
- Imitation interference generator or relay type (with full reproduction of the structure of the code of the original signal);
- Intelligent with a full imitation of information contained in the navigation signal.

Table 3. Extremely dangerous distances of interference sources (jammers) (Korotonoshko and Perunov, 2007)

Channel		GPS L1	GPS L1, L2C	GLONASS L1	GLONASS L1, L2	GPS+ GLONASS
Interference / Jamming type	**Power, W**	**Distance to the jammer (km)**				
Noise aiming	0.1	85	70	60	40	38
	1.0	280	230	200	132	125
	10.0	850	700	600	400	380
	100.0	2800	2300	2000	1320	1250
Harmonic	0.1	280	230	87	49	46
	1.0	829	736	278	157	147
	10.0	2800	2300	870	490	460
	100.0	8290	7360	2780	1570	1740
Imitative generated	0.1	900	740	391	252	236
	1.0	2880	2368	1251	803	755
	10.0	9000	7400	3921	2520	2360
	100.0	28800	23680	12510	8030	7550

On the basis of the above information, it is possible to propose the following possible measures to increase the GNSS jamming immunity:

1. Improvement of the beam pattern of reception antenna at small elevation angles;
2. Control of the antenna's beam pattern to reduce the sensitivity in the direction of the jamming source;
3. An antenna array with signal polarization;
4. Improved signal processing in the receiver;
5. Combination of GNSS receivers with the Inertial Navigation Systems (INS);
6. Use of dual-frequency receivers L1, L2;
7. Use of multi-frequency receivers.

The possible gain in interference immunity, shortening the distance of suppression and possibilities of proposed measures are presented at the Tables 4 and 5.

Thus, estimating the possible gain in the stability of navigation GNSS receivers immunity to interferences, the most promising method is beam pattern control in the receiving antenna (reducing sensitivity or setting the "0" (nulls) in the beam pattern in the interference source direction). It can be realized in the adaptive antenna compensators of interferences. The advantages of this approach are the following:

- The gain in interference immunity can be very significant;
- There is no need to adjust the GNSS receiver.

Table 4. Means for interference immunity at the consumer equipment (Korotonoshko and Perunov, 2006)

	Means of interference immunity	**The possible benefit with regards to the standard GNSS receivers, dB**	**Possible cost accretion with regards to the standard GNSS receivers, %**	**Notations**
1	Improvement the beam pattern in the receiving antennas at low elevation angles	10 – 15	30	Real, in all consumer systems
2	Control of antenna's beam pattern, reducing sensitivity in the interference direction	20 – 25	Up to 100	Practically efficient for one interference source, it's necessary to know the direction of the interference source.
3	Antenna array with signal polarization	10 – 15	До 50	Acts not in all environment conditions
4	Improvement of signal processing in a receiver	До 20	5 – 10	Requires additional research regarding implementation methods
5	Combining GNSS receiver with INS	10 – 15	10 – 300	The cost is defined by the level of INS and has the tendency to decrease
6	Using double frequency receivers L1, L2	5	20 – 30	
7	Using multifrequency receivers	8	40 – 50	

Table 5. Estimation of the efficiency of protection measures when using GNSS receiver in a local zone under terroristic suppression that uses 50 W ("basic") receiver (Korotonoshko and Perunov, 2006)

	Receiver	An increase of interference immunity with regards to the basic variant, dB	The distance of suppression (interference) coverage from the "basic" receiver, km (antenna of the transmitter is at the height of 1 m)
1	Standard receiver GPS L1 or GLONASS L1	–	57,0
2	Receiver GPS L1 or GLONASS L1 with improved antenna's beam pattern at low elevation angles	15	17,5
3	Receiver GPS L1 or GLONASS L1 with improved antenna beam pattern at low elevation angles, integrated with INS	10	10,0
4	The receiver from the 3rd row, but with additional frequency L2 (double frequency)	5	6,0
5	The receiver from the 3rd row, but with additional frequencies L2, L3 (three frequency)	8	4,1
6	The receiver from the 4th row, with additional integration GPS/GLONASS	5	3,3
7	The receiver from the 5th row, with additional integration GPS/GLONASS	5	2,2

Adaptive interference compensators are based on antenna arrays and adaptive methods to control the beam pattern. Very often adaptive interference compensators are called adaptive antenna arrays.

Adaptive antenna arrays have been studied for several decades already. The main directions of adaptive antenna arrays development are radiolocation systems and radio communication systems because the main task was increasing the interference immunity from lateral petals. The greatest attention is paid to the issues of processing of radio location information in the context of correlated interferences, which involves not only the accumulation of useful signals, but also the compensation of interfering signals (Widrow and Stearns, 1985; Losev, 1988; Ratynskiy, 2003; Monzingo and Miller, 2004; Dzhigan, 2013).

Starting from 2004 – 2007 the papers devoted to interference immunity of GNSS systems started to appear in open sources, see for example (Brown and Gerein, 2001; Brown and Mathews, 2007; Gheethan, Herzig and Mumcu, 2013; Magiera and Katulski, 2015; Gao et.al., 2016). According to the structure of signals and interferences, the GNSS differ from radio location systems and radio communication systems. Therefore, when using adaptive methods for compensation of interferences in the channels of GNSS, it should take into account a number of important features, which often complicate the implementation of adaptive interference compensator. Thus, unlike radio location and communication systems, in GNSS systems the frequency-time structure of a useful signal is not known in advance, which excludes the possibility of using a number of widely used methods of adaptive interference compensation with the use of a reference signal. The continuous nature of the time structure of the GNSS signal significantly impedes the ability to distinguish the compensating voltage from the interference and to exclude the effect of a useful signal on the adaptation chain. This largely limits the possibility of using GNSS multichannel spatial-temporal processing devices with adaptive compensation of correlated noise.

There is a large variety of methods and tools for constructing adaptive interference compensators for the GNSS receivers. They are determined by a combination of efficiency criteria and adaptive control algorithms. Basically, all developments in this sphere exploit the solutions developed for in radio location and communication systems, so-called beamforming-systems (Monzingo and Miller, 2004). The compensation channel in the beamforming system uses the LMS or RMS adaptive algorithm based on the criterion of the least squares, which in turn is based on the matrix inversion and QR decomposition.

The disadvantages of beamforming-systems include the slow convergence of LMS or RMS adaptive algorithms (Monzingo and Miller, 2004), as well as the need for a priori data referring to the interference and useful signal (Dzhigan, 2013), therefore the beamforming-systems operate in two stages:

- Estimation of the direction (angle) of interference arrival, using high-resolution MUltiple SIgnal Classification (MUSIC) or Estimation of Signal Parameters via Rotational Invariance Technique (ESPRIT) (Oumar, Siyau and Sattar, 2012);
- Using measured angle, calculation of weight coefficients and beam pattern formation.

Unfortunately, the maximum damping factor in such systems does not exceed 25 - 30 dB. However, the development of beamforming-systems in our time is going due to their main advantage - the possibility of using an antenna array with a large aperture, the step between individual elements of the antenna array.

At the same time, beamforming systems do not use all the possibilities of solving the Wiener-Hopf equation (1), which assumes that all information about sources of interference, namely, its angular position in space, is in the interference correlation matrix (Widrow and Stearns, 1985; Dzhigan, 2013).

$$\mathbf{W} = \mathbf{R}^{-1}\mathbf{S}, \tag{1}$$

where $\mathbf{W}$ is a vector of weight coefficients, $\mathbf{R}^{-1}$ is the inverse of the interference correlation matrix, $\mathbf{S}$ – column vector characterizing the amplitude-phase distribution of the signal through the receiving channels.

To calculate the weight coefficients vector by the expression (1) it is necessary to perform direct inversion of the correlation matrix. However, in practice, this correlation matrix is unknown. Therefore, the ones may calculate the most plausible estimate of the correlation matrix L using temporal samples of random amplitudes of the input process. Thus, the adaptive processing of the signal reduces to the finding the estimate $\hat{\mathbf{W}}$ of the vector W:

$$\hat{\mathbf{W}} = \hat{\mathbf{R}}^{-1}\mathbf{S} \tag{2}$$

The expression (2) is a simple implementation of the weight coefficients vector calculation. It is only necessary to calculate the estimate of the inverse correlation matrix $\mathbf{R}^{-1}$, i.e. to use the direct method of calculating the inverse correlation matrix. However, if a weighing vector is evaluated by formula (2), two problems may arise. First, when $L < N$ (short sample size), the correlation matrix becomes singular and therefore has no inverse, and when $L \geq N$ it is poorly conditioned (Widrow and Stearns, 1985; Monzingo and Miller, 2004; Dzhigan, 2013).

However, the direct methods of the interference correlation matrix inverse calculation $\mathbf{R}^{-1}$ provide the following important advantages:

- Small time of obtaining the estimate of weight coefficients vector;
- High coefficient of interference suppression;
- There is no necessity in a priory data about the parameters of interference and useful signal, which is quite important for practical applications.

The analysis of different publications devoted to the development of adaptive compensators of interferences (*Losev*, 1988), (*Ratynskiy*, 2003), (*Dzhigan*, 2013) together with own experience (Shvets, 2013), (Shvets V. and Shvets O., 2013), (Shvets and Kharchenko, 2017) shows that it is possible to increase the level of spatial interference suppression (more than 30 dB) with implementation of the scientifically substantiated technical solutions (Figure 1)

From the above materials it can be concluded that at present it is urgent and necessary to develop scientific methods of direct computation of a correlation or inverse correlation matrix of interference that will enable to increase the level of spatial suppression of interferences, reduce the time of adaptation to the interferences, reduce the time to calculate weight coefficients (Ratynskiy, 2003; Dzhigan, 2013; Shvets and Kharchenko, 2017).

MATHEMATICAL MODELS OF INTERFERENCES TO AERO NAVIGATION SYSTEMS

To solve the problems of increasing the interference immunity of air navigation systems, it is necessary to have the information signals arriving at the input of the digital signal processor (correlator) after the path of the high frequency of the satellite receiver. The availability of such information to the developers allows them to make the optimal design of equipment, which increases the interferences immunity of the satellite navigation systems.

Since as the real "signal/interference" mixture (combinations) obtaining might be difficult, it is more convenient to use the mathematical models.

The interferences may affect both the radio channel and the digital signal processor (correlator) of satellite navigation receivers, which is substantiated by the structure of radiation.

Figure 1. Technical means to achieve high efficiency in interferences attenuation in adaptive interference compensators

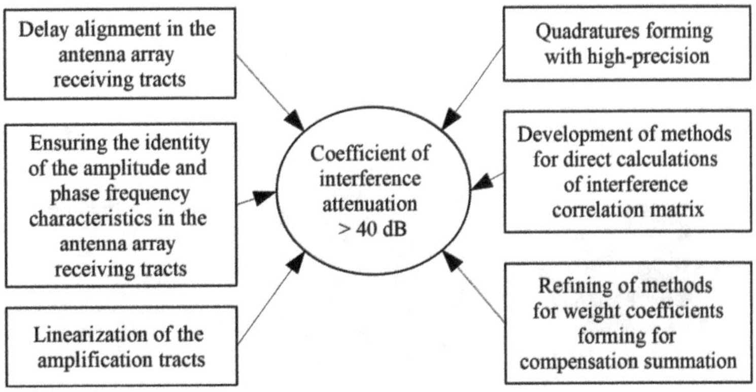

The current level of electronics development allows us to receive signals with any specific features and statistical parameters. Therefore, it would be more convenient to systematize the interferences according to their statistical characteristics, for their further modeling. The systematization results are presented in Figure 2.

The process of radio interferences modeling is carried out in the following stages:

- Modeling of non-correlated radio interference with a given distribution density;
- Modeling of correlated radio interference with a given correlation function and distribution density.

Random Variables Modeling With Specified Probability Distribution Law

For the modeling of random variables whose possible values are beyond the limits of certainly limited interval (a, b) (random variables with truncated distribution laws), as well as random variables whose distribution laws can be approximated by truncated, the Neumann method is sufficiently universal (Bykov, 1971; Shvets, Kasperovych and Ostrovskiy, 2010, 2011).

The method of Neumann includes the following steps:

Using random numbers generator, we obtain a pair of numbers, uniformly distributed on an interval (0, 1) – x_1 and x_2.

By means of transformations (by inverse function methods) the two numbers X^*_1 and X^*_2 are obtained, which are distributed on intervals (a,b) and $(0,W_{max})$:

$X^*_1=a+x_1(b-a)$ i $X^*_2=x_2W_{max}$, де $W_{max}=\max f(y)$

Figure 2. Statistical properties of radio interferences

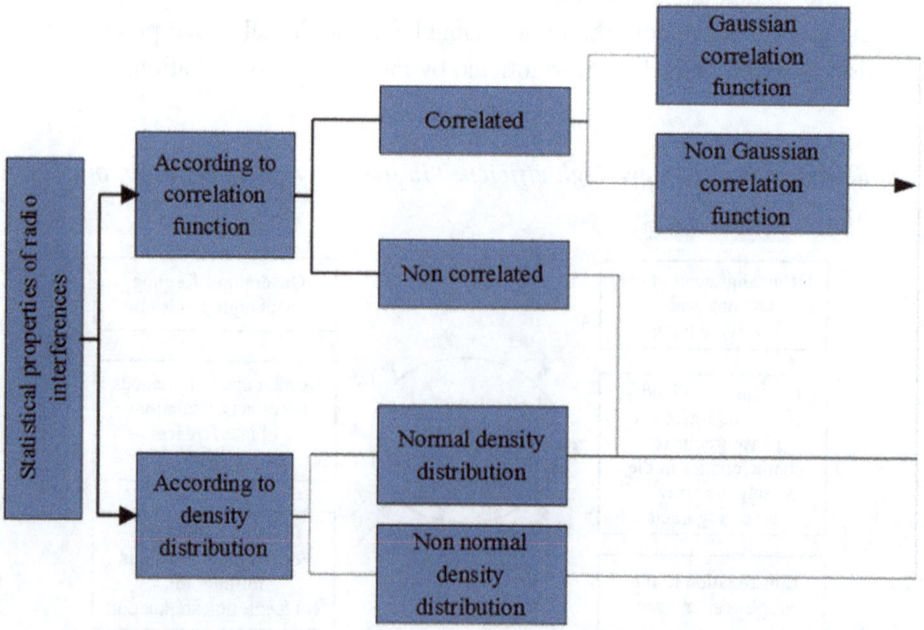

(Figure 3).

As the realization of a random variable, such number of X^*_1 is taken (accepted) from the pair of X^*_1 and X^*_2, satisfying the inequality: $X^*_2 < f(X^*_1)$.

Illustrative Example of Random Variables Modeling With Specified Probability Distribution Function

Figure 4 demonstrates the interface of the program for random variables modeling with specified probability distribution type together with the result for Rayleigh probability distribution function. The uniformly distributed random number, generated by the special generator, (Figure 4, 1st plot), by means of Neumann's method, is transformed into a random variable with the Rayleigh distribution law (Figure 4, 2nd plot). The 3rd plot at Figure 15 shows the desired probability distribution function (i.e., calculated by analytical expression), while the 4th plot here reflects the distribution density of generated random variable according to the desired distribution law.

The given method allows modeling the stochastic processes of any complexity. A prerequisite here is a presence of an analytical expression of the distribution density (probability distribution function) of the modeled random process.

The fragment of a program code implementing Neumann method in Matlab

```
clear all
% taking input data (digitized data array)
f=uigetfile('*.dat');
fid = fopen(f); [A,len]=fscanf(fid,'%f'); fclose(fid);
fid = fopen(f);clear A; [A]=fscanf(fid,'%f',[2,len/2]); fclose(fid);
A=A'; x=A(:,1); y=A(:,2);
%  approximation by cubic spline
s=csapi(x,y);
% calculation of an integral under curve for the further normalizing procedure
Si=fnint(s);
I=fnval(Si,max(A(:,1)))-fnval(Si,min(A(:,1)));
% calculation of the normalizing factor
```

Figure 3. Choice of random numbers by Neumann method (Bykov, 1971)

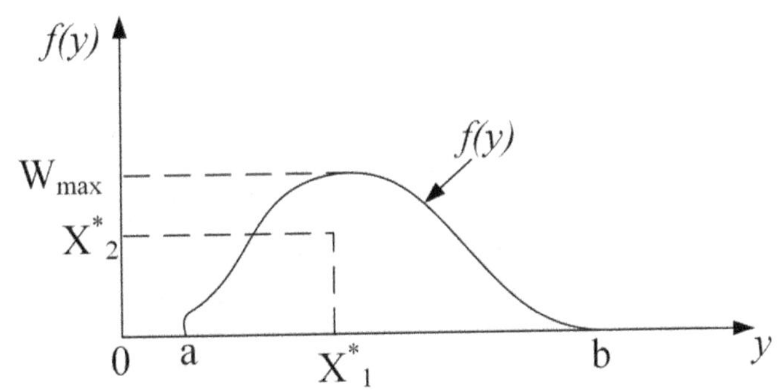

Figure 4. The result of random process modeling with Rayleigh density distribution

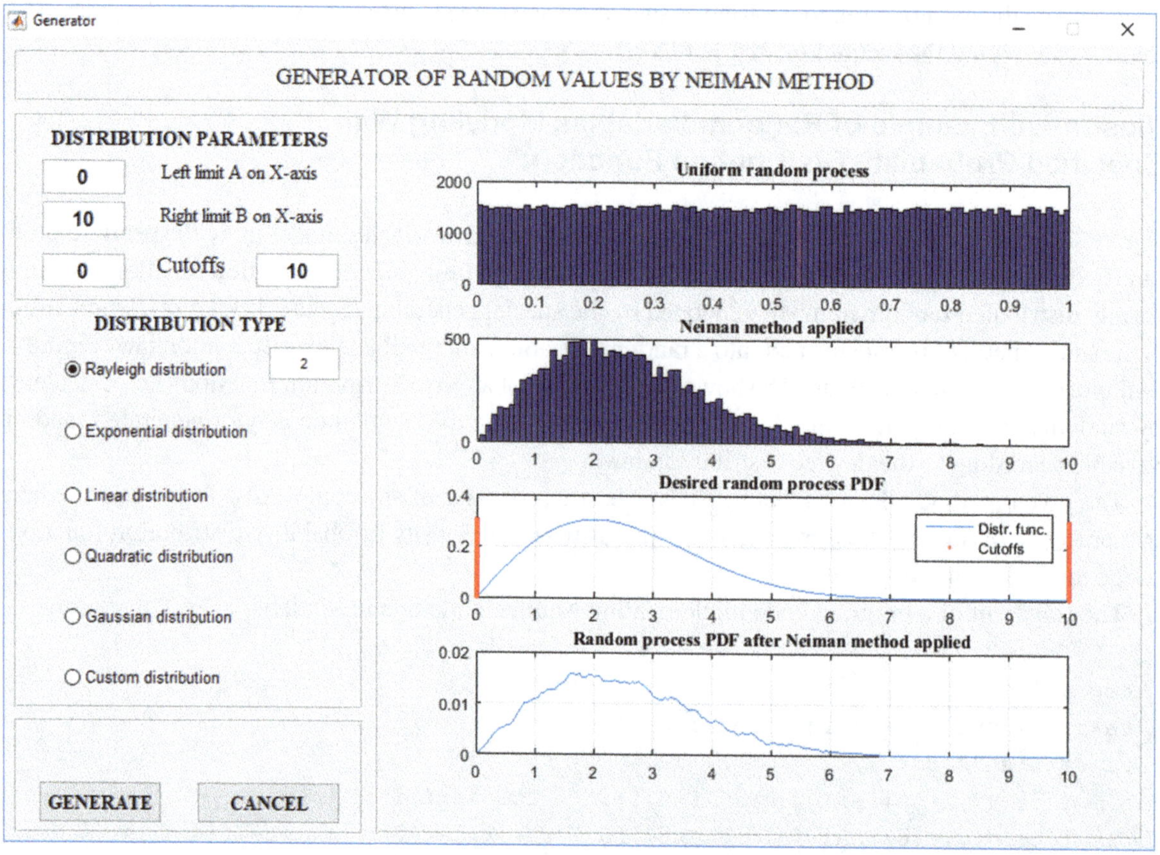


```
knorm=1/I;
% data normalizing. The purpose is to have the area under curve to be equal to
1.
y1=y.*knorm;
plot(x,y,'r x',x,fnval(s,x),'b'); grid on
% approximation of normalized values by cubic spline
% getting the values corresponding to the norm conditions
% this obtained spline is used for further interpolation
s=csapi(x,y1); clear A; clear len; clear y; clear f; clear fid;
% plotting graphs of distribution densities
hold on; plot(min(x):0.3:max(x),fnval(s,min(x):.3:max(x)),'g'); grid on
xlabel('Interval of random values);
ylabel('Probability density distribution');
% Implemenation of Neumann method
Pmax=max(y1); a=min(x); b=round(max(x)); p=1;
while p<=20000
```

```
    x1=rand; x2=rand; x11=a+(b-a)*x1; x22=Pmax*x2;
    if x22<=fnval(s,x11)
        rn(p)=x11; p=p+1;
    end
end
clear p;
[pp,xout]=hist(rn,12); p1=sum(pp); pp=pp./p1;
plot(xout,pp,':');
legend(' digitized values ',' approximated values ',' interpolated values ','
generated values ',0);
file=uiputfile('genrnd.dat', ' Save file with name ');
[F]=fopen(file,'wt'); fprintf(F,'%f\n',rn); fclose(F);
```

Modeling of Correlated Stationary Random Processes

The problem of modeling random processes can be seen as a problem of synthesis of discrete linear shaping filter that converts discrete white noise in a correlated discrete stochastic process with the desired correlation and spectral characteristics.

Obtaining weight coefficients by decomposition of the spectral density function in the Fourier series. The approach to finding the weighting coefficients c_k means of the formula is proposed:

$$c_k = \frac{1}{\omega}\int_0^{\omega_0}\left[\frac{\omega_0}{\pi}G(\omega)\right]^{\frac{1}{2}}\cos\frac{k\pi\omega}{\omega_0}d\omega, \qquad \omega_0 = \frac{\pi}{\Delta t}, \tag{3}$$

We might see that the weighting coefficients here are the Fourier series coefficients under its decomposition in a series of cosines of the spectral density function $G(\omega)$ modeled process, put in the power ½.

The expression (3) can be obtained based on the following considerations. The continuous stationary centered normal random process ξ(t) needed to be modeled is specified by some energy spectrum:

$$G(\omega) = \int_{-\infty}^{\infty} R(\tau)e^{-j\omega\tau}d\tau, \tag{4}$$

The usual spectral density $G(\omega)$ decreases at sufficiently large ω, and becomes negligible starting with a certain frequency ω_c. Then the random process ξ (t) can be replaced with sufficient accuracy by the process $\xi_0(t)$ with the energy spectrum

$$G_0(\omega) = \begin{cases} G(\omega), & |\omega| \le \omega_c \\ 0, & |\omega| > \omega_c \end{cases}. \tag{5}$$

We will consider the random process $\xi_0(t)$ as the result of a continuous white noise $x(t)$ influence with a limited frequency ω_c spectrum on a continuous linear system, whose transfer function is determined by the relation

$$G_0 \left| K(j\omega) \right|^2 = G_0(\omega), \tag{6}$$

where G_0 is white noise spectral density (See Figure 5). The relation (6) reflects the known from the theory of random processes fact: the energy spectrum of a noise at the output of a linear system is equal to the product of the energy spectrums of the input noise per square of the modulus of the system transfer function (complex frequency characteristics).

The infinite set of linear systems having different phase-frequency characteristics, which are the arguments of the complex function $K(j\omega)$, satisfy the condition (6). The authors choose one of those systems (system K_0) with a phase-frequency characteristic, equal to zero. To satisfy the condition (3), it should have the following form:

$$K_0(j\omega) = \sqrt{\left[\frac{1}{G_0} G_0(\omega) \right]}. \tag{7}$$

The impulse response characteristic, corresponding to the transfer function (7), is equal

$$h(t) = \frac{1}{2\pi} \int_{-\infty}^{\infty} K_0(j\omega) e^{j i \varpi t} d\omega = \frac{1}{\pi} \int_0^{\varpi_c} \sqrt{\left[\frac{1}{G_0} G_0(\omega) \right]} \cos \omega t d\omega \tag{8}$$

Using the presented above methods of modeling the random processes with given distribution density and correlation function, a software product has been created that allows receiving a "signal/interference" mixture for further research.

The program is implemented in the MatLab programming environment. The results of the random processes modeling with a specified correlation function according to the equations (4) – (8) (linear one, with the correlation coefficient equal to 1) are presented in Figure 6. Additionally, the figure demon-

Figure 5. The energy spectrum of white noise at the output of a linear system (Bykov, 1971)

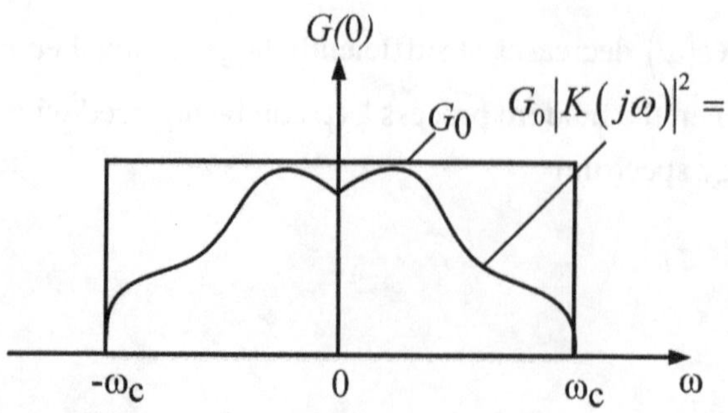

strates the program interface, according to which it is possible to choose the type of correlation function together with the correlation coefficient. It is possible also to input here either normally distributed noise or with custom specified distribution type.

INCREASING THE INTERFERENCE IMMUNITY OF SATELLITE AERO NAVIGATION SYSTEMS

One of the most efficient ways to eliminate the interferences impact is an implementation of adaptive antenna arrays (Dzhigan, 2013; Shvets, 2013; Shvets V. and Shvets O., 2013).

In general case the task of interference immunity ensuring consists in passing the mix of useful and interference signal through the filter, in which the interference is maximally suppressed and useful signal is almost not changed.

The design of such filters is based on the theory of optimal filtration. Such filters might have constant parameters or to be adaptive. The design of filters with constant parameters is based on the apriori knowledge of characteristics of useful signal and interference. When there is minimal knowledge about those signals features, the adaptive filtration principles are generally used (Widrow and Stearns, 1985).

Adaptive devices of signal processing act by the principle of closed contours. The input signal *x(n)* is filtered (or weighted) in a programmable filter to obtain the output signal $\hat{y}(n)$, which is then compared with the standard or training signal *y(n)* to obtain the error signal *e(n)*. The error signal is used to correct the filter parameters (weight coefficients) usually using an iterative method with the aim of gradual error minimization.

Adaptive devices could be subdivided into two big subgroups: adaptive filters and adaptive antennas. Actually, adaptive antennas are filters as well, but they are the spatial ones. The principle of adaptive compensation could be demonstrated in the simplest example (Losev, 1988), (Monzingo and Miller, 2004). The structural scheme is shown in the Figure 7.

The useful signal S_i and interference n_0 come to the main channel of the device. Let's assume that the useful signal S_i and interference n_0 are non-correlated. The base channel receives only interference $n_{1,}$ non-correlated with the signal. The interference signals n_0 and n_1 are correlated with each other, but the correlation function is unknown apriori. After passing through the adaptive filter the interference n_1 will be transformed at the output in noise signal *y*. The parameters of the filter should be chosen in

Figure 6. Results of modeling of random processes with specified correlation function

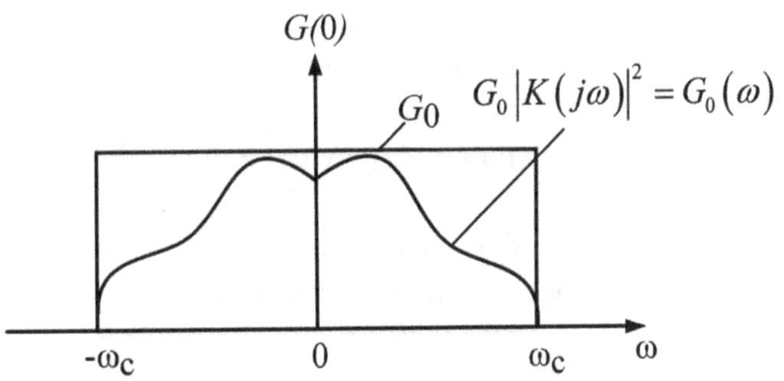

Figure 7. Adaptive filter structural scheme (Widrow and Stearns, 1985)

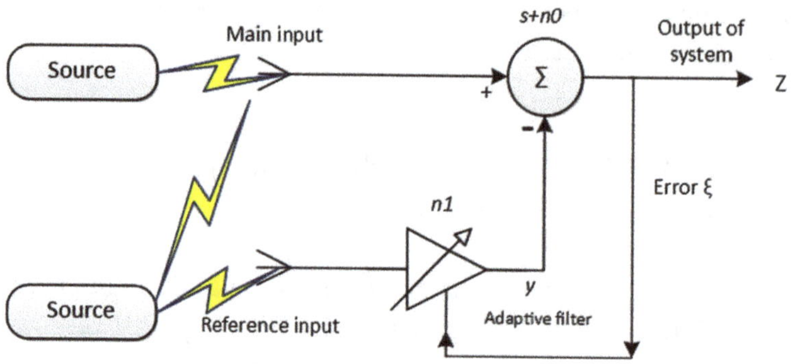

such a way that the voltage y differ from the voltage of interference $n_{0.}$ At the adder's output the signal is the following:

$$Z = S + n_0 - y.$$

With well-known characteristics of the channels, to which interferences are coming to the main and reference inputs, it would be possible to construct a filter with constant parameters, which will turn the interference n1 into n0. Then after the subtraction, we would receive a signal S.

In practice, such filters can not be used, since they require very precise adjustment. And most importantly, the characteristics of the interference channels are a priori unknown. Therefore, an adaptive filter is used, the parameters of which are automatically adjusted depending on the signal-interference environment. The algorithms of the functioning of the adaptive device depend on the particular application and are synthesized according to different criteria.

An adaptive antenna system is based on the antenna arrays and, due to sufficient degrees of freedom, allows performing interferences detection and nulls forming in the beam pattern in the direction of each interference source. To do this, it is necessary that the number of elements of the antenna array was greater on one than the number of interference sources.

The adaptive antenna system represents a set of electronic devices that solve the problems of receiving and converting microwave signals to the digital representation, spatial and temporal processing, quasi-optimal weighting, and adaptive process control.

With a well-developed theory of such systems, there is a large variety of ways of constructing them using various methods of approaching optimal solutions depending on the constraints due to the specific application of the adaptive antenna system.

Such equipment should be designed taking into account clear limits on the weight and dimensions while ensuring high quality of operation because the latter is connected with the safety of flights. This imposes its peculiarities when considering solutions.

The choice of methods and techniques for building AAS is:

- Due to the expediency of using adaptive antenna systems for satellite navigation;
- Analysis of methods and means of constructing adaptive antenna system, including performance criteria, adaptive algorithms, generalized structures.

An antenna array adopts electromagnetic radiation from navigation signal and interference sources, converting them into electrical signals. These signals are separated and fed into the subsystem of the transformation of PPS signals. The purpose of PPS is to convert microwave signals into digital samples.

The main element of the adaptive antenna system is the adaptive processor. As a result of synthesis according to the chosen criterion, an adaptive control algorithm for AAS, implemented in the adaptive processor, is determined.

Calculated in AP weight factors through the PC control device are fed into the chart-creating subsystem of the chipboard. The purpose of the DSP is to form the amplitude-phase distribution in the channels of the AR in such a way that in the resulting beam pattern the nulls are formed in the direction of interferences arrival, and, therefore, signals are received with minimal energy losses.

For a certain class of adaptive algorithms, a feedback signal may be required that separates from the output of the AAS and is converted by the PPS device to the type perceived by the AP.

The antenna array beam pattern may be presented in the following form (Monzingo and Miller, 2004):

$$g\left(\alpha\right)=f\left(\alpha\right)\cdot h\left(\alpha\right), \tag{9}$$

where $f\left(\alpha\right)$ is a beam pattern of the elementary transmitter, $h\left(\alpha\right)$ is an array factor, α is an angle of electromagnetic wave income.

For the flat equidistant antenna array the array factor is described by the expression:

$$h\left(\alpha,\beta\right)=\sum_{l=0}^{L-1}\sum_{m=0}^{M-1}W_{lm}^{*}\exp\left(\frac{2\pi l d_x}{\lambda}\cos\alpha+\frac{2\pi m d_y}{\lambda}\cos\beta\right), \tag{10}$$

where d_x is a distance between antenna array elements along the x axis, d_y - along y axis, λ – wavelength of received oscillation, α, β – angles of oscillation input, W_{lm}^{*} - weight multipliers in elements (l, m) of the array.

In factorized form it looks like the following:

$$h\left(\alpha,\beta\right)=h_1\left(\alpha\right)\cdot h_2\left(\beta\right). \tag{11}$$

The weight coefficients vector is factorized as well:

$$W_{lm}=W_{xl}\cdot W_{ym} \tag{12}$$

In such case, we obtain the simplest solution of adaptive antenna array beam pattern control.

The main task of the adaptive processor is to calculate the weight coefficients vectors W_{xl} and W_{ym} based on the results of analysis of the signal-interference environment.

Optimal Adaptive Methods to Form the Weight Coefficients

It is possible to use several different efficiency criteria (performance measures) when making a synthesis of an adaptive processor (Figure 3.2), which adjusts the weight coefficients for each element of the antenna array. Those criteria are applied using the optimal weight processing of the signal and interference mixture if the vector of weight coefficients is calculated by the Wiener-Hopf formula (1) (Widrow and Stearns, 1985), (Dzhigan, 2013).

Depending on the inverse correlation matrix of interference $\mathbf{R}^{-1}$ determination, it is possible to synthesize optimal methods for calculating weight coefficients.

There are four typical criteria (Monzingo and Miller, 2004):

1. Mean square error (MSE).
2. Signal to noise ratio (SNR).
3. Maximum likelihood (ML).
4. Minimum noise variance (MV).

The results of mathematical modeling of the above-mentioned criteria for plotting the beam pattern of the adaptive antenna array are presented below.

Illustrative Example for the Mean Square Error Criterion

The details on this criterion are provided in (Monzingo and Miller, 2004), chapter 3.3.1. The authors have created the corresponding functions in MatLab.

Function ***w_plargrad(m,n,dxnal,dynal)*** calculates and plots the beam pattern of adaptive antenna array according to the equation (10). Input paramters: *m* – quantity of elements on X axis; *n* – quantity of elements on Y axis; *dxnal, dynal* – ratios *dx/λ, dy/λ*.

```
function w_plargrad(m, n, dxnal, dynal)
% Author V. Shvets, 2010
A=0; A1=0; i=sqrt(-1);
az=0:1:360;
inp1=fopen('weith.txt','r'); inp2=fopen('faz.txt','r');
w1=fscanf(inp1,'%f'); fz1=fscanf(inp2,'%f');
fclose(inp1); fclose(inp2);
p=0;
for k=1:m
    for b=1:n
        p=p+1; W(k,b)=w1(p);
        betha(k,b)=fz1(p)*pi/180;
    end
end
if n==1
    az=zeros(size(az));
end
```

```
if m==1
    az=ones(size(az));
    az=90*az;
end
um=0:1:90;
kx=2*pi*(dxnal);
ky=2*pi*(dynal);
[AZ UM]=meshgrid(az, um);
for u=1:m
    for v=1:n
         A=A+W(u,v)*exp(i*betha(u,v))*exp(i*((u-1)*kx.*sind(UM).*cosd(AZ)+(v-
1)*ky.*sind(UM).*sind(AZ)));
    end
end
F=abs(A);
x=F.*sind(UM).*cosd(AZ);
y=F.*sind(UM).*sind(AZ);
z=F.*cosd(UM)
figure(1);
mesh(x,y,z);
figure(2)
contour(AZ,UM,z,150); grid on; colorbar; colormap(hot);
figure(3);
surf(AZ,UM,z); shading interp; colorbar;
figure(4)
contour(x,y,z,64); grid on; colorbar; colormap(hot);
```

Function ***model_sign_sko(m,n,dx,dy,az,um1,kol,fi0,sigma,AZs,UMs,fis)*** calculates reference signal in adaptive antenna array, amplitude-phase distribution of interference in adaptive antenna array channels, correlation matrix of interference, weight coeffitients to control the directed diagram of the adaptive antenna array. Input parameters: *m* – quantity of elements on X axis; *n* – quantity of elements on Y axis; *dxnal, dynal* – ratios dx/λ, dy/λ; *az* – azimuth of interference signal source; *um*1 – elevation angle of interference signal source; *kol* – samples quantity for the interference and useful signal; *fi*0 – initial phase shift of interference; *sigma* – variance of interference signal; *AZs* – azimuth of useful signal; *UMs* – elevation angle of useful signal; *fis* – initial phase shift of useful signal.

```
function [weith,faz]=model_sign_sko(m,n,dx,dy,az,um1,kol,fi0,sigma,AZs,UMs,f
is)
% Author V. Shvets, 2010
clc;
j=sqrt(-1);  s90=sind(90-az);
fi0=pi*fi0/180; fis=pi*fis/180;
rr=randn(kol,1); ri=randn(kol,1); rnd=complex(rr,ri);
% interference modeling part
```

```
for a=1:kol
    Si(a,1)=exp(j*(2*pi*a*10/128+fi0));
end
Si=Si+sigma.*rnd;
p=0;
for ii=1:n
    for k=1:m
        p=p+1;
        for a=1:kol
            S(a,p)=Si(a,1)*exp(j*(sind(um1)*2*pi*((k-1)*dx*sind(az)+(ii-
1)*dy*s90)));
        end
    end
end
N=S'; Ns=sum(N); Ns=Ns'; clear N;
for i=1:p
    N(:,i)=Ns;
end
R1=corrcoef([S N]);
R=R1(1:p,1:p);
% modeling of amplitude-phase distribution of signal
p=0;
for ii=1:n
    for k=1:m
        p=p+1;
        v(p)=randn*exp(j*(sind(UMs)*2*pi*((k-1)*dx*sind(AZs)+(ii-1)*dy*sind(90-
AZs))));
    end
end
v=v';
w=R\v;
weith=abs(w)/max(abs(w)); faz=angle(w)*180/pi;
[F]=fopen('weith.txt','wt');[F1]=fopen('faz.txt','wt');
fprintf(F,'%f\n',weith); fprintf(F1,'%f\n',faz);
fclose(F); fclose(F1);
```

Main function ***AAR_sko(m,n,dx,dy,az1,um1,kol,fio,sigma,AZs,UMs,fis)*** combines two above presented ones.

```
function AAR_sko(m,n,dx,dy,az1,um1,kol,fio,sigma,AZs,UMs,fis)
% Author V. Shvets, 2010
 [weith,faz]=model_sign_sko(m,n,dx,dy,az1,um1,kol,fio,sigma,AZs,UMs,fis);
w_plargrad(m,n,dx,dy);
```

The results of modeling are presented at the Figure 8 for the following input parameters: interference azimuth 241°, interference elevation angle 28°, matrix of adaptive antenna array 2×2, *dx, dy* = 0.25, interference variance – 1, azimuth of useful signal 110°, elevation angle of useful signal 18°, phase shift of useful signal and interference 0°.

It can be seen from the Figure 8 that null is formed in the interference direction.

However, this criterion has the following disadvantages:

- To apply it in GNSS, there is a requirement for the receiver side to have full information about expected useful signal;
- For optimization of weight coefficients, it is necessary to foresee the device, which would exclude the influence of reference signal when calculating the inverse correlation matrix of interference $\mathbf{R}^{-1}$ (Losev, 1988)

Illustrative Example for the Maximum Likelihood Criterion

The details on this criterion are provided in (Monzingo and Miller, 2004), chapter 3.3.3. The corresponding functions in Matlab have been created by the authors.

Figure 8. The beam pattern of adaptive antenna array using the MSE criterion for weight coefficients calculation

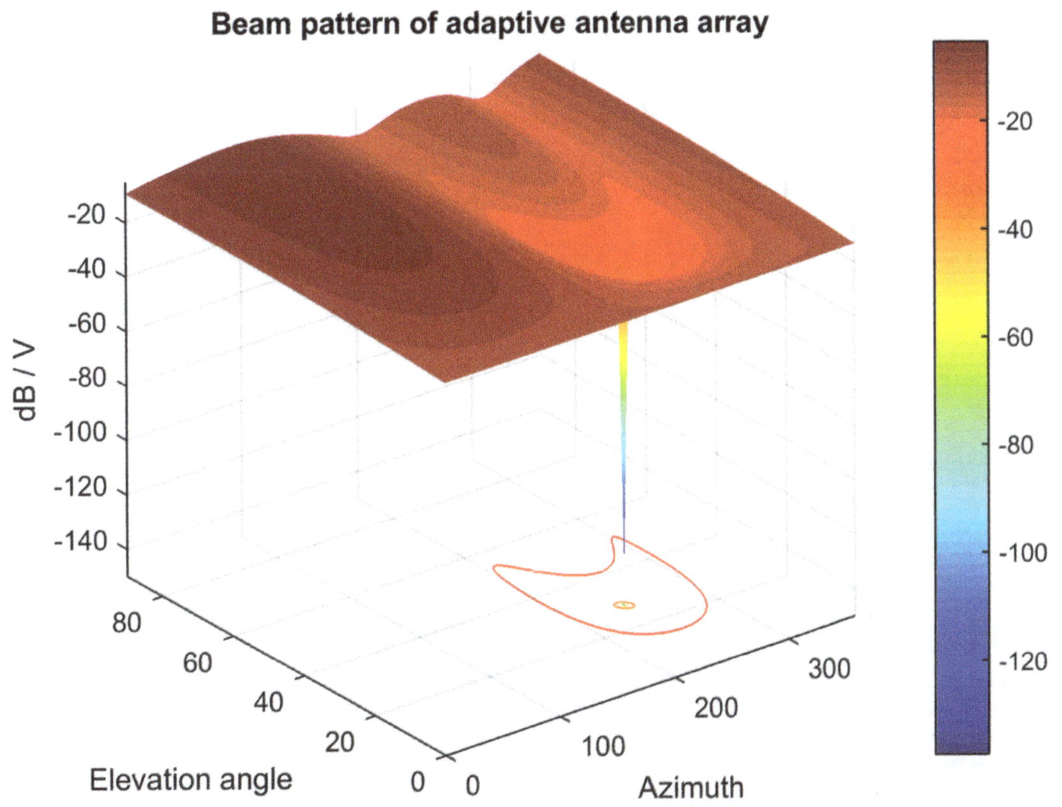

Function ***model_sign_max_trust*** calculates an amplitude-phase distribution of partial diagrams of the adaptive antenna array, amplitude-phase distribution of interference in the adaptive antenna array channels, the correlation matrix of interference, the weight coefficients to control the directed diagram of the adaptive antenna array.

Input parameters: *m* – quantity of elements on X axis; *n* – quantity of elements on Y axis; *dxnal, dynal* – ratios *dx/λ, dy/λ*; *az* – azimuth of interference signal source; *um*1 – elevation angle of interference signal source; *kol* – samples quantity for the interference and useful signal; *fi*0 – initial phase shift of interference; *sigma* – variance of interference signal; *AZs* – azimuth of useful signal; *UMs* – elevation angle of useful signal; *fis* – initial phase shift of useful signal.

```
function [weith,faz]=model_sign_max_trust(m,n,dx,dy,az,um1,kol,fi0,sigma,AZs,U
Ms,fis)
% Author V. Shvets, 2010
clc;
j=sqrt(-1);  s90=sind(90-az);
fi0=pi*fi0/180; fis=pi*fis/180;
rr=randn(kol,1); ri=randn(kol,1); rnd=complex(rr,ri);
% interference modeling part
for a=1:kol
    Si(a,1)=exp(j*(2*pi*a*10/kol+fi0));
end
Si=Si+sigma.*rnd;
p=0;
for ii=1:n
    for k=1:m
        p=p+1;
        for a=1:kol
            S(a,p)=Si(a,1)*exp(j*(sind(um1)*2*pi*((k-1)*dx*sind(az)+(ii-
1)*dy*s90)));
        end
    end
end
N=S'; Ns=sum(N); Ns=Ns'; clear N;
for i=1:p
    N(:,i)=Ns;
end
R1=corrcoef([S N]);
R=R1(1:p,1:p);
% modellign of amplitude-phase distribution of the signal
p=0;
for ii=1:n
    for k=1:m
        p=p+1;
        v(p)=exp(j*(sind(UMs)*2*pi*((k-1)*dx*sind(AZs)+(ii-1)*dy*sind(90-
```

```
AZs))));
     end
end
vs=v';
R=inv(R);
znam1=v*R; znam=znam1*vs;
chisl=R*vs;
w=chisl/znam;
weith=abs(w)/max(abs(w)); faz=angle(w)*180/pi;
[F]=fopen('weith.txt','wt');[F1]=fopen('faz.txt','wt');
fprintf(F,'%f\n',weith); fprintf(F1,'%f\n',faz);
fclose(F); fclose(F1);
```

Main function ***AAR_max_trust(m,n,dx,dy,az1,um1,kol,fio,sigma,AZs,UMs,fis)*** combines the presented above function related with correcponding weigh coefficients calculations accoding to the Maximum Likelihood criterion and presented in previous illustrative example function for calculation and plotting the beam pattern of adaptive antenna array.

```
function AAR_max_trust(m,n,dx,dy,az1,um1,kol,fio,sigma,AZs,UMs,fis)
% Author V. Shvets, 2010
clc
[weith,faz]=model_sign_max_trust(m,n,dx,dy,az1,um1,kol,fio,sigma,AZs,UMs,fis);
w_plargrad(m,n,dx,dy);
```

The results of modeling are presented at the Figure 9 for the following input parameters (*the same as in previous illustrative example*): interference azimuth 241°, interference elevation angle 28°, matrix of adaptive antenna array 2×2, *dx, dy* = 0.25, interference variance – 1, azimuth of useful signal 110°, elevation angle of useful signal 18°, phase shift of useful signal and interference 0°.

It can be seen from the Figure 9 that null is formed in the interference direction, i.e. the results coincide with the results from the previous illustrative example (when modeling was done for the Mean Square Error criterion).

However, this criterion has such a disadvantage that it should be available full information about wavefront signal, i.e., it is necessary to know in advance the direction on the source of the useful signal.

Illustrative Example for the Minimum Noise Variance Criterion

The details on this criterion are provided in (Monzingo and Miller, 2004), chapter 3.3.4. The corresponding functions in Matlab have been created by the authors.

Function ***model_sign_min_sigma*** calculates the amplitude-phase distribution of interference in adaptive antenna array channels, the correlation matrix of interference, weight coefficients to control the directed diagram of the adaptive antenna array.

Input parameters: *m* – quantity of elements on X axis; *n* – quantity of elements on Y axis; *dxnal, dynal* – ratios dx/λ, dy/λ; *az* – azimuth of interference signal source; *um*1 – elevation angle of interfer-

Figure 9. The beam pattern of adaptive antenna array using Ml criterion for weight coefficients calculation

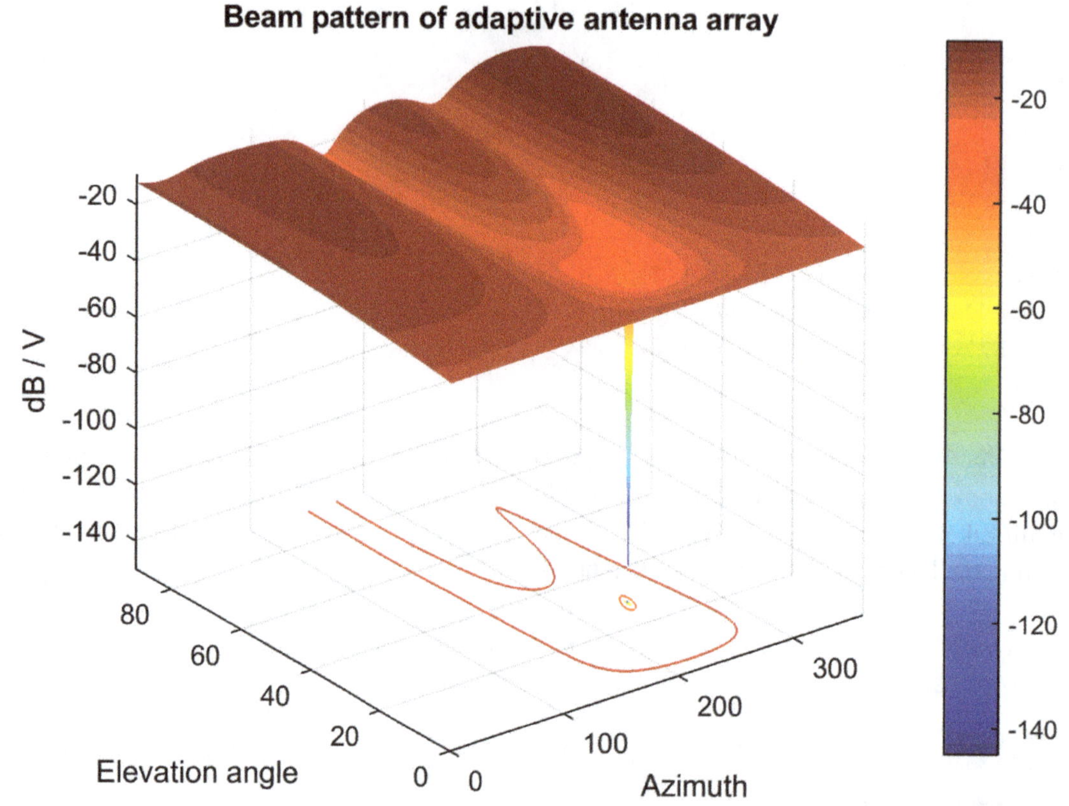

ence signal source; *kol* – samples quantity for the interference and useful signal; *fi0* – initial phase shift of interference; *sigma* – variance of interference signal.

```
function [weith,faz]=model_sign_min_sigma(m,n,dx,dy,az1,um1,kol,fi0,sigma)
% Author V. Shvets, 2010
clc;
j=sqrt(-1);  s90=sind(90-az);
fi0=pi*fi0/180;
rr=randn(kol,1); ri=randn(kol,1); rnd=complex(rr,ri);
% interference modeling part
for a=1:kol
    Si(a,1)=exp(j*(2*pi*a*1/kol+fi0));
end
Si=Si+sigma.*rnd;
p=0;
for ii=1:n
    for k=1:m
        p=p+1;
        for a=1:kol
```

```
            S(a,p)=Si(a,1)*exp(j*(sind(um1)*2*pi*((k-1)*dx*sind(az)+(ii-
1)*dy*s90)));
        end
    end
end
N=S'; Ns=sum(N); Ns=Ns'; clear N;
for i=1:p
    N(:,i)=Ns;
end
R1=corrcoef([S N]);
R=R1(1:p,1:p);
% unity vector modeling
p=0;
for ii=1:n
    for k=1:m
        p=p+1;
        v(p)=1;
    end
end
vs=v';
R=inv(R);
znam1=v*R; znam=znam1*vs;
chisl=R*vs;
w=chisl/znam;
weith=abs(w)/max(abs(w)); faz=angle(w)*180/pi;
[F]=fopen('weith.txt','wt');[F1]=fopen('faz.txt','wt');
fprintf(F,'%f\n',weith); fprintf(F1,'%f\n',faz);
fclose(F); fclose(F1);
```

Main function ***AAR_min_sigma(m,n,dx,dy,az1,um1,kol,fio,sigma)*** combines the presented above function related with correcponding weigh coefficients calculations accoding to the Minimum Noise Variance criterion and presented in previous illustrative example function for calculation and plotting the beam pattern of adaptive antenna array.

```
function AAR_min_sigma(m,n,dx,dy,az1,um1,kol,fio,sigma)
% Author V. Shvets, 2010
clc
[weith,faz]=model_sign_min_sigma(m,n,dx,dy,az1,um1,kol,fio,sigma);
w_plargrad(m,n,dx,dy);
```

A disadvantage of this method relies upon the necessity to receive the useful signal by means of the beam pattern. The results of modeling for Minimum Noise Variance Criterion calculation are presented at the figures below.

It can be seen from the Figure 10 that null is formed in the interference direction, i.e. the results coincide with the results from the previous illustrative examples.

Theoretical Background and Illustrative Example for the Proposed Adaptive Method of Weight Coefficients Calculations Without Knowledge on Reference Useful Signal

The above criteria for the weight coefficients calculation imply that the direction to the useful signal source is known, whether the amplitude-phase signal distribution is known in the channels of the antenna array. Therefore, these algorithms implementation requires a priori knowledge about useful signal or reference signal formation in the receiving channel. The reference signal can be formed from a received mixture of the signal with interference.

Analytically the reference signal is calculated according to the expression (1.26) from (Losev, 1988), and the weight coefficients from the expression (1.27) from (Losev, 1988). As it follows from those expressions, to ensure the stability of functioning of the adaptive device it is necessary to choose correctly the value of $1/\gamma$. If this is not done correctly, then the increase of weight coefficients will cause the system failure. As well to form the reference signal it is proposed in (Losev, 1988) to transfer service combinations, however, it is not possible to do in GNSS systems (GLONASS, GPS, and GALILEO

Figure 10. The beam pattern of adaptive antenna array using MNV criterion for weight coefficients calculation

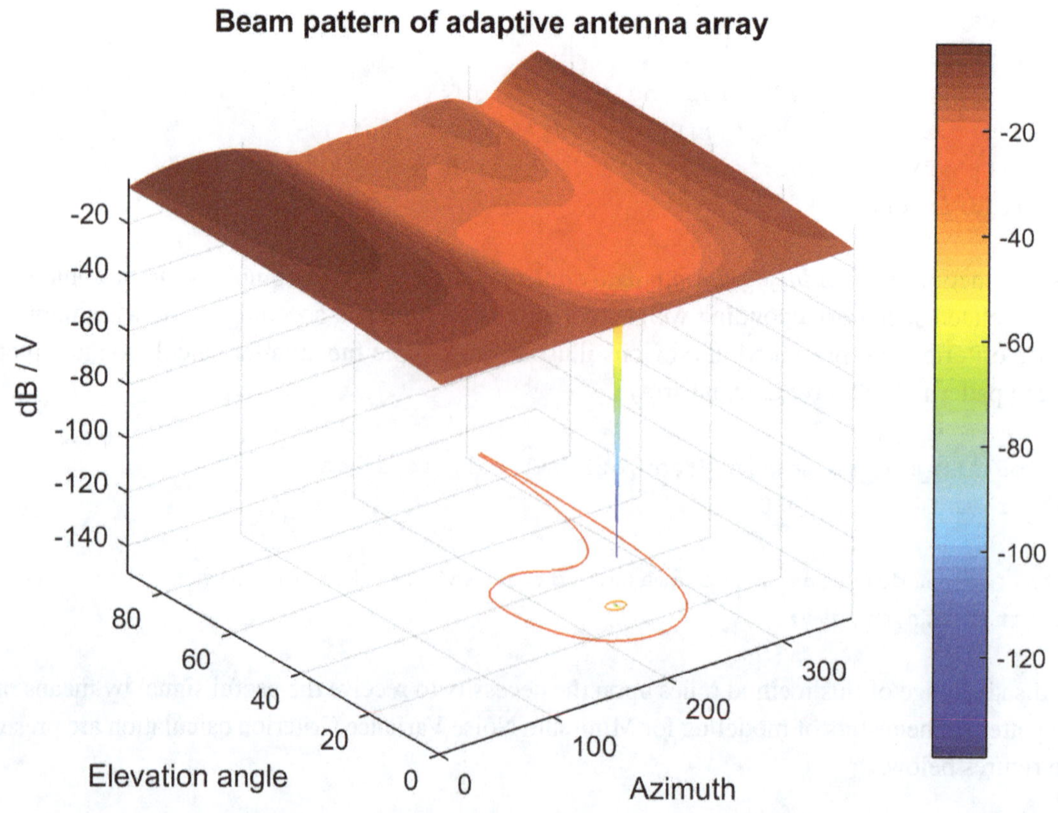

do not foresee transfer of any additional information). Therefore to ensure the stability of functioning of the adaptive array system it is necessary to eliminate the dependence of weight coefficients vector from the reference signal or to remove the signal component from the signal and interference mixture.

Now we analyze the influence of signal component on the estimation of the interference correlation matrix, which has an impact on the accuracy of interference coordinates determination.

In contradistinction to the previous methods, there is no possibility to form the reference channel in the antenna array for GNSS. Therefore useful signal and interference will be received by the same antennas and processed by the same receiving channel. And due to this fact, the mutual correlation of useful signal and interference will appear, which will cause the changes in interference correlation matrix, and this, in turn, will cause the change of the weight coefficient vector.

Also, to decrease the influence of non identity of adaptive antenna array channels (non identity of the frequency characteristics of the receivers, constructive non identity of the adaptive antenna array elements, the difference of quadrature channels in amplification gain and phase shift, and other factors) on the calculation of the weight coefficients, it is necessary to exclude the signal component from the calculation of the interference correlation matrix. To do this it is recommended to exclude the signal from the signal/interference mixture (Grigoriev, 2010) by means of filtration. However, after a detailed analysis of this method, it has been noticed the significant change in interference correlation function. As well this change of interference correlation function depends on the filter delay-bandwidth.

To eliminate (remove) this dependence, the method for calculation of weight coefficients without reference signal has been developed, which is described by the functional scheme (Figure 11). The condition of this method implementation is that the random vectors of signal and interference are non-correlated **cov**$(u_c,u_п)=0$.

The frequency shifter (converter) transforms the signal with the frequency f_1 in a signal with frequency f_2, which are constant every time. Then the rejection filter cuts the signal component and feeds the voltage of interference with the inverse sign to the adder, at the output of which we get the voltage of the useful signal. The voltage of interference we get from the subtraction of signal voltage from the "signal/interference" mixture. The double subtraction is proposed with the aim not to introduce changes in the interference correlation matrix.

To calculate the weight coefficients vector the authors use the criterion of minimal mean square error.

An error signal is calculated as a difference:

$$\varepsilon = w^T u_{sj} - S, \tag{13}$$

where $\mathbf{u}_{sj}$ is a voltage, mixture of the useful and jamming signal, S - reference (useful) signal copy.

Square of an error:

$$\varepsilon^2 = S^2 - 2Sw^T u_{sj} + w^T u_{sj}^* u_{sj}^T w .$$

After finding mathematical expectation ε^2, it is possible to find the expression for the mean square error:

Figure 11. Adaptive antenna system without the reference signal (Shvets V. and Shvets O., 2013)

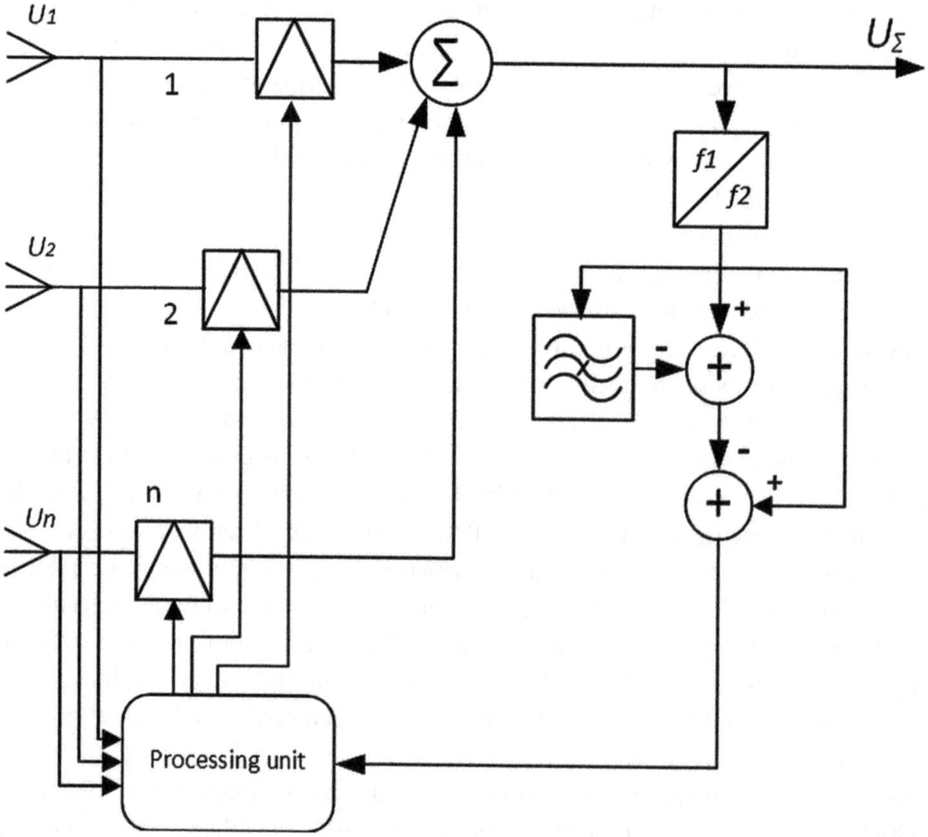

$$E\left[\varepsilon^2\right] = S^2 + w^T COV_{sj} w - 2w^T r_{sj,s}, \tag{14}$$

where $\mathbf{r}_{\mathrm{sj,s}} = \overline{\mathbf{u}^*_{\mathrm{sj}} S}$, E – mean value, $\mathbf{COV}_{\mathrm{sj}}$ – covariance matrix of the useful signal and interference mixture.

Covariance matrix $\mathbf{COV}_{\mathrm{sj}}$ will be equal to the sum of covariance matrices of interferences $\mathbf{COV}_{\mathrm{j}}$ and useful signal $\mathbf{COV}_{\mathrm{s}}$, $\mathbf{COV}_{\mathrm{sj}} = \mathbf{COV}_{\mathrm{j}} + \mathbf{COV}_{\mathrm{s}}$. When forming the error signal the covariance matrix is calculated as the following:

$$\mathbf{COV}_{\mathrm{sj}} = \mathbf{COV}_{\mathrm{j}} + \mathbf{COV}_{\mathrm{s}} - \mathbf{COV}_{\mathrm{s}} = \mathbf{COV}_{\mathrm{j}}.$$

The vector $\mathbf{r}_{\mathrm{sj,s}}$ in expression (14) is a correlation vector of voltages $\mathbf{u}^*_{\mathrm{sj}i}$ and S (i=1,2, …, N), at $S^*_i = 0$, i.e., when the reference signal is abcent $\mathbf{r}^{\mathrm{T}}_{\mathrm{sj,s}} = \left[\overline{\mathbf{u}^*_{\mathrm{sj1}}}\, \overline{\mathbf{u}^*_{\mathrm{sj2}}} \ldots \overline{\mathbf{u}^*_{\mathrm{sj}N}}\right]$, where $\mathbf{u}^*_{\mathrm{sj}i} = \mathbf{u}^*_{\mathrm{j}i}$.

Otherwise, $\mathbf{r}^{\mathrm{T}}_{\mathrm{sj,s}} = \mathbf{r}^{\mathrm{T}}_{\mathrm{j}}$.

Then the expression (14) will have the following form:

$$E\left[\varepsilon^2\right] = w^T COV_j w - 2w^T r_j, \tag{15}$$

Optimization of expression (15) is performed by minimization due to an appropriate choice of weight coefficients **w**. Since as mean square error (15) is a quadratic function from vector **w,** it has the single minimum, which could be found when differentiating expression (15) and put it to be equal to zero

$$\frac{dE\left[\varepsilon^2\right]}{dw} = 0 \text{. Since as } \frac{dE\left[\varepsilon^2\right]}{dw} = 2COV_j w - 2r_j, \tag{16}$$

Then the vector of weight coefficients will be defined according to the following expression

$$\mathbf{w} = \mathbf{COV}_j^{-1}\mathbf{r}_j. \tag{17}$$

Since as $\mathbf{COV}_j = \sigma_j \mathbf{R}_j$, then the expression (17) could be written in the following form

$$\mathbf{w} = \sigma_j \mathbf{R}_j^{-1}\mathbf{r}_j.$$

The expression (17) corresponds to the Wiener-Hopf expression regarding calculation of the weight coefficients in the adaptive antenna array.

The results of mathematical modeling are presented below.

Function ***model_sign_min_ bez_opsign*** calculates the amplitude-phase distribution of interference in adaptive antenna array channels, weight coefficients to control the directed diagram of the adaptive antenna array.

Input parameters: *m* – quantity of elements on X axis; *n* – quantity of elements on Y axis; *dx, dy* – ratios *dx/λ, dy/λ*; *az* – azimuth of interference signal source; *um*1 – elevation angle of interference signal source; *kol* – samples quantity for the interference and useful signal; *fi*0 – initial phase shift of interference; *sigma* – variance of interference signal.

```
function [weith,faz]=model_sign_bez_opsign(m,n,dx,dy,az,um1,kol,fi0,sigma)
% Author V. Shvets, 2012
clc;
j=sqrt(-1);
fi0=pi*fi0/180;
rr=randn(kol,1); ri=randn(kol,1); rnd=complex(rr,ri);
% modeling the interference
for a=1:kol
    Si(a,1)=exp(j*(2*pi*a*1/kol+fi0));
end
Si=Si+sigma.*rnd;
p=0;
for ii=1:n
```

```
    for k=1:m
        p=p+1;
        for a=1:kol
          S(a,p)=Si(a,1)*exp(j*sind(um1)*2*pi*((k-1)*dx*sind(az)+(ii-
1)*dy*sind(90-az)));
        end
    end
end
R=corrcoef(S);
% modeling the interference distribution vector in adaptive antenna array
channels
v=randn(1,m*n);
v=v';
R=inv(R); w=R*v;
weith=abs(w)/max(abs(w)); faz=angle(w)*180/pi;
[F]=fopen('weith.txt','wt');[F1]=fopen('faz.txt','wt');
fprintf(F,'%f\n',weith); fprintf(F1,'%f\n',faz);
fclose(F); fclose(F1);
function AAR_bez_opsign(m,n,dx,dy,az1,um1,kol,fio,sigma)
% Author V. Shvets, 2012
 [weith,faz]=model_sign_bez_opsign(m,n,dx,dy,az1,um1,kol,fio,sigma);
w_plargrad(m,n,dx,dy);
```

The results of modeling are presented at the Figure 12 for the following input parameters: interference azimuth 241°, interference elevation angle 28°, matrix of adaptive antenna array 2×2, *dx, dy* = 0.25, interference variance – 1, phase shift of useful signal and interference 0°.

It can be seen from the Figure 12 that null in the beam pattern is formed in the interference direction, i.e. the results coincide with the results from the previous illustrative examples.

Theoretical Background and Illustrative Example for the Proposed Direct Estimation Method of Correlation Matrix Using Pulse Characteristics of Spatial Filter

The considered above methods in the illustrative examples (Mean Square Error, Maximum Likelihood, Minimum Noise Variance criterions) imply some apriori knowledge about signal or interference characteristics. In practical cases, the interference characteristics are not known in advance and are constantly changing. Therefore, it is necessary to update values of the weight vector to be able to track the change in signal-interference environment. It results in a constant estimation of signal vector s and interference matrix R.

Therefore, it is proposed to apply direct methods of interference correlation matrix estimation (Gilmore, 1998). However, the main difficulty in this method is in the inversion of poorly conditioned interference matrix R.

Let us suppose that the source of the interference is a white noise of some power. Then the interference spectrum will form the following elements (See Figure 13), i.e. the filter F and the interference amplifier

Figure 12. The beam pattern of adaptive antenna array using proposed method for weight coefficients calculation without the reference signal

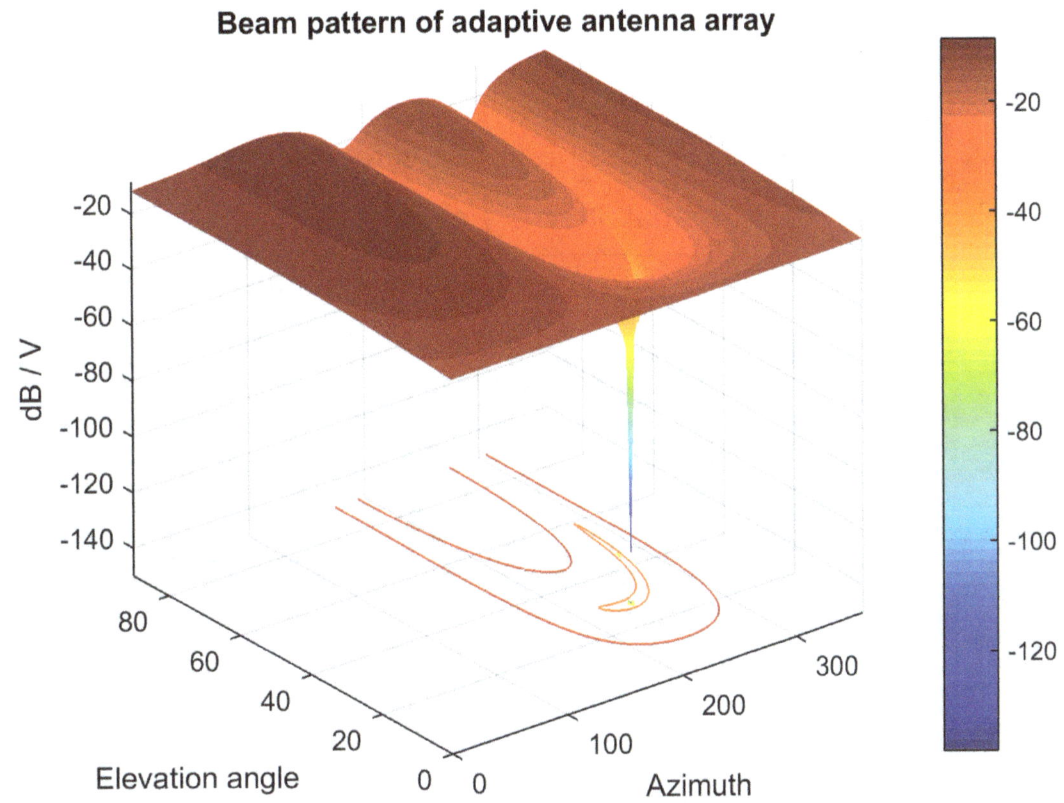

U, the propagation medium, A1, A2, A3, A4-antennas and the radio path of the adaptive antenna array. All these elements will constitute a spatial filter with certain impulse characteristics.

The spectral power of the interference density is determined by expression (4). Usually, the spectral density $G(\omega)$ becomes negligible starting with certain high-frequency ω_c. Then the interference $\xi(t)$ can be rather accurately replaced with the process $\xi_0(t)$ with energy spectrum according to the expression (5). As it was mentioned earlier, the random process $\xi_0(t)$ is considered as a result of continuous white noise influence $x(t)$ with a limited by frequency ω_c spectrum on a continuous linear system with transfer function determined by the relation (6). The transfer function satisfying all necessary conditions is determined by the relation (7), and its impulse response characteristic is defined by relation (8).

To express the impulse characteristics of the linear system (filter) in terms of the interference correlation function we substitute (4) in square brackets of relation (8), and get the following:

$$h(t) = \frac{1}{\pi}\int_0^{\omega_c}\left[\frac{1}{G_0}\int_0^{\tau} R(\tau)e^{-j\omega\tau}d\tau\right]^{\frac{1}{2}} e^{i\omega t}d\omega\,. \tag{18}$$

Expression (18) can be written in a matrix form for discrete signals

Figure 13. Elements comprising a spatial filter (Shvets and Kharchenko, 2017)

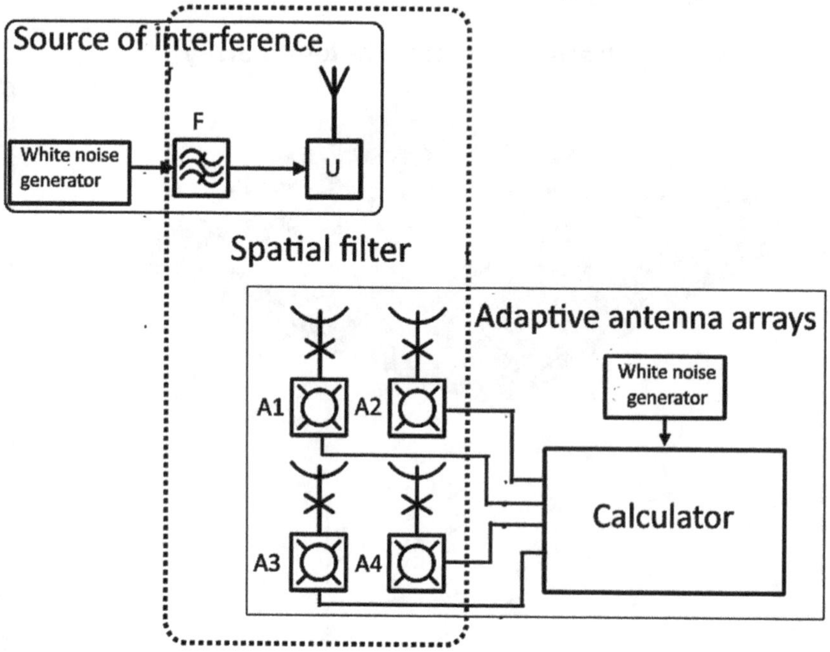

$$\mathbf{h} = \mathbf{W}^* \left[\mathbf{Wr}\right]^{\frac{1}{2}}, \tag{19}$$

where $\mathbf{W} = e^{-j\frac{2\pi kn}{N}}$ is a matrix of turning factors of the discrete Fourier transform, **W*** is acomplex conjugate matrix to **W**, **r** is a correlation function vector, **h** is an impulse characteristics vector.

Proceeding from the fact that AAA is an element of a spatial filter, expression (19) can be written for a two-dimensional form of dimension $N \times N$

$$\mathbf{H} = \mathbf{W}^* \left[\mathbf{WR}\right]^{\frac{1}{2}}, \tag{20}$$

where **R** – interference correlation matrix.

Let us find the inverse correlation matrix from (20). After taking the squaring of (20), we obtain

$$\mathbf{H}^2 = \mathbf{W}^{*2}\mathbf{WR}. \tag{21}$$

Let us multiply both elements from the right side of (21) by $\mathbf{R}^{-1}$

$$\mathbf{H}^2\mathbf{R}^{-1} = \mathbf{W}^{*2}\mathbf{WRR}^{-1} = \mathbf{W}^{*2}\mathbf{WI} = \mathbf{W}^{*2}\mathbf{W}. \tag{22}$$

Let us multiply both elements from the left side of (22) by $(\mathbf{H}^2)^{-1}$

$$\begin{aligned}&\left(\mathbf{H}^2\right)^{-1}\mathbf{H}^2\mathbf{R}^{-1}=\left(\mathbf{H}^2\right)^{-1}\mathbf{W}^{*2}\mathbf{W},\\&\mathbf{I}\mathbf{R}^{-1}=\left(\mathbf{H}^2\right)^{-1}\mathbf{W}^{*2}\mathbf{W},\\&\mathbf{R}^{-1}=\left(\mathbf{H}^2\right)^{-1}\mathbf{W}^{*2}\mathbf{W}.\end{aligned} \tag{23}$$

where: $\mathbf{R}^{-1}$ — inverse correlation matrix of interference, **H** – impulse characteristics matrix of a spatial filter, **I** – unit matrix. Thus, (23) is an expression for determining the inverse correlation matrix of the interference.

In expression (23), the matrix **H**, like the matrix **R**, is unknown. In order to get the impulse characteristics of the filter, we should use the *system identification* method. By the term *system identification,* we mean the definition of the impulse response *h (n)* if it is not known.

The output of the system is connected with its input by the following relation (convolution)

$$y(n)=\sum_{m=0}^{n}h(m)x(n-m)=h(n)\otimes x(n)\,, \tag{24}$$

where: $y(n)$ – output sequence (interference with correlation function $\boldsymbol{r}$), $x(n)$ – input sequence (normal white noise), $h(n)$ – the desired impulse characteristics of the shaping filter.

Thus, having a white noise generator from the input sequence (i.e. from the outputs of the intermediate frequency amplifiers) embedded in AAA, the system identification method determines the impulse characteristics of the spatial filter necessary to calculate the inverse correlation matrix of the interference (See Figure 13).

Figure 14 shows the impulse characteristics of the four AAA channels obtained by the system identification method, since the phase front of the interference depends on the direction of its arrival (angles θ and φ), so the impulse characteristics of the AAA channels will differ and carry the information about the correlation function of the interference (18).

Figure 15 shows the AAA beam pattern obtained using the expression (23).

Figure 15 demonstrates that the null is formed on the direction of the interference source in the beam pattern correctly.

In order to increase the interference immunity of NSS while being a part of the GNSS equipment, it is possible to use AAAs, which allow to significantly reducing the signal reception coefficient from the direction where the source of interference is located.

To achieve the result it is proposed to use a qualitatively new method of direct calculation of the inverse correlation matrix based on the impulse response of the spatial filter (AAA). The proposed method makes it possible to reduce computational costs, due to the predetermined matrices of the turning multipliers of the Fourier transform. The matrix of the impulse characteristics is well-conditioned, so it is possible to calculate the inverse matrix of the impulse characteristics from it. Using the proposed method, it is possible to compensate for instrumental errors introduced by the design parameters of the AAA, which in turn reduces the error while measuring the direction of the source of interference.

Figure 14. Adaptive antenna array channel impulse characteristics view

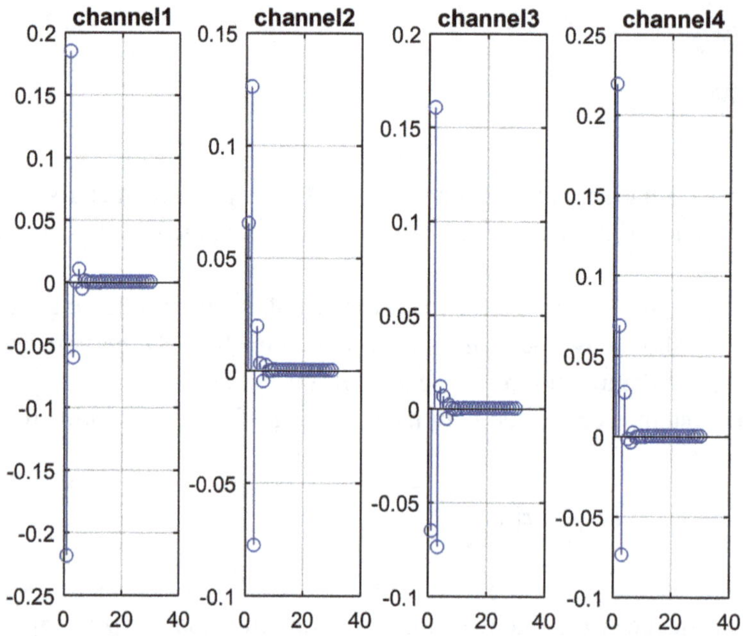

Figure 15. AAA beam pattern with generated null in the direction of the interference source

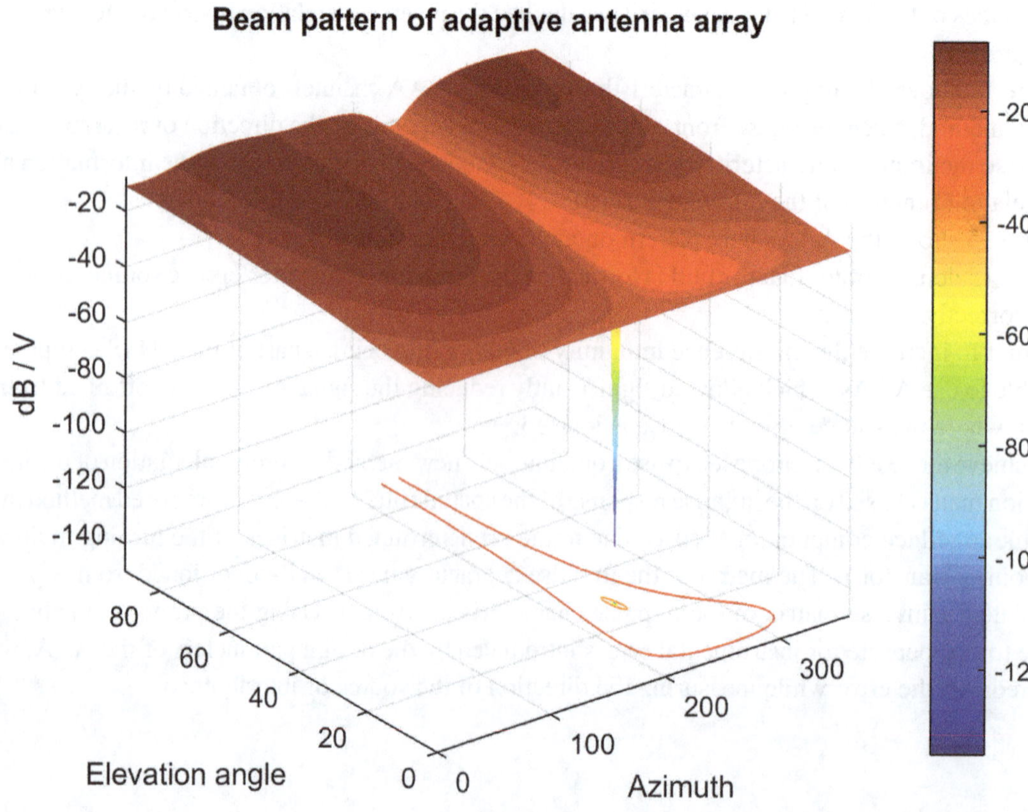

FUTURE RESEARCH DIRECTIONS

Based on the method of pulse characteristics of the spatial filter the model of adaptive interferences compensator for GNSS could be further developed. Another direction of research is a creation of an adaptive antenna array model, consisting of three and four (2x2) patch antennas. Correspondently it would be necessary to create a model of radio receiving channels (for 3 and 4 elements). The next step of research might include investigation of adaptive compensator properties (like the accuracy of determination of interference source direction, interference attenuation gain, etc) depending on constructive parameters of the antenna array and radio receiving channels.

It would be interesting as well to investigate the influence of different types of interferences on the above-mentioned adaptive compensator properties. Finally, the prototype of the adaptive compensator of GNSS interferences could be developed using the software-defined radio technologies.

CONCLUSION

Many of the safety-critical applications rely on position, navigation and (or) time information, coming from GNSS. The navigation receivers are potentially exposed to deliberate interferences due to the very weak signal power near the Earth's surface. Recently, the GNSS users concern more on intentional interferences (jamming attacks), since at the modern level of radio-electronic components development it becomes quite easy to create own jamming device, and there are already a lot of such devices available on the market. To ensure necessary requirements to the accuracy, integrity, and availability of GNSS under influence of unintentional and deliberate GNSS interferences and increase the jamming immunity of consumer receivers it has been proposed to implement the adaptive antenna arrays.

Unfortunately, the available beamforming technologies successfully used in radiolocation and radio communication could not be directly applied for GNSS, since they have the different structure of the signals. In a contrast to radio location and communication systems, in GNSS systems the frequency-time structure of a useful signal is not known in advance, which excludes the possibility of using a number of widely used methods of adaptive interference compensation with the use of a reference signal. Therefore, the authors have developed and presented the methods for weight coefficients calculations to form the nulls in the beam pattern of the adaptive antenna array in the direction of the interference source.

The proposed by authors methods have been compared with the known optimal methods for calculating weight coefficients (mean square error, maximum likelihood, minimum noise variance). The comparative analysis of methods for the equal test conditions approved the efficiency of proposed methods. The illustrative examples, including fragments of MatLab codes with modeling results, for all methods are presented in a chapter. It could be seen from the examples that the null in the beam pattern is formed correctly and equally by all considered methods. The proposed by authors methods provide some additional benefits as well. One of them is the absence of the requirement of knowledge about GNSS signal parameters (direction of arrival, signal structure, etc). Another one is in an elimination of the problem of poor conditioning or even singularity of interference correlation matrix in the process of weight coefficients calculations for beam pattern control.

All considered methods have been tested for different interferences models using the approaches for random variables modeling with the specified probability distribution and correlation functions, presented in a chapter.

REFERENCES

Bond, L. (1998). Overview of GPS Interference Issues. In *GPS Interference Symposium* (pp. 341-352). Boston: Volpe National Transportation System Center.

Boynton, F. (2014). *GNSS interference / denials, needs and challenges*. Retrieved from https://gpsworld.com/wp-content/uploads/2014/05/Loctronix-2014-GNSS-Webinar-140521-final.pdf

Brown, A., & Gerein, N. (2001). *Test Results of a Digital Beamforming GPS Receiver in a Jamming Environment*. Retrieved from: https://pdfs.semanticscholar.org/cde4/47bd0435360900e16c9c15e417067db27443.pdf

Brown, A., & Mathews, B. (2007). *Test Results from a Novel Passive Bistatic GPS Radar Using a Phased Sensor Array*. Retrieved from: https://pdfs.semanticscholar.org/481e/1afcc5652ac831492d66556301938f6a603d.pdf

Bykov, V.V. (1971). *Tsifrovoe modelirovaniye v statisticheskoy radiotekhnike* [Digital modeling in statistical radio technique]. Moscow: Sovetskoye radio.

Colby, G. (1997). Test Results of the Joint FAA/DoD Investigation of GPS Interference. *Proceedings of the 10th International Technical Meeting of the Satellite Division of the Institute of Navigation, GPS-97*, 217-228.

Corrigan, T. M., Hartranft, J. F., Levy, L. J., Parker, K. E., Pritchett, J. E., Pue, A. J., ... Thompson, T. (1999). *GPS Risk Assessment Study: Final Report*. The John Hopkins University Applied Physics Laboratory.

Dzhigan, V. I. (2013). *Adaptivnaja fil'tracija signalov: teorija i algoritmy* [Adaptive filtering of signals: theory and algorithms]. Moscow: Tehnosfera.

Federal Aviation Administration. (2013a). *Global positioning system directorate systems engineering and integration*. Interface Specification IS-GPS-800.

Federal Aviation Administration. (2013b). *GPS, "NAVSTAR GPS Control Segment to User Support Community Interfaces", ICD-GPS-870, rev. B*. Author.

Forssel, B., & Olsen, T. B. (2003). Jamming Susceptibility of Some Civil GPS Receivers. *GPS World*, *1*, 54–58.

Gao, G. X., Sgammini, M., Lu, M., & Kubo, N. (2016). Protecting GNSS Receivers from Jamming and Interference. *Proceedings of the IEEE*, 1327–1338. 10.1109/JPROC.2016.2525938

Gheethan, A. A., Herzig, P. A., & Mumcu, G. (2013). Compact 2 × 2 coupled double loop GPS antenna array loaded with broadside coupled split ring resonators. *IEEE Transactions on Antennas and Propagation*, *61*(6), 3000–3008. doi:10.1109/TAP.2013.2253539

Gilmore, S. (1998). The Impact of Jamming on GPS. In *Symposium on GPS Interference and Mitigation Technique*, (pp. 467-478). Boston: Volpe National Transportation System Center.

Grigoriev, L. N. (2010). *Tsifrovoe formirovaniye diagrammy naprvlennosti v fazirovannikh antennikh reshetkakh* [Digital forming of directional diagram in phased-array antennas]. Moscow: Radiotekhnika.

Hansen, A. J. (1997). *WAAS Precision Approach Metrics: Accuracy*. Integrity, Continuity and Availability.

Huang, J., Lo Presti, L., Motella, B., & Pini, M. (2016). GNSS spoofing detection: Theoretical analysis and performance of the Ratio Test metric in open sky. *ICT Express. Elsevier*, *2*(1), 37–40. doi:10.1016/j.icte.2016.02.006

International Civil Aviation Organization. (2000). *Validated ICAO GNSS Standards and Recommended Practices*. SARPS.

International Civil Aviation Organization. (2014). *Aeronautical Telecommunications Annex 10 to the Convention on International Civil Aviation, Volume I (Radio Navigation Aids)*. Author.

Key, E. L. (1995). *Technique to Counter GPS Spoofing*. International Memorandum, MITRE Corporation.

Korotonoshko, A.N., & Perunov, Y.M. (2006). *Stoykost' i radiotekhnicheskiy immunitet transportnykh sistem, osnovannykh na tochnoy sputnikovoy navigatsii* [Persistence and radio technical immunity of transport systems relying on precise satellite navigation]. Moscow: Novosti navigatsii.

Korotonoshko, A.N., & Perunov, Y.M. (2007). *Imitatsionniye radiotekhnicheskiye pomekhi sistemam spitnikovoy navigatsii* [Imitating radio technical interferences to the satellite navigation systems]. Moscow: Novosti navigatsii.

Konin V. & Harchenko V. (2010) *Sistemy sputnikovoj radionavigacii* [Satellite radionavigation systems]. Kyiv: Holtech Reference.

Kuusniemi, H. (2012). *Effects of GNSS jammers and potential mitigation approaches*. Retrieved from: http://www.unoosa.org/pdf/sap/2012/un-latvia/ppt/3-14.pdf

Littlepage, R. S. (1998). The Impact of Interference on Civil GPS. *Proceedings ION, GPS-98*, 821–828.

Liu, Y., Li, S., Fu, Q., & Liu, Z. (2018). Impact assessment of GNSS spoofing attacks on INS/GNSS integrated navigation system. *Sensors (Switzerland)*, *18*(5), 1433. doi:10.339018051433 PMID:29734715

Losev, Y.I. (1988). *Adaptivnaya compensatsiya pomekh v kanalakh svyazi* [Adaptive compensation of interferences in communication channels]. Moscow: Radio i svyaz.

Lyusin, S. V., Sazonov, L. B., Khazanov, I. G., & Komarov, A. S. (1998). Combined GPS/GLONASS Receiver with High Antijamming Performance. *Proceedings ION, GPS-98*(September), 775–782.

Magiera, J., & Katulski, R. (2015). Detection and mitigation of GPS spoofing based on antenna array processing. *Journal of Applied Research and Technology*, *13*(1), 45–57. doi:10.1016/S1665-6423(15)30004-3

Radio Technical Commission for Aeronautics. (1998). *Minimum Aviation Performance Standards for the Automatic Dependent Surveillance Broadcast (ADS-B), RTCA DO-247*. Author.

Monzingo, R. A., & Miller, T. W. (2004). *Introduction to adaptive arrays*. SciTech Publishing, Inc.

Oumar, O. A., Siyau, M. F., & Sattar, T. P. (2012). Comparison between MUSIC and ESPRIT direction of arrival estimation algorithms for wireless communication systems. *The First International Conference on Future Generation Communication Technologies*, 99-103. 10.1109/FGCT.2012.6476563

Ovcharenko, L. A., & Poddybniy, V. N. (1992). *Pomekhoustoychivost priema fazomanipulirovannykh signalov na fone naiboleye neblagopriyatnykh pomekh* [Interference immunity of receiving phase-shift keying signals with the most unfavorable interferences]. Moscow: Radiotekhnika.

Pinker, D., & Walker, C., (1998). Jamming the GPS signal. *Proceedings ION-98*, 829-837.

Psiaki, M. L., & Humphreys, T. E. (2016). GNSS Spoofing and Detection. *Proceedings of the IEEE, 104*(6), 1258–1270. doi:10.1109/JPROC.2016.2526658

The Commission. (2001). Report of the Commission to Assess United States National Security Space Management and Organization.

Ratynskiy, M.V. (2003). *Adaptatsiya i sverkhrazresheniye v antennikh reshetkakh* [Adaptation and superresolution in antenna arrays]. Moscow: Radio i svyaz.

Rodgers, C. (1991). Development of a Low Cost PC Controlled GPS Satellite Signal Simulator. In *Proceedings of the 15th Biennial Guidance Test Symposium.* Holloman AFB.

Radio Technical Commission for Aeronautics. (2008). *Assessment of radio frequency interference relevant to the GNSS L1 frequency band RTCA/DO-235B.* Author.

Russian Institute of Space Device Engineering. (2008). *GLONASS Interface Control Document*, Edition 5.1. Author.

Rügamer, A., & Kowalewski, D. (2015). *Jamming and Spoofing of GNSS Signals – An Underestimated Risk?!* Academic Press.

Shvets, V.A. (2012). *Eksperymentalni doslidzhennya zavadostiykosti system GPS* [Experimental studies of impedance of GPS systems]. Kyiv: Visnyk inzhenernoyi akademiyi Ukrayiny.

Shvets V.A. (2013). *Formuvannia vahovykh koefitsiientiv priamymy metodamy v antennykh reshitkakh system GPS* [Formation of weight coefficients by direct methods in antenna arrays of GPS systems]. Kyiv: Visnyk inzhenernoyi akademiyi Ukrayiny.

Shvets, V.A., Shvets, O.V. (2013). *Formuvannya vahovykh koefitsiyentiv bez opornoho syhnalu v adaptyvnykh antennykh reshitkakh system GPS, HALILEO, GLONASS* [Formation of weight factors without reference signal in adaptive antenna arrays of GPS, GALILEO, GLONASS systems]. Kyiv: Visnyk inzhenernoyi akademiyi Ukrayiny.

Shvets, V. A., & Kharchenko, V. P. (2017). Pulse characteristics of network satellite systems adaptive antenna for assessing correlation interference matrix. *Proceedings of the National Aviation University, 4*(73), 30 – 35. Retrieved from: http://er.nau.edu.ua:8080/handle/NAU/32567

Shvets, V.A., Kasperovych, N.O., & Ostrovskiy, Y.P. (2010). Certificate about copyright on work No 36059 from 14.12.2010. Computer program. *Modelyuvannya zavad suputnykovykh aeronavigatsiynykh system iz vyznachenoyu corelyatsiynoyu funktsiyeyu* [Modeling the interferences to satellite aeronavigation systems with defined correlation function].

Shvets, V.A., Kasperovych, N.O., Ostrovskiy, Y.P. (2011). Certificate about copyright on work No 36677 from 25.01.2011. Computer program. *Modelyuvannya zavad suputnykovykh aeronavigatsiynykh system* [Modeling the interferences to satellite aeronavigation systems].

Sushych, O.P. (2012). *Eksperymentalna otsinka vplyvu navmysnykh zavad na aparaturu spozhyvacha hlobalnoi navihatsiinoi suputnykovoi systemy* [Experimental assessment of the effect of deliberate interference on the equipment of the consumer of the global navigation satellite system]. Kyiv: Visnyk inzhenernoyi akademiyi Ukrayiny.

US Department of Defense. (1999). *US Department of Transportation*. Federal Radionavigation Plan.

John, A. (2001). *Vulnerability Assessment of the Transportation Infrastructure Relying on the Global Positioning System*. Cambridge, MA: Volpe National Transportation Systems Center. Retrieved from https://www.navcen.uscg.gov/pdf/vulnerability_assess_2001.pdf

Ward, P. (1994). GPS Receiver RF Interference Monitoring, Mitigation and Analysis Technique. *Journal of the Institute of Navigation*, *41*(4).

Journal Telesatellite International. (2010). Retrieved from: www.Tele-satellite.com

Widrow, B., & Stearns, S. (1985). *Adaptive signal processing*. Englewood Cliffs, NJ: Prentice-Hall.

Winer, B. (1996). GPS Receiver Laboratory RFI Tests. *Proceedings of the Institute of Navigation National Technical Meeting*.

Wallis, S. (1999). GPS Open Air Testing – Jamming at Woomera. *Proceedings of 1999 Technical Meeting & 19th Biennial Guidance Test Symposium*.

Wildemeersch, M. (2010). *Impact Study of Unintentional Interference on GNSS Receivers*. Tech. Rep. EUR-24742-EN. Publications Office of the European Union. doi:10.2788/57794

ADDITIONAL READING

Dovis, F. (2015). GNSS Interference Threats and Countermeasures. Edited by F. Dovis. Boston|London: ARTECH HOUSE, INC.

Isoz, O. (2012) *Interference Detection and Localization in the GPS L1 Frequency Band*. Lule°a: Universitetstryckeriet, Luleå 2012. Available at: https://www.diva-portal.org/smash/get/diva2:990261/FULLTEXT01.pdf

Kaplan, E., & Hegarty, C. (2006). *Understanding GPS Principles and Applications* (2nd ed.). London: ARTECH HOUSE, INC; Available at http://d1.amobbs.com/bbs_upload782111/files_33/ourdev_584835O21W59.pdf

Parkinson, B. W., & Spilker, J. J. (1996) *Progress In Astronautics and Aeronautics: Global Positioning System: Theory and Applications*. American Institute of Aeronautics & Astronautics (Geografia (E-libro)). Available at: https://books.google.com.ua/books?id=t0eGFpSwN0wC

Sklar, J. R. (2003). Interference Mitigation Approaches for the Global Positioning System, *LINCOLN LABORATORY JOURNAL*, 14(2), pp. 167–180. Available at: http://citeseerx.ist.psu.edu/viewdoc/download?doi=10.1.1.70.4732& rep=rep1&type=pdf

Vagle, N., Broumandan, A., & Lachapelle, G. (2016) 'Analysis of Multi-Antenna GNSS Receiver Performance under Jamming Attacks', *Sensors*. Multidisciplinary Digital Publishing Institute, 16(11), pp. 1–21. doi: . Available at: http://www.mdpi.com/1424-8220/16/11/1937 doi:10.339016111937

Ward, K. D. (2006) *Development of Smart Antenna Technology Final Report, Smart Antenna Technology*. Kidlington. Available at: https://www.ofcom.org.uk/__data/assets/pdf_file/0014/36014/smartpres1.pdf

KEY TERMS AND DEFINITIONS

Antenna Array (or Array Antenna): A set of multiple connected antennas, which work together as a single antenna to transmit or receive radio waves.

Antenna Pattern (or Radiation Pattern): The directional (angular) dependence of the strength of the radio waves from the antenna or other source.

Beamforming: A method used to create the radiation pattern of the antenna array by adding constructively the phases of the signals in the direction of the targets/mobiles desired, and nulling the pattern of the targets/mobiles that are undesired/interfering targets.

Global Navigation Satellite System (GLONASS): The global navigation satellite system provided by the Russian Federation. The nominal baseline constellation of GLONASS comprises 24 Glonass-M satellites that are uniformly deployed in three roughly circular orbital planes at an inclination of 64.8° to the equator. The altitude of the orbit is 19,100 km. The orbit period of each satellite is 11 hours, 15 minutes, 45 seconds. The orbital planes are separated by 120° right ascension of the ascending node. Eight satellites are equally spaced in each plane with 45° argument of latitude. Moreover, the orbital planes have an argument of latitude displacement of 15° relative to each other.

Global Positioning System (GPS): The global navigation satellite system provided by the United States of America. The GPS baseline constellation consists of 24 slots in six orbital planes, with four slots per plane. Three of the slots are expandable and can hold no more than two satellites. Satellites that are not occupying a defined slot in the GPS constellation occupy other locations in the six orbital planes. Constellation reference orbit parameters and slot assignments as of the defined epoch are described in the fourth edition of the GPS Standard Positioning Service Performance Specification, dated September 2008. As of that date, the GPS constellation had 30 operational satellites broadcasting healthy navigation signals: 11 in Block IIA, 12 in Block IIR and 7 in Block IIR-M.

Jamming (Electromagnetic): The deliberate use of unintended radiation, reradiation or reflection of electromagnetic energy for the purpose of preventing or reducing the effective use of a signal.

Meaconing: Re-broadcast of GPS signals in such a way as to create a stronger, but erroneous fix.

Spoofing: Broadcast of signals, which can appear to be genuine GNSS signals.

Chapter 15
Design Features of High-Performance Multiprocessor Computing Systems

Gennady Shvachych
National Metallurgical Academy of Ukraine, Ukraine

Nina Rizun
Gdansk University of Technology, Poland

Olena Kholod
Alfred Nobel University, Ukraine

Olena Ivaschenko
National Metallurgical Academy of Ukraine, Ukraine

Volodymyr Busygin
National Metallurgical Academy of Ukraine, Ukraine

EXECUTIVE SUMMARY

The chapter analyzes the ways of development of high-performance computing systems. It is shown that a real breakthrough in mastering parallel computing technologies can be achieved by developing an additional (actually basic) level in the hierarchy of hardware capacities of multiprocessor computing systems of MPP-architecture, the personal computing clusters. Thus, it is proposed to create the foundation of the hardware pyramid of parallel computing technology in the form of personal computing clusters. It is shown that on the basis of multiprocessor information systems processing and control, the control systems are implemented for many industries: space industry, aviation, air defense and anti-missile defense systems, and many others. However, the production of multiprocessor information processing and control systems is hampered by high cost at all its stages. As a result, the total cost of the system often makes it as an inaccessible tool. The use of modern multiprocessor cluster systems would reduce the costs of its production.

DOI: 10.4018/978-1-5225-7588-7.ch015

BACKGROUND

The paper analyzes the ways of development of high-performance multiprocessor computing systems. It is shown that a real breakthrough in mastering parallel computing technologies can be achieved by the development of an additional (actually, basic) level in the hardware hierarchy of multiprocessor computing systems of the *MPP* architecture - personal computing clusters. Thus, it is proposed to create the pyramid foundation of parallel computing technology hardware in the form of personal computing clusters. The scope of such systems is mastering parallel computing technologies, creation and debugging of parallel programs, including problem-oriented packages and libraries, as well as a simulation run of the software developed.

It is shown that on the basis of multiprocessor information and control systems, the latter are implemented for many industries: space industry, aviation, anti-aircraft and anti-missile defense systems, and many others. However, the production of multiprocessor information and control systems is hampered by high operational cost at all its stages. As a result, the overall cost of a system often makes it an inaccessible tool. Using modern multiprocessor clustered computing systems would reduce the cost of its production.

Currently, significant interest in the construction of parallel multiprocessor computing systems (MCS) is determined by the use of standard public technologies and components (Bashkov et al., 2011; The Blackford MultiCore cluster specification, http://www.mvs.icc.ru/cluster_info.html; Ivaschenko et al., 2013; Latsys, 2003). This is due to a number of factors. Let's emphasize the main ones. First, according to market needs, the performance growth of the standard network technologies such as *GI(Gigabit Ethernet), FC (Fiber Channel)* and *IB (InfiniBand)* allows them to be seen as a communication medium for multiprocessor computing systems made by the *NUMA (Non-Uniform Memory Access)* (Non-Uniform Memory Architecture (NUMA)). Secondly, an essential factor was the growing popularity of the *Linux* freely distributed operating system. At the first stage of its use, it was positioned as the *UNIX* platform for the platforms based on Intel architecture, but relatively fast versions appeared for other popular microprocessors, including the leader in performance over the past few years, the *Alpha* microprocessors.

Depending on the problems features and the budget, the project system can have all kinds of configuration options. The most affordable standard motherboard is the *Core LGA1155 Z77* motherboard, which includes two free *PCI Express* 3.0 8 line slots on the Core i7-4790K 4 GHz platform and *Gigabit Ethernet* network adapters and *FDR InfiniBand*, which allows increasing performance, reducing latency and reducing network power boards consumption. The cluster nodes are interconnected by *Gigabit Ethernet* switch and *InfiniBand*, which is designed for the corresponding number of ports. The number of nodes in the system and their configuration depends on the requirements for computing resources in the form of specific problems, and on the user's financial possibilities.

In addition, the sharp competition among manufacturers of computer technology leads to the fact that the situation when components market prices change rather dynamically, especially with regard to the release of new products models. A wide range of modern electronic products allows to certainly state that by using standard components it is possible to build high-performance computing systems of general purpose in a very short time with the maximum full consideration of the needs and capabilities of different users.

In regards to the mentioned above, let's emphasize several important features of the multiprocessor systems design. Today, making supercomputers include serial microprocessors, equipped with each own local memory, which are interconnected with some communication medium (Voevodin & Zhumaty,

2007). The overwhelming majority of functioning supercomputing systems is, in fact, multiprocessor parallel computing systems of *MPP* architecture (*Massively Parallel Processing*) (Ivaschenko et al., 2013; Massive-parallel architecture, http://www.myshared.ru). The multiprocessor computing systems built on the basis of local networks have begun to be called "the cluster systems" or simply "clusters". This is due to the fact that the logically mentioned *MPP* system is slightly different from the usual local area network. This paper considers multiprocessor computing systems of "blades" server solutions architecture, where similar maternal modules are installed in a single package. Practice shows that the "blade" systems are more compact and easy to maintain, and their implementation is not much more expensive compared to the multiprocessor computing systems. The organization of the "blade" cluster is to merge the blades into a single computer network by a switch that is installed in the same place. For that, there may be enough one hard disk, which is the image of the boot system, while using the Network boot. When you turn on the Master system, the node distributes the IP addresses for all nodes in the cluster through the network switch, that is, the initial initialization takes place, and the cluster is ready for operation.

It is in view of the latest advances in designing multiprocessor computing systems and the module of a high-performance multiprocessor computing system with an expanding computing domain (Pat. 57663 Ukraine, 2011). In the cluster configuration, *N* blades and the modular principle of its implementation were selected. This provides, if necessary, the expansion of computing capabilities by installing additional modules. Each node operates under the control of its own copy of the operating system, and most often use standard operating systems: *Linux, NT, Solaris* and the like. The nodes composition and power of the described cluster may vary, which allows for the making of heterogeneous systems. The switching network connects the processors to each other. This architecture has many advantages: if necessary, one can add processors, increasing the performance of such a cluster; If one has limited financial possibilities or know the necessary computing power in advance, one can easily select the necessary system configuration. The name of such systems underlines theoretically unlimited scalability of devices of such a class.

Note that the development of such systems is a relevant problem. This is due not only to the fundamental limitation of the maximum possible performance of ordinary serial computers but also to the almost constant existence of computational problems, which solution requires additional capabilities of existing computer facilities. Those include, for instance, not only problems of aviation, but also, for example, numerical simulation of processes of hydrodynamics and metallurgical thermophysics (Ivaschenko et al., 2008; Shvachych, 2006; Shvachych, 2007), problems of image recognition, optimization problems with a large number of parameters, climate simulation, computations in genetic engineering, designing integrated circuits, analysis of environmental pollution (Milenin et al., 2006), as well as the computation of a wide range of multidimensional nonstationary problems (Roosta, 2000), etc.

This paper, for the built-in multiprocessor computing system, considers the following problem. There is a difference grid of *N* dimension; the computation time of a problem being solved by a single-processor system is a parameter *t*, but it is not decisive. An essential aspect here is the increase in the grid size, and in size that would be larger than the size that can be processed in the memory of a single computer. This procedure is crucial for a more detailed computation or for obtaining some new effects of the investigated processes. It is necessary to study the computations' features in the conditions of the considered class of problems based on the application of the network interface *InfiniBand* in multiprocessor computing systems.

ORDEVAL APPROACH

Constructive Implementation and Access Modes to the Multiprocessor Computing System

The purpose of these researches is to highlight the processes of making a multiprocessor computing system module, which real performance would be the peak in solving highly related problems with an expanding computational domain. In addition, this system should have the increased reliability and high performance. The proposed device blocks must be completed by mass-produced computing devices. The MCS module block diagram is shown in Figure 1.

So, the module contains one Master node (*PM*001) and *N* computing slave-nodes (*PN*001, *PN*002, ..., *PN*00n), two managed switches (*SW*1, *IB*1), a reconfigurable network for data interchange between computing nodes, the local reservation and intermediate computing system (*TCA Controller Storage System*), a key component backup mechanism, and also provides nodes downloading into *GI* network (*Gigabit Ethernet*) via the *SW*1 switch. The master-node and slave-hosts use the same components (motherboards, processors, *Gigabit Ethernet* network cards, *InfiniBand ConnectX-3 HCA* external bipolar network cards (*MCX354A-FCBT*)). In particular, the master-node is further equipped with a hard drive Disk (*SSD*) and *DVD*. The multiprocessor computing system dial-up network supports four configuration modes: star, ring, line, and lattice. They focused on implementation of a boundary data interchange that reflects the peculiarities of problems to be solved by such a multiprocessor computing system.

Setting up the *IB*1 switch and its configuring are run by the master-node using two *Gigabit Ethernet* ports (*IB1GI.i*1 - control, *IB1GI.i*2 - scaling). The *TCA Controller Storage System* local reservation and intermediate computing system are connected to Port 16 of the Managed Switch *IB*1.

As a constructive one was chosen the single case, which is the cell of the computer cabinet. This is due to the fact that on the one hand, if necessary, several modules can be placed in a single housing, and on the other hand, this approach provides compactness, successful cooling and easy access to the nests and card components that are being adjusted. The computing system includes a vertical, parallel to each other arrangement of the motherboard, which corresponds to the idea of "blade" servers.

The high-performance multiprocessor computing system module block diagram with an expandable computational domain is shown in Figure 2. Consider the sequence of steps for initializing the operation of the multiprocessor computing system. After the power supply to the *ATX* unit of the master node and the external *PUSK* signal from the *P*01 control panel, the start and initialization of the master-node of the multiprocessor system begin. Direct download of the operating system can be either from the hard disk or from a *CD/DVD* device. The operating system boot is performed by a specially-oriented configuration script that configures the *DHCP* server. In addition, the number of computer nodes in the system is determined here, if necessary, the access to the Internet environment or to the external network is configured. Here are the basic settings and settings. Due to the consistent *ATX* power supply and slave-node initialization, the required power of the *UPS* unit is reduced, all computing slave-nodes are started up and operating systems get downloaded. Downloading and configuring all computing nodes in the cluster completes the script operation. The system is ready for parallel computing.

The master-node (*PM*001) through the *KGI SW*1 switch provides a data stream related to control and diagnostics. In its turn, the slave-nodes, in accordance with the algorithm for solving problems and processes, implement the mode of necessary computing. The data interchange among computing nodes and the loading of problem conditions is carried out on a separate network, which is organized

Figure 1. The high-performance multiprocessor computing system module block diagram

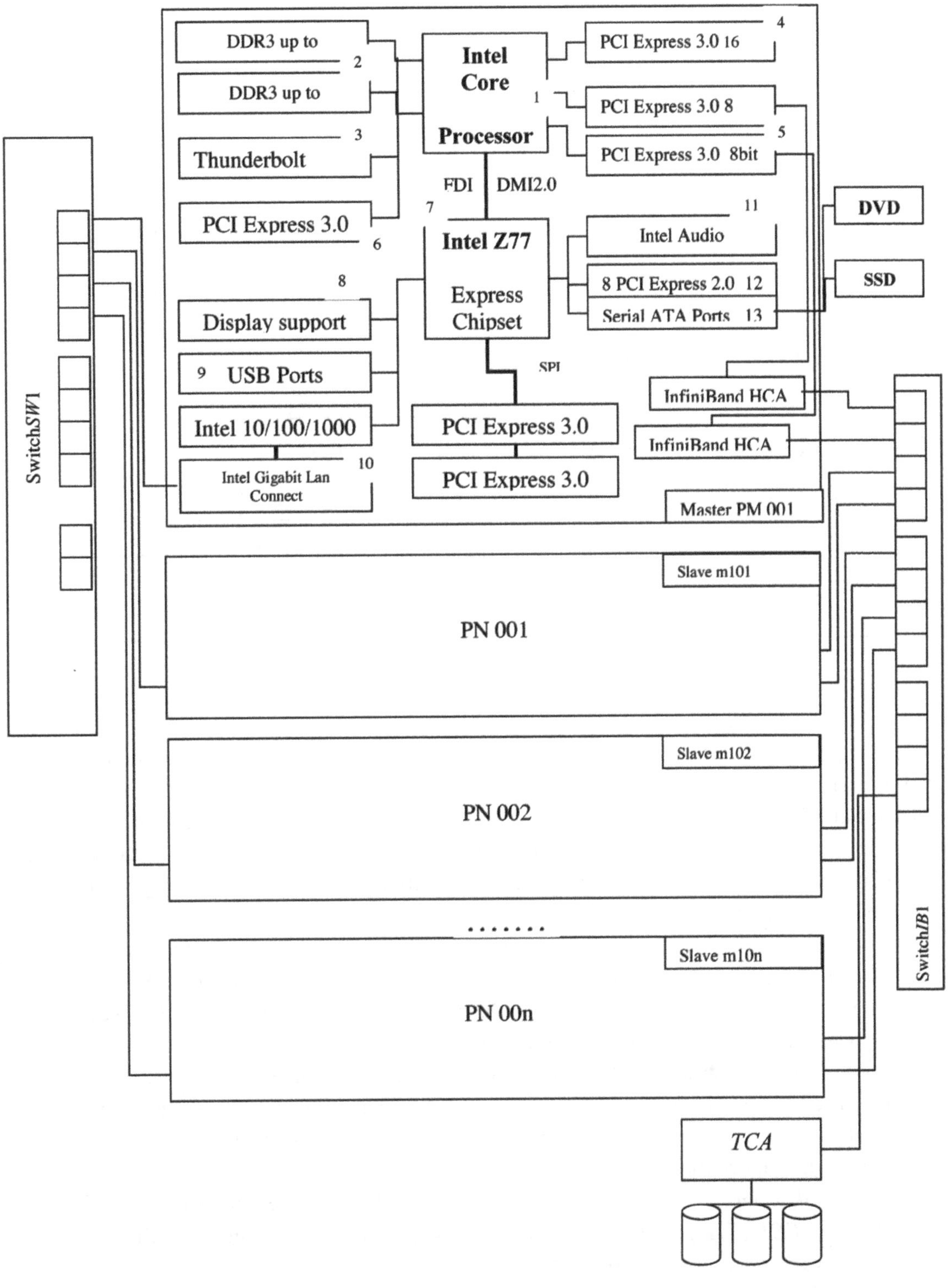

Figure 2. High-performance multiprocessor computing system module block diagram with the distributed computational domain

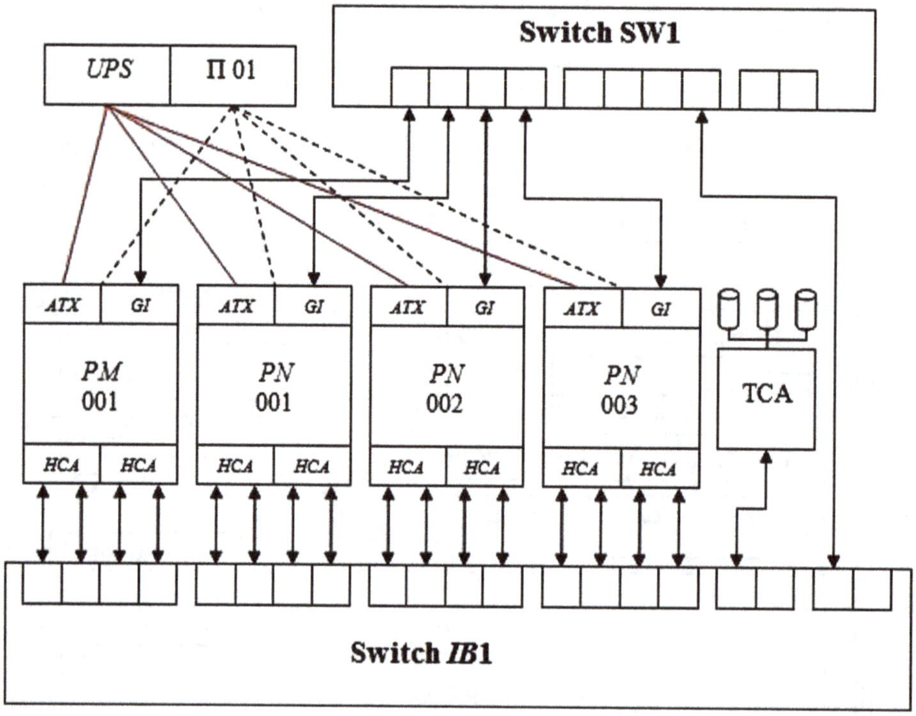

by a managed switch *KIB IB*1. In order to achieve maximum performance of the cluster system, single or duplex adapters of *InfiniBand* are used and the reconfiguring process of the structure of the second network is carried out in accordance with the specifics of the problems to be solved. The results of the intermediate and final computations are transmitted to the master node via the *InfiniBand KIB*'s managed switch. In this case, the control and transfer of the corresponding data from the slave-nodes are carried out by *HCA* network adapters. Directly this computation storage for further processing is carried out by the *TCA* network adapter.

The technical result achieved after the introduction of this system is that the data interchange among the computing nodes is transferred to a separate network using *InfiniBand* technology. This increased the data interchange speed and significantly reduced the network latency (channel download) that connects the cluster nodes. The use of a dial-up environment in a data-sharing network with point-to-point connections, unlike the previous variants of the networks that used the bus connection, allowed to significantly increase the speed of data transmission among the nodes of the multiprocessor computing system and reduce the latency in the data packet transfer medium. This is due to the fact that all transfers begin and end on the *HCA*-adapter of the channel.

Entering the *QDR (Quad Data Rate)* mode in the two-port *MCX353A-FCBT* (maximum 2 x 56 Gbps data transfer rate, *FDR (Fourteen Data Rate)* mode, has agreed on computing capabilities of the *CPU* and *PCI Express Interface* (2 x 32 = 64 Gb/s).

Using the *RDMA (Remote Direct Memory Access)* principle allows one transferring data without additional buffering and does not require active OS, as well as libraries or applications. This contributes to a significant reduction in the load on the processor system during data transmission, resulting in the release of processor computing resources and reduced latency in the data transfer medium. Network system download and the introduction of the mechanism of key components backup, as well as the use of power supplies for each blade of the multiprocessor computing system, can increase the system functioning reliability as a whole.

The cause-and-effect relationship among the set of essential features of such a system and the technical result that is achieved is that the introduction of subnets for system download, diagnostics, and data interchange dumped the computing system network, improved its availability and performance.

The setting and configuration of computing nodes software get simplified due to the network boot. At the same time, there are no network drives in the computing nodes, and the download, the configuration, the diagnostics, and the control occurs through the *SW*1 first switch network. These approaches allow flexibly reconfiguring the setting, update and adapt the software to a specific problem.

The network loading of the multiprocessor computing system module, the module key components backup, as well as a significant reduction in the number of system components, allows increasing the operation reliability.

To ensure high reliability of the system power supply of multiprocessor computing system, the voltage is fed via an uninterruptible power supply (*UPS*), which is connected to the control unit P01, from it through the power network interfaces (splitters) the current flows into the power supply units of the main module and system slave-nodes. Thus, in each blade of the multiprocessor computing system module, there are similar types of power supplies. This approach reduces voltage jumps when switching on power supplies, increases the system reliability, realizes the optimal loading mode and reduces power consumption of the computer system as a whole. The described power supply interface of the computer nodes of the multiprocessor system module facilitated the simplification of the structure of this operation, eventually, a significant reduction in the cost of the computing system by using one *UPS* to the entire module.

This energy consumption mode introduction allowed abandoning specialized integrated air conditioning systems, which also reduced the system cost as a whole. However, the use of the same components of the power supply system and the backup mode allowed increasing the system's operation reliability.

The data interchange among computing nodes has been transferred to a separate network using *InfiniBand* technology, which has increased the system's overall performance and significantly reduced the network latency (channel download) that connects the cluster nodes.

Using the *RDMA* principle allows transferring data without additional buffering and does not require active operation of the OS, libraries or applications, which significantly reduces the load on the system processors during the data transmission, resulting in the release of processor computing resources and reducing latency in the data transfer medium.

The use of two-port *HCA*-adapters due to the 4-aggregation mode of network channels of the link interface allows changing the network configuration by improving its bandwidth. The reconfigurable network implementation promotes the cluster system efficiency, adapting its network structure to solve each specific type of problem.

Network system download and the introduction of the mechanism of key components backup, as well as the use of power supplies for each blade of the multiprocessor computing system, can improve the functioning reliability of the system module. Possible communication modes of working PC with the multiprocessor computing system are shown in Figure 3.

Figure 3. The communication mode of a working PC with a multiprocessor computing system

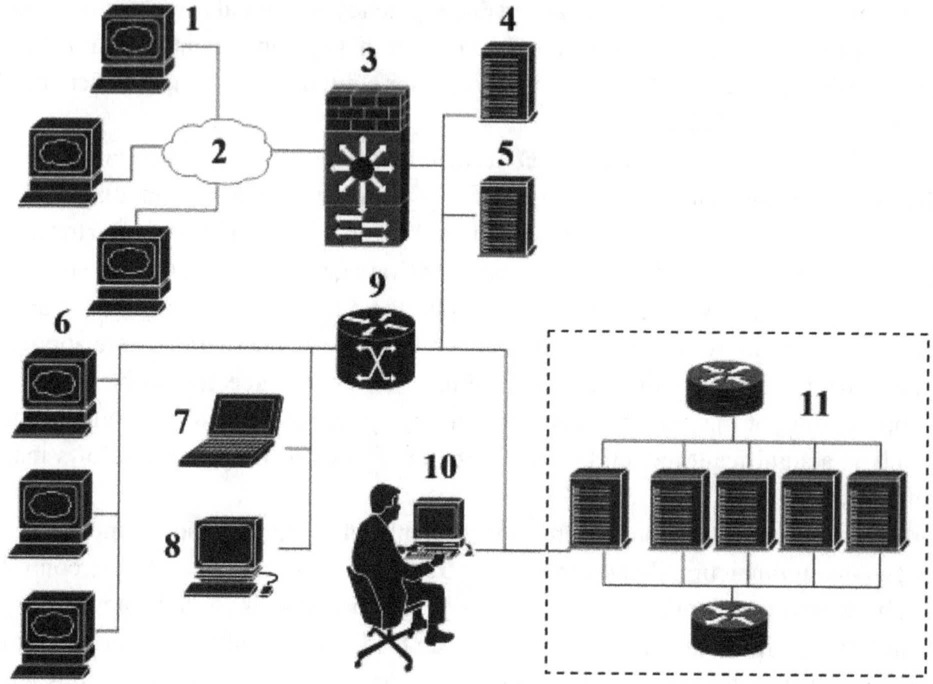

Thus, remote access to the resources of a multiprocessor computing system can be provided by workstations (1) connected to the *INTERNET* network (2) through the personal firewall (3) of the university network. Internal access to the resources of the multiprocessor computing system is provided through the fixed stations of the working groups and laboratories of the university units (6), personal mobile laptop-stations (7), experimental laboratories (8) and through the operator personal workplace of the multiprocessor computing system (10).

Research on the Energy Efficiency of Processors and Network Interface

The high-quality cluster power supply is an extremely important factor. It depends on both the system durability and the extent of its availability. In addition, the issue of electricity consumption for the cluster may be very significant. Note that the power consumption of the cluster is far greater than the amount of electricity used by as many ordinary personal computers, as the computing loads are higher. In addition, the power of server equipment is almost always greater than the capacity of personal machines. In addition, such systems use additional equipment and almost always work in the round-the-clock mode.

At the same time, not only the power supply but also its reliability and guaranteed quality is an essential factor. This is especially true for round-the-clock systems.

In order to ensure high reliability of the power supply system of the multiprocessor computing system, the voltage is fed through the uninterruptible power supply (*UPS*), which is connected to the *P*01 control unit, from it through the power network interfaces (splitters) the current enters the power supply units of the main module (*ATX*) and the systems' (*ATX*) slave-nodes. Thus, in each module blade of the multiprocessor computing system, there are similar types of power supplies. This approach reduces voltage jumps when switching on power supplies, increases the reliability of the system, realizes the mode of the optimal loading and reduces the power consumption of the computer system as a whole. The described power supply interface of the module computer nodes of the multiprocessor system allowed to simplify the structure of its operation and eventually significantly reduce the cost of the computer system by using the similar type of mass production units using one *UPS* to the entire module. Each MCS blade uses a standard power supply (*PS*). Note that tested different PSs, in particular, *CHIEFTEC-500W*, *COOLERMASTER-550W*, and *TARGA-400W*. They all showed a fairly good stock of power. For constructive reasons, the *CHIEFTEC-500* type power supply was chosen.

Each MCS blade must be switched on via *UPS*. Such sources are able to smooth out slight voltage fluctuations in the power grid to ensure stable operation of the whole computer system during short-term interruptions of electricity supply. A more serious situation may occur when an emergency power is turned off. In this case, *UPS* can provide uninterrupted system operation for 5-10 minutes. It is possible to compute the power consumption for the *UPS*, based on the power of the equipment powered. Extremely important is the *UPS* status control. Most modern uninterruptible power supplies are able to report their current state via *COM*-port, *USB*, or over the network through *SNMP (Simple Network Management Protocol*, a protocol designed to control and manage equipment on the network).

For the processors used in the research system (*Core i7 - 4790K* 4 GHz), peak computing power is 512 (256) *GFlops* at a power consumption of 88 *W*, respectively, the energy efficiency is 5.8 (2.9) *GFlops/W*. At the same time, we note that the actual efficiency will be less because in this case, the efficiency of power supplies, uninterruptible power supply systems, and so on was not taken into account. We also note that for the *Tsubame-KFC* supercomputer developed by the *GSIC (Global Scientific Information and Computing Center)* at the Tokyo Institute of Technology, showed the energy efficiency index at 4.5 *GFlops/W*. For the proposed *GPU-CPU* processors, measurements of peak performance and power consumption are performed; the research data are presented in Table 1.

To increase the energy efficiency of a multiprocessor computing system, it is recommended to use the 400 volts direct current industry standard power supply. That reduces the amount of energy conversion from source to consumer. Note that the transformation (from alternating current to constant, then vice versa and so on) significantly reduces the system efficiency. The proposed approach enables: first, to reduce actual power consumption by about 15% (the claimed efficiency of *Emerson Electric* power supplies is 92%), and second, to lower the required cooling capacity by the same 15% (which is very significant), third, to provide a higher installation density due to using cables of smaller diameter.

Moreover, the computing device can be imagined in the form of an object that converts energy into the computation, and the gross indicator of a well-designed object is *PUE (Power Usage Effectiveness)* ratio, the percentage of the total power consumption of the installation to the power consumed by the

Table 1. Energy efficiency of GPU-CPU type processors

Name	Type	Computing productivity, GFlops	Power, W	Energy efficiency, GFlops/W
Core i7-4790K GHz	CPU	512 / 256	88	5.8 (3)
	GPU	432	88	4.9
	CPU-GPU	512 + 432	88	10.7
Radeon HD 7970	GPU	3800 / 947	250	15/3.8

computing equipment. Note that the ideal value of the *PUE* ratio is1.0. At present, a ratio of 1.2 and less is considered good, and for existing data centers (data processing centers): - 1.6 ... 3.0. The reasons for high *PUE* ratios are inefficient cooling, administrative costs, and the use of non-adapted buildings and premises as compared to cloud-based data centers.

In addition, to increase the energy efficiency of the multiprocessor computing system, a set of *GPU-CPU* processors is used. In this case, the integrated *IntelHD Graphics 4600 (GT2)* processor has an output of 432 *GFLOPS (Peak Pixel Fill Rate: 5.4 GPixels/s. Peak Texel Rate: 10.8 GTexels/s. Peak Polygon Rate: 675 MPolys/s).* The external video adapter bandwidth, the *Radeon HD 7970 GPU*, is about *3800 GFLOPS* at 32-bit mathematical computations and *947 GFLOPS* for double-precision *GFLOPS* operations when consuming 250 *Watts* of power.

In addition, in general, a significant contribution to the energy consumption of the system makes the network equipment and data input/output operations. In order to reduce this indicator, the transition from *Gigabit Ethernet* network technology to *FDR InfiniBand* technology was implemented. This allowed increasing the data transfer speed, reducing latency and power consumption (compared to 1/10/40 *Gigabit Ethernet* technology). For the *Gigabit Ethernet* and *InfiniBand* network interfaces performance and power measurements have been made, the research data are presented in Table 2.

Thus, due to the system energy-efficient components use and the introduction of this energy consumption mode, it was possible to abandon the specialized integrated air conditioning systems, which also reduced the cost of the system as a whole and reduced the energy efficiency of *PUE.*

However, the same components use of the power supply system and its backup mode contributed to improving the reliability of the operation of the MCS as a whole.

The Conjugation Features of Multiprocessor Computing Systems

In the configuration of the proposed multiprocessor computing system (Shvachych et al., 2007) *N* blades are selected. Note that their number for solving a certain problem is limited by the dimensions of the computing rack. But its use allows expanding the computing power of the system by installing additional modules. Figure 4 shows the interface scheme for two modules of multiprocessor computer systems. The diagram depicts a master module that contains one master node (*PM*001) and *N* computing nodes (*PN*001, *PN*002, ..., *PN*00*N*), as well as a module with an expansion function (*PM*011 Backup Wizard node and *PN*011, PN012, ..., PN01*N* computing nodes). In this case, the switch *SW*1 forms a network for controlling, downloading and diagnosing the extended computing system as a whole. All integrated master-node and slave-node interfaces are connected to the inputs/outputs of this switch.

Table 2. Energy efficiency of the multiprocessor computing system network interface

Name	Type	Speed GBit	Power W	Energy efficiency, GBit/W
Gigabit Ethernet	*Gigabit*	1	8	0,12
Gigabit Ethernet 10	*Gigabit*	10	25	0,4
Gigabit Ethernet 40	*Gigabit*	40	120	0,33
InfiniBand 5463642	*InfiniBand*	56	50	1,25

The use of energy efficient equipment for multiprocessor modules enabled each blade to be fed from the *ATX* personal power supply, thereby reducing energy consumption and increasing the reliability of the entire computer system. The master node is powered by an individual *ATX* power supply. The start and main module initialization of the multiprocessor computing system are run via the control panel (*P*01). The master-node and computing slave-nodes use the same components (motherboards, processors, external network cards). In addition, the master-node is additionally equipped with an *SSD* hard disk, *DVD-ROM*, drives. The same configuration also has an extension module.

The computing complex operation begins after the voltage supply to the master unit (*ATX*) and the external *PUSK* signal. From the control panel *P*01, the start and initialization of the master-node of the multiprocessor computing system begin. Direct *OS* download can be made from an *SSD* hard drive, or from a *CD/DVD* device, a *USB* device.

In order to implement the network boot system among the ports 0, 1, 2, 3, 4, 5, 6, 7, 8 and managed switch *SW*1, a virtual local area network (*VLAN*) is created. The master unit module contains an optional external network card for administration and network management of the multiprocessor computing system. In this case, the *PM*001 master node connects through the input/output of the internal bi-directional interface of the *Gigabit Ethernet* (*PM*001.i1) network card with the port 0 of the managed switch *SW*1.

In the data interchange, the two-port network adapter *InfiniBand* 1 of the first interface *HCA*1.1 connected to the port of the managed switch *IB*1 00, the network adapter *HCA*1.2 second interface - with the port of the managed switch *IB*1 01, another two-port network adapter *InfiniBand* 2 of the first interface *HCA*2.1 - with the port of the managed switch *IB*1 02, the network adapter *HCA*2.2 second interface - with the port of the managed switch *IB*1 03.

In this case, the *PN*001 computing unit is connected by the input/output of the internal bi-directional interface *GI* (*PN*001.i1) to port 1 of the managed switch *SW*1. In the data interchange, the two-port network adapter *InfiniBand* 1 connects to the first interface *HCA*1.1 - with the port of the managed switch *IB*1 04, the second interface of the network adapter *HCA*1.2 - with the port of the managed switch *IB*1 05, the other two-port network adapter *InfiniBand* 2 of the first interface *HCA*2.1 - with the port of the managed switch *IB*1 06, the second interface of the network adapter *HCA*2.2 - with the port of the managed switch *IB*1 07.

In this case, the *PN*002 computing unit is connected by the input/output of the internal bi-directional interface *GI* (*PN*001.i1) to the port 2 of the managed switch *SW*1. In the data interchange, the two-port network adapter *InfiniBand* 1 is connected by the first interface *HCA*1.1 with the port of the managed switch *IB*1 08, the second interface of the network adapter *HCA*1.2 - with the port of the managed switch *IB*1 09, two-port network adapter *InfiniBand* 2 first interface *HCA*2.1 - with the port of the managed switch *IB*1 10, the second interface of the network adapter *HCA*2.2 - with the port of the managed switch *IB*1 11.

In this case, other computing nodes of the system are combined in the same way. *TCA (Controller Storage System)* is connected to the port of a managed switch *IB*1 16, with the *IB*1 (*Gigabit Ethernet IB1GI.i1*) configuration interface set to port 7 of managed switch *SW*1.

After the OS downloading, a specially oriented configuration script is launched configuring the operation of the *DHCP* server. In addition, it is necessary to determine the number of system computer nodes, if necessary, to configure access to the Internet environment or to the external network. Here are the basic settings and configurations are set. Serial supply to the power supply (*ATX*) and the initialization of slave-nodes reduce the power required by the *UPS* unit, run all computing slave-nodes, and downloading the OS. Downloading and configuring all computing nodes in the cluster completes the script operation. The system is ready for parallel computing.

Figure 4. Structural conjugation scheme of the multiprocessor computing system modules network interface

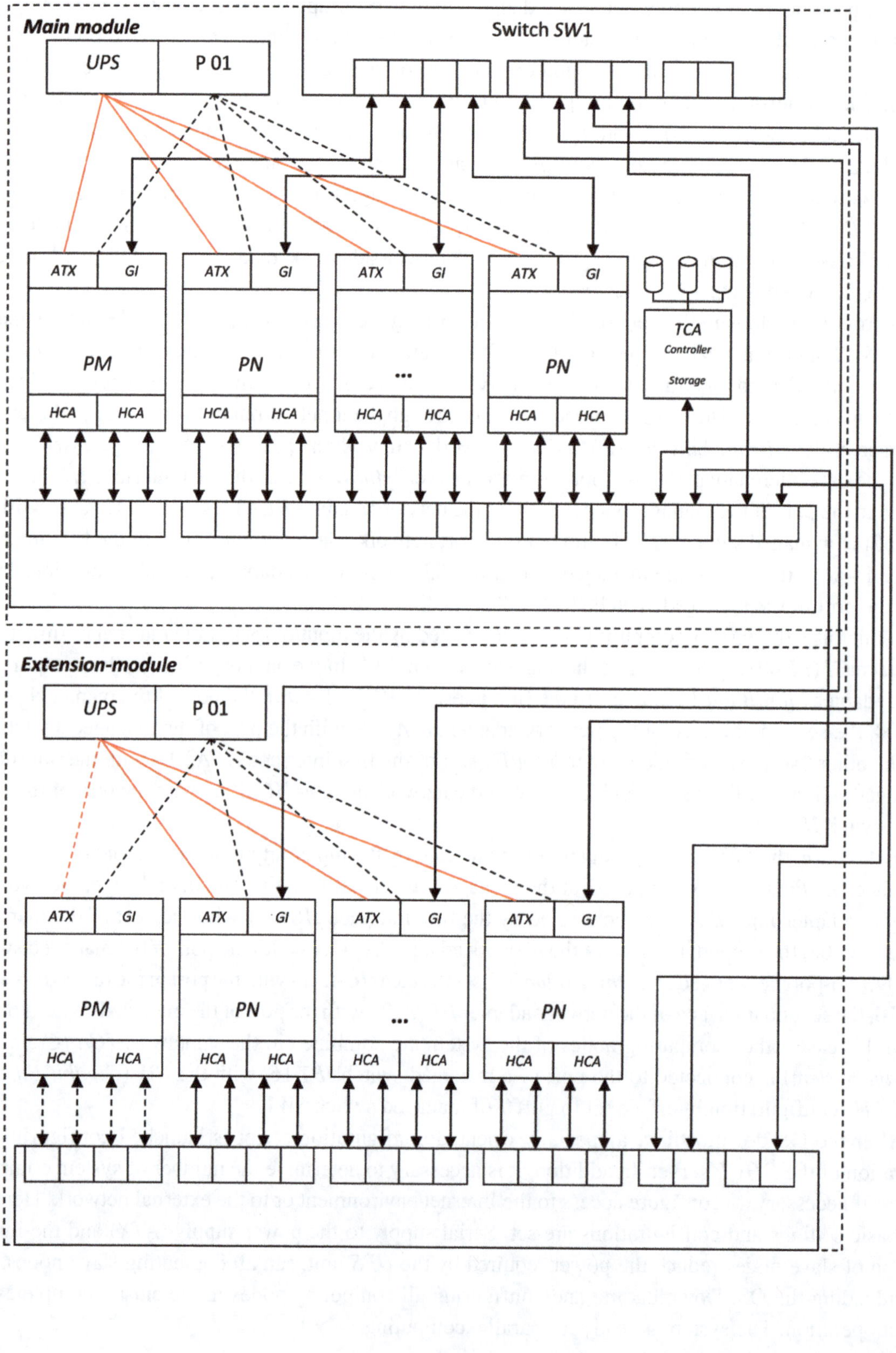

The master node (*PM*001) through the *SW*1 managed switch provides a data stream related to the control and diagnostics of the computing system. In its turn, slave-nodes, in accordance with the algorithm for solving problems and processes, implement the necessary computing mode. The data interchange among the computing nodes and the conditions of the problems downloading is carried out on a separate network, which is organized with a managed switch *IB*1. In order to achieve maximum performance of the cluster system, single or duplex adapters of *InfiniBand* are used and the process of reconfiguring the structure of the second network according to the specifics of the problems to be solved. The results of the intermediate and final computations are transmitted to the master node via the *InfiniBand IB* managed switch. In this case, the control and transfer of relevant data from the slave-nodes are done by the *HCA (Host Channel Adapters)* network adapter. Directly stored computing data for further processing is performed by a *TCA (Target Channel Adapters)* network adapter.

The conjugation mode of the main module and the expansion module is realized by the managed switches *IB*1 and *IB*2, with the port 18 *IB*1 connected to port 18 *IB*2, port 20 *IB*1 - with port 20 *IB*2, and port 07 *SW*1 - with port 19 *IB*2. This approach allows combining the main module and extension module control networks to provide the mode for downloading, diagnosing and controlling the whole multiprocessor computing system.

The described approach allows connecting data interchange networks among multiprocessor modules without loss of throughput. Using a configurable computing complex, a number of computational experiments were conducted, which results confirmed the effectiveness of the proposed approach. This is especially true for solving problems with non-stationary thermal conductivity, simulation problems for research of the thermophysical properties of materials, forecast simulation of ecological systems under in natural and anthropogenic factors, simulation of the temperature regime for thermal processing of long-length products.

Illustrative Example

As these studies show, nowadays the multiprocessor computing systems are developing quite dynamically. For this reason, parallel computations have become available to many researchers. The creation of parallel computing systems required the development of a new approach for arranging of appropriate computational algorithms. It is quite obvious that such an approach is aimed at the vectorization procedure implementation for computations. Then, under such conditions, the description of the corresponding models leads to rather complicated dependencies. For their comprehension and description, one has to turn to the newest achievements of various fields of applied mathematics. The methods of solving applied problems, which have already become classical, are not suitable from the algorithmic structure point of view. In addition, they do not always satisfy the accuracy and speed of computation requirements. Thus, it is quite obvious that the problems discretization is distinguished by those equations systems that have a large number of unknowns. For this reason, new ideas and approaches in computational mathematics are emerging, which are aimed towards the new methods developed for realizing numerical experiments. In this paper, the example of solving the thermal problems of the aviation complex shows the procedure for the formation of parallel computational algorithms and an example of their implementation on multiprocessor computing systems.

Among the known algorithms for recursive decomposition of solutions of systems of equations, it is obvious that only the algorithm of cyclic reduction of permutations admits the highest degree of its vectorization (Ivaschenko et al., 2015; Ivaschenko, et al., 2014). It is shown (Shvachych & Shmukin, 2014) that in the algorithm of such "odd-even" reduction there are many possibilities for parallelization. Therefore the algorithm is based on the method of parallelizing a mathematical problem. Note that its application is a very promising direction of applied mathematics because the problems of linear algebra represent a significant part of all problems of numerical analysis (according to some estimates, it is about 75%). It is obvious that the application of graphs for describing and researching the general mapping problem of the above algorithm on the architecture of parallel computing systems makes this procedure vivid.

The systems parallelization of the tridiagonal structure was realized by a numerical-analytical approach, which predetermined their maximally parallel algorithmic form. This approach contributes to the minimum possible time for the implementation of the developed algorithm on parallel computing systems. In addition, during the parallel computations of arithmetic expressions, in the developed algorithm, the error in the output data is separated from the rounding operations. Such a procedure became possible because with this approach the recursive structure of solution vectors computation is excluded, which usually predetermines the accumulation of rounding errors.

In addition, the necessary smoothness of the original solution and input data for a particular problem was taken into account, which allowed adapting the computational algorithm to the singularities of the numerical solution. This procedure was carried out by additional nodes located among the nodes of a given grid area. This allows you to accurately display the computational algorithm. In this case, computations in the construction of graphs or isolines can be performed in parallel and simultaneously (Shvachych & Shmukin, 2014). The carried out computational experiments showed high efficiency of the proposed approach.

So, the approach to the solution of the thermal problem of the aviation complex is realized in the form of a procedure for mathematical modeling using a multiprocessor computing system. In this case, consider one of the options for its implementation.

Test Problem

Let the boundary conditions of the first kind be given on the aircraft body surface. For definiteness, the boundary function will be considered as known, but it will be the desired function in solving the heat conduction problem. In the quasilinear approximation of the boundary Dirichlet problem in the grid domain there associated a system of differential effects that take the form of a system of ordinary differential equations, that is,

$$T^{/}_{p,n+1}(\varepsilon_t) = \{A_p(n+4)(n+2)T_{p,n+3}(\varepsilon_t) + B_p(n+1)T_{p,n+2}(\varepsilon_t)\}, \quad (1)$$

for $n = 0,1,2,...$, in this case

$$A_p = Dt1\frac{\lambda_{p,1}}{CV_{p,1}Dx1^2}, \text{ if } p = \overline{1,2m-1}; \tag{2}$$

$$B_p = Dt1\frac{\lambda_{p,2}}{CV_{p,1}Dx1^2}.$$

Assuming an implicit in time scheme of the first order approximated and defining the closing connections, we obtain central differences of the following form:

$$\left.\begin{array}{l} T_{p,2} = \frac{1}{2}T\left(T_{p+1,1} - T_{p-1,1}\right) \\ T_{p,3} = \frac{1}{2}\left(T_{p+1,1} + T_{p-1,1} - 2T_{p,1}\right) \end{array}\right\}, \tag{3}$$

Next, we obtain a system of equations of the tridiagonal structure, i.e.

$$C_p T_{p+1,1} - T_{p,1} + D_p T_{p-1,1} = f_{p,1}, \tag{4}$$

in this case

$$C_p = \frac{\left(A_p + B_p / 2\right)}{\left(1+2A_p\right)},\ D_p = \frac{\left(A_p + B_p / 2\right)}{\left(1+2A_p\right)},\ f_{p,1} = -\frac{TO_{p,1}}{\left(1+2A_p\right)}, \text{ if } p = \overline{1,2m-1}. \tag{5}$$

The solution of the system of equations (4) is realized by the sweep method. To solve a direct problem in the system of equations (4), (5), we must assume that $T_{0,1} = TW\left(t_j\right)$, and $T_{2m,1} = TL\left(t_j\right)$. When $TW\left(t_j\right)$ is a given and $TL\left(t_j\right)$ is the required boundary function, then, solving the heat conduction problem, it is necessary to select the function in such a way that in the solutions of the systems of equations (4), (5) the functional with the mean-square discrepancy is minimal. The entire time interval, where the desired function $T\left(t,x\right)$ is determined by setting the initial and boundary conditions, is divided into time intervals *Dt1* according to the sampling scheme. The same scanning step during the solution of the boundary heat conduction problem does not have to coincide with the sampling step *Dt1* adopted in the controlled system of equations (4), (5). The algorithm for solving the boundary heat conduction problem begins with the first step of the scan. Suppose that as a zeroth approximation a certain value is taken, when *R* = *TL*, it is known from a priori information and is clearly greater than the unknown boundary function. This circumstance is verified experimentally by the method of mathematical simulation. After substituting the value of R into the solution of the controlled mathematical simulation (4), (5), the calculated value of the temperature function $T_5\left(t_j\right)$ is obtained at the same location along the coordinate, where a function $T_5(t_j)$ that is known from the experiment is given. This allows finding the numerical value of the residual ($T_P(t_j) - T_5(t_j)$). If the discrepancy ($T_P(t_j) - T_5(t_j)$)> 0, then the value of R decreases by some amount ΔT. In other cases, it increases by the same amount (ΔT). Further, this process gets developed until the discrepancy changes its sign to the opposite. The establishment of this fact means that the required approximate solution of the heat conduction problem lies in the interval [*a*, *b*], where a and b are denoted numbers at which the values of the functions at these points have different signs. By computing the value of the discrepancy at the midpoint, which lies between *a* and *b*, we can further refine the value of the unknown boundary function by an interpolation formula. Repeating this iterative procedure as many times as necessary, the value of the control parameter is computed with any predetermined accuracy. Naturally, it should be borne in mind here that the function $T_5(t_j)$ is known by the results of the experiment and has some error ε. Thus, the required solution of the heat conduction problem occurs through scanning steps and, assuming a linear approximation in the time layers, write that

$$TL_{v+\varepsilon_t,1}(\varepsilon_t) = TL_{v,1} + \varepsilon_t TL_{v,2} \tag{6}$$

in this case

$$\left.\begin{aligned} TL_{v,2} &= (TL_{v+1,1} - TL_{\nu,1}) \\ \varepsilon_t &= \frac{t - t_v}{t_{v+1} - t_\nu} \in [0,1] \end{aligned}\right\}, \text{ if } \nu = 1,2,\ldots, . \tag{7}$$

Then it is possible to solve the heat conduction problem on a full-time interval, when $t \in [t_{0,} t_k]$ at *ν*-th time layers. This possibility arises, firstly, because the sweep algorithm realizes the desired nonstationary solution on any time layer. The transition in the systems of equations (1) - (7) to the time variable ε_t unifies the algorithm with respect to any time interval by means of automatic accounting of the initial values. Figure 5 shows the results of the test problem solution of thermal conductivity for the exact values of the experimental function.

The simulation process was implemented on a personal computing cluster. From the analysis of simulation results, it follows that the developed method for solving the heat conduction problem sufficiently effectively minimizes the residuals. We also note that the algorithm developed in this formulation is absolutely stable. The carried out computational experiments showed high efficiency of the proposed approach.

The Results Analysis of the Test Problem

In this paper, we present a method for solving the thermal problems of an aviation complex using the most parallel computational forms. It is shown that in the algorithm of "odd-even" reduction there are hidden many parallelization possibilities. This is because the permutations algorithm solves the system

Figure 5. The results graph of restoring external boundary conditions

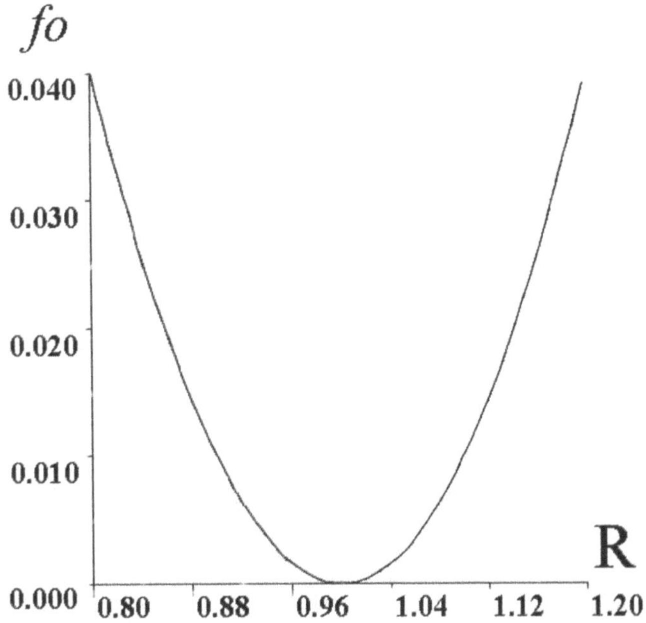

of equations simultaneously from both ends. Therefore, its use for solving tridiagonal systems on parallel systems can be very promising.

Parallelizing tridiagonal systems based on numerical-analytical sampling methods, firstly, does not impose any restrictions on the topology of the grid nodes of the computational domain. And, secondly, with respect to the parallel computation of arithmetic expressions, the error in the initial data from the rounding operations inherent in a real PC is separated in it. With this approach, the recurrent structure of the computation of the desired decision vectors is excluded, which, as a rule, leads to the accumulation of rounding errors. The parallel form of the algorithm made in this way is maximal, and, therefore, has the minimum possible time for implementing the algorithm on parallel computing systems.

CONCLUSION

In modern conditions, the multiprocessor computing systems based on standard, commonly used technologies and components are becoming increasingly essential. Especially this problem is relevant to the aviation industry. Due to the high demand and suggestions on blade configurations in scientific practice, it is proposed the "blade" cluster computing complex for solving problems with the distributed computational domain. When designing a cluster, its making is of great importance. Right at the stage of designing a multiprocessor computing system, it is necessary to predict the possibility of its expansion or modification in the future. Note that the most successful solution is the placement of a multiprocessor system in the rack. Such an arrangement proved to be appropriate even for a small computing system. Inside the rack, there are nodes, hardware for effective components connection, means of control of the internal network system and so forth. Each blade works by managing its copy of the standard operating system. The composition and power of the nodes of the multiprocessor computing system may be different. This paper considers a homogeneous system. The interaction among the nodes of the computer system is established by specialized libraries.

Introduction to the multiprocessor computing system of the separate *InfiniBand* data interchange computing network, *VLAN* support, especially organized data interchange modes in the *KIB* managed switch network, as well as the mode of network downloading of processors and the mechanism of the module key components backup allowed:

- Firstly, due to the use of *InfiniBand* technology, set the following priorities: low latency, scalability, backup, the ability to select the required speeds from a given range, which in its turn helps to use the developed system to solve tightly coupled problems and problems with expanding computational domain;
- Secondly, through the terminal or the *WEB*-interface, to change the configuration of the computing network, adapting its structure to solve each specific type of problem;
- Thirdly, due to the use of *RDMA* tools, the *InfiniBand* technology provides direct data interchange among the nodes *RAM* of a multiprocessor computing system, which improves the speed of computing when solving tightly coupled problems, provides high-speed access to the cluster node's memory and data interchange among them, unloads the *CPU* when data is interchanged and reduces the channel loading that passes among the cluster nodes;
- Fourthly, due to the use of *ConnectX* adapters to provide new connectivity capabilities with different computing environments, which increases the productivity of the entire computer system and helps unload the *CPU* (*InfiniBand* traffic maintaining);
- Fifthly, due to the modular construction principle, to simplify the design, build-up or replacement of failing computing nodes, as well as the operation of the entire system.

This section explores the processors' energy efficiency and network interface. The high-quality cluster power supply is an extremely essential factor in the operation of such systems. It depends on both the lifetime of the multiprocessor system and the extent of its availability. In addition, for the cluster, the issue of electricity consumption may be very significant. The developed power supply interface of the MCS module computing nodes allowed simplifying the system structure, and ultimately, significantly reducing its cost by applying mass-produced same-type units using one *UPS* to the entire module. For *GPU-CPU* research processors, measurements of peak performance and power consumption have been performed, and the PUE energy efficiency ratio has been determined in a designed system. The proposed approach allowed: firstly, to reduce actual power consumption by about 15% (the efficiency of *Emerson Electric* power supplies is 92%), and secondly, to reduce the required cooling capacity to the same 15% (which is very significant), thirdly, to provide a higher mounting density by using cables of smaller diameter. For *Gigabit Ethernet* and *InfiniBand* network interfaces there were performance measurements and power consumption checked. At the same time, due to system energy efficient components and implementation of the energy consumption advanced mode, it was possible to abandon specialized integrated air conditioning systems, which also reduced the system cost as a whole and reduced the *PUE* energy efficiency. At the same time, the application of components of the same type of the power supply system and the backup mode allowed improving the system functioning reliability as a whole.

In addition, in this research section shows that for solving many of the tightly coupled problems with modular MCS in its greatest productivity mode, there may not be enough suggested blades. With this regard, it is recommended to develop a new qualitative direction of designing multiprocessor computing systems - the conjugation of several modules into a single computing complex. This paper reveals the modules conjugation features of multiprocessor systems to improve their computational possibilities. In this case, an option has been developed, which connects two modules of the multiprocessor computing system - the main module and the module-extension. The connection feature of two modules interfaces of multiprocessor computing systems and data networks was investigated. So, in order to avoid the mutual influence during the data transfer/receipt among the computer nodes, the *VPN* network among the ports of the managed switches *IB*1 and *IB*2 was put into operation. Experimental computations were carried out to test the developed computing complex. The proposed approach effectiveness is confirmed by the successful resolution of a wide range of problems.

REFERENCES

Bashkov, Y. O., Ivaschenko, V. P., & Shvachych, G. G. (2011). High-performance multiprocessor system based on the personal computing cluster. *Scientific Papers of Donetsk National Technical University, 9*(179), 312–324.

Ivaschenko, V. P., Shvachych, G. G., & Kholod, E. G. (2014). *Extreme algorithms for solving problems with higher order accuracy. In Applied and fundamental research* (pp. 157–170). St. Louis, MO: Publishing House Science and Innovation Center, Ltd.

Ivaschenko, V. P., Shvachych, G. G., & Shmukin, A. A. (2008). Parallel computations and applied problems of metallurgical thermophysics. *System Technologies: The Regional Collection of the Proceedings, 3*(56), 123–138.

Ivaschenko, V. P., Shvachych, G. G., & Shmukin, A. A. (2013). Investigation of the components selection problem of a modular multiprocessor system in the problem of metallurgy thermal processing. *System Technologies: The Regional Collection of the Proceedings, 2*(85), 78–92.

Ivaschenko, V. P., Shvachych, G. G., & Sobolenko, O. V. (2015). The system of automated control of temperature regimes of thermal treatment of steel products. *Metallurgical and Mining Industry, 1*, 142–147.

Ivaschenko, V. P., Shvachych, G. G., & Tkach, M. A. (2013). Features of using multi-core processors in multiprocessor computing systems. Newest computer technologies, 11, 132–135.

Latsys, A. O. (2003). *How to build and use a supercomputer*. Moscow: Bestseller, 240. Retrieved from http://www.myshared.ru/ slide/

Milenin, A., Dyya, H., Muskalsky, Z., & Pylyarchik, Ya. (2006). Simulation with the finite element method of the wire drawing process in roller coils. *Metizy, 2*(12), 30–33.

Non-Uniform Memory Architecture (NUMA): Investigating the subsystem memory of AMD Opteron dual-processor platforms using the RightMark Memory Analyzer. (n.d.). Retrieved from http://www.ixbt.com/cpu/rmma-numa.shtml

Patent 57663 Ukraine, IPC G06F 15/16. (2011). *The module of the high-efficiency multiprocessor system of high readiness*. National Metallurgical Academy of Ukraine.

Roosta, S. H. (2000). *Parallel Processing and Parallel Algorithms: Theory and Computation*. Springer-Verlag. doi:10.1007/978-1-4612-1220-1

Shvachych, G. G. (2006). On the design of parallel computing in the simulation of environmental identification problems. *Mathematical Simulation*, *2*(14), 23–34.

Shvachych, G. G. (2007). On parallel computer technologies of cluster type solutions of multidimensional non-stationary problems. *Vĕdeckэ potenciбl svĕta - materiбly IV mezinбrodnн vĕdecko-praktickб conference. Technickŭvĕdy. Mathematics Fyzika. Moderni informaun technologie. Vэstavbaaarchitektura.* Prague: Publishing House "Education and Science" s.r.o.

Shvachych, G. G., Kolpak, V. P., & Sobolenko, M. A. (2007). Mathematical simulation of thermal processing high-speed modes of long-length articles. *Theory and Practice of Metallurgy: State Scientific and Technical Journal*, *4-5*(59-60), 61–67.

Shvachych, G. G., & Shmukin, A. A. (2014). Features of the design of parallel computing algorithms for a computer in problems of heat and mass transfer. *East-European Magazine of Advanced Technologies, 3,* 42–47.

Shvachych, G. G., & Shmukin, A. A. (2014). On the concept of unbounded parallelism in the problems of thermal conductivity. *East-European Journal of Advanced Technologies*, *3*(9), 81–84.

The Blackford MultiCore cluster specification: Institute of systems dynamics and control theory of the SB RAS. (n.d.). Retrieved from http://www.mvs.icc.ru/cluster_info.html

Voevodin, V. V., & Zhumaty, S. A. (2007). Computation and cluster systems. Moscow: Publishing House of Moscow University.

ADDITIONAL READING

Ivaschenko, V. P., Shvachych, G. G., & Ivaschenko, O. V. (2017). The numerical-analytical concept of solutions of applied problems on the basis of schemes of the raised order of accuracy. *Collection of scientific works "Computer modeling: analysis, management, optimization", 1(1),* 85–90.

Ivaschenko, V. P., Shvachych, G. G., & Semenov, S. G. (2017). Efficient parallelization algorithms of the applied tasks in multiprocessor computing systems. *System technologies. Regional interuniversity collections of scientific works, 2(109)*, 57–66.

Semenov, S. G., & Shvachych, G. G. (2017). Model for calculating the time limits for software development projects. *System management, navigatsii zvoyu, 1 (41)*, 43–50.

Shvachych, G. G., Kholod, E. G., Ivaschenko, E. V., & Busygin, V. V. (2018). Visualization of the decisions of applied problems in the multiprocessor system. Natural and Technical Sciences, 1(17) (February), Budapest, 65–69.

Shvachych, G. G., Pobochii, I. A., Ivaschenko, E. V., & Busygin, V. V. (2018). Research on the problem of compatibility in the multi-processing compound systems. *Science Review, 2(9)*, (February), Warsaw, 19–23.

KEY TERMS AND DEFINITIONS

Blade-Server Solutions: In this chapter, the authors consider multiprocessor systems where several similar motherboards are installed in the same package. Practice proves that "blade" systems are more compact and easy to maintain, and their implementation is not much more expensive compared to multiprocessor computing systems.

Cluster Computing Systems: Multi-processor computing systems built on the basis of local networks, known as "cluster computing systems."

Communication Environment: For the construction of supercomputers are taken serial microprocessors, each with its own local memory and connected through some communication environment.

InfiniBand Technology: The exchange of data among computing nodes of a multiprocessor system is transferred to a separate network using the *InfiniBand* network interface, which increased the system's overall performance and significantly reduced the latency (channel download) of the network that connects nodes of the cluster computing system.

"Odd-Even" Reduction Algorithm: Among the known algorithms for recursive decomposition of solutions of systems of equations, only the algorithm of cyclic reduction of permutations allows the highest degree of its vectorization. In the algorithm of such an "odd-even" reduction, there are many possibilities for parallelization. In this connection, this approach is based on the method of parallelizing a mathematical problem.

Supernumerary Devices: In fact, here we are talking about multiprocessor parallel computing systems of MPP architecture (massively parallel processing).

Vectorization of Computations: The creation of parallel computing systems required the development of a new approach for making of appropriate computational algorithms. For this reason, new ideas and approaches are emerging in computational mathematics, which is aimed towards the development of new methods for realizing numerical experiments based on the computations parallelization.

Chapter 16
Realization Features of System Software of Multiprocessor Computing Systems

Boris Moroz
University of Customs and Finance, Ukraine

Eugene Fedorov
Donetsk (Pokrovsk) National Technical University, Ukraine

Ivan Pobochii
National Metallurgical Academy of Ukraine, Ukraine

Dmytro Kozenkov
National Metallurgical Academy of Ukraine, Ukraine

Larisa Sushko
Dnipro State Agrarian and Economic University, Ukraine

EXECUTIVE SUMMARY

The chapter is aimed at the problem of use and adjustment of system software of multiprocessor computing systems. The main principles of the Linux operating system were considered, which were necessary when constructing a multiprocessor computing system. These studies also cover new ways to remotely access the memory of processor systems through the use of RDMA technology for InfiniBand technology. Thus, it has been shown that the RDMA principle, together with the formation of a separate computing network in the data interchange environment, and the implementation of VLAN mechanisms, allowed the data transmission among nodes memory of the multiprocessor computing system without additional buffering. This approach does not require the active OS operation, libraries, or applications on those nodes of the system which memory is requested.

DOI: 10.4018/978-1-5225-7588-7.ch016

BACKGROUND

Nowadays there is a rapid increase in the number of multiprocessor computing systems and their total productivity in the world (Bashkov et al., 2013; Shvachych et al., 2008; Latsys, 2003; Voevodyn & Zhumatyy, 2007). This is due to the fact that such systems became publicly available and cheap hardware platforms for high-performance computing (Ivaschenko et al., 2017; Shvachych et al., 2018; Ivaschenko et al., 2018; Shvachych et al., 2018). At the same time, the interest has sharply increased in the problems of the computer networks (*GRID*) and awareness of the fact that their implementation will have a huge impact on the development of human society, is comparable to that of the emergence of unified electric networks at the beginning of XX century. In this regard, considering the problems of mastering multiprocessor computing systems, it should be taken into account that they are the first step in the creation of such computing networks.

The more so the practice puts a variety of problems before the application scientists, the full solution of which in most cases is possible only by multiprocessor computing systems (Alishov et al., 2016; Ivaschenko et al., 2005). Obviously, today, based on multiprocessor information and management systems, control systems are being implemented for many industries: the space industry, aviation, anti-aircraft and anti-missile defense systems, and many others. However, the production of multiprocessor information and management systems is hampered by the high costs at all its stages. As a result, the overall cost of a system often makes it an inaccessible tool. The use of modern multiprocessor cluster systems would reduce the cost of its production (Ivashchenko et al., 2011; Bashkov et al., 2011).

For the aviation industry, the development and implementation of new materials are relevant. And here an important component of the reliability of aircraft is the interaction with metallurgical production. At the same time, in the metallurgical industry, there is a lot of diverse and interconnected processes. First and foremost, these are the technologies of smelting and casting of iron-carbon alloys, heating, rolling and heat treatment of metal products (Shvachych et al., 2003; Ivaschenko et al., 2003), as well as the operation of auxiliary equipment, which includes filling machines, buckets, bowls, and so forth. Industrial practice shows that neither the intensification of the processes of metallurgical production nor the constructive improvement of various metallurgical equipment is possible without studying and analyzing the phenomena of heat and mass transfer (Ivaschenko et al., 2005; Shvachych et al., 2007). At the same time solving these problems by well-known standard approaches is a complex problem, the overcoming of which is possible only through the use of modern multiprocessor computing technologies. At the same time, one of the main features of the introduction of such technologies is to increase the speed and computing productivity (Shvachych et al., 2018; Shvachych et al., 2017). High-performance computing allows to solve multi-dimensional problems, including those that require a large amount of processor time (Ivaschenko et al., 2018; Shvachych et al., 2018). Speed control makes allows to efficiently manage technological processes, or even create the preconditions for developing new promising technological processes.

The relevance of this type of research is determined, first of all, by the need for theoretical substantiation and integrated practical consideration of the peculiarities of the creation of modular multiprocessor computing systems for solving applied problems of industry; is to formulate a methodology for studying the effectiveness of these systems, as well as to research prospects for the application of modern network interfaces for the design of multiprocessor computing systems. The solution of the mentioned aviation and metallurgical problems is also the development of a family of numerical-analytical methods that harmoniously fit into the problem of constructing modern distributed algorithms (Ivaschenko et al., 2017;

Shvachych et al., 2018; Alishov et al., 2016). It is known that the complexity of developing applications for multiprocessor computing systems with distributed memory is a major obstacle to their widespread implementation. Thus, it is possible to overcome the difficulties encountered in the application of multiprocessor computing technologies to the solution of industrial applications.

At the same time, specialized software plays an important role in the design of modular multiprocessor computing systems (Official site of MPI-forum, http://www.mpi-forum.org; Official page of mpich1 package, http://www-unix.mcs.anl.gov/mpi/mpich1), that when using any technology of parallel programming, eventually, one receives software complexes. With this regard, consideration of the development tools of parallel program complexes should be separated from consideration of the means of their communication and execution. Summarizing this fact, one can conclude that the operation of multiprocessor computing systems is provided by four types of specialized system software, namely:

- Operating systems;
- Means of communication (interaction) among the nodes of a multiprocessor computing system;
- Means of development of parallel applications;
- Means for administering multiprocessor computing systems.

So, let's proceed with the consideration and configuration of the *HPC MPI* system. All configurations will be executed on the basis of the *Scientific Linux* OS based on the family of *RedHat* OS. The choice of this distribution is due to the fact that there is an *Intel Fortran Compiler* in the standard configuration, which is necessary for the system as a whole. This OS includes the necessary programs and some other specialized packages.

First of all, it should be noted that any program that is not part of the kernel can be executed in such an operating system only in the form of a *process*. On the one hand, this concept is basic for this type of OS. On the other hand, there are many definitions and interpretations of the process concept. They are usually quite cumbersome. In this regard, we can give some informal definition, which, however, differ by the simplicity and clarity of the interpretation. So, the process is a special way to execute an application, organized in such a way that it runs on a separate processor.

Let's consider what properties of processes in the *Linux* OS fit this definition. First of all, processes in *Linux* are pseudo-parallel, that is, from a logical point of view, this occurs in such a way that the execution of each of them gets engaged in a separate processor. Since the switching rhythm of a single physical processor among processes is not provided, the only reasonable model of relative progression of processes in time is the assumption that each of them moves in its own rhythm independently. Any process has its own address space: since none can, even if it wants to, "rub the memory" of another. This fact is fully consistent with the informal definition of the process.

Each node of the multiprocessor computing system runs on its own copy of the standard operating system. The composition and power of the nodes can be different within a single module of the multiprocessor computing system, but in this case, a homogeneous multiprocessor computing system is being considered, which hardware configuration is given in Table 1.

Let's consider some additional explanations to the Table. 1. Only the controlling master-node has a hard drive, slave-nodes are diskless. Another important detail is that *Gigabit Ethernet* (*eth*0) and *InfiniBand* (*eth*1,2) communication technologies are used in the computing system.

Creating a set of programs for implementation on a computer system can be divided into two main stages:

Table 1. Hardware configuration of a multiprocessor computing system

Type	Processor	Cores	RAM Gb	Quantity
Controlling master-node	Intel Core i7-4790K 4 GHz	4	8	1
Computing slave-node	Intel Core i7-4790K 4 GHz	4	8	3

- The sequential algorithm is subjected to decomposition (parallelization), that is, it is divided into independently working branches; for mutual interaction, there are introduced two additional non-mathematical operations: data reception and transmission;
- A parallel algorithm is written in the form of a program where the operations of data reception and transmission are expressed by the terms of a specific communication system among the branches.

ORDEVAL APPROACH

High-Performance, Multi-Processor High-Reliability System

The paper proposes a concept for making a multiprocessor computing system designed to solve the aviation industry's problems, which would have peak real efficiency and productivity. In addition, this system should be characterized by increased reliability and high energy efficiency. Blocks of the proposed system were implemented with the help of mass production computing facilities. The FAWN technology has further developed.

There was grounded the idea of choosing the form factor of computing nodes, for which there are several traditional solutions today. The projected cluster system is based on blade technologies. For these reasons, it is a densely packed module with blade type processors installed in the rack. Inside the rack, there are nodes, equipment for efficient connection of components, equipment for controlling the system's internal network, etc. Each cluster blade operates under the control of its copy of the standard operating system. The composition and power of the nodes can be different within a single module, but in this case, a homogeneous module was considered. The interaction between nodes of the cluster system is established using the programming interface, that is, specialized libraries of functions. But there was a problem with the number of blades. Some developers of such systems are convinced that the number of computing nodes of a multiprocessor computing system completely depends on the class of problems solved. For example, the chosen mathematical models for the research of metallurgical production are based on differential equations in partial derivatives. In view of these circumstances, it was decided to limit the six-node module to the multiprocessor module, and if necessary, a cluster building extension option would be provided. It was also found that fewer nodes may not produce the desired result. Thus, in making a multiprocessor computing system, special attention was focused on the possibility of expansion or modification of the cluster in the future.

For this purpose, there was carried out an analysis of the choice of architecture and parameters of the computing nodes of the cluster system, since they largely determine its characteristics. It was at the cluster design stage that the main parameters of the motherboard nodes were provided, taking into account the specifics of problems that would be solved by such a system. The multiprocessor module includes one master node (MNode001) and five computing slave nodes (NNode001, NNode002,

NNode003, NNode004, NNode005), three managed switches (SW1, SW2, SW3), intermediate switch memory buffers reconfigured a network for interchanging data between computing nodes, virtual local area networks, a mechanism for redundancy of key components, and also provides network boot nodes. Its block diagram is demonstrated in Figure 1.

A series of experimental computation was conducted on the configured computing system. The effectiveness of the proposed approach is confirmed by the successful resolution of broad-based problems. This work is aimed at disclosing the problems of development of specialized software developed by the modular computing system.

Some Features of the OS Implementation for Computing Systems With Distributed Memory

Due to the fact that the personal computing cluster is made from the finished hardware component, its operating system, for the most part, consists of ready-made software products of general purpose. So, on the cluster nodes and in the control machine there could operate their local *Linux* OS (Ozerov, 2005).

However, the OS of multiprocessor computing system also includes specialized components, for instance, a communication library and a basic system for launching parallel programs. For the cluster

Figure 1. Block diagram of a multiprocessor computing system

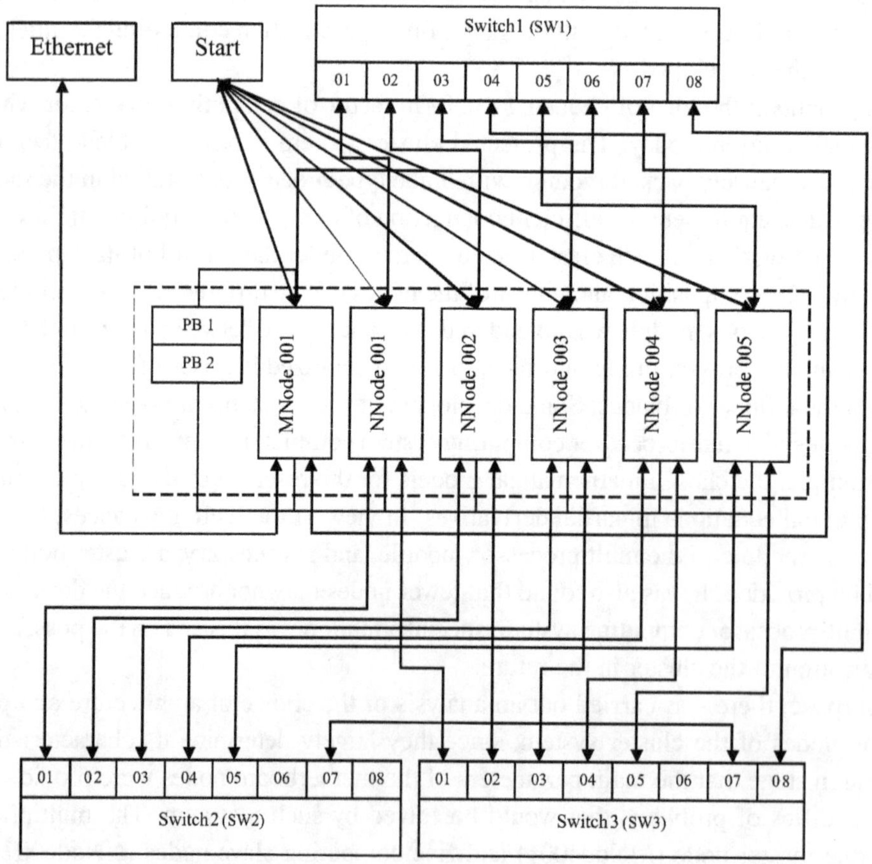

work, one needs the following network services: *DNS* server, *NFS* server, *TFTP* server, *NTP* server, *DHCP* server, *SSH* server are the main services for multiprocessor computing system operation.

In the first step, consider the configuration of multiprocessor computing system networks. It is carried out both during OS installation and after the procedure. Configuring network interfaces can be done by both graphical tools and console utilities. Below is an example of setting up networks by direct editing configuration files. Files with settings for network interfaces are in */etc/sysconfig/networking-scripts* directory for *Scientific Linux* distribution. The names of the files are *ifcfg-eth*0, *ifcfg-eth*1 and *ifcfg-eth*2 correspond to first, second, and third network interface respectively.

File listing ifcfg-eth0 (internal interface):

```
cat/etc/sysconfig/networking-scripts/ifcfg-eth0
DEVICE=eth0
BOOTPROTO=none
BROADCAST=192.168.31.255
IPADDR=192.168.31.1
NETMASK=255.255.255.0
NETWORK=192.168.31.0
ONBOOT=yes
TYPE=Ethernet
```

The file listing with an external interface is not given due to the lack of relevance in this context and the very slight differences with previous code.

The *Domain Name System (DNS)* server (*bind*-9.2.4-27.0.1.el4) of the main module performs direct and inverse transformation of node names into *IP* addresses. Reverse conversion is used by some software packages, for example, *torque-pbs (*problem breakdown system*)*. The *DNS* server is configured as a caching and serving the internal network of the multiprocessor computing system.

The main configuration file looks like this:

```
options {
        directory "/var/named";
        dump-file "/var/named/data/cache_dump.db";
        statistics-file "/var/named/data/named_stats.txt";
        // query-source address * port 53;
};
controls {
        inet 127.0.0.1 allow { localhost; } keys { rndckey; };
};
zone "." IN {
        type hint;
        file "named.ca";
};
zone "localdomain" IN {
        type master;
        file "localdomain.zone";
```

```
        allow-update { none; };
};
zone "localhost" IN {
        type master;
        file "localhost.zone";
        allow-update { none; };
};
zone "0.0.0.127.in-addr.arpa" IN {
        type master;
        file "named.local";
        allow-update { none; };
};
// Remark, in which file the description of the direct zone is
         // Cluster and the fact that the server is the main for this zone
zone "cluster" IN {
        type master;
        file "cluster";
        allow-update { none; };
};
 // Remark, in which file the description of the inverse transform is
         // For the cluster zone
zone "31.168.192.in-addr.arpa" {
        notify no;
        type master;
        file "192.168.31";
        allow-update { none; };
};
include "/etc/rndc.key";
```

The description file for the cluster zone is located in the */var/named/chroot/var/* directory. The contents of the cluster file are as follows:

```
$TTL    86400
@               IN SOA  cluster. root.parallel.cluster (
                                  2007042700      ; serial (d. adams)
                                  3H              ; refresh
                                  15M             ; retry
                                  1W              ; expiry
                                  1D)            ; minimum
                IN NS        parallel.cluster.
parallel        IN A         192.168.31.1
master          IN A         192.168.31.1
m01             IN A         192.168.31.11
```

```
...
m03                IN A               192.168.31.13
```

The file with the description of inverse transformations for the cluster zone is in the same directory (*/var/named/chroot/var/named*); it also has a "soft" link, which is in the */var/named* directory. The name of the file with the data in the back zone is 192.168.3.1.

The contents of this file are as follows:

```
$TTL     86400
@        IN       SOA      cluster. root.parallel.cluster.   (
                                     2007042600 ; Serial
                                     28800       ; Refresh
                                     14400       ; Retry
                                     3600000     ; Expire
                                     86400)     ; Minimum
         IN       NS       parallel.cluster.
 1        IN       PTR      parallel.cluster.
 11       IN       PTR      m01.cluster.
...
13       IN       PTR      m03.cluster.
```

TFTP (Trivial File Transfer Protocol) is a simple file transfer protocol. It is mainly used for initial downloading of diskless *slave*-nodes. This protocol does not include authentication capabilities and uses the transport protocol as *UDP*. The package containing this server is called *tftp*-server. If it is not in the system, then it needs to be installed. The *tftp* service works through *xinetd* super servers to use the following commands to run them:

```
/sbin/chkconfig --level 345 xinetd on
/sbin/chkconfig --level 345 tftp on
```

They configure the launch of *tftp* and *xinetd* services during execution of protocols with numbers from 3 to 5. Their launch is performed by the following command:

```
/sbin/service tftp start
/sbin/service xinetd start
```

Dynamic Host Configuration Protocol (DHCP) protocol is network-based and is required for automatic distribution of *IP* addresses and other network parameters of slave-nodes. Each node connects to a dedicated *DHCP* server that returns its network settings, including *IP* address, gateway address, and *DNS* servers. The configuration file for */etc/dhcpd.conf* cluster contains following data:

```
default-lease-time 600;
max-lease-time 7200;
authoritative;
```

```
ddns-update-style none;
option domain-name "cluster";
option routers 192.168.31.1;
option domain-name-servers 192.168.31.1;
option swap-path code 128 = string;
option swap-size code 129 = integer 32;
subnet 192.168.31.0 netmask 255.255.255.0 {
        range  192.168.31.11   192.168.31.21
        option subnet-mask              255.255.255.0;
        option broadcast-address        192.168.31.255;
        group {
          filename "linux-install/pxelinux.0";
          option root-path "/diskless/root";
          next-server 192.168.31.1;
          host m01 {
              option host-name "m01.cluster";
              hardware ethernet 00:0E:0C:4D:16:95;
              fixed-address 192.168.31.11;
          }
# The full configuration file is not provided
          # given the irrelevance and the same setting of the setting
     ... ... ... ... ...
host m12 {
              option host-name "m12.cluster";
              hardware ethernet 01:0E:0C:4D:19:95;
              fixed-address 192.168.31.21;
          }
        }
}
allow booting;
allow bootp;
```

In addition, the following line has been added to */etc/sysconfig/dhcpd*:

```
DHCPDARGS=eth0
```

In addition, this research section shows that parallel programs, which require high-speed, process, as a rule, solid amounts of data. For the local conservation of results and intermediate calculations, the *TCA* system is used. The main advantage of such an approach is the simplicity of implementation and substantial unloading of the master node. The results of the intermediate and final computations are transmitted to the *TCA* via a managed *Infiniband* switch. In this case, the control and transfer of the corresponding data from the *slave*-nodes are carried out using the *HCA* network adapters.

So, we describe the modes of using disk space in a designed multiprocessor computing system. The files with the output and the results of the parallel program computations are placed on a single file server. To organize access to these files from the blades, *NFS (Network file system)* is used. This server is required for exporting file systems or individual directories for diskless *slave*-nodes. The /etc/exports configuration file looks like this: *# export filesystems for dumb nodes*:

```
# Exports the root file system, which is read-only
/diskless/root    192.168.31.0/24 (ro,sync,no_root_squash)
```

When building a high-speed network inevitably there is a need for *RDMA*, the direct access to remote memory. That is, instead of the traditional packets interchange over the network, servers turn directly to the memory of each other. Of course, this requires specific network equipment that supports *RDMA*. There are two commonly used alternatives: the *Infiniband* and 10*G Ethernet*. It should also be noted that if one makes a software *RDMA* code, following only standard *API*s, it will work the same way on both *Infiniband* and 10*G Ethernet*. Physically, under the same *API*, there is a completely different hardware, the *Infiniband* or 10 *Gigabit Ethernet*. Therefore, the choice is the *Infiniband* technology, it can be separated from the applications development (http://tslab.com.ua/en/publication/category/3/50).

At significant differences in time setting on the various blades of the multiprocessor computing system there may be interruptions in the parallel program's operation. Therefore, to synchronize the time, the system uses an *NTP* server. At the first stage, the time is set on the main node, and then on the *slave*-nodes using the synchronization command. The *ntp* server configuration file looks like this:

```
restrict 192.168.31.0 mask 255.255.255.0  nomodify notrap
restrict 127.0.0.1
restrict 127.127.1.0
restrict 127.127.27.0 nomodify
server  127.127.1.0     # local clock
fudge   127.127.1.0 stratum 10
driftfile /var/lib/ntp/drift
broadcastdelay  0.008
keys     /etc/ntp/keys
```

Software Communication Among the Nodes of the Multiprocessor Computing System

The interaction among the nodes of the multiprocessor computing system occurs through the so-called programming interface, that is, specialized libraries of functions. The communication system, in turn, consists of two components: software and hardware.

The software component contains two basic methods, therefore, the data can be transmitted:

- Through the shared memory, and the branches access synchronization to it occurs by semaphores;
- In the form of messages.

The first method of the simulation of interprocess communication programming through shared memory with synchronization of semaphores is basic for *SMP* and single processor computers. *Shared memory* (*SHM*) functions are part of each multitasking operating system. Within the same computer, the rest of interprocess communication is implemented by the *SHM* team.

The second method of the interprocessor programming model is typical for all types of networks. The interaction among the nodes of the computing cluster is realized by the programming interface *MPI* (Official site of MPI-forum, http://www.mpi-forum.org). *MP* is a technology for creating parallel running programs, which is based on the transmission of messages between processes (the processes themselves can be executed both on a single and on different computing nodes). *MPI* technology is a typical example of the imperative (based on the principle of complete programmer control by a sequence of computations, data sharing and information interchange among processes) technology for creating programs for parallel execution (Rajkumar et al., 1999; Massive-parallel architecture, http://www.myshared.ru/slide/641118; Pat. 61944 Ukraine; http://www.mpi-forum.org).

Each data transmission simulation has its advantages and disadvantages. Let's analyze them. The advantages of the shared memory (*SHM*) simulation versus *MPI* can be described as follows. In a shared memory simulation, the data used is stored in a single copy without creating local copies for each branch of application. Branches that work with such data spend little time synchronizing data access. In general, the communication speed ratio and computing speed in a simulation with shared memory are higher than in a messaging simulation. Consequently, the class of problems suitable for parallelization in *SHM* technology is in principle broader than *MPI*.

However, some programmers point out that in terms of data rates, the *MPI* simulation is not worse than *SHM*. This is due to the following circumstances:

- A well-parallel application for interaction among branches (data transfer and synchronization) takes a little time, a few percents of the total operation period; Thus, the transfers slowing down, say, twice, does not mean the same general decline in productivity, since it will decrease by a few percents, which can be considered quite acceptable;
- Means available in *MPI* technology (asynchronous/permanent communications) and especially in MPI-2 (remote access memory) provide a data rate comparable to that of *SHM*.

It is noteworthy that a certain part of the programmers notes other properties of the *MPI* simulation, which makes it more acceptable compared to the *SHM* simulation. Among these, we will name the main:

- All possible implementations of *SHM* or, in principle, are not transferred beyond SMP machines, or are not sufficiently effective or not sufficiently distributed;
- Cost of computing and communications in *SMP* machines decreases more slowly than in *MPP* complexes and networks;
- The method of data interchange in the form of messages from the very beginning is characterized by greater visibility, and therefore convenient for the programmer;
- Unlike *MPI*, the *SHM* simulation usually only implements the most basic operations.

Thus, the *MPI* programming simulation, on the one hand, reduces the gap in the speed of programs execution, and on the other - increases the gap in the speed of their making.

The main features of *MPI* technology are:

- Providing communication among branches of a parallel program;

- *SPMD* programming (*Single Program - Multiple Data*) with message transmission;
- Implementation as a library for programming languages *C* and *Fortran* and downloading applications;
- Support for heterogeneous computations;
- Interface mobility and, consequently, the mobility of the created programs.

An application that employs *MPI* technology is easier to configure (narrowing the space for stereotyped parallel programming errors) and is more likely to be transferred to other platforms (ideally, by simple recompiling).

There are three generations of *MPICH* (Official site of the MPI Forum, http://www.mpi-forum.org; Official page of the lam-mpi package, http://www.lam-mpi.org; Official page of mpich1 package, http://www-unix.mcs.anl.gov/mpi/mpich1) related to the development of the *ADI* method. The first version of *ADI-1* was designed for computers characterized by mass parallelism, where the mechanism of interchange among processes belonged to the system. This method provided good performance at low overhead costs. This version of *MPICH* has been installed on parallel computers: *Intel iPSC/860, Delta, Paragon, nCUBE.*

The *ADI-2* has been introduced to achieve great flexibility in the implementation and effective maintenance of communication mechanisms with a large number of functions.

The *ADI-3* is designed to provide greater compatibility with new networks that have remote access, as well as a multi-threaded operating environment and support for *MPI-2* operations (including remote access to memory and dynamic process management). The *ADI-3* is the first version of *MPICH*, which, during the design, did not set the problem of close correspondence with other libraries with messaging, since *MPI* has supplanted the majority of systems for the interchange of messages in scientific computations. The *ADI-3*, like previous ones, is designed to facilitate the transfer of *MPICH* to new platforms and communication methods.

Consequently, *ADI* is a set of functions (which can be implemented as functions or as macros) from a custom *MPI* set. This is exactly what creates protocols that distinguish *MPICH* from other *MPI* implementations. In particular, the *ADI* level includes procedures for picketing messages and connecting header information, managing buffering policy, and also designed to match the reception of incoming messages, etc.

To illustrate the essence of the MPICH technology work, consider, for instance, the implementation of simple functions of sending and receiving *MPI_Send* and *MPI_Recv*. The procedures of this group are forming the basis of the processes interaction. Recall that the basic mechanism of communication among processes in the *MPI* is the sending and receiving of messages. The main functions that provide the "point-to-point" connection are the *send* and *receive* operators. For this purpose two protocols can be used: *Eager* and Rendezvous.

Let's describe the *Eager* protocol action. When sending data, the *MPI* along with the buffer address must enter the user's information about the tag, the device, as well as the length of the message, its source, and the recipient. It is called the 'envelope'. The sent message consists of a shell that accompanies the data. The data sending method along with the shell is called the *eager* ("greedy") protocol. When a message arrives, there are two possible cases: either the corresponding reception procedure is started or not. In the first case, the input for data is provided. In the second case, the situation is more complex. The reception process should remember that the message was received, and therefore it is necessary to save it somewhere. The first requirement to perform is relatively easy by tracking the incoming message queue. When the program executes the command *MPI_Recv*, it first checks this queue. If the message is received, then the operation is run and completed. But there may be a problem with the data. What, for instance, happens when a lot of subordinate processes will almost simultaneously send their long messages to the main process (for example, more than 100 MB in size each) and the space in its memory for placement is not enough? A similar situation arises, for example, with the multiplication of matrices. The *MPI* standard requires that data reception takes place under such conditions and not refused. All this leads to buffering.

The *Rendezvous* Protocol. In order to solve the problem of delivering a large amount of intended data, it is necessary to control the quantity and time of receiving the data by the recipient process. A simple solution to the problem is to send only the shell to the recipient process. When one needs to receive data (and there is a place for the placement), it sends a message to the sender with the phrase "now send data". They are received. This method is called the "rendezvous" protocol: the recipient reserves a space in memory only for storing shells of a small size, rather than storing the data themselves.

It should be noted that the interchange of messages between the two processes is characterized by three main important properties:

1. The exchange is bilateral. If the data is to be transmitted, then one process must execute the "transmit" operation, and the other one is the "take" operation.
2. The interchange has the properties of a "rendezvous", that is, a means of not only data transfer but also synchronization. When transmitting, the programmer absolutely must not guess whether the recipient at this time can accept the data, because, in the case of unwillingness, the communication library will automatically slow down the transmitter until the receiver goes into readiness. If the receiver is in a working condition prior to the appearance of messages, then the communication library will slow down the acceptance operation.
3. The interchange and synchronization are built on the basis of messages, rather than a continuous byte stream. Each transmission of a message must correspond to one and only one acceptance of the same length.

Thus, it becomes apparent that the basic communication library is the implementation of the mechanism of a "bilateral rendezvous" among processes that operate on different cluster blades. However, the OS blade must have mechanisms of interprocess interaction in the network version, which allows implementing a two-way rendezvous.

Thus, we note that the command for launching a parallel program is a rather complicated operation, which is always tied to a specific communication library. In this regard, the effective organization of the communication network is the most essential condition for the functioning of the computer system. The key role is played by software. Therefore, it is necessary to consider the features of setting up the system software of the computing cluster. The main node and *slave*-nodes have the package *lam*-7.0.6-5 included in the *Scientific Linux* distribution. Its configuration file is located in the directory at */etc/lam/lam-bhost.def.* The contents of the configuration file are as follows:

```
m03 cpu=4
m02 cpu=4
m01 cpu=4
master cpu=4
```

The main work with making the *MPI* standard and the building of its implementations was carried out at the Argonne National Laboratory of the United States (http://www.anl.gov/). Here, *MPI*s, known as *MPICH* (the CH adding is taken from the name of the *Chameleon* package, which was previously used for messaging systems since many of its packets were included in *MPICH*), have been prepared and widely promoted. In this version of the multiprocessor computing system, the *MPICH* version 1.2.7*p*1 is used, the package was assembled from the source code with the following options:

```
-rsh=ssh -with-device=ch_p4 --with-comm=shared --refix=/usr/local/gmpich-
1.2.7p1-f90=gfortran
```

The */usr/local* directory creates a "soft" link from the *mpich* program library to the *gmpich*-1.2.7*p1* directory. The *machines.LINUX* configuration file is located in the directory at */usr/local/mpich-1.2.7p1/share*. The contents of the configuration file is as follows:

```
m03:4
m02:4
m01:4
parallel:4
```

Each line contains the name of the *slave*-node or *IP* address after the colon indicates the number of kernels in the *slave*-node of the multiprocessor computing system.

Configuring the System Software of a Multiprocessor Computing System and the Features of Its Operation

Let's turn from the issues related with the OS properties and communications media of the multiprocessor computing system (MCS) to the consideration of the system configuration and operation. It should be noted that the MCS must be considered as one multiprocessor computer that does not use the time distribution system and works in mono-user mode. Only under this condition, its work will be productive. In each case, one needs own approach to security issues, application software, administration, download and use of the file system. Most *Linux* distributions are capable of maintaining security at a very high level. However, it should be borne in mind that in the cluster, it can 'come into conflict' with the computing and administrative settings.

Note that a parallel program is essentially a set of processes that are executed in different MCSs and periodically interact with each other. Let's dwell on the basic means of organizing such interaction.

At the first stage, we will consider implementation in the MCS the remote command execution. To run a parallel program, it is necessary to start the process on another cluster blade available on the network. The main basic tool used for this purpose in the vast majority of parallel programs is the *rsh* remote run command.

So, typing while managing the cluster on the blade A the command:

```
rsh B Is -alF /home/mydir,
```

the user will see an answer from the *Is alf home mydir* command performed on the *B* blade. It is clear that it is thus fairly easy to "split" the list of nodes in the parallel branch program that are to be executed on those nodes. It is clear that in the example given, the *Is* command will be executed on the *B*-blade as a process, and it must belong to a user. Therefore, it becomes obvious that the *rsh* command must have an *authorization system*, that is, somewhere the information about the blades and users should be stored, and their authority to execute the *rsh* command.

Consequently, on the basis of the considered considerations, it is possible to formulate general implementation principles in a cluster system of the commands remote execution:

1. External access rights are given on the domain to which this access is granted;
2. Access permissions are given in the form of configuration files, access to which should be allowed only to those on whose behalf they allow access from the outside; At the same time, note that the root user can at any time change the permissions from the outside.

The *slave*-nodes are configured directly according to the instructions (http://www.lam-mpi.org). So, the *diskless/snapshot* directory was previously created, which acts as root */diskless/root* for *slave*-nodes, where all the files to be replaced will be located. So, to configure *slave*-nodes, one needs to copy the root file system with the command:

```
rsync -a -e / /diskless/root/
```

As a constructive one the single case was chosen, which is the cell of the computing cabinet. This is due to the fact that, on the one hand, if necessary, several modules can be placed in a single housing, and on the other hand, this approach provides compactness, successful cooling and easy access to the nests and card components that are being adjusted. The computing system includes the vertical, parallel to each other, the location of the motherboard, which corresponds to the idea of "*Blade*" servers.

After the power supply to the *ATX* unit of the master unit and the external *PUSK* signal from the *P*01 control panel, the startup and initialization of the master unit of the multiprocessor computing system begin. Direct OS download can be made from a hard drive or from a *CD/DVD* device. After the OS downloading, a specially oriented configuration script is launched that configures the *DHCP* server operation. In addition, the number of computing nodes in the system is determined, if necessary, it is configured to access the Internet environment or the external network. Here the basic settings and configurations are set. Serial supply to the power supply (*ATX*) and the initialization of *slave*-nodes reduce the required power of the *UPS* block, run all computing *slave*-nodes, and their OS download. Downloading and configuring all computing nodes completes the corresponding script operation. The system is ready for parallel computing.

The master node (*PM*001) through the *KGI SW*1 switch provides the diagnostic flow of data related to controls. In its turn, the *slave*-nodes, in accordance with the algorithm for solving problems and processes, implement the necessary computing mode. The data interchange among computing nodes and the problem conditions loading is carried out on a separate network, which is organized by a managed switch *KIB IB*1. In order to achieve maximum system performance, one or two-port *Infiniband* adapters are used and the process of reconfiguring the structure of the second network according to the specifics of the problems to be solved. The results of intermediate and final computations are transmitted to the master node via the *Infiniband KIB*'s managed switch. In this case, the control and transfer of the corresponding data from the *slave*-nodes are carried out by *HCA* network adapters. The direct storage of these computations for further processing is performed by the *TCA* network adapter.

Experimental Research and Processing

The developed multiprocessor computing system was applied to the solution of the thermal problem of the aviation complex. The proposed approach was implemented as a procedure for mathematical simulation. At the same time, an important stage of the research was the development of a package of applied programs for the solution of coefficient heat conduction problems by mathematical simulating methods. The package was created taking into account the requirements of object-oriented programming. The simulation procedure was implemented based on the use of a multiprocessor computing system. The package of applied programs is designed for processing thermophysical experiments by inverse methods for studying the temperature regime of the aircraft body. The main goal, which was put at its creation, was to provide practical assistance to the researcher at all stages of the processing of experimental data.

To solve the problem, additional conditions are considered that allow us to divide the researched problem into two - temperature and flux. The first of them allows solving the coefficient problems in the whole given range of temperature variation with the control parameter in the form of the thermal diffusivity coefficient (model 1), the second one in the form of the coefficients of thermal conductivity or heat capacity (model 2). This approach corresponds to the classical methods of technical thermophysics. The research of mathematical models 1 and 2 was carried out by the method of lines. Moreover, model 1 (for example, algebraic or functional) and model 2 (gradient) allow solving the coefficient problem in an extreme formulation. As a test problem, it was proposed to determine the thermophysical properties of a specific industrial material (http://www.mpi-forum.org;). The thermal properties of the aircraft body were researched. For this, the temperature field of a sample having the shape of a cylinder was simulated. When solving such a coefficient problem, the following initial data were used: the thermal diffusivity coefficient. The results of simulation performed by means of a multiprocessor computing system are presented in Figure 2. The solution of the coefficient problem was carried out by control over the dimensionless coefficient of thermal diffusivity at $R = a / a_0$. The simulation results analysis (Figure 2) demonstrate that the minimum of the residual corresponds to the value of the parameter $R \approx 1$. The exact value of the control parameter $R = 1$. For the heat conduction problem from table data $\lambda = 0.16$. The identification of such a parameter is provided in Figure 3.

So, the features of the developed approach are as follows. Its version with the use of accurate input data absolutely coincides with the exact result of the analytical solution, and the errors in the results of computation the recoverable causal characteristics, which take into account the error of the input data, approximately equal the errors in the output data.

Analysis of Experimental Data Results

The solution of the inverse coefficient problem in the proposed formulation reduces to the direct determination of sequences of values of functionals on mathematical models and the computation of the minimal carriers $J_{\min}$ inside them. The determination procedure can be implemented by simple sorting or by changing the sign $J^*(a) * J^*(b)$ on the segment of R = a, R = b, where for the linear value of the functional $(a < b)$. It is clear that $J^*(R) = 0$ on a separated segment $R \in [a,b]$ will have a root. The

Figure 2. Graph of the computation results of the coefficient problem with the control parameter relative to the thermal diffusivity coefficient

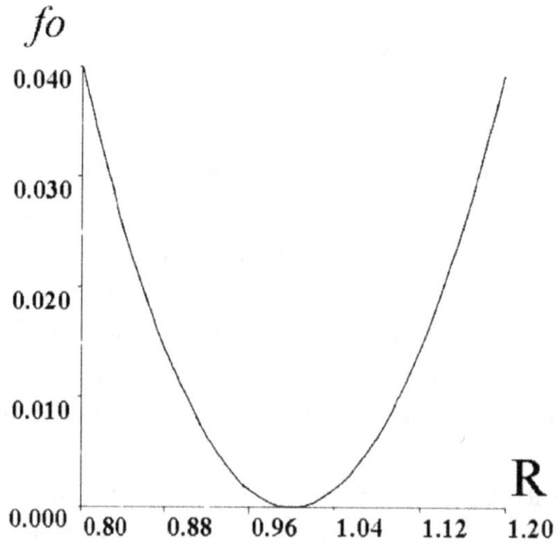

Figure 3. Graph of the computation results of the coefficient problem with the control parameter relative to the thermal conductivity coefficient

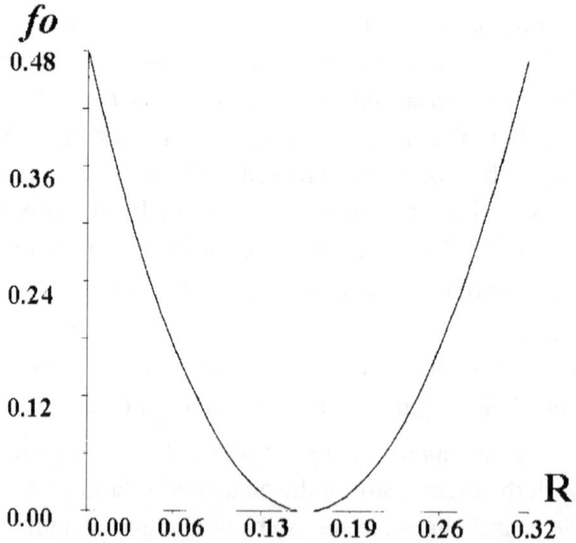

refinement of the values of this root can be realized with any predetermined accuracy from known dependencies.

It should be noted that the partition of the total time interval into independent intervals with the solution of inverse problems in each of them according to the scheme indicated above allows determining the value of the unknown parameters as functions of temperature $T_{p,1}(T)$. Therefore, the subsequent stage of processing experimental data consists in constructing temperature dependences $\lambda_{p,1}(T)$, $C\nu(T)$ in the form of certain polynomial expansions of one degree or another by the method of mathematical planning and regression analysis. At this stage, to verify and establish the adequacy, it is advisable to apply a discrete nonlinear mathematical model on a full space-time interval.

CONCLUSION

The research considers the problems of use and adjustment of the system software of the multiprocessor computing system. For the designed MCS, the OS *Linux* was selected. It is now considered standard for computing cluster systems. There are at least three explanations for this fact. The first is purely technical. The *MS Windows*, as the most widespread, could be an alternative, if it were not for its initial orientation human communication with the computer. However, by analyzing the basic system mechanisms that are mainly used in clusters, one can conclude that there is absolutely no user intervention. The second and perhaps the main explanation is the infrastructure peculiarity. With the independent cluster compilation and operation, these considerations are the foreground. In this sense, it's obvious that the *Linux* OS has no alternatives. Thirdly, a very significant explanation is that the use of *Linux* can be introduced for free. In these researches, the main features of *Linux* were considered, which is necessary when designing a multiprocessor computing system. First of all, we note that any program that is not part of the kernel can be executed in such an operating system only as a process. So, here are considered the main properties of processes in the *Linux* OS.

In addition, the section discusses the use of communication tools (interactions) among the nodes of the multiprocessor computing system. It is shown that the most widespread in our time parallel programming technology with partitioned memory is *MPI* technology. The main way of interacting parallel processes in such systems is acknowledged by the transmission of messages. This is reflected in the technology name, the *Message Passing Interface*. The *MPI* standard offers an interface that should be adhered to as an *MPI* programming system as well as a user when creating own programs. It is shown that an application with *MPI* technology is easier to set up (narrowing the space for stereotyped parallel programming errors) and is more likely to be transferred to other platforms (ideally, by simple recompiling). The *MPI* syntax makes it easy to create applications in the *SPMD* (*single program multiple data)* simulation since one program runs on different processors with its own data. One and the same function can be called on the source node and on the receiver nodes, and the type of operation (transmission or reception being performed) in this case is determined by the software parameter. This syntax makes *SPMD* programs much more compact, although more complex to understand.

These researches highlight new means of remote access to the processor systems memory due to the use of *RDMA* mechanisms of *InfiniBand* technology. Thus, the *RDMA* principle, along with the formation of a separate computer network in the data interchange environment, and the *VLAN* mechanisms implementation, allowed the data transmission among the node's memory in a multiprocessor computing system without additional buffering. This approach does not require the active OS operation, libraries or applications on those system nodes, which memory is requested. The proposed mechanisms allowed to increase the computing speed when solving tightly-coupled problems, provide high-speed access to the system nodes memory and data interchange among them, unload the *CPU-GPU* in the data interchange and reduce the channel load that is passing among the computing nodes.

In addition, it is shown that parallel programs that require high-speed performance typically handle solid data volumes. For the local conservation of results and intermediate computations, the *TCA* system is used. The main advantage of such an approach is the implementation simplicity and significant unloading of the master node. The results of intermediate and final computations are transmitted to the *TCA* via a managed *Infiniband* switch. In this case, the control and transfer of the corresponding data from the slave-nodes are carried out by HCA network adapters.

REFERENCES

Alishov, A. N., Shvachych, G. G., Semenov, S. G., & Fedorov, E. E. (2016). Efficient of extreme algorithms of high order accuracy for solving applied tasks. *Mathematics and Computer Science: Journal of Qafqaz University, 1*, 3–11.

Bashkov, E. O., Ivaschenko, V. P., & Shvachych, G.G. (2011). High-efficient multiprocessor system based on personal computing cluster. *Problems of Simulation and Automatic Projecting, 9*(179), 312–324.

Ill'chenko, K. D., Chechenev, V. A., Ivashchenko V. P., & Tereschenko, V. S. (1999). *Teplofyzyches-kye svoystva promyshlennykh materyalov: Spravochnyk*, 152.

Ivaschenko, V. P., Shvachych, G. G., Ivaschenko, E. V., & Busygin, V. V. (2018). Effective algorithms for solving coefficient problems of high accuracy. *System Technology, 4*(117), 95–108.

Ivaschenko, V. P., Shvachych, G. G., & Semenov, S. G. (2017). Efficient parallelization algorithms of the applied tasks in multiprocessor computing systems. *System Technology, 2*(109), 57–66.

Ivaschenko, V. P., Shvachych, G. G., & Shmukyn A. A. (2005). Some aspects of the problem of mathematic simulation of the metallurgical thermophysical tasks based on parallel computing system of cluster type. *Modern Problem of the Metalurgy, 7*, 23–30.

Ivashchenko, V. P., Bashkov, E. A., & Shvachych, G. G. (2011). Informational systems and technology: monography. Krasnoyarsk: Academic Press.

Ivashchenko, V. P., Shvachych, G. G., Sobolenko, A. V., & Protopopov, D. V. (2003). Informational system of intellectual support of making solutions for the metal rolling. *Vostochno-evropeyskyy zhurnal peredovykh tekhnolohyy, 3*, 4–9.

Latsys, A. O. (2003). How to build and use supercomputer. Moscow: Academic Press.

Massive-Parallel Architecture. (n.d.). Retrieved from http://www.myshared.ru/ slide / 641118 /

Moving to a Balanced Computer Architecture. (n.d.). Retrieved from http://tslab.com.ua/en/publication/category/3/50

Official Page of the LAM-MPI Package. (n.d.). Retrieved from http://www.lam-mpi.org

Official Page of the mpich1 Package. (n.d.). Retrieved from http://www-unix.mcs.anl.gov/mpi/mpich1

Official Site of the MPI Forum. (n.d.). Retrieved from http://www.mpi-forum.org

Official Website of the Laboratory. (n.d.). Retrieved from http://www.anl.gov

Ozerov, S., & Kalinichenko, A. (n.d.). Parallel computations: clusters. *Computerra: The Site of Computer Magazine*. Retrieved from http://offline.computerra.ru/2005/618/244051

Rajkumar, B. (1999). *High-Performance Cluster Computing* (Vol. 1). Architectures and Systems.

Red Hat Enterprise Linux 4 System Administration Guide. (n.d.). Retrieved from http://www.rhd.ru/docs/manuals/enterprise/RHEL-4-Manual/sysadmin-guide/ch-diskless.html

Shvachych, G. G., Kholod, E. G., Ivaschenko, E. V., & Busygin, V. V. (2018). Visualization of the decisions of applied problems in multiprocessor system. Natural and Technical Sciences, 1(17), 65–69.

Shvachych, G. G., Kolpak, V. P., & Sobolenko, M. A. (2007). Mathematic simulation of the speed modes of thermo processing of the long products. *Theory and Practice of the Metallurgy, 4–5*(59–60), 61–67.

Shvachych, G. G., Moroz, B. I., Pobochii, I. A., Ivaschenko, E. V., & Busygin, V. V. (2018). Maximally parallel forms of distributed simulation of the dynamic system. World Science, 4(32), 12–20.

Shvachych, G. G., Sobolenko, A. V., & Sobolenko, M. A. (2003). Research of the temperature regime of operation of annealing furnaces of mine type of wires. *Theory and Practice of the Metallurgy, 1,* 59–62.

Shvachych, G. G., Sobolenko, O. V., & Ivashchenko, O. V. (2017). Network Interface link aggregation of multiprocessor computing systems. *Strategy of Quality in Industry and Education: Proceedings from XII International Conference*, 236–241.

Shvachych, G. G., Tkach, M. A., & Shcherbyna, P. A. (2008). Supercomputers and high-efficient computing. *Proceedings from B″deshcheto problemyte na svetovnata nauka: materyaly za 4-a mezhdunarodna praktychna konferentsyya. V. 21. S″vremenyy tekhnolohyy na ynformatsyy*, 22–27.

Voevodyn, V. V., & Zhumatyy, S. A. (2007). Computing and cluster systems. Moscow: Academic Press.

ADDITIONAL READING

Alishov, A. N., Shvachych, G. G., & Tkach, M. A. (2013). Multiprocessor modeling technologies for the applied Monte Carlo tasks. *Mathematics and Computer Science: Journal of Qafqaz University*, *1*, 3–10.

Bertsekas, D. P., & Tsitsiklis, J. N. (1989). *Parallel and Distributed Computation* (p. 718). New Jersey: Numerical Methods.

Chandra, R., Menon, R., Dagum, L., Kohr, D., Maydan, D., & McDonald, J. (2000). *Parallel Programming in OpenMP* (p. 231). New Jersey.

Rajkumar, B. (1999). *High-Performance Cluster Computing* (Vol. 1, p. 453). New Jersey: Architectures and Systems.

Roosta, S. H. (2000). *Parallel Processing and Parallel Algorithms: Theory and Computation* (p. 600). NY: Springer-Verlag. doi:10.1007/978-1-4612-1220-1

Shvachych, G. G. (2008). Prospects of construction highly productive computers systems on the base of standard technologies. *Strategy of Quality in Industry and Education: IV International Conference*, Bulgaria, 815–819.

Shvachych, G. G., Konovalenkov, V. S., & Tkach, M. A. (2013). Model-based parallel construction of a numerical-analytical scheme. Science and Education a New Dimension: Natural and Technical Science, V. 8, Budapest, 76–78.

Xu, Z., & Hwang, K. (1998). *Scalable Parallel Computing Technology, Architecture* (p. 557). Boston: Programming.

KEY TERMS AND DEFINITIONS

Blade-Server Solutions: In this chapter, the authors consider multiprocessor computing systems with several similar motherboards installed in the same package. Practice shows that "blade" systems are more compact and easy to maintain, and their implementation is not much more expensive compared to multiprocessor computing systems.

Cluster Computing Systems: Multi-processor computing systems based on local networks, known as "cluster computing systems."

Communication Environment: For making supercomputers there are used serial microprocessors, each with its own local memory and connected via some communication environment.

DNS Server: Domain name system server (bind-9.2.4-27.0.1.el4) of the main module performs direct and inverse transformation of node names into IP addresses.

InfiniBand Technology: The data interchange among computing nodes of a multiprocessor computing system is transferred to a separate network using the InfiniBand network interface, which increased the system's overall performance and significantly reduced the latency (channel download) of the network that connects nodes of the cluster computing system.

Supernumerary Devices: In fact, this is about multiprocessor parallel computing systems of MPP architecture (massively parallel processing).

TFTP Protocol: Trivial file transfer protocol is a simple file transfer protocol. It is mainly used for initial downloading of diskless slave nodes.

Vectorization of Computations: The creation of parallel computing systems also required the development of a new approach for the making appropriate computing algorithms. For this reason, new ideas and approaches in the field of computational mathematics are now emerging, which are aimed towards the development of new methods for the implementation of numerical experiments based on the computations' parallelization.

Chapter 17
Critical Aviation Information Systems:
Identification and Protection

Sergiy Gnatyuk
National Aviation University, Ukraine

Zhengbing Hu
Central China Normal University, China

Viktoriia Sydorenko
National Aviation University, Ukraine

Marek Aleksander
State Higher Vocational School in Nowy Sącz, Poland

Yuliia Polishchuk
National Aviation University, Ukraine

Khalicha Ibragimovna Yubuzova
Satbayev University, Kazakhstan

EXECUTIVE SUMMARY

This chapter is devoted to developing formalization methods for identification and security objects of critical information infrastructure (CII) in civil aviation. The analysis of modern approaches to the CII identification was carried out that gave a possibility to determine weaknesses and to formalize a scientific researches task. As a result, the unified data model was developed for formalizing the process of a list of CII objects forming in certain field and at the state level. Moreover, the specialized technique was developed. Besides, the identification method was proposed, and it gives a possibility to determine elements of CII field, mutual influences, and influence on functional operations of critical aviation information system. Furthermore, special software was developed and implemented that can be useful for CII elements identification and also for determining its influences on functional operations. Also, the basic aspects of cybersecurity ensuring for identified critical aviation information system were described in this chapter.

DOI: 10.4018/978-1-5225-7588-7.ch017

INTRODUCTION

Present-day trends in information and communication technologies (ICT) have caused phenomenal dependence the society form services which different infrastructure proposed. For now, quality and accessibility of these services are the main points of infrastructure development of the state. According to that, the ensuring their protection and stability are the most essential and mandatory part of State security in developed countries. Increased concentration of facilities and resources for protection the different types of electronic infrastructure had necessitated ranking of infrastructure objects, choosing the most important of them and creating the "critical infrastructure" (CI) definition. Typically, this category relates to energy and transmission line, oil and gas line, seaports, high-speed and government communications channels, life-saving systems of megacities, high-tech enterprises and enterprises of the military-industrial complex, and also the central government authority. The civil aviation (CA), given the need to ensure sustained communication and cooperation between ground-based and aircraft systems, are required special attention. Therefore, identifying the objects which are critical for ensuring the system information continuing operation is the first priority. Nevertheless, an unlimited number of objects and system parameters that constantly varied and unforeseen behavior of objects with lots of interlinkages are the main reason for difficulties with the identified objects of state critical infrastructure. The basic component of critical infrastructure is an informational part – so-called CII (critical information infrastructure). The main reasons for the CII importance are the widespread usage in all areas of human activity of ICT, dependence on them of citizens, society and the state, as well as increasing vulnerabilities and potential threats of different nature. Moreover, in some countries a strong focus on the CI importance for the nation (even the CII definition is critical national information infrastructure). In case of Ukraine, the legislative framework for regulating the protection of CI still in an early development stage, particularly, continuing the process of identifying the objects of state CI in different fields (with no great success, unfortunately). The problematic of CII protection (CIIP) in various fields was exploring by the Ukrainian and foreign scientists such as: H. Alcaraz, D. Biryukov, D. Bobro, D. Gritsalis, O. Dovgan, E. Yeliseyev, A. Kondratiev, M. Merabti, L. Romano, H. Sitarlis, I. Fovino, V. Kharchenko and others. However, the most of researches are not systemic: 1) mostly researches oriented on development and implement preventive and countermeasures for protection particular CI or CII objects; 2) not enough focus making on the mechanism of formulating the list of state CII, and known approaches (according to the international standards and recommended practices), which are not formalized that make it difficult to use them on the state level, especially in CA. According to that, a development the unified data model and the methods of identifying the objects of CII is the actual scientific task that has theoretical and practical values.

State of the Art in CIIP

Modern society totally depends on ICT, the dysfunction and breakdown of which may lead to chaos, significant financial losses and even mass deaths of people.

The truth is, much of mankind inclined to take the most important services (in particular, their quality) as a matter of course until something or someone breaks their work. To determinate and consolidate the most important and vulnerable assets of the state, relatively recently, the term of the CI was introduced into international law. It is known, that establishment of the regulatory framework in the field of CIIP a long process. Most developed countries formulated their own policy (concept) and have developed practical guidelines for the protection CI objects, paying particular attention to the objects of CII (information systems and networks).

The analysis of legislative framework shows the field of CII protection in our State is in an early stage of development. The current legislation still does not define an exhaustive list of the state CII objects, and only the objects of separate fields are specified (Biriukov, Kondratov & Sukhodolia, 2015; Dovgan, 2013), which require protection from the state: enterprises which have important meaning for economics and State security; the most essential power industry objects; especially important objects of the oil and gas industry; important state objects, including points of management of state authorities and local self-government bodies; objects of possible terrorist attacks; objects that are subject for protection and defense in the event of emergencies and during a special period; objects that must be protect by units of the State Protection Service by contracts; objects of high danger (including the list of especially dangerous enterprises, the termination of which requires special measures to prevent damage to life and health of citizens, property, buildings, the environment), objects that are included in the State register of potentially hazardous objects, radiation hazardous objects for which the object design threat is being developed; objects classified as civil defense categories; objects owned by economic entities, the project which is carried out taking into account the requirements of engineering and technical measures of civil protection; the emergency-dispatching system of emergency assistance to the population for the only free emergency call number 112; rescue services; the national system of confidential communication; payment systems; immovable objects of cultural heritage objects.

According to (Biriukov &Kondratov 2012), for now, in Ukraine are continuing the development of a proposal about forming the list of ITS objects of state CI. Unfortunately, this list has not been formed in any sector of the CI. Moreover, the lack of clearly defined terminological basis complicated integration our state to the global information space.

It is established that in order to protect the most important CII objects, it is necessary to identify these objects according to certain criteria or critical parameters (Figure 1, block 1).

The analysis of criteria by which it is possible to choose or identify the CII objects was conducted in (Lyadovska, 2014). It was found that one of the firsts criteria for identification of the CI was specified in the EU Directive (Council Directive 2008). Accordingly, each state should identify potential CI that meet the requirements of the two main groups of criteria – intersectoral and sectoral. Intersectoral criteria should include: 1) criteria of losses (estimated from the perspective of the potential number of dead or injured); 2) criteria of economic consequences (assessed in terms of the significance of economic losses and / or degradation of products and services, including potential environmental impacts); 3) criteria of impact on the public (estimated from the point of view of influence on public trust, physical suffering and the violation of everyday life, including the loss of important services). Sectoral criteria should take

Figure 1. The general scheme of the state CIIP stages

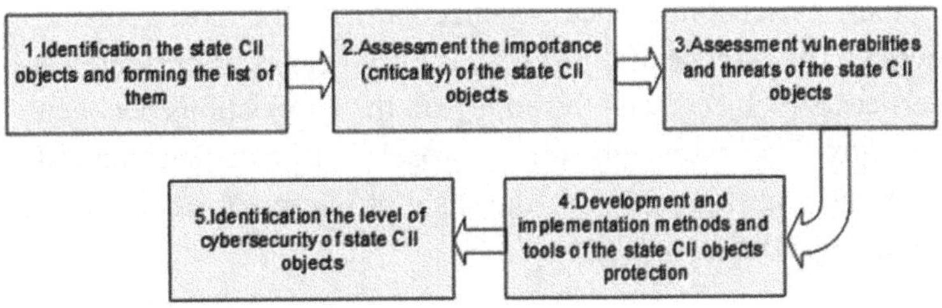

into account the characteristics of individual sectors of CI. They determine the features or functions of objects included in the CII objects.

In the USA, regarding to (Tsigichko, Smolyan, &Chereshkin, 2006), adapted to divide the CI into those that related to international organizations (energy, transport, banking and financial system, ICT objects) and those that are not related to them (for example, water supply, rescue services, public administration). In accordance with (Lyadovska, 2014), CI objects are classified according to the categories of impacts in different directions and sectors: economy, finance, environment, health and safety, technological environment, duration of impact. Moreover, the criticality can be described by three general characteristics (Fekete, 2011): critical share, critical time and critical quality.

In Ukraine just one list of criteria exists (Resolution of the Cabinet of Ministers of Ukraine from 23.08.2016, N°563), which can be used for identification CII objects is the List of negative effects that a cyberattack could cause to the ITS, which includes: 1) the emerging the disaster situation of anthropogenic origin and / or negative impact on the ecological protection condition of the state; 2) negative impact on the energy protection condition of the state; 3) negative impact on the energy protection condition of the state; 4) negative impact on the defense capacity, ensuring national protection and national law order condition; 5) negative influence on the state management system condition; 6) negative influence on the socio-political situation in the state; 7) negative influence on the state image; 8) violation of the stable functioning of the state financial system; 9) violation of the sustainable functioning of the state transport infrastructure; 10) violation of the sustainable functioning of the state's ICT infrastructure, including its interaction with the relevant infrastructures of other states.

With regard to the approaches of CII identification, according to (Kondratiev, 2012; Mattioli & Levy-Bencheton, 2014), there are in the developed countries are known a small number of methods and models that can provide managers of the relevant management units with the opportunity to make a reasonable and correct decision on the CI protection. However, the issue of feasibility and effectiveness of the application of these methods for revealing the CII objects is still not explored and open. To find and formulate the universal, clearly formalized approach that will be applied in various fields, in (Gnatyuk, Sydorenko & Duksenko, 2015) was conducted a multicriteria analysis of approaches to the choosing and identification the CII objects. Each of the approaches to the identification of CII objects in accordance with (Gnatyuk, Sydorenko & Duksenko, 2015; Gnatyuk &Sydorenko,2015; Sydorenko, Aleksander & Nagi, 2016) is described in details below.

1. **K. Clausewitz theory for network architectures:** The basic idea of this theory is to find the "central point" (core) of the system, where its key forces and powers are concentrated (Kondratiev, 2012).
2. **The self-organizing networks A. Barabasi theory:** The essence of this theory is any unstructured (Poisson's) network under the influence of a set of well-known rules and laws, primarily of an economic and social nature, after a certain time (after some number of iterations) takes an appropriate structure without any external influence organized around the most valuable or important nodes.
3. **The graphs theory:** The essence of this theory is the CI can be represented as a weighted oriented graph whose vertices are objects, and the edges are the connections between them.
4. **Priority asset model:** The essence of this approach is to calculate the risk index of an object, which depends on the object rating using the scale of the factor category and the significance of this factor (Kondratiev, 2012).

5. **Identification of the CI objects based on categorization:** The essence of this approach is to identify the dangerous infrastructure objects by comparing the value of the integral criterion with the value of "unacceptable damage" $K_{\text{Н}} : K_{\text{ІНТ}} = K_{\pi} + K_{\text{екОН}} + K_{\delta a\pi} + K_{\text{екО}\pi}$, where K_{π} – the financial loss, determined by the number of dead and injured people, in the case of realization the attack on the objects., $K_{\text{екОН}}$ – the financial lossas a result of failure the most vulnerable elements of the object,, $K_{\delta a\pi}$ – the inventory building value (or cost of reconstruction), $K_{\text{екО}\pi}$ – the value of the expected environmental damage in the case of a terrorist attack on the object. Accordingly, "unacceptable damage" $(K_{\text{Н}})$, is the lower level of damage, after which the object should be classified as hazardous (Stislavsky, 2009). The process of identification ends with creating the list of dangerous objects (for which running the condition $K_{\text{ІНТ}} \geq K_{\text{Н}}$); after that all objects should be categorized.
6. **Simulation:** The simulation that based on software reproduction of the time-wise process of functioning the system becomes a real tool for understanding and the CI full-fledged research. The purpose of simulation (in accordance with (Bratushka, 2009; Dudenhoeffer et al., 2006)) is creating a simulation model of CI objects and to carry out an imitation experiment over them.
    6.1. **CI modeling system:** The Critical Infrastructure Interdependency Modeling System (CIMS) (Dudenhoeffer et al., 2006)is a discrete events simulation that modeling and imitates infrastructures and interdependences which exists between them at the level relevant to the situation. The CIMS was developed to study the relationship between infrastructure networks, or rather the variability of system behavior, which manifests itself when one or more nodes in the system are out of order. Moreover, the system provides a highly visual and interactive environment for monitoring cascading effects and impacts on infrastructure.
    6.2. **"Afina" simulation model:** The "Afina" model – is the software tool that was developed for the big systems analysis (including politic, military, economical and information sectors) as well as for detection the interdependence and interrelated elements (Kondratiev, 2012).

To assess the feasibility and effectiveness of application of analyzed approaches for identify the CII, it is need to analyzed them according to the following basic criteria (Table 1): 1) clarity of formalization (clarity and comprehensibility of mathematical calculations); 2) simplicity of implementation (absence of complicated procedures); 3) flexibility and versatility (the possibility of changing certain parameters as needed and applied in various fields of human activity); 4) accuracy (high degree of approximation of the true value of a particular parameter); 5) efficiency (the ability to correctly and quickly perform calculations); 6) information component (taking into account the features of building IS and networks, cyberspace architecture); 7) objectivity (the possibility of independent evaluation).

In accordance with Table 1 and (Gnatyuk, Sydorenko & Duksenko, 2015), the multicriteria analysis of these approaches represented that the most successful (in terms of CII application) are approaches based on the graphs theory and the simulations (CI modeling system and "Afina" simulation model), which, like many other approaches, are based on the graphs theory. In addition, the knowledge of the self-organizing networks A. Barabasi theory and the identification of the CI objects based on categorization are widely used. Furthermore, some aspects of these approaches can also be used for the CII objects.

Table 1. Analysis of approaches to identify the CII objects

N°	The known approach to identification	Core criteria						
		1	2	3	4	5	6	7
1.	K. Clausewitz theory for network architectures	+	-	-	-	-	-	-
2.	The self-organizing networks A. Barabasi theory	+	+	-	+	-	-	-
3.	The graphs theory	+	-	+	+	-	-	+
4.	Priority asset model	-	+	-	-	-	-	-
5.	Identification of the CI objects based on categorization	+	-	+	+	-	-	-
6.1	CI modeling system	+	-	+	+	+	-	+
6.2	"Afina" simulation model	+	-	+	+	+	-	+

The Unified Data Model for CII Defining

The state CA, as a CI field, requires the special protection. In accordance with the guidance documents in this area (in particular (Doc 30, 2010; Doc 8973 ICAO, 2014)), it is necessary to identify and protect the so-called critical aviation information system (CAIS). It is obvious that unauthorized interference with the CAIS work can lead to significant economic losses, human casualties and the destruction of national infrastructure. However, none of the ICAO or ECAC guiding documents on the protection of international CA includes a complete list of CAIS, which complicates the development of effective methods for protecting CAIS from various types of cyber threats. In view of this, a unified data model was developed to forming a list of state objects CII (Resolution of the Cabinet of Ministers of Ukraine from 23.08.2016, N°563).

According to (Resolution of the Cabinet of Ministers of Ukraine from 23.08.2016, N°563), a full set of CII systems categories in a particular field was introduced:

$$\mathbf{S} = \{\bigcup_{i=1}^{n} \mathbf{S}_i\} = \{\mathbf{S}_1, \mathbf{S}_2, \ldots, \mathbf{S}_n\}, \tag{1}$$

where $\mathbf{S}_i \subseteq \mathbf{S}$ $(i = \overline{1,n})$ are systems categories in some CI field, n is the total number of system categories. Important, that each set can be represented in three views: a set with a (I) index, for instance $\mathbf{S}_1$; a set with a (IIO) object name index, for instance, S_{ISAO} and a set with (IO) object name (IO), for instance **ISAO**.

The set of $\mathbf{S}_i$ categories can be represented as the set of systems:

$$\mathbf{S}_i = \{\bigcup_{j=1}^{m_i} \mathbf{S}_{ij}\} = \{\mathbf{S}_{i1}, \mathbf{S}_{i2}, \ldots, \mathbf{S}_{im_i}\}, \tag{2}$$

where $\mathbf{S}_{ij} \subseteq \mathbf{S}_i$ $(i=\overline{1,n}, j=\overline{1,m_i})$ are systems of i category, m_i is the number of system i categories. Accordingly, with (2), expression (1) can be presented as:

$$\mathbf{S} = \{\bigcup_{i=1}^{n} \mathbf{S}_i\} = \{\bigcup_{i=1}^{n} \{\bigcup_{j=1}^{m_i} \mathbf{S}_{ij}\}\} = \{\{\mathbf{S}_{11}, \mathbf{S}_{12}, \ldots, \mathbf{S}_{1m_1}\}, \{\mathbf{S}_{21}, \mathbf{S}_{22}, \ldots, \mathbf{S}_{2m_2}\}, \ldots, \{\mathbf{S}_{n1}, \mathbf{S}_{n2}, \ldots, \mathbf{S}_{nm_n}\}\}, (i=\overline{1,n}, j=\overline{1,m_i}) \quad (3)$$

The set of $\mathbf{S}_{ij}$ systems can be represented as a set of subsystem:

$$\mathbf{S}_{ij} = \{\bigcup_{k=1}^{r_{ij}} S_{ijk}\} = \left\{S_{ij1}, S_{ij2}, \ldots, S_{ijr_{ij}}\right\}, \quad (4)$$

where $S_{ijk} \subseteq \mathbf{S}_{ij}$ $(i=\overline{1,n},\ j=\overline{1,m_i},\ k=\overline{1,r_{ij}})$ is a set of a subsystem of $\mathbf{S}_{ij}$ the system, r_{ij} is a number of a subsystem of ij the system. Accordingly (4), expression (3) can be presented as:

$$\mathbf{S} = \{\bigcup_{i=1}^{n} \mathbf{S}_i\} = \{\bigcup_{i=1}^{n} \{\bigcup_{j=1}^{m_i} \mathbf{S}_{ij}\}\} = \{\bigcup_{i=1}^{n} \{\bigcup_{j=1}^{m_i} \{\bigcup_{k=1}^{r_{ij}} S_{ijk}\}\}\} = \{\{\{S_{111}, S_{112}, \ldots, S_{11r_{11}}\}, \{S_{121}, S_{122}, \ldots, S_{12r_{12}}\}, \ldots, \{S_{1m_1 1}, S_{1m_1 2}, \ldots, S_{1m_1 r_{nm_1}}\}\}, \{\{S_{211}, S_{212}, \ldots, S_{21r_{21}}\}, \{S_{221}, S_{222}, \ldots, S_{22r_{22}}\}, \ldots, \{S_{2m_2 1}, S_{2m_2 2}, \ldots, S_{2m_2 r_{nm_2}}\}\}, \ldots, \{\{S_{n11}, S_{n12}, \ldots, S_{n1r_{n1}}\}, \{S_{n21}, S_{n22}, \ldots, S_{n2r_{n2}}\}, \ldots, \{S_{nm_1 1}, S_{nm_1 2}, \ldots, S_{nm_n r_{nm_n}}\}\}\} \quad (5)$$

The sets of subsystems of S_{ijk} the system can be presented as the subsystem of the system:

$$\mathbf{S}_{ijk} = \{\bigcup_{p=1}^{v_{ijk}} S_{ijkp}\} = \left\{S_{ijk1}, S_{ijk2}, \ldots, S_{ijkv_{ijk}}\right\}, \quad (6)$$

where $S_{ijkp} \subseteq \mathbf{S}_{ijk}$ $(i=\overline{1,n},\ j=\overline{1,m_i},\ k=\overline{1,r_{ij}},\ p=\overline{1,v_{ijk}})$ is a subset of $\mathbf{S}_{ijk}$ the system, v_{ijk} is a number of subset of ijk subsystem.

Due to (6), expression (3) can be presented as:

$$
\begin{aligned}
S &= \{\bigcup_{i=1}^{n} S_i\} = \{\bigcup_{i=1}^{n}\{\bigcup_{j=1}^{m_i} S_{ij}\}\} = \{\bigcup_{i=1}^{n}\{\bigcup_{j=1}^{m_i}\{\bigcup_{k=1}^{r_{ij}} S_{ijk}\}\}\} = \{\bigcup_{i=1}^{n}\{\bigcup_{j=1}^{m_i}\{\bigcup_{k=1}^{r_{ij}}\{\bigcup_{p=1}^{v_{ijk}} S_{ijkp}\}\}\}\} = \\
&= \{\{\{\{S_{1111}, S_{1112},\dots,S_{111v_{111}}\},\{S_{1121},S_{1122},\dots,S_{112v_{112}}\},\dots,\{S_{11r_{11}1},S_{11r_{11}2},\dots,S_{11r_{11}v_{11r_{11}}}\}\}, \\
&\{\{S_{1211},S_{1212},\dots,S_{121v_{121}}\},\{S_{1221},S_{1222},\dots,S_{122v_{122}}\},\dots,\{S_{12r_{12}1},S_{12r_{12}2},\dots,S_{12r_{12}v_{12r_{12}}}\}\},\dots, \\
&\{\{S_{1m_111},S_{1m_112},\dots,S_{1m_11v_{1m_11}}\},\{S_{1m_121},S_{1m_122},\dots,S_{1m_12v_{1m_12}}\},\dots,\{S_{1m_1r_{1m_1}1},S_{1m_1r_{1m_1}2},\dots,S_{1m_1r_{1m_1}v_{1m_1r_{nm_1}}}\}\}\}, \\
&\{\{\{S_{2111},S_{2112},\dots,S_{211v_{211}}\},\{S_{2121},S_{2122},\dots,S_{212v_{212}}\},\dots,\{S_{21r_{21}1},S_{21r_{21}2},\dots,S_{21r_{21}v_{21r_{21}}}\}\}, \\
&\{\{S_{2211},S_{2212},\dots,S_{221v_{221}}\},\{S_{2221},S_{2222},\dots,S_{222v_{222}}\},\dots,\{S_{22r_{22}1},S_{22r_{22}2},\dots,S_{22r_{22}v_{22r_{22}}}\}\},\dots, \\
&\{\{S_{2m_211},S_{2m_212},\dots,S_{2m_21v_{2m_21}}\},\{S_{2m_221},S_{2m_222},\dots,S_{2m_22v_{2m_22}}\},\dots,\{S_{2m_2r_{2m_2}1},S_{2m_2r_{2m_2}2},\dots,S_{2m_2r_{2m_2}v_{2m_2r_{nm_2}}}\}\}\},\dots, \\
&\{\{\{S_{n111},S_{n112},\dots,S_{n11v_{n11}}\},\{S_{n121},S_{n122},\dots,S_{n12v_{n12}}\},\dots,\{S_{n1r_{n1}1},S_{n1r_{n1}2},\dots,S_{n1r_{n1}v_{n1r_{n1}}}\}\}, \\
&\{\{S_{n211},S_{n212},\dots,S_{n21v_{n21}}\},\{S_{n221},S_{n222},\dots,S_{n22v_{n22}}\},\dots,\{S_{n2r_{n2}1},S_{n2r_{n2}2},\dots,S_{n2r_{n2}v_{n2r_{n2}}}\}\},\dots, \\
&\{\{S_{nm_n11},S_{nm_n12},\dots,S_{nm_n1v_{nm_n1}}\},\{S_{nm_n21},S_{nm_n22},\dots,S_{nm_n2v_{nm_n2}}\},\dots,\{S_{nm_nr_{nm_n}1},S_{nm_nr_{nm_n}2},\dots,S_{nm_nr_{nm_n}v_{nm_nr_{nm_n}}}\}\}\}\}
\end{aligned}
\tag{7}
$$

Depending on the detailing possibility of CII field categories, the S_{ijkp} subsystem can be presented in the form of subsets with an in-depth detailing level. Therefore, it's necessary to represent a complete set of categories of systems in the CII field ($\mathbf{S}$) in general view as follows:

$$
\mathbf{S} = \{\bigcup_{i_1=1}^{n_0}\{\bigcup_{i_2=1}^{n_{i_1}}\{\dots\{\bigcup_{i_l=1}^{n_{i_1,i_2,\dots,i_{l-1}}} \mathbf{S}_{i_1,i_2,\dots,i_l}\}\}\}\}, \tag{8}
$$

where

$$
\mathbf{S}_{i_1,i_2,\dots,i_l} \subseteq \mathbf{S}\ (i_1 = \overline{1,n_0},\ i_2 = \overline{1,n_{i_1}},\ i_l = \overline{1,n_{i_1,i_2,\dots,i_{l-1}}})
$$

are levels of detailing the category of $\mathbf{S}$ systems, $\mathbf{S}_{ij}$ is a number of levels of detailing the category of systems.

On the base of the unified data model was created the technique which allows forming the list of CII objects in certain field and at the state level. The technique realizes using the following steps: 1) Identification the categories of the system of CII certain field or state, in general, in accordance with (1); 2) Identification of the systems and the subsystems according to (2) - (7); 3) Defining the list of objects for CII field or state on a general view on the basis of (8).

THE PROPOSED METHOD FOR CII IDENTIFICATION

Moreover, the method of identifying the CII objects in CA was developed. This method is implemented in the following stages: 1) defining of CII elements; 2) defining the possible factors of influence on the

CII elements; 3) identifying the extent of damage and the weight of the factor's influence on the CII elements; 4) defining the functions of influence of CII elements; 5) the graph-analytical mapping of the functional processes of the CII system; 6) assessment of the CII system functioning quality.

Input data: 1) the structural-functional schemes of the analyzed system; 2) the information about the infrastructure elements and their functional operations of the analyzed system; 3) a detailed description of all possible factors and their parameters that may affect on the system functioning.

Output data:1) the set of identified infrastructure elements; 2) the number of influencing factors on the infrastructure element and a description of their parameters; 3) the table of extent of damage of the infrastructure elements for each factor of influence; 4) the scales table of influence of elements of infrastructure for each factor of influence; 5) the elements list of the infrastructure for which the influence ratios are established and the value of the influence function is calculated; 6) the graph of the CII system functional process; 7) the influence matrix of the infrastructure elements on their functional operations and a list of ranked by the order of importance of the CII system infrastructure elements.

Below is described in detail stages of realization the developed method. The procedure of forming the CII elements implemented at the system level (when $l = 2$ in accordance to (8) and refers only to elements of the information infrastructure (EII), which fully reflect the structure of the selected detailing level.

The method of identifying the CII objects is implemented for a particular system, defined (1) the set $\mathbf{S} = \{\bigcup_{i=1}^{n} S_i\} = \{S_1, S_2, \ldots, S_n\}$, $S_i \subseteq \mathbf{S}$, $(i = \overline{1,n})$ are CII systems, n is the total number of systems.

Stage 1: Defining of CII Elements

Step 1.1: *Forming the possible CII. For forming the possible CII of the particular system* S_i, each expert from $\mathbf{E}$ the set of $\mathbf{E}$ ($\mathbf{E} = \{\bigcup_{j=1}^{N} E_j\} = \{E_1, E_2, \ldots, E_N\}, E_j \subseteq \mathbf{E}$ $(j = \overline{1,N})$ are experts in CII field, N is total number of experts), forming all possible EII ($\mathbf{L}^j = \{\bigcup_{k=1}^{h} L_k^j\} = \{L_1^j, L_2^j, \ldots, L_h^j\}$, $L_k^j \subseteq \mathbf{L}^j$ $(k = \overline{1,h})$ are the possible EII which formed by j experts, h is the number of elements which characterize the system S_i). As a result of Step 1.1, the matrix (9) was formed, which shows all possible CII elements of the particular system S_i. The formed EII $\{L_1^j, L_2^j, \ldots, L_h^j\}$ can have the different length, so according to described approach in (Shershakov, Trahtengerts & Kamaev, 2015), these elements complemented by an empty element ω_0 so that all the rows of the matrix L have the same length.

$$L = \begin{pmatrix} L_1^1 & L_2^1 & \ldots & L_k^1 & \omega_0 \\ L_1^2 & L_2^2 & \ldots & L_k^2 & \omega_0 \\ & & \ldots & & \\ L_1^N & L_2^N & \ldots & L_k^N & \omega_0 \end{pmatrix} \tag{9}$$

Step 1.2: *Identification of unique EII.* For identification of unique EII, from obtaining the possible L_k^j, it is necessary to identify the particular set $\mathbf{F}$ ($\mathbf{F} = \{\bigcup_{ai=1}^{e} F_{ai}\} = \{F_1, F_2, \ldots, F_e\}$, $F_{ai} \subseteq \mathbf{F}$ $(ai = \overline{1,e})$ are unique EII, e is the number of unique EII).

Step 1.3: *EII coherence.* Forming the coherence EII from obtained unique EII F_{ai} is accomplished by establishing the matching of the F_{ai} elements in L the matrix in the form of the matrix $\mathbf{V}$ ($\mathbf{V} = \{\bigcup_{bi=1}^{d} V_{bi}\} = \{V_1, V_2, \ldots, V_d\}$, $V_{bi} \subseteq \mathbf{V}$ $(bi = \overline{1,d})$ is the matching of the F_{ai} elements in the L matrix, d is the number of matching corresponding F_{ai} in matrix L). The matching of EII elements occurs according to the expression: $V_{bi} > \frac{N}{2}$, the obtained values d and their respective EII, are proposed for approval as a coherent set of CII elements: $\mathbf{a} = \{\bigcup_{m=1}^{b} a_m\} = \{a_1, a_2, \ldots, a_b\}$, where $a_m \subseteq \mathbf{a}$ $(m = \overline{1,b})$ are CII elements, which fully reflect the structure of the CII system, b is the total number of these elements.

Step 1.4: *Defining the representation graph of identifiable CII elements.* The graph-analytical mapping of identifiable EII is presented by unoriented graph (10):

$$\Gamma\left(\{a_m\}, \{p_{m\,m'}\}\right), \tag{10}$$

where vertices a_m $(m = \overline{1,b})$ correspond to identified CII elements, and edges $p_{m\,m'}$ are links between elements a_m, where $m = \overline{1,b}$, $m' = \overline{1,b}$, $m \neq m'$ (see Figure 2).

Stage 2: Defining the Possible Factors of Influence on the CII Elements

Step 2.1: *Defining the set of influence zone.* On this step, forming the set of influence zone $\mathbf{Z} = \{\bigcup_{ci=1}^{v} Z_{ci}\} = \{Z_1, Z_2, \ldots, Z_v\}$, where $Z_{ci} \subseteq \mathbf{Z}$ $(ci = \overline{1,v})$ is influence zone on EII, v is the total

Figure 2. The graph-analytical mapping of identifiable a_m

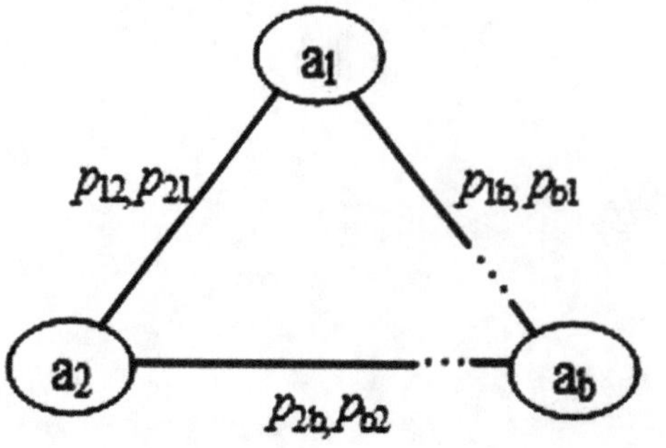

number of influence zone, which meets the following condition: in each zone Z_{ci} gets one CII element a_m (a set of zones that are part of the CII territory system can be entered).

Step 2.2: *Defining the influence factors on CII elements.* For forming the possible influences factors the corresponding set should be introduced $\Phi = \{\bigcup_{di=1}^{s} \Phi_{di}\} = \{\Phi_1, \Phi_2, \ldots, \Phi_s\}$, where $\Phi_{di} \subseteq \Phi$ $(di = \overline{1,s})$ is influence factor on EII a_m, s is total number of the influence factors. Each factor of Φ_{di} for particular zone Z_{ci} can be introduced as the set of parameters $\mathbf{O}^{\Phi_{dV}} = \{\bigcup_{ei=1}^{z} O_{ei}^{\Phi_{dV}}\} = \{O_1^{\Phi_{dV}}, O_2^{\Phi_{dV}}, \ldots, O_z^{\Phi_{dV}}\}$, where $O_{ei}^{\Phi_{dV}} \subseteq \mathbf{O}^{\Phi_{dV}}$ $(ei = \overline{1,z})$ are parameters of influence factor Φ_{di} on EII a_m, z is the total number of factors parameter Φ_{di}, that formed on the basis of expert knowledge and can be described by text message or contain the quantitative indicators. As a result of those operations, the set of possible influence factors Φ_{di}, in which each factor is a set of $\Phi_{di}\left(Z_{ci}, O_{ei}^{\Phi_{dV}}\right)$ is forming.

Stage 3: Identifying the Extent of Damage and the Weight of the Factor's Influence on the CII Elements

For each determined influence factor Φ_{di} and EII element, according to (Shershakov, Trahtengerts & Kamaev, 2015), the values of two quantities are fixed $d_{gi}(a_m, \Phi_{di})$ and $\varphi_{gi}(a_m, \Phi_{di})$ – the extent of element a_m damage and the weight of the factor's influence Φ_{di} on EII respectively, where $gi = \overline{1, f}$, $f = b \cdot s$. For assessment the extent of element $d_{gi}(a_m, \Phi_{di})$ damage the linguistic scale used is as follows: *Damage absent* – "0" (EII has not been influenced or influenced insignificantly); *Middle damage* – "1" (the impact on EII caused significant damage to equipment); *Complete failure* – "2" (the impact on EII led to complete destruction). The exact values of linguistic variables *Damage absent*, *Middle damage*, *Complete failure* is established for each CII system individually.

Further, each expert E_j determining $d_{gi}(a_m, \Phi_{di})$ and $\varphi_{gi}(a_m, \Phi_{di})$ for all elements a_m for all factors Φ_{di}. In addition, for any EII a_m must fulfill the condition $\sum_{di=1}^{s} \varphi_{di}(a_m) = 1$. After that, the data is processed from all the experts E_j as an agreed assessment of the extent of element damage $d_{gi}(a_m, \Phi_{di})$ and the weight of influence $\varphi_{gi}(a_m, \Phi_{di})$ on CII elements the values are taken in accordance with (11), where the values of the threshold values t_0, t_1 get the condition $0 < t_0 < t_1 \leq 2$ are set in advance and can be reviewed depending on the CII system.

$$d(a_m, \Phi_{di}) = \begin{cases} 0, \text{ if } \frac{1}{N}\sum_{gi=1}^{f} d_{gi}(a_m, \Phi_{di}) < t_0; \\ 1, \text{ if } t_0 \le \frac{1}{N}\sum_{gi=1}^{f} d_{gi}(a_m, \Phi_{di}) < t_1; \\ 2, \text{ if } t_1 \le \frac{1}{N}\sum_{gi=1}^{f} d_{gi}(a_m, \Phi_{di}); \end{cases} \quad (11)$$

$$\varphi(a_m, \Phi_{di}) = \frac{1}{N}\sum_{di=1}^{s} \varphi_{di}(a_m, \Phi_{di})$$

Stage 4: Defining the Functions of the Influence of CII Elements

Step 4.1: *Define the relations of influence between CII elements.* At this step, the existence the relations of influence between the CII elements are determined and agreed by the following rule: two CII elements a_m and $a_{m'}$ $(m = \overline{1,b}, m' = \overline{1,b}, m \neq m')$ related to the ratio of influence if the damage to the element a_m causes damage to the element $a_{m'}$.

Thus, for each possible pair of CII elements $(a_m, a_{m'})$ each expert indicates the value of the ratio of influence r as follows $(r \in [-;+])$: if there is influence, then put "+", if there is no influence – put "-". The following is the processing of the data received from each E_j, where the value $K_{m\,m'}$ is equal to the number of "+" in the line corresponding to the pair $(a_m, a_{m'})$, and the value r_w is agreed score, takes value "+", if in the appropriate line is performed the inequality: $K_{m\,m'} > \beta N$, "-" – if is not performed. The value of the score $0 < \beta < 1$ is pre-determined and can be reviewed depending on the CII system.

Step 4.2: *Define functions of mutual influence between CII element pairs.* For defining the value of influence the damage elements on the other CII elements, on the basis of the proposed approach in (Shershakov, Trahtengerts & Kamaev, 2015), each expert E_j fixes the value of the function of influence – $h_{m\,m'}(d_{gi})$. Definition of the latter is carried out as follows: for EII pairs $(a_m, a_{m'})$, for which is established in step 4.1 the ratio of influence r_w = "+", it is necessary to specify a value $h_{m\,m'}$ which shows the influence degree on the element $a_{m'}$, if the element is damaged a_m (definition $h_{m\,m'}$ conducted relative to two levels of influence degree *Middle damage* – "1", and *Complete failure* – "2"). Further, received data from all experts E_j is processed – as an agreed assessment of the function of influence $h_{m\,m'}(d)$ for pair $(a_m, a_{m'})$ values are taken according to (12), where $y = \overline{1,u}$. The limit value τ fulfills the conditions $0 < \tau \le 2$,is pre-determined and can be reviewed depending on the CII system.

$$h_{mm'}(d) = \begin{cases} 1, \text{if } \frac{1}{N}\sum_{y=1}^{u} h_{mm'}^{y}(d) \le \tau, \\ 2, \text{if } \tau < \frac{1}{N}\sum_{y=1}^{u} h_{mm'}^{y}(d), \ d = 1,2 \end{cases} \tag{12}$$

Stage 5: The Graph-Analytical Mapping of the Functional Processes of the CII System

The mapping of the functional processes of the CII system based on the approach described in (Shershakov, Trahtengerts & Kamaev, 2015), can be represented by an oriented acyclic graph (13):

$$G(\{B_{ii}\} \cup \{A_q\} \cup \{C_{ji}\}, \{P_{qq'}\}), \tag{13}$$

where vertices $A_q, (q = \overline{1,x})$ are functional operations performed by one EII a_m (x is total number of functional operation), vertices B_{ii} $(ii = \overline{1,g})$ and C_{ji} $(gi = \overline{1,w})$ are input and output data the system operations resp. (regarding this system, where g is total number of input data, w is the total number of output data), and edges $P_{qq'}$ are connections between operations $A_q, A_{q'}$, where $q = \overline{1,x}$, $q' = \overline{1,x}$, $q \ne q'$.

The availability of the oriented edge $P_{qq'}$ in the graph, which comes from the vertices A_q to vertices $A_{q'}$, means that the operation A_q is performed initially and then is performed $A_{q'}$ operation. The edge that comes from B_{ii} to A_q, called input for this system and symbolizes that on the input of the operation A_q received the data that is the result of the operation B_{ii}.In turn, the edge, that comes from A_q to C_{ji}, called output for this system and symbolizes that on the input of the operation C_{ji} received the data that is the result of the operation A_q (see Figure 3).

Such representation allows mapping in a convenient form the formalized description of the functional stages of operations and links between them, as well as the corresponding input and output data.

Figure 3. The graph-analytical mapping of the functional processes of the CII system

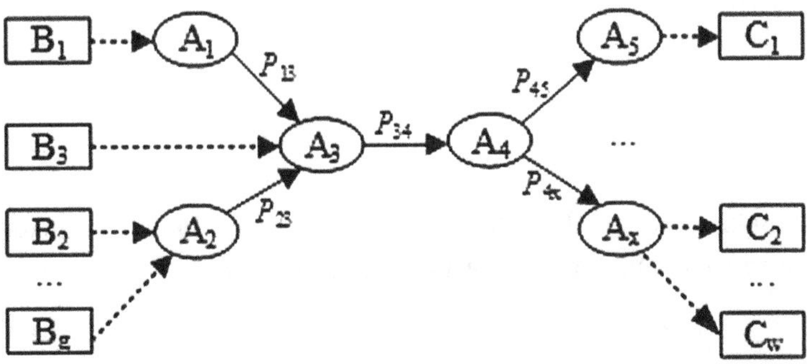

Stage 6: Assessment of the CII System Functioning Quality

The result of the influence on EII may be a reduction of the quality of performance the functional operations. To evaluate the quality of execution of functional operations $Q(A_q)$ is introduced in following linguistic scales: *Normal* – "0" (the operation is performed in accordance with the functional rules); *Deviation* – "1" (the operation is carried out, but there are significant deviations from the functional regulation); *Interruption* – "2" (operation is not performed). The influence matrix $Q_d^q(d(a_m))$ shows the execution of a functional operation A_q provided that EII a_m has a corresponding damage $d(a_m)$ and it is formed consistently by each expert E_j for all EII and their respective operations, where is the top index expression Q_d^q corresponds to the number of the operation and the lower index of expression corresponds to the value of the damage $d \in \{0,1,2\}$. Then, data is processed, received from each expert E_j,as a coordinated assessment of the performance the functional operation, the value is taken in accordance with (14), where the limit value s_0, s_1 fulfills the conditions $0 < s_0 < s_1 \le 2$ (established by the head of the relevant infrastructure and experts pre-determined and can be reviewed depending on the CII system).

$$Q(A_q) = \begin{cases} 0, \text{if } \frac{1}{N} \sum\limits_{q,d,m} Q_d^q(d(a_m)) < s_0; \\ 1, \text{if } s_0 \le \frac{1}{N} \sum\limits_{q,d,m} Q_d^q(d(a_m)) < s_1; \\ 2, \text{if } s_1 \le \frac{1}{N} \sum\limits_{q,d,m} Q_d^q(d(a_m)) \end{cases} \tag{14}$$

Ranging the agreed qualities of the functioning of the CII system by all E_j owing occurs by comparing the quantitative values obtained in the influence matrix. The sum of the quantitative indicators of quality $Q(A_q)$ obtained for one EII a_m, is compared with the sum of the quantitative indicators of quality $Q(A_q)$, obtained for other EII and ranging as follows; $\{VEI_1 > VEI_2 >,...,> VEI_o\}$, where the set of

$$\mathbf{VEI} = \{\bigcup_{li=1}^{o} VEI_{li}\} = \{VEI_1, VEI_2, ..., VEI_o\},$$

$$VEI_{li} \subseteq \mathbf{VEI}, \ (li = \overline{1,o})$$

is ranked in order of importance for EII system, o is the number in order of ranked EII relative to the sum of quality indicators (it should be noted that the number $o = b$).

The Experimental Study for Identifying CII Objects in CA

On the basis of the proposed unified data model, the technique was developed. Using this technique, the list of CII objects in CA was formed, accordingly (Gnatyuk & Vasyliev, 2016; Gnatyuk, Sydorenko & Seilova, 2017; Sydorenko, Zhmurko, Polishchuk & Gnatyuk, 2017; Gnatyuk, Aleksander & Sydorenko, 2018), as a result of what at detailing level $l = 4$, are determined 3 sets of categories, 17 sets of systems, 97 subsets of subsystems, 125 subsystems of the CAIS. A fragment of the formed list is given in the Table 2.

For the pilot study of the CII object identification method experiment, a specialized software tool was developed that allows identifying the CII objects in any field and determining their influence on functional operations.

The technology platform provides objects (data and metadata) and objects management mechanisms. Objects (data and metadata) are described as configurations. When automating any activity (software development), it consists of its own configuration of objects, which is a complete application. The con-

Table 2. The formed list of CII objects in CA

Detailing Level	Parameters	Systems / Subsystems of CAIS
$l = 1$	$n = 3$	ISAO, BSPS, ISAA
$l = 2$	$m_1 = 5,\ m_2 = 7,\ m_3 = 5.$	**SAE, RZZP, SSP, SOD, SMZ, SPS, SZV, NAVS, SSPZ, OSL, SVI, ABSK, CRS, GDS, IDS, BSP, DCS**
$l = 3$	$r_{1.1} = 5,\ r_{1.2} = 4,\ r_{1.3} = 9,$ $r_{1.4} = 5,\ r_{1.5} = 3,\ r_{2.1} = 4,$ $r_{2.2} = 3,\ r_{2.3} = 8,\ r_{2.4} = 4,$ $r_{2.5} = 2,\ r_{2.6} = 5,\ r_{2.7} = 4,$ $r_{3.1} = 2,\ r_{3.2} = 18,\ r_{3.3} = 8,$ $r_{3.4} = 8,\ r_{3.5} = 5.$	**SAPE, SANE, ZAR,** *SASZ, MTM, NDB, VOR, DME, ILS, PSR, SSR, MSSR, RADS, SMR, WRAD, MLAT, ADS, DF,* **ASYPR,** *SPPP,* **ESAN, SOPD,** *SOPA, SCMAU, KRAMS, SADIS,* **DPPT,** *DZP,* **TPT,** *POP, BRS, CPDLS, AKARS, SNS, INS, ARK, RV, BVOR, BD, BILS, DVKZ, TRA, TCAS, SRPZ, BMR, OBCH, MCDU,* **DBSVS, DSVS, OSVS,** *NSVSO, ISVS, APIL, SAU, PILS, PNK, DELTM, PANAM, AMDS,* **TGDS,** *SAB, TRES, APSS, ABCS, ACA, AXS, IBE, KUI, MER, NAV, PATH, RAD, AKF, TTI, WSMS, SIR, BKNG, OKT, EXP, ORB, HRS, TRAV, HOT, PRLN, STD, SAF, ODOC, ZVPR, PROCO, SABZ, PPKK, POV, SITA,* **TAIS,** *SAMDS, JKCS, HCS.*
$l = 4$	$v_{1.1.1} = 3,\ v_{1.1.2} = 6,\ v_{1.1.3} = 2,$ $v_{1.4.1} = 7,\ v_{1.4.3} = 2,\ v_{1.4.4} = 1,$ $v_{2.1.1} = 3,\ v_{2.1.3} = 2,\ v_{2.6.1} = 5,$ $v_{2.6.2} = 5,\ v_{2.6.3} = 3,\ v_{3.2.2} = 3,$ $v_{3.5.2} = 3.$	*NRPZ, CPDLC, ACARS, ZPRZZ, SCGZ, ZRZ, AFTN, AMHS, MOD, VOLM, ATIS, ODSS, OPD, MKS, ZVI, KGZ, PPR, ZBP, ARTAS, SDDS, IFPS, PST, PDT, KPPT, STATL, DYNL, BBSB, RVM, BMRL, BMDX, BSIB, PFD, ND, HUD, HMD, IDIS, OSZ, OGPP, OVP, APLL, GALL, WSPN, TCRS, TDCS, TTSH.*

figuration is created in the special operation software mode called “Configurator”, and then the operating mode called “1C: Enterprise” is started, in which the user gets access to the basic functions implemented in this application (configuration). The platform itself is not a software product for end users but serves as a foundation for the development and operation of application solutions.

In (Gnatyuk, Sydorenko & Kinzeriavyi, 2017), the pilot study of the EII object identification method in CA was carried out based on the system S_{SNS} are satellite navigation systems (SHS), level of system detail $l = 2$, and the adequacy of the method response to change the input data has been proved.

This software implements the following features: entry of input parameters; defining the set of CII identifiable elements; defining the possible factors of influence on the CII element and description of their parameters; creation of tables of extent of damage and the weight of the factor’s influence on the CII elements; determination the list of CII element pairs, for which the influence ratios and their calculated values of the influence function are established; allocation of functional stages of operations, connections between them, corresponding input and output data and construction a graph of functional processes; defining the influence matrix of CII elements on their functional operations and selecting the list of ranked by the importance order of the CII elements.

Stage 1: Defining of CII Elements

For system S_{SNS} (Soloviev, 2000), on stage 1, at $N = 3$, the matrix of the possible EII was formed, accordingly (9):

$$L = \begin{pmatrix} L_1^1 & L_2^1 & L_3^1 & L_4^1 & L_5^1 \\ L_1^2 & L_2^2 & L_3^2 & L_4^2 & \omega_0 \\ L_1^3 & L_2^3 & L_3^3 & L_4^3 & L_5^3 \end{pmatrix},$$

where L_1^1 is artificial satellite, L_2^1 is control station, L_3^1 is additional station, L_4^1 is observation station, L_5^1 are receivers; L_1^2 is artificial satellite, L_2^2 is control and observation station, L_3^2 are additional stations, L_4^2 are receivers; L_1^3 is artificial satellite, L_2^3 is control and observation station, L_3^3 are additional stations, L_4^3 are SPS- receivers, L_5^3 are PPS- receivers.

After that a set of unique EIIs is allocated, at $e = 8$,

$$\mathbf{F}_{\text{SNS}} = \{\bigcup_{ai=1}^{8} F_{ai}\} = \{F_1, F_2, \ldots, F_8\},$$

where F_1 is the artificial satellite, F_2 is control station, F_3 is the additional station, F_4 is observation station, F_5 are receivers; F_6 is control and observation station, F_7 are SPS-receivers, F_8 are PPS-receivers.

Then, a set of coincidences, at $N = 3,\ d = 8$,

$$\mathbf{V}_{\text{SNS}} = \{\bigcup_{bi=1}^{8} V_{bi}\} = \{V_1, V_2, \ldots, V_8\} = = \{3,1,3,1,2,2,1,1\},$$

and an agreed EII set are allocated:

$$\mathbf{a}_{\text{SNS}} = \{\bigcup_{m=1}^{4} a_m\} = \{a_1, a_2, \ldots, a_4\},$$

where a_1 is the artificial satellite, a_2 is control and observation station, a_3 is the additional station, a_4 are receivers, accordingly (Soloviev,2000; Golub & Sheremet, 2016).

Results of the implementation stage 1 are shown in Figure 4.

For the system S_{SNS}, at $b = 4$, accordingly (10), the graph vertices Γ is a_1 the artificial satellite, a_2 is control and observation station, a_3 is the additional station, a_4 are receivers, and the links between these elements are edges:

$$p_{12}, p_{21}, p_{13}, p_{31}, p_{14}, p_{41}, p_{23}, p_{32}, p_{24}, p_{42}, p_{34}, p_{43}$$

(see Figure 5).

Stage 2: Defining the Possible Factors of Influence on the CII Elements

For system S_{SNS}, at, $b = 4$ and $v = 2$, accordingly (Soloviev,2000), the zone set presented as:

$$\mathbf{Z}_{\text{SNS}} = \{\bigcup_{ci=1}^{2} Z_i\} = \{Z_1, Z_2\},$$

Figure 4. The stages of CII elements defining

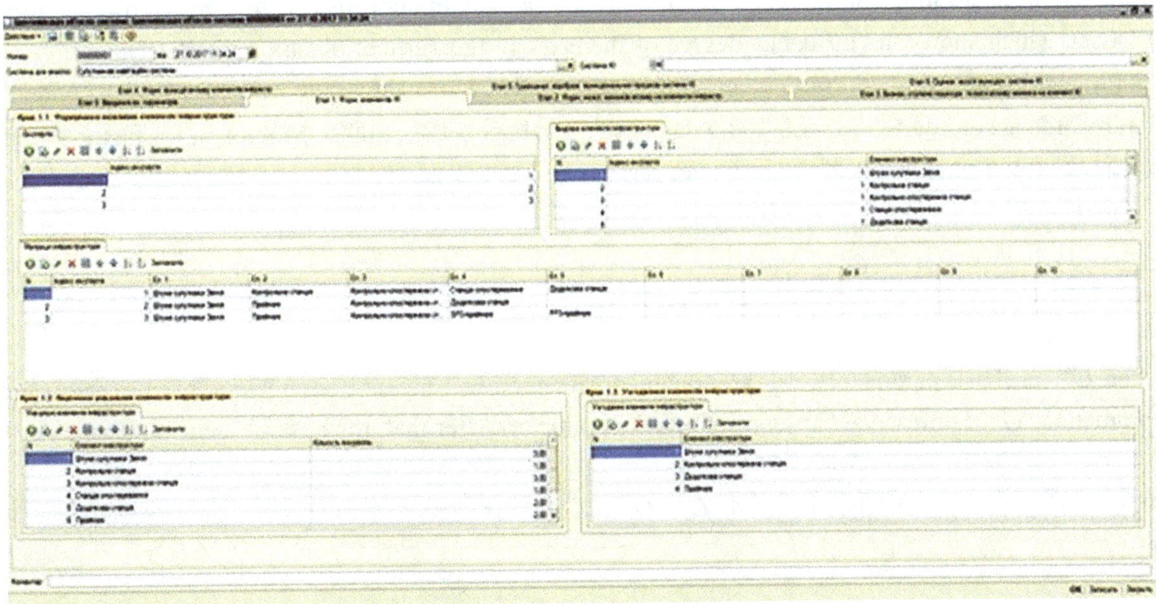

Figure 5. The graph-analytical mapping of CII elements at $b = 4$ *for* S_{SNS}

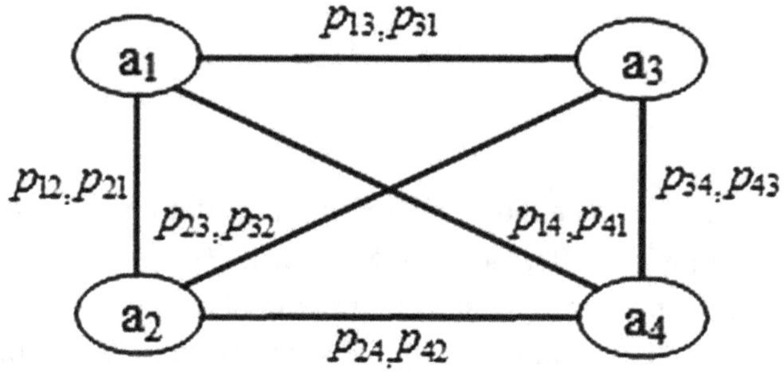

where Z_1 is space or orbital zone, Z_2 is ground management and control zone. For S_{SNS}, at, $b = 4$ and $s = 7$, accordingly (Snovidov, 2015; Maksimenko, 2016)the set of factors of influences can be presented as:

$$\Phi_{SNS} = \{\bigcup_{di=1}^{7} \Phi_{di}\} = \{\Phi_1, \Phi_2, \ldots, \Phi_7\},$$

where Φ_1 is geometric factor (GDOP), which indicates the state of the influence of pseudo-range metrics errors (the last one characterizes the measure of remoteness of the consumer from the GPS- satellite)of hours for accuracy of coordinate calculation; Φ_2 is horizontal factor (HDOP), which shows the influences degree of the accuracy the definition of the horizontal on the calculation of coordinates error; Φ_3 is relative factor (RDOP), dimensionless index describing the effect on the accuracy of determining the coordinates of the pseudo-range error; Φ_4 is time factor (TDOP), is equal to the factor of reduction of the accuracy normalized for the period of 60 s; Φ_5 is vertical factor (VDOP), describes the degree of influence of the accuracy of the hours metrics on the accuracy of coordinates; Φ_6 are situation factors (PDOP), which shows the influence degree of the error in the vertical plane on the determination of coordinates accuracy; Φ_7 is communication factor (CDOP), which shows the value of network connection records according to the NLS-KDD database (Tavallaee, et al., 2009). Moreover, for factor Φ_7, at $z = 5$, the set of parameters of the influence factor represented as:

$$\mathbf{O}^{\Phi_7} = \{\bigcup_{ei=1}^{5} O_{ei}^{\Phi_7}\} = \{O_1^{\Phi_7}, O_2^{\Phi_7}, \ldots, O_5^{\Phi_7}\},$$

where $O_1^{\Phi_7}$ are basic parameters; $O_2^{\Phi_7}$ are content parameters; $O_3^{\Phi_7}$ are time parameters; $O_4^{\Phi_7}$ are hardware parameters; $O_5^{\Phi_7}$ is presence / absence of attack parameter. After that, the possible sets of parameters

$$\Phi_7\left(Z_1, O_1^{\Phi_7}, O_2^{\Phi_7}, O_3^{\Phi_7}, O_4^{\Phi_7}, O_5^{\Phi_7}\right); \Phi_7\left(Z_2, O_1^{\Phi_7}, O_2^{\Phi_7}, O_3^{\Phi_7}, O_4^{\Phi_7}, O_5^{\Phi_7}\right)$$

to form for a factor Φ_7.

Stage 3: Identifying the Extent of Damage and the Weight of the Factor's Influence on the CII Elements

For system S_{SNS} at $b=4$ and $s=7$, accordingly (Soloviev, 2000; Maksimenko, 2016) resp., agreed by the experts in accordance with (11) the values of the extent of damage and the weight of the factors are indicated in the Table 3 (the value of limit score $t_0=1$ and $t_1=1{,}5$).

Stage 4: Defining the Functions of Influence of CII Elements

For the specified system S_{SNS}, at $b=4$, accordingly (Soloviev, 2000), the possible EII pairs are formed and the influence between these elements is estimated (the value of score). The processed value presented in Table 4 (the value $\beta=0{,}5$), where, by gray color, are marked pairs, for which established the ratio of influence.

For pairs for which, according to (12), established the ratio of influence (Table 4), should be define the value of the function of influence and displayed it's in the Table 5 ($\tau=1$).

Stage 5: The Graph-Analytical Mapping of the Functional Processes of the CII System

For the studied system, accordingly (Golub & Sheremet, 2016), at $x=4$, $g=2$, $w=2$, to display the scheme of the functional process using the graph (13),in which the vertices A_q correspond to functional operations (A_1 is satellite segment, A_2 is control and observation, segment A_3 is additional segment, A_4 is user segment), vertices B_1, B_2, C_1, C_2 are correspond to the input and output data of operations A_q, and edges $P_{12}, P_{14}, P_{21}, P_{23}, P_{24}, P_{34}$ are links between elements $A_q, A_{q'}$ (was installed in step 4.1) (Figure 6).

Table 3. An example of damage degree value and influence weight on CII elements

	Φ_1		Φ_2		Φ_3		Φ_4		Φ_5		Φ_6		Φ_7	
	d_1	φ_1	d_2	φ_2	d_3	φ_3	d_4	φ_4	d_5	φ_5	d_6	φ_6	d_7	φ_7
a_1	1	0,2	1	0,1	0	0,1	1	0,1	1	0,2	0	0	2	0,3
a_2	1	0,2	1	0,2	1	0,2	1	0,2	1	0	1	0	1	0,2
a_3	1	0,1	0	0,1	0	0,1	0	0,1	0	0,1	1	0,1	1	0,4
a_4	1	0,2	0	0,1	0	0,1	0	0,1	0	0,1	0	0,1	1	0,3

Table 4. The ratio of influence between CII elements

The pair $(a_m, a_{m'})$	The result			The number of "+" $(K_{mm'})$	The agreed score (r_w)
	1	2	3		
(a_1, a_2)	+	+	+	*3*	+
(a_1, a_3)	-	-	+	*1*	-
(a_1, a_4)	+	-	+	*2*	+
(a_2, a_1)	+	+	+	*3*	+
(a_2, a_3)	+	+	+	*3*	+
(a_2, a_4)	+	-	+	*2*	+
(a_3, a_1)	-	-	-	*0*	-
(a_3, a_2)	-	-	-	*0*	-
(a_3, a_4)	+	+	+	*3*	+
(a_4, a_1)	-	+	-	*1*	-
(a_4, a_2)	+	-	-	*1*	-
(a_4, a_3)	+	-	-	*1*	-

Stage 6: Assessment of the CII System Functioning Quality

For system S_{SNS} accordingly (14), we will construct the agreed influence matrix of all EII on all functional system operations ($q_0 = 0{,}5$, $q_1 = 1{,}5$) will be constructed. Moreover, will form, at $b = o = 4$, a set of ranked by the importance order for the EII system: $\mathbf{VEI}_{\text{SNS}} = \{a_1, a_2, a_3, a_4\}$, where a_1 is artificial satellite, a_2 is control-observation station, a_3 is additional station, a_4 are receivers.

Results of the implementation stage 6 are shown in the Table 6.

For assessment the adequacy of proposed method, its response to the change in input data was checked. For the studied system S_{SNS} , the number of EIIs and CII elements of KII are changed, which respectively indicated a change in the output data. Below is a verification of the developed method. Consequently, an experimental study proved the possibility of using the developed method for identifying elements of the CII field, determining the interaction and impact on CAIS functional operations.

Table 5. Evaluating the functions of impact

The pair $(a_m, a_{m'})$	The result			The agreed score $\left(h^y_{mm'}(d)\right)$
	1	2	3	
(a_1, a_2)	$h^1_{12}(1) = 2,$ $h^1_{12}(2) = 2$	$h^2_{12}(1) = 1,$ $h^2_{12}(2) = 2$	$h^3_{12}(1) = 2,$ $h^3_{12}(2) = 2$	$h^y_{12}(1) = 2,$ $h^y_{12}(2) = 2$
(a_1, a_4)	$h^1_{13}(1) = 1,$ $h^1_{13}(2) = 1$	$h^2_{13}(1) = 1,$ $h^2_{13}(2) = 2$	$h^3_{13}(1) = 1,$ $h^3_{13}(2) = 2$	$h^y_{13}(1) = 1,$ $h^y_{13}(2) = 2$
(a_2, a_1)	$h^1_{21}(1) = 0,$ $h^1_{21}(2) = 2$	$h^2_{21}(1) = 1,$ $h^2_{21}(2) = 1$	$h^3_{21}(1) = 0,$ $h^3_{21}(2) = 2$	$h^y_{21}(1) = 1,$ $h^y_{21}(2) = 2$
(a_2, a_3)	$h^1_{23}(1) = 1,$ $h^1_{23}(2) = 1$	$h^2_{23}(1) = 1,$ $h^2_{23}(2) = 1$	$h^3_{23}(1) = 1,$ $h^3_{23}(2) = 1$	$h^y_{23}(1) = 1,$ $h^y_{23}(2) = 1$
(a_2, a_4)	$h^1_{24}(1) = 1,$ $h^1_{24}(2) = 2$	$h^2_{24}(1) = 1,$ $h^2_{24}(2) = 2$	$h^3_{24}(1) = 1,$ $h^3_{24}(2) = 2$	$h^y_{24}(1) = 1,$ $h^y_{24}(2) = 2$
(a_3, a_4)	$h^1_{34}(1) = 1,$ $h^1_{34}(2) = 1$	$h^2_{34}(1) = 1,$ $h^2_{34}(2) = 1$	$h^3_{34}(1) = 1,$ $h^3_{34}(2) = 2$	$h^y_{34}(1) = 1,$ $h^y_{34}(2) = 2$

Figure 6. The stage of mapping of the functional processes of the CII system

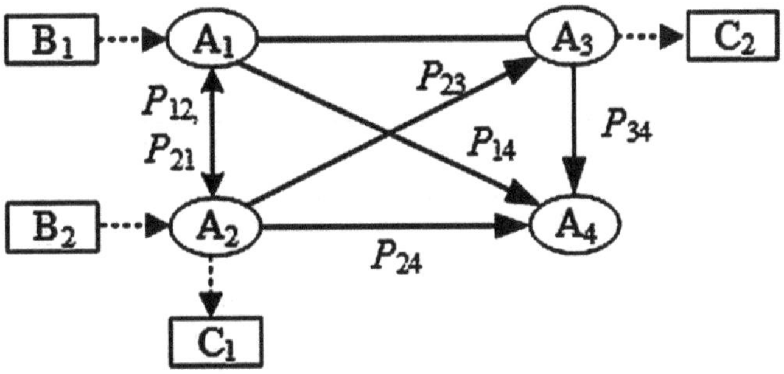

Requirements for Cybersecurity and CIIP in CA

Last attacks and terrorist acts in cyberspace demonstrated that its reality had clearly expressed political tint and more evident in cyber influence at the international level. Studying parallel the ICT development, it should be notice that the main causes of cyberterrorism is, above all, a sharp increase of productivity and simultaneously cheapening of contemporary computing facilities, making them accessible and considerably expands the set of potential cyberthreats (potential cause of an unwanted incident, which may result in harm to a system, individual or organization), and the lack of clear boundaries in cyberspace

Table 6. The influence matrix of CII elements on functional operations

Operation (A_q)	The chart of degrees of elements damage to elements a_m		
	$d(a_1)=0$	$d(a_1)=1$	$d(a_1)=2$
A_1	0	2	2
A_2	0	1	2
A_3	0	1	2
A_4	0	1	1
	$d(a_2)=0$	$d(a_2)=1$	$d(a_2)=2$
A_1	0	1	2
A_2	0	2	2
A_3	0	1	2
A_4	0	1	1
	$d(a_3)=0$	$d(a_3)=1$	$d(a_3)=2$
A_1	0	0	1
A_2	0	0	1
A_3	0	1	2
A_4	0	1	2
	$d(a_4)=0$	$d(a_4)=1$	$d(a_4)=2$
A_1	0	0	1
A_2	0	1	1
A_3	0	0	1
A_4	0	1	2

that eliminates the distinction between external and internal sources of threats for state's cybersecurity. In addition, cyberspace provides an opportunity for attackers to manipulate the information and its society perception at its discretion, and allows realized terroristic acts with unprecedented efficiency and make the task of identifying intruders very difficult (Gnatyuk, 2016).

Today we know many examples of attacks on CA throughout the world (e.g. Malaysia, Turkey et al), but most of these cases are not advertised. This is the purpose of hidden blocking vulnerabilities, but appearance of unwanted consequences can make the world community in a loud voice to talk about cyberterrorism in CA and respond in the short term. The control aviation security documents declare following requirements to ensure CAIS security against cyberthreats. European control document (Doc 30, 2010) declares that measures addressing cyberthreats to CA have been included in the National Civil Aviation Security Program, the National Quality Control Program and the National Civil Aviation Security Training Program. Security controls consist of below measures: a) implementation of effective measures to protect CAIS; b) include the CAIS in their threats assessment processes; c) separation the CAIS networks from public; d) responsibility for securing CAIS is allocated by operators to a properly selected, recruited and trained individual; e) security measures are considered in the design, implementation, operation and disposal of new CAIS; f) supply chain security measures for hardware and software should be applied to CAIS; g) remote access to CAIS is only permitted under pre-arranged and secure conditions; h) cyberattack incidents must be recorded for future evaluation and counter & preventive measures efficiency increasing.

Cybersecurity is actual and very important problem of CA. This problem is complex and multilevel (Figure 1), it solving needs step-by-step multi-face approach. Mentioned problems are relevant for Ukraine by virtue of its participation as a full member in ICAO, ECAC and other communities. To solve the most of CA security problems from viewpoint of security against cyberthreats Ukraine (and other similar developing states) must implement existed or develop cybersecurity methods, techniques and tools.

CONCLUSION

The analysis of modern approaches to the state CII objects identification was carried out. As a result, it was established that today in Ukraine there is no exhaustive list of CII objects and effective mechanisms for its defining. It is also determined that well-known approaches to the CI objects identification are oriented, as a rule, to economic, environmental, technogenic and other state security domains, and do not take into account the CII characteristics. The analysis allowed to formalize the tasks of scientific research – methods and models development for formalization, identification and cybersecurity ensuring of the CII objects in CA.

A unified data model is developed that allows formalizing the process of forming a list of CII objects. The list of CII objects in CA was formed with detailing level $l = 4$, and 3 sets of categories, 17 sets of systems, 97 subsets of subsystems and also 125 subsystems of the CAIS were determined. The indicated results can be used by the relevant state authorities to formulate a list of CII objects in order to apply adequate security methods and tools.

The method of identification was developed, which makes it possible to define elements of the CII field, their mutual influence and influence on CAIS functional operations. Also a special software application that can be used to identify the CII elements and determine their impact on functional operations, both in aviation and in other areas of state CI is developed and implemented.

Also the basic aspects of cybersecurity ensuring for identified CAIS in accordance were described in this chapter. Based on these, effective methods and techniques for identified objects criticality assessment as well as vulnerabilities and threats estimation can be proposed and implemented. Also, new improved cybersecurity methods and tools can be developed and implemented for CAIS cybersecurity ensuring.

REFERENCES

Biriukov, D., & Kondratov, S. (2012). Critical infrastructure protection: problems and prospects of implementation in Ukraine: analytical report. Academic Press.

Biriukov, D., Kondratov, S., & Sukhodolia, O. (2015). Green paper on critical infrastructure protection in Ukraine. Academic Press.

Bratushka, S. (2009). Simulation as a tool to study complex economic systems. *Scientific Bulletin NLTU Ukraine*, *8*, 22-28.

Doc 30 ECAC (2010). *Policy Statement in the Field of Civil Aviation Security* (Restricted), 13, 138 p.

Doc 8973 ICAO (2014). *Guidance on Aviation Security* (Restricted), 9, 818 p.

Dovgan, O. (2013). Critical infrastructure as an object of protection against cyberattacks. *Information security: the challenges and threats of our time,* 17-20.

Dudenhoeffer, D., Permann, M., & Manic, M. (2006). CIMS: a framework for infrastructure interdependency modeling and analysis. *Proceedings of the winter simulation conference WSC 2006*, 478-485. 10.1109/WSC.2006.323119

Fekete, A. (2011). Common criteria for the assessment of critical infrastructures. *International Journal of Disaster Risk Science*, *2*(1), 15–24. doi:10.100713753-011-0002-y

Gnatyuk, S. (2016). *Meeting Security Challenges Through Data Analytics and Decision Support*. NATO Science for Peace and Security Series – D: Information and Communication Security. IOS Press.

Gnatyuk, S., Aleksander, M., & Sydorenko, V. (2018). Unified data model for defining state critical information infrastructure in civil aviation. *The 9th IEEE International Conference on Dependable Systems, Services and Technologies (DESSERT-2018),* 37-42.

Gnatyuk, S. & Sydorenko, V. (2015). An overview of methods for evaluating critically important objects. *Problems and prospects for the development of aviation and cosmonautics*, 110.

Gnatyuk, S., Sydorenko, V., & Duksenko, O. (2015). Modern approaches to critical infrastructure objects detection and identification. *Ukrainian Scientific Journal of Information Security*, *21*(3), 269–275.

Gnatyuk, S., Sydorenko, V. & Kinzeriavyi, V. (2017). Method of object identification of critical information infrastructure in aviation. *Information Technology and Security*, *5*(2), 27-39.

Gnatyuk, S., Sydorenko, V., & Seilova, N. (2017). Universal data model for the formation of the critical information infrastructure of the state objects list. *Ukrainian Scientific Journal of Information Security*, *23*(2), 80–91.

Gnatyuk, S., & Vasyliev, D. (2016). Modern critical aviation information systems. *Ukrainian Scientific Journal of Information Security*, *22*(1), 51–57.

Golub, O. & Sheremet, S. (2016). *Satellite navigation systems in transport.* Academic Press.

Kondratiev, A. (2012). Modern trends in the study of critical infrastructure in foreign countries. *Foreign Military Review*, *1*, 19–30.

Lyadovska, V. (2014). Methods and criteria for the identification of objects of the state's critical infrastructure. *Integrated Intelligent Robotic Complexes,* 356-358.

Maksimenko, Y. (2016). *Investigation of the parameters of ground and onboard receivers using the GPS system*. Academic Press.

Mattioli, R., & Levy-Bencheton, C. (2014). Methodoľogies for the identification of critical information infrastructure assets and services. ENISA Report.

On the identification and designation of European critical infrastructures and the assessment of the need to improve their protection: Council Directive 2008/114/EC.

Resolution on approval of the procedure for the formation of the list of information and telecommunication systems of critical infrastructure objects of the state. Resolution of the Cabinet of Ministers of Ukraine of 23.08.2016, N°563.

Review of the new platform "1C Enterprise 8.3". (n.d.). Retrieved from http://erp–project.com.ua/index.phpuk/novini/item/206–obzor–novoj–platformy–1s–predpriyatie–8–3

Shershakov, V., Trahtengerts, E., & Kamaev, D. (2015). *Network-centric methods of computer support for the management of emergency response*. LENAND.

Snovidov, I. (2015). Factors affecting measurement accuracy. *Bulletin of the State Hydrography*, *1*(48).

Soloviev, Y. (2000). *Satellite navigation systems*. Eco-Trendz.

Stislavsky, A. (2009). Building a methodology for ensuring transport security based on categorization. *Bulletin of the National University of Water Management and Nature Management*, *3*(47), 155-165.

Sydorenko, V., Aleksander, M., & Nagi, A. (2016). Modern approaches to the definition and identification of critical information infrastructure. *Current issues of cyber security and information security*, 134-137.

Sydorenko, V., Zhmurko, T., Polishchuk, Yu., & Gnatyuk, S. (2017). *Data model for forming critical infrastructure objects and determining its connectivity. In Inzynier XXI wieku* (pp. 329–350). Monografia, Poland: ATH.

Tavallaee, M., Bagheri, E., Lu, W., & Ghorbani, A. (2009). A detailed analysis of the KDD CUP 99 data set. *Second IEEE Symposium on Computational Intelligence for Security and Defense Applications (CISDA)*. 10.1109/CISDA.2009.5356528

Tsigichko, V., Smolyan, G. & Chereshkin, D. (2006). *Ensuring the safety of critical infrastructures in the USA*. Academic Press.

KEY TERMS AND DEFINITIONS

Critical Information Infrastructure: The part of information infrastructure, a set of information, telecommunication, and information and telecommunication systems, the malfunctioning of which may lead to an accident or emergency and the state's inability to perform its functions.

Critical Information Infrastructure Object: The communication or technological system of CI object, the cyber-attack on which will directly affect the stable functioning of such CI object.

Critical Infrastructure: The totality of state infrastructure objects that are the most critical for the economy and industries, the society functioning and the population protection, and the decommissioning or destruction of which may have an impact on national security and defense, the natural environment, and can lead to significant financial losses and human casualties.

Critical Infrastructure Protection: A complex of measures implemented in the regulatory, organizational, technological tools aimed at ensuring the security and stability of CI.

Compilation of References

Achtelik, M., Bachrach, A., He, R., Prentice, S., & Roy, N. (2009, April). Stereo vision and laser odometry for autonomous helicopters in GPS-denied indoor environments. In *Unmanned Systems Technology XI* (Vol. 7332, p. 733219). International Society for Optics and Photonics. doi:10.1117/12.819082

Adding a Google Earth Overlay. (n.d.). Retrieved from https://help.gpsinsight.com/docs/about-maps/using-3d-maps/adding-a-google-earth-overlay/

Aghabozorgi, S., Shirkhorshidi, A. S., & Wah, T. Y. (2015). Time-series clustering–A decade review. *Information Systems, 53*, 16–38. doi:10.1016/j.is.2015.04.007

Aircraft Pitch Control System Using Lqr And Fuzzy Logic Controller. (2016). International Journal Of Advance Engineering and *Research for Development, 3*(5). doi:10.21090/ijaerd.030564

Alamdari, F., & Fagan, S. (2017). Impact of the adherence to the original low-cost model on the profitability of low-cost airlines. In *Low Cost Carriers* (pp. 73–88). Routledge.

Alishov, A. N., Shvachych, G. G., Semenov, S. G., & Fedorov, E. E. (2016). Efficient of extreme algorithms of high order accuracy for solving applied tasks. *Mathematics and Computer Science: Journal of Qafqaz University, 1*, 3–11.

Allam, A., Bakeir, M., & Abo-Tabl, E. (2008). Some Methods for Generating Topologies by Relations. *Bulletin of the Malaysian Mathematical Sciences Society, 2*(31), 35–45.

Allianz Global Corporate & Specialty, Embry-Riddle Aeronautical University. (2014). *Global Aviation Safety Study: A Review of 60 Years of Improvement in Aviation Safety*. Authors.

Almubayed, A., Hadi, A., & Atoum, J. (2015). A Model for Detecting Tor Encrypted Traffic using Supervised Machine Learning. *International Journal Of Computer Network And Information Security, 7*(7), 10–23. doi:10.5815/ijcnis.2015.07.02

Alvarez-Taboada, F., Paredes, C., & Julian-Pelaz, J. (2017). Mapping of the Invasive Species Hakea sericea Using Unmanned Aerial Vehicles (UAV) and WorldViev-2 Imagery and an Object-Oriented Approach. *Remote Sensing, 9*(5), 913–931. doi:10.3390/rs9090913

Amine Chatti, M. (2012). Knowledge management: A personal knowledge network perspective. *Journal of Knowledge Management, 16*(5), 829–844. doi:10.1108/13673271211262835

Arefin Md, T., & Amin Md, R. (2010). Congestion Control and Buffering Technique for Video Streaming over IP. *International Journal of Latest Trends in Computing., 1*(2), 133–137.

Arkhangelskiy, V., Bohaenko, I., Grabowskiy, G., & Ryumshyn, N. (1999). *Neural networks in systems of automation*. Kiev: Techniques.

Aviation Accident Statistics. (2018). *National Transportation Safety Board.* Retrieved from www.ntsb.gov/aviation/aviation.htm

Aviation Human Factors Related Industry News 1Parts of this section are compiled from "Aviation Human Factors Industry News" and reproduced with permission of Roger Hughes. (2015). *Aviation Psychology and Applied Human Factors, 5*(1), 67-69. doi:10.1027/2192-0923/a000077

Avila, M., Funk, M., & Henze, N. (2015). Drone Navigator: Using Drones for Navigating Visually Impaired Presons. *Assets, 15*(10), 1–2.

Barba-Guaman, L., Calderon-Cordova, C., & Quezada-Sarmiento, P. A. (2017). Detection of moving objects through color thresholding [Detección de objetos en movimiento a través de la umbralización del color]. *Iberian Conference on Information Systems and Technologies.*

Barlow, R. E., & Proschan, F. (1965). *Mathematical Theory of Reliability.* New York: John Wiley and Sons.

Bashkov, E. O., Ivaschenko, V. P., & Shvachych, G.G. (2011). High-efficient multiprocessor system based on personal computing cluster. *Problems of Simulation and Automatic Projecting, 9*(179), 312–324.

Bashkov, Y. O., Ivaschenko, V. P., & Shvachych, G. G. (2011). High-performance multiprocessor system based on the personal computing cluster. *Scientific Papers of Donetsk National Technical University, 9*(179), 312–324.

Bauer, F. H. (2015). *GPS Space Service Volume (SSV) Ensuring Consistent Utility Across GPS Design Builds for Space Users.* Retrieved from https://www.gps.gov/governance/advisory/meetings/2015-06/bauer.pdf

Bay, H., Ess, A., Tuytelaars, T., & Gool, L. V. (2009). Speeded-Up Robust Features. *Proceedings of the 9th European Conference on Computer Vision,* 404-417.

Bay, H., Ess, A., Tuytelaars, T., & Van Gool, L. (2008). SURF: Speeded Up Robust Features. *Computer Vision and Image Understanding, 110*(3), 346–359. doi:10.1016/j.cviu.2007.09.014

Beard, R. W., & McLain, T. W. (2012). *Small Unmanned Aircraft: Theory and Practice.* Princeton University Press.

Beard, R. W., McLain, T. W., Goodrich, M., & Anderson, E. P. (2002). Coordinated target assignment and intercept for unmanned air vehicles. *IEEE Transactions on Robotics and Automation, 18*(6), 911–922. doi:10.1109/TRA.2002.805653

Benallegue, A., Mokhtari, A., & Fridman, L. (2006). Feedback linearizationand high order sliding mode observer for a quadrotor UAV. *Proceedings of the 2006 International Workshop on Variable Structure Systems,* 365. 10.1109/VSS.2006.1644545

Ben-Asher, Y., Feldman, S., Gurl, P., & Feldman, M. (2008). Distributed Decision and Control for Cooperative UAVs using Ad-Hoc Communication. *IEEE Transactions on Control Systems Technology, 16*(3), 511–516. doi:10.1109/TCST.2007.906314

Benchoff, B. (2016). *A Brief History of 'Drone'.* Available at: https://hackaday.com/2016/09/26/a-brief-history-of-drone/

Berbeka, J. (2016). Fotografia powietrzna jako innowacyjne narzędzie współtworzenia przeżyć turystów. *Prace naukowe Uniwersytetu Ekonomicznego we Wrocławiu, 473(30),* 69-76.

Berner, B., & Chojnacki, J. (2016). Wykorzystanie dronów do transportu towarów [The use of drones for goods transporting]. *Autobusy, R17*(6), 233–235.

Berner, B., & Chojnacki, J. (2016a). Wykorzystanie dronów w rolnictwie precyzyjnym [The use of drones in precision farming]. *Technika Rolnicza i Leśna, 3,* 19–21.

Berner, B., & Chojnacki, J. (2017). Zastosowanie bezzałogowych statków powietrznych do opryskiwania upraw rolniczych [The use of unmanned aerial vehicles for spraying agricultural crops]. *Technika Rolnicza Ogrodnicza Leśna*, *2*, 23–25.

Beshelev, S. D., & Gurvich, F. G. (1973). *Expert assessment*. Moscow: Science.

Bezruk, V., Chebotareva, D., Ivanenko, M., & Jo, S. (2014). Multicriteria Optimization in Planning of Mobile Communication Networks. *Proceeding of the 20th International Conference on Microwaves, Radar and Wireless Communications*, 633–639. 10.1109/MIKON.2014.6899974

Bielecki, K. (2015). *Zastosowanie dronów do inspekcji sieci ciepłowniczych* [Application of drones for inspection of heating networks]. Available at: http://www.inzynierbudownictwa.pl/technika,materialy_i_technologie,artykul,zastosowanie_dronow_do_inspekcji_sieci_cieplowniczych,8704

Biriukov, D., & Kondratov, S. (2012). Critical infrastructure protection: problems and prospects of implementation in Ukraine: analytical report. Academic Press.

Biriukov, D., Kondratov, S., & Sukhodolia, O. (2015). Green paper on critical infrastructure protection in Ukraine. Academic Press.

Blösch, M., Weiss, S., Scaramuzza, D., & Siegwart, R. (2010, May). Vision based MAV navigation in unknown and unstructured environments. In Robotics and automation (ICRA), 2010 IEEE international conference on (pp. 21-28). IEEE.

Bobek, S. (1992). A framework for integrating decision support systems into office information systems. *Decision Support Systems*, *8*(3), 211–216. doi:10.1016/0167-9236(92)90015-H

Bond, L. (1998). Overview of GPS Interference Issues. In *GPS Interference Symposium* (pp. 341-352). Boston: Volpe National Transportation System Center.

Bonin-Font, F., Ortiz, A., & Oliver, G. (2008). Visual navigation for mobile robots: A survey. *Journal of Intelligent & Robotic Systems*, *53*(3), 263–296. doi:10.100710846-008-9235-4

Borisov, A., Krumberh, O., & Fedorov, I. (1990). *Decision Making Based on Fuzzy Models: Examples of Using*. Riga: Zynatne.

Borovikov, V. P. (2008). *Neural Networks. STATISTICA Neural Networks: Methodology and Technology of Modern Data Processing* (2nd ed.). Moscow: Hotline-Telecom.

Boynton, F. (2014). *GNSS interference / denials, needs and challenges*. Retrieved from https://gpsworld.com/wp-content/uploads/2014/05/Loctronix-2014-GNSS-Webinar-140521-final.pdf

Bratushka, S. (2009). Simulation as a tool to study complex economic systems. *Scientific Bulletin NLTU Ukraine*, *8*, 22-28.

Breese, J. S., Heckerman, D., & Kadie, C. (1998, July). Empirical analysis of predictive algorithms for collaborative filtering. In *Proceedings of the Fourteenth Conference on Uncertainty in artificial intelligence* (pp. 43-52). Morgan Kaufmann Publishers Inc.

Brown, A., & Gerein, N. (2001). *Test Results of a Digital Beamforming GPS Receiver in a Jamming Environment*. Retrieved from: https://pdfs.semanticscholar.org/cde4/47bd0435360900e16c9c15e417067db27443.pdf

Brown, A., & Mathews, B. (2007). *Test Results from a Novel Passive Bistatic GPS Radar Using a Phased Sensor Array*. Retrieved from: https://pdfs.semanticscholar.org/481e/1afcc5652ac831492d66556301938f6a603d.pdf

Brunner, R. J., & Kim, E. J. (2016). Teaching data science. *Procedia Computer Science*, *80*, 1947–1956. doi:10.1016/j.procs.2016.05.513

Bujnowski, M. (2017). Integracja bezzałogowych statków powietrznych z unijnym systemem lotnictwa cywilnego [Integration of unmanned aerial vehicles into the EU civil aviation system]. *Kwartalnik Antymonopolowy i Regulacyjny*, *6*(2), 74–97.

Burdziakowski, P. (2016). Przegląd budowy i funkcjonalności współczesnych bezzałogowych statków powietrznych do celów fotogrametrycznych [Review of the construction and functionality of modern unmanned aerial vehicles for photogrammetric purposes]. *Biuletyn WAT*, *65*(4), 69–91. doi:10.5604/12345865.1228956

Bykov, V.V. (1971). *Tsifrovoe modelirovaniye v statisticheskoy radiotekhnike* [Digital modeling in statistical radio technique]. Moscow: Sovetskoye radio.

Capolupo, A., Pindozzi, S., Okello, C., Fiorentino, N., & Boccia, L. (2015). Photogrammetry for environmental monitoring: The use of drones and hydrological models for detection of soil contaminated by copper. *The Science of the Total Environment*, *514*(9), 298–306. doi:10.1016/j.scitotenv.2015.01.109 PMID:25668282

Castillo, P., Lozano, R., & Dzul, A. (2005). *Experimental implementation of linear and nonlinear control laws*. IEEE Control Systems Magazine.

Chao, X., Zeping, L., Guangquan, X., Zhiyong, F., Hongyan, T., & Haifeng, Z. (2015). *3D Reconstruction of Tree-Crown Based on the UAV Aerial Images* (Vol. 2015). Mathematical Problems in Engineering.

Charles F. Kettering. (2018). Available at: https://en.wikipedia.org/wiki/Charles_F._Kettering

Chen, H., Chiang, R. H., & Storey, V. C. (2012). Business intelligence and analytics: From big data to big impact. *Management Information Systems Quarterly*, *36*(4), 1165–1188. doi:10.2307/41703503

Chen, J., Zhang, X., Xin, B., & Fang, H. (2016). Coordination Between Unmanned Aerial and Ground Vehicles: A Taxonomy and Optimization Perspective. *IEEE Transactions on Cybernetics*, *46*(4), 959–972. doi:10.1109/TCYB.2015.2418337 PMID:25898328

Chernyak, L. (2012). *SDN - from concept to market*. Open Systems.

Chiliński, M. T., Markowicz, K. M., & Markowicz, J. (2016). Observation of vertical variability of black carbon concentration in lower troposphere on campaigns in Poland. *Atmospheric Environment*, *137*(4), 155–170. doi:10.1016/j.atmosenv.2016.04.020

Chojnacki, J. (2017). Status operatorów bezzałogowych statków powietrznych w świetle międzynarodowego prawa humanitarnego [Status of operators of unmanned aircraft in the light of international humanitarian law]. *Witelona w Legnicy*, *22*(1), 137–147.

Choset, H., & Burdick, J. (2000). Sensor-based exploration: The hierarchical generalized Voronoi graph. *The International Journal of Robotics Research*, *19*(2), 96–125. doi:10.1177/02783640022066770

Choset, H., Lynch, K. M., Hutchinson, S., Kantor, G., Burgard, W., Kavraki, L., & Thrun, S. (2005). *Principles of Robot Motion: Theory, Algorithms, and Implementation*. Cambridge, MA: MIT Press.

Christensen, B. R. (2015). Use of UAV or remotely piloted aircraft and forward-looking infrared in forest, rural and wildland fire management: Evaluation using simple economic analysis. *New Zealand Journal of Forestry Science*, *45*(16), 1–9.

Chyrkov A. (2016). Poshuk pidozrilykh ob'jektiv na video z kamery bezpilotnogho litaljnogho aparatu na osnovi analizu ghistoghram [Finding suspicious objects on a video from an unmanned aircraft camera based on histogram analysis]. *Problems of creation, testing and operation of complex information systems: Collection of scientific work of Zhytomyr Military Institute named after S. P. Korolev*, *13*, 126–135.

Chyrkov, A., & Prystavka, P. (2018). *Algorithm for Suspicious Object Search on Video Stream from Aircraft Camera and an Example of its Application. In Advances in Computer Science for Engineering and Education. ICCSEEA 2018. Advances in Intelligent Systems and Computing* (Vol. 754, pp. 340–348). Cham, Switzerland: Springer.

Cisco. (2004). *Internetworking technologies*. Cisco Press.

Claesson, A., Fredman, D., Svensson, L., Ringh, M., Hollenberg, J., Nordberg, P., ... Ban, Y. (2016). Unmanned aerial vehicles (drones) in out-of-hospital-cardiac-arrest. *Resuscitation and Emergency Medicine*, *24*(124), 1–9. PMID:27729058

Cleveret, T. (2016, December 5). *Google map overlayer.* Retrieved from http://cleveret.net/?page_id=161

Clothier, R. A., Greer, D. A., Greer, D. G., & Mehta, A. M. (2015). Risk Perception and the Public Acceptance of Drones. *Risk Analysis*, *35*(6), 1167–1183. doi:10.1111/risa.12330 PMID:25689883

Colby, G. (1997). Test Results of the Joint FAA/DoD Investigation of GPS Interference. *Proceedings of the 10th International Technical Meeting of the Satellite Division of the Institute of Navigation, GPS-97*, 217-228.

Collins, K. (2017). *London police deploy drones to search for missing people.* Available at: https://www.cnet.com/news/london-police-deploy-drones-to-search-for-missing-people/

Colomina, I., & Molina, P. (2014). Unmanned aerial systems for photogrammetry and remote sensing: A review. *ISPRS Journal of Photogrammetry and Remote Sensing*, *92*(4), 79–97. doi:10.1016/j.isprsjprs.2014.02.013

Corrigan, T. M., Hartranft, J. F., Levy, L. J., Parker, K. E., Pritchett, J. E., Pue, A. J., ... Thompson, T. (1999). *GPS Risk Assessment Study: Final Report.* The John Hopkins University Applied Physics Laboratory.

D'Andrea, R. (2014). Guest Editorial Can Drones Deliver? *IEEE Transactions on Automation Science and Engineering*, *11*(3), 647–648. doi:10.1109/TASE.2014.2326952

Dang, A. D., & Horn, J. (2015). Formation Control of Leader-Following UAVs to Track a Moving Target in a Dynamic Environment. *Journal of Automation and Control Engineering*, *1*, 1–8.

Das, A. K., Fierro, R., Kumar, V., Ostrowski, J. P., Spletzer, J., & Taylor, C. J. (2002). A Vision-Based Formation Control Framework. *IEEE Transactions on Robotics and Automation*, *18*(5), 813–825. doi:10.1109/TRA.2002.803463

Datasheet DHT11. (2010). Retrieved April 11, 2018, from: https://ru.scribd.com/document/351339794/Datasheet-DHT11

Datasheet Preview for BMP085 by Bosch. (2011). Retrieved April 4, 2018, from: https://www.datasheets.com/datasheet/bmp085-bosch-43859722.html

Davenport, T. H., & Patil, D. J. (2012). Data scientist: The sexiest job of the 21st century. *Harvard Business Review*. PMID:23074866

De Havilland DH82B Queen Bee. (2018). Available at: https://www.dehavillandmuseum.co.uk/aircraft/de-havilland-dh82b-queen-bee/

De Sousa, J., Girard, A., & Hedrick, J. (2004). Elemental maneuvers and coordination structures for unmanned air vehicles. *Proc. of the 43th IEEE Conf. on Decision and Control (CDC)*, *5*, 608–613. 10.1109/CDC.2004.1428711

Dhillon, B. S. (2005). *Reliability, Quality, and Safety for Engineers*. Boca Raton, FL: CRC PRESS.

Dhillon, B. S. (2006). *Maintainability, Maintenance, and Reliability for Engineers*. New York: Taylor & Francis Group. doi:10.1201/9781420006780

Dias, M., Zlot, R., Kalra, N., & Stentz, A. (2006). Market-Based Multirobot Coordination: A Survey and Analisys. *Proceedings of the IEEE*, *94*(7), 1257–1270. doi:10.1109/JPROC.2006.876939

Diggelen, F. (2007). *Update: GNSS Accuracy: Lies, Damn Lies, and Statistics.* Retrieved from: http://gpsworld.com/gpsgnss-accuracy-lies-damn-lies-and-statistics-1134/

DJI spark review: a smart selfie drone that's fun and frustrating. (2017). Available at: http://planforcrisis.com/dji-spark-review-a-smart-selfie-drone-thats-fun-and-frustrating/

Dobryakova, L., & Ochin, E. (2016). W sprawie klasyfikacji systemów transportowych w tym bezzałogowych [Regarding the classification of transport systems, including unmanned vehicles]. *Autobusy*, *R17*(6), 1680–1687.

Doc 30 ECAC (2010). *Policy Statement in the Field of Civil Aviation Security* (Restricted), 13, 138 p.

Doc 8973 ICAO (2014). *Guidance on Aviation Security* (Restricted), 9, 818 p.

Dovgan, O. (2013). Critical infrastructure as an object of protection against cyberattacks. *Information security: the challenges and threats of our time,* 17-20.

Drone-assisted water rescue demo. (2017). Available at: https://www.airmedandrescue.com/video/1581

Dudek, B., & Czapaj-Atłas, R. (2016). Rozwój dronów i ich zastosowań w energetyce [The development of drones and their applications in the energy sector]. *Energetyka (Warszawa)*, *1*, 21–26.

Dudenhoeffer, D., Permann, M., & Manic, M. (2006). CIMS: a framework for infrastructure interdependency modeling and analysis. *Proceedings of the winter simulation conference WSC 2006*, 478-485. 10.1109/WSC.2006.323119

Dzhigan, V. I. (2013). *Adaptivnaja fil'tracija signalov: teorija i algoritmy* [Adaptive filtering of signals: theory and algorithms]. Moscow: Tehnosfera.

Edwards, T., & Titchenal, S. (2009, November 19). *Creating image overlays for online mapping services.* Retrieved from http://railsandtrails.com/mapoverlays/imageoverlays.htm

Elsey, R. M., & Trosclair, P. L. III. (2016). The Use of an Unmanned Aerial Vehicle to Locate Alligator Nests. *Southeastern Naturalist (Steuben, ME)*, *15*(1), 76–82. doi:10.1656/058.015.0106

European GNSS Agency. (2017, June 9). *GNSS Market Report Issue 5.* Retrieved from https://www.gsa.europa.eu/system/files/reports/gnss_mr_2017.pdf

European Organisation for the Safety of Air Navigation (Eurocontrol). (2000). *Simulations Facilities for Air Traffic Control Training. HUM.ET1.ST07.3000-REP-02.* Brussels, Belgium: Author.

European Organisation for the Safety of Air Navigation (Eurocontrol). (2004a). ATM Services' Personnel (2nd ed.). ESARR 5. Belgium, Brussels: Author.

European Organisation for the Safety of Air Navigation (Eurocontrol). (2004b). *EATM Training Progression and Concepts.* Brussels, Belgium: Author.

European Organisation for the Safety of Air Navigation (Eurocontrol). (2015). *Specifications for the ATCO Common Core Content Initial Training* (2nd ed.). Brussels, Belgium: Author.

European Space Agency. (2012). *Galileo and Space Service Volume.* Retrieved from: http://www.unoosa.org/pdf/icg/2012/icg-7/wg/wgb2-2.pdf

Evenden, G. I. (2003). *Cartographic Projection Procedures for the UNIX Environment — A User's Manual.* Open-File Report 90-284. U.S. Geological Survey. Retrieved from http:// trac.osgeo.org/proj

Everett, H. R. (2015). *Unmanned Systems of World Wars I and II.* MIT Press.

Eyerman, J., Letterman, C., Pitts, W., Holloway, J., Hinke, K., Schanzer, D., ... Kaydos-Daniels, S. C. (2013). *Unmanned Aircraft and the Human Element: Public Perceptions and First Responder Concerns. Research Brief, Post-Project Report.* Institute for Homeland Security Solutions.

Fazli, S., Pour, H. M., & Bouzari, H. (2009). Particle Filter Based Object Tracking with Sift and Color Feature. *Second International Conference on Machine Vision*, 89–93. 10.1109/ICMV.2009.47

Federal Aviation Administration. (2013a). *Global positioning system directorate systems engineering and integration.* Interface Specification IS-GPS-800.

Federal Aviation Administration. (2013b). *GPS, "NAVSTAR GPS Control Segment to User Support Community Interfaces", ICD-GPS-870, rev. B.* Author.

Fekete, A. (2011). Common criteria for the assessment of critical infrastructures. *International Journal of Disaster Risk Science*, *2*(1), 15–24. doi:10.100713753-011-0002-y

Fellner, R. (2015). Bezpieczeństwo i bezzałogowe statki powietrzne w dyskursie naukowo-prawniczym [Safety and unmanned aerial vehicles in the scientific-legal discourse]. *Seria Transport*, *95*, 197–212.

Felzenszwalb, P.F., Girshick, R.B., McAllester, D., & Ramanan, D. (2009). Object detection with discriminatively trained part based models. *IEEE PAMI, 99.*

Feng, F., Li, Y., & Leoreanu-Fotea, V. (2010). Application of level soft sets in decision making based on interval-valued fuzzy soft sets. *Computers & Mathematics with Applications (Oxford, England)*, *60*(6), 1756–1767. doi:10.1016/j.camwa.2010.07.006

Flammini, F., Pragliola, C., & Smarra, G. (2016). Railway Infrastructure Monitoring by Drones. *2016 International Conference on Electrical Systems for Aircraft, Railway, Ship Propulsion and Road Vehicles & International Transportation Electrification Conference (ESARS-ITEC)*, 1-6.

Floreano, D., & Wood, R. (2015). Science, technology and the future of small autonomous drones. *Nature*, *521*(5), 460–466. doi:10.1038/nature14542 PMID:26017445

Forssel, B., & Olsen, T. B. (2003). Jamming Susceptibility of Some Civil GPS Receivers. *GPS World*, *1*, 54–58.

Fortz, B., & Throup, M. (2002). Optimizing OSPF / IS-IS weights in a changing world. *IEEE Journal on Selected Areas in Communications*, *6*, 1–31.

France: Demonstrating the use of drones for railway maintenance and service. (2015). Available at: https://uic.org/com/uic-e-news/464/article/france-demonstrating-the-use-of?page=iframe_enews

Frenkel, I. B., Karagrigoriou, A., Lisnianski, A., & Kleyner, A. V. (2013). *Applied reliability engineering and risk analysis: probabilistic models and statistical inference.* Wiley. doi:10.1002/9781118701881

Gan, X., Tang, X., & Gao, J. (2013). Missile Autopilot Design Based on Fuzzy Self-Tuning Immune PID Control. *Advanced Materials Research*, *739*, 550–554. . doi:10.4028/www.scientific.net/AMR.739.550

Gao, G. X., Sgammini, M., Lu, M., & Kubo, N. (2016). Protecting GNSS Receivers from Jamming and Interference. *Proceedings of the IEEE*, 1327–1338. 10.1109/JPROC.2016.2525938

Gardner, M. W., & Dorling, S. R. (1998). Survey and Critique of Techniques for Extracting Rules from Trained Artificial Neural Networks. *Atmospheric Environment*, *32*(14–15), 2627–2636. doi:10.1016/S1352-2310(97)00447-0

George, G., Osinga, E. C., Lavie, D., & Scott, B. A. (2016). *Big data and data science methods for management research.* Academic Press.

Gettys, J. (2011). Bufferbloat: Dark Buffers in the Internet. Internet Computing. *IEEE Computer Society*, *3*(15), 96–95.

Gheethan, A. A., Herzig, P. A., & Mumcu, G. (2013). Compact 2 × 2 coupled double loop GPS antenna array loaded with broadside coupled split ring resonators. *IEEE Transactions on Antennas and Propagation*, *61*(6), 3000–3008. doi:10.1109/TAP.2013.2253539

Gilmore, S. (1998). The Impact of Jamming on GPS. In *Symposium on GPS Interference and Mitigation Technique*, (pp. 467-478). Boston: Volpe National Transportation System Center.

Gis-Lab. (2010). *The format of the geographic registration file*. Retrieved from http://gis-lab.info/qa/tfw.html

Gnadinger, F., & Schmidhalter, U. (2017). Digital Counts of Maize Plants by Unmanned Aerial Vehicles (UAVs). *Remote Sensing*, *9*(5), 544–559. doi:10.3390/rs9060544

Gnatyuk, S. & Sydorenko, V. (2015). An overview of methods for evaluating critically important objects. *Problems and prospects for the development of aviation and cosmonautics*, 110.

Gnatyuk, S. (2016). *Meeting Security Challenges Through Data Analytics and Decision Support*. NATO Science for Peace and Security Series – D: Information and Communication Security. IOS Press.

Gnatyuk, S., Aleksander, M., & Sydorenko, V. (2018). Unified data model for defining state critical information infrastructure in civil aviation. *The 9th IEEE International Conference on Dependable Systems, Services and Technologies (DESSERT-2018),* 37-42.

Gnatyuk, S., Sydorenko, V. & Kinzeriavyi, V. (2017). Method of object identification of critical information infrastructure in aviation. *Information Technology and Security*, *5*(2), 27-39.

Gnatyuk, S., Sydorenko, V., & Duksenko, O. (2015). Modern approaches to critical infrastructure objects detection and identification. *Ukrainian Scientific Journal of Information Security*, *21*(3), 269–275.

Gnatyuk, S., Sydorenko, V., & Seilova, N. (2017). Universal data model for the formation of the critical information infrastructure of the state objects list. *Ukrainian Scientific Journal of Information Security*, *23*(2), 80–91.

Gnatyuk, S., & Vasyliev, D. (2016). Modern critical aviation information systems. *Ukrainian Scientific Journal of Information Security*, *22*(1), 51–57.

Goedeking, P. (2010). *Networks in Aviation: Strategies and Structures*. Springer. doi:10.1007/978-3-642-13764-8

Goldberg, D., Corcoran, M., & Picard, R. G. (2014). Remotely Piloted Aircraft Systems & Journalism. *Reuters Institute for the Study of Journalism*, *6*, 1–41.

Golub, O. & Sheremet, S. (2016). *Satellite navigation systems in transport*. Academic Press.

Goncharenko, A. V. (2015). Applicable aspects of alternative UAV operation. In *Proceedings of IEEE 3rd International Conference on Actual Problems of Unmanned Air Vehicles Developments (APUAVD 2015)*, (pp. 316 – 319). Kyiv, Ukraine: IEEE. 10.1109/APUAVD.2015.7346630

Goncharenko, A. V. (2017). Optimal UAV maintenance periodicity obtained on the multi-optional basis. In *Proceedings of IEEE 4th International Conference on Actual Problems of Unmanned Air Vehicles Developments (APUAVD 2017)*, (pp. 65 – 68). Kyiv, Ukraine: IEEE. 10.1109/APUAVD.2017.8308778

Goncharenko, A. V. (2018). Multi-optional hybrid effectiveness functions optimality doctrine for maintenance purposes. In *14th International Conference on Advanced Trends in Radioelectronics, Telecommunications and Computer Engineering (TCSET 2018)*, (pp. 771 – 775). Lviv-Slavske, Ukraine: Academic Press. 10.1109/TCSET.2018.8336313

Goodwin, E. (1975). A statistical study of ship domain. *Journal of Navigation*, *28*(03), 328–344. doi:10.1017/S0373463300041230

Google Maps Overlay. (n.d.). Retrieved from https://www.w3schools.com/graphics/google_maps_overlays.asp

Gorakala, S. K., & Usuelli, M. (2015). *Building a Recommendation System with R*. Packt Publishing Ltd.

Gorbunov, I. E. (2006). Methodology of analysis and synthesis of reconfigurable topologies of mobile communication networks. *Mathematical Machines and Systems*, *2*, 48–59.

Gosiewski, Z., & Ambroziak, L. (2013). UAV Autonomous Formation Flight Experiment with Virtual Leader Control Structure. *Diffusion and Defect Data, Solid State Data. Part B, Solid State Phenomena*, *198*, 254–259. doi:10.4028/www.scientific.net/SSP.198.254

GOST R 51794-2008. (2009). *Globalnye navigatsionnye sputnikovye sistemy. Sistemy koordinat. Metody preobrazovaniya koordinat opredelyaemykh tochek* [Global navigation satellite systems. Coordinate systems. Methods for converting the coordinates of defined points]. Moskow: standardinform. Retrieved from http://ingeo-pro.ru/upload/SNIP/gost-r-51794-2008.pdf (in Russian)

Gowtham, S., Nazeer, A., & Saranya, G. (2017). Survillence Drones imparted with Optimized Cloud Server Techniques. IEEE, 17, 1-3.

GPS Directorate. (2013, September 24). *Interface Specification IS-GPS-200 (rev. H)*. Retrieved from https://www.gps.gov/technical/icwg/IS-GPS-200H.pdf

Grabbe, S., Sridhar, B., & Mukherjee, A. (2010). Evaluation of an Integrated Traffic Flow Management Decision Making Approach. *Air Traffic Control Quarterly*, *18*(4), 331–358. doi:10.2514/atcq.18.4.331

Grancharova, A., Grøtli, E. I., & Johansen, T. A. (2013). Rotary-Wing UAVs Trajectory Planning by Distributed Linear MPC with Reconfigurable Communication Network Topologies. *IFAC Proceedings Volumes*, *46*(27), 198 – 205.

Grigoriev, L. N. (2010). *Tsifrovoe formirovaniye diagrammy naprvlennosti v fazirovannikh antennikh reshetkakh* [Digital forming of directional diagram in phased-array antennas]. Moscow: Radiotekhnika.

Guiraud, M., & Nachescu, M. (2012). Drones as a tool of sustainable development - consequences of their usage and new forms of risk management. *Economic and Environmental Studies*, *12*(4), 467–479.

Gulevitch, S. P., Veselov, Yu. G., Pryadkin, S. P., & Tyrnov, S. D. (2012). Analysis of factors affecting safety of flight of unmanned aerial vehicles (UAV). Causes of accidents of UAVs and methods of preventing them. Science and education Russia. Bauman Moscow State Technical University.

Gupta, L., Jain, R., & Vaszkun, G. (2015). Survey of Important Issues in UAV Communication Networks. *IEEE Communications Surveys and Tutorials*, 1 – 32.

Hahn, G. J., & Shapiro, S. S. (1967). *Statistical models in engineering*. New York: John Wiley & Sons, Inc.

Hahsler, M. (2011). *Recommenderlab: A Framework for Developing and Testing Recommendation Algorithms*. Southern Methodist University.

Haidari, L. A., Brown, S. T., Ferguson, M., Bancroft, E., Spiker, M., Wilcox, A., ... Lee, B. Y. (2016). The economic and operational value of using drones to transport vaccines. *Vaccine*, *34*(24), 4062–4067. doi:10.1016/j.vaccine.2016.06.022 PMID:27340098

Haigh, K., & Shewchuk, J. (1994). Geometric Similarity Metrics for Case-Based Reasoning. *Working Notes from the AAAI-94 Workshop on Case-Based Reasoning*, 182–187.

Hajiyev, C., & Caliskan, F. (2013). *Fault diagnosis and reconfiguration in flight control systems* (Vol. 2). Springer Science & Business Media.

Haleplidis, E. (2014). *For CES Applicability to SDN-enhanced NFV*. Proc. Euro. Wksp. Software Defined Networks.

Haleplidis, E. (2015). *Software-Defined Networking (SDN)*. Layers and Architecture Terminology.

Hallsten, J. (2017). *The essential guide to building your Amazon drone delivery pad.* Available at: http://www.hallsteninnovations.com/drone-delivery-pad/

Hammer, J., Piper, G., Thorp, O., & Watkins, J. (2004). *Investigating Virtual Structure Based Control Strategies for Spacecraft Formation Maneuvers*. Providence, RI: AIAA Guidance, Navigation, and Control Conference and Exhibit. doi:10.2514/6.2004-4898

Hansen, A. J. (1997). *WAAS Precision Approach Metrics: Accuracy*. Integrity, Continuity and Availability.

Hassas Yeganeh, S., & Ganjali, Y. (2012). Kandoo: a framework for efficient and scalable offloading of control applications. *Proceedings of the first workshop on Hot topics in software defined networks*, 19–24. 10.1145/2342441.2342446

Hauger, S., Tso, K. H., & Schmidt-Thieme, L. (2008). Comparison of recommender system algorithms focusing on the new-item and user-bias problem. In *Data Analysis, Machine Learning and Applications* (pp. 525–532). Berlin: Springer. doi:10.1007/978-3-540-78246-9_62

Heller, S., & McKeown. (2012). The controller placement problem. *Proceedings of the first workshop on Hot topics in software defined networks, HotSDN.*

Hird, J. N., Montaghi, A., McDermid, G. J., Kariyeva, J., Moorman, B. J., Nielsen, S. E., & McIntosh, A. C. S. (2017). Use of Unmanned Aerial Vehicles for Monitoring Recovery of Forest Vegetation on Petroleum Well Sites. *Remote Sensing*, *9*(5), 413–433. doi:10.3390/rs9050413

Hnot, T. V., & Nehrey, M. V. (2017). Data Science Algorithms in Modeling Business Processes. *Economics and Society.*, *12*, 187.

Hodgson, A., Kelly, N., & Peel, D. (2013). Unmanned Aerial Vehicles (UAV) for Surveying Marine Fauna: A Dugong Case Study. *PLoS One*, *8*(11), 1–15. doi:10.1371/journal.pone.0079556 PMID:24223967

How drones are already being deployed by railways around the world. (2014). Available at: https://www.smartrailworld.com/how_drones_are_already-being-used-by-railways-around-the-world

Hoyland, A., & Rausand, M. (1994). *System reliability theory*. New York: John Wiley & Sons, Inc.

Huang, J., Lo Presti, L., Motella, B., & Pini, M. (2016). GNSS spoofing detection: Theoretical analysis and performance of the Ratio Test metric in open sky. *ICT Express. Elsevier*, *2*(1), 37–40. doi:10.1016/j.icte.2016.02.006

Hui, W., Hu, J., & Shyue, K. (2008). Role of coordinates in computational fluid dynamics. *International Journal of Computational Fluid Dynamics*, *22*(1-2), 3–9. doi:10.1080/10618560701737112

Hwang, I., Kim, J., & Tomlin, C. (2007). Protocol-Based Conflict Resolution for Air Traffic Control. *Air Traffic Control Quarterly*, *15*(1), 1–34. doi:10.2514/atcq.15.1.1

Ill'chenko, K. D., Chechenev, V. A., Ivashchenko V. P., & Tereschenko, V. S. (1999). *Teplofyzyches-kye svoystva promyshlennykh materyalov: Spravochnyk*, 152.

Immerzeel, W. W., Kraaijenbrink, P. D. A., Shea, J. M., Shrestha, A. B., Pellicciotti, F., Bierkens, M. F. P., & de Jong, S. M. (2014). High-resolution monitoring of Himalayan glacier dynamics using unmanned aerial vehicles. *Remote Sensing of Environment, 150*(5), 93–103. doi:10.1016/j.rse.2014.04.025

International Civil Aviation Organization (ICAO). (1999). *Manual of technical provisions for the aeronautical telecommunication network (ATN)*. Retrieved from https://www.icao.int/safety/acp/repository/_%20Doc9705_ed2_1999.pdf

International Civil Aviation Organization. (2000). *Validated ICAO GNSS Standards and Recommended Practices*. SARPS.

International Civil Aviation Organization. (2005). *Doc 9849 AN/457 Global Navigation Satellite System (GNSS) Manual*. Retrieved from https://www.icao.int/Meetings/PBN-Symposium/Documents/9849_cons_en[1].pdf

International Civil Aviation Organization. (2007). Air Traffic Management (15th ed.). Doc. ICAO 4444-ATM/501. Montreal: Author.

International Civil Aviation Organization. (2007). Procedures for Air Navigation Services. Air Traffic Management (15th ed.). Doc. ICAO 4444-ATM/501. Montreal: Author.

International Civil Aviation Organization. (2008). *Threat and Error Management (TEM) in Air Traffic Control. Cir. ICAO 314-AN/178*. Montreal, Canada: Author.

International Civil Aviation Organization. (2014). *Aeronautical Telecommunications Annex 10 to the Convention on International Civil Aviation, Volume I (Radio Navigation Aids)*. Author.

Isermann, R. (2006). *Fault-diagnosis systems: an introduction from fault detection to fault tolerance*. Springer Science & Business Media. doi:10.1007/3-540-30368-5

Issariyakul, T., & Hossain, E. (2006). Channel-quality-based opportunistic scheduling with ARQ in multi-rate wireless networks: Modeling and analysis. *IEEE Transactions on Wireless Communications, 4*(5), 796–806. doi:10.1109/TWC.2006.1618929

Ivaschenko, V. P., Shvachych, G. G., & Semenov, S. G. (2017). Efficient parallelization algorithms of the applied tasks in multiprocessor computing systems. *System Technology, 2*(109), 57–66.

Ivaschenko, V. P., Shvachych, G. G., & Shmukin, A. A. (2008). Parallel computations and applied problems of metallurgical thermophysics. *System Technologies: The Regional Collection of the Proceedings, 3*(56), 123–138.

Ivaschenko, V. P., Shvachych, G. G., & Shmukin, A. A. (2013). Investigation of the components selection problem of a modular multiprocessor system in the problem of metallurgy thermal processing. *System Technologies: The Regional Collection of the Proceedings, 2*(85), 78–92.

Ivaschenko, V. P., Shvachych, G. G., & Shmukyn A. A. (2005). Some aspects of the problem of mathematic simulation of the metallurgical thermophysical tasks based on parallel computing system of cluster type. *Modern Problem of the Metalurgy, 7*, 23–30.

Ivaschenko, V. P., Shvachych, G. G., & Tkach, M. A. (2013). Features of using multi-core processors in multiprocessor computing systems. Newest computer technologies, 11, 132–135.

Ivaschenko, V. P., Shvachych, G. G., Ivaschenko, E. V., & Busygin, V. V. (2018). Effective algorithms for solving coefficient problems of high accuracy. *System Technology, 4*(117), 95–108.

Ivaschenko, V. P., Shvachych, G. G., & Kholod, E. G. (2014). *Extreme algorithms for solving problems with higher order accuracy. In Applied and fundamental research* (pp. 157–170). St. Louis, MO: Publishing House Science and Innovation Center, Ltd.

Ivaschenko, V. P., Shvachych, G. G., & Sobolenko, O. V. (2015). The system of automated control of temperature regimes of thermal treatment of steel products. *Metallurgical and Mining Industry, 1*, 142–147.

Ivashchenko, V. P., Bashkov, E. A., & Shvachych, G. G. (2011). Informational systems and technology: monography. Krasnoyarsk: Academic Press.

Ivashchenko, V. P., Shvachych, G. G., Sobolenko, A. V., & Protopopov, D. V. (2003). Informational system of intellectual support of making solutions for the metal rolling. *Vostochno-evropeyskyy zhurnal peredovykh tekhnolohyy, 3*, 4–9.

Jadbabaie, A., Lin, J., & Morse, A. (2003). Coordination of groups of mobile autonomous agents using nearest neighbor rules. *IEEE Transactions on Automatic Control, 48*(6), 988–1001. doi:10.1109/TAC.2003.812781

Jagadish, H., Ooi, B.-Ch., Tan, K.-L., Yu, C., & Zhang, R. (2005). iDistance: An Adaptive B+-Tree Based Indexing Method for Nearest Neighbor Search. *ACM Transactions on Database Systems, 30*(2), 364–397. doi:10.1145/1071610.1071612

Jaidee, U., Munoz-Avila, H., & Aha, D. W. (2013). Case-based goal-driven coordination of multiple learning agents. *Proc. of the 21st Int. Conf. on Case-Based Reasoning*, 164–178. 10.1007/978-3-642-39056-2_12

Jain, A., & Kumar, A. (2017). Traffic Congestion Management System: An Approach with Machine Learning. *IMS Manthan, 12*(01). doi:10.18701/imsmanthan.v12i01.10338

James, G., Witten, D., Hastie, T., & Tibshirani, R. (2013). *An introduction to statistical learning* (Vol. 112). New York: Springer. doi:10.1007/978-1-4614-7138-7

Jeffery, L., & Rhodes, G. (2011). Insights into the development of face recognition mechanisms revealed by face aftereffects. *British Journal of Psychology, 102*(4), 799–815. doi:10.1111/j.2044-8295.2011.02066.x PMID:21988385

John, A. (2001). *Vulnerability Assessment of the Transportation Infrastructure Relying on the Global Positioning System.* Cambridge, MA: Volpe National Transportation Systems Center. Retrieved from https://www.navcen.uscg.gov/pdf/vulnerability_assess_2001.pdf

Journal Telesatellite International. (2010). Retrieved from: www.Tele-satellite.com

Junaid, A. B., Konoiko, A., Zweiri, Y., Sahinkaya, M. N., & Seneviratne, L. (2017). Autonomus Wireless Self-Charging for Multi-Rotor Unmanned Aerial Vehicles. *Energies, 10*(6), 803–817. doi:10.3390/en10060803

Juneja, P., & Kashyap, R. (2016). Optimal approach for CT image segmentation using improved energy based method. *International Journal of Control Theory and Applications, 9*(41), 599–608.

Kaliaev, I., Gaiduk, A., & Kapustian, S. (2009). *Models and Algorithms of Collective Control in Groups of Robots.* Moscow: Fizmatlit.

Kaminski, M. E. (2013). Drone Federalism: Civilian Drones and the Things They Carry. *California Law Review Circuit, 4*(5), 57–74.

Karoly, G., Leizer, K., & Tokody, D. (2017). Radiofrequency identification by using drones in railway accidents and disaster situations. *Interdisciplinary Description of Complex Systems, 15*(2), 114–132.

Karpenko, A. V. (2010). *Complex «LEER» with Unmanned Aerial Vehicle «ORLAN-10»*. Retrieved from http://bastion-opk.ru/orlan-10/

Karpov, D. (2012). *Sshivka izobrazheniy, poluchennykh v rezultate aerofotosemki* [Cross-linking of images obtained as a result of aerial photography] (PhD thesis). St. Petersburg: St. Petersburg National Research University of Information Technologies, Mechanics and Optics. Retrieved from https://archive.li/o/fiaQi/is.ifmo.ru/projects/2012/karpov/description.pdf (in Russian)

Kashyap, R., & Piersson, A. (2018a). Big Data Challenges and Solutions in the Medical Industries. In Handbook of Research on Pattern Engineering System Development for Big Data Analytics. IGI Global. doi:10.4018/978-1-5225-3870-7.ch001

Kashyap, R., Gautam, P., & Tiwari, V. (2018). Management and Monitoring Patterns and Future Scope. In Handbook of Research on Pattern Engineering System Development for Big Data Analytics. IGI Global. doi:10.4018/978-1-5225-3870-7.ch014

Kashyap, R. (2018). Object boundary detection through robust active contour based method with global information. *International Journal Of Image Mining*, *3*(1), 22. doi:10.1504/IJIM.2018.093008

Kashyap, R., & Gautam, P. (2013). Microarray image segmentation using improved GOGAC method. *International Journal on Computer Science and Engineering*, *2*(4), 67–74.

Kashyap, R., & Gautam, P. (2015). Modified region based segmentation of medical images. *International Conference on Communication Networks (ICCN)*, 209–216. 10.1109/ICCN.2015.41

Kashyap, R., & Gautam, P. (2016). Fast level set method for segmentation of medical images. In *Proceedings of the International Conference on Informatics and Analytics (ICIA-16)*. ACM. 10.1145/2980258.2980302

Kashyap, R., & Gautam, P. (2017a). Fast Medical Image Segmentation Using Energy-Based Method. *Biometrics. Concepts, Methodologies, Tools, and Applications*, *3*(1), 1017–1042. doi:10.4018/978-1-5225-0983-7.ch040

Kashyap, R., & Gautam, P. (2017b). *Fast Medical Image Segmentation Using Energy-Based Method.* In *Pattern and Data Analysis in Healthcare Settings*. Medical Information Science Reference.

Kashyap, R., & Piersson, A. (2018b). *Impact of Big Data on Security. In Handbook of Research on Network Forensics and Analysis Techniques* (pp. 283–299). IGI Global. doi:10.4018/978-1-5225-4100-4.ch015

Kashyap, R., & Tiwari, V. (2017). Energy-based active contour method for image segmentation. *International Journal of Electronic Healthcare*, *9*(2–3), 210–225. doi:10.1504/IJEH.2017.083165

Kaur, P., Singh, D., Singh, G., & Singh, N. (2011). Analysis, comparison and performance evaluation of BNP scheduling algorithms in parallel processing. *International Journal of Information Technology and Knowledge Management*, *1*(4), 279–284.

Keating, M. (2016). Georgia country police use drones to investigate serious traffic wrecks. *American City & Country.* Available at: http://americancityandcounty.com/public-safety/georgia-county-police-use-drones-investigate-serious-traffic-wrecks-related-video

Keesey, L. (2010). *Navigator Technology Takes GPS to a New High.* Retrieved from https://www.nasa.gov/topics/technology/features/navigator-gps.html

Kelso, T. S. (2014). *Definition of a Yuma Almanac* Retrieved from https://celestrak.com/GPS/almanac/Yuma/definition.asp

Key, E. L. (1995). *Technique to Counter GPS Spoofing.* International Memorandum, MITRE Corporation.

Kharchenko, V., Chynchenko, Yu., & Raychev, S. (2007). Threat and Error Management in Air Traffic Control. *Proceedings of the National Aviation University, 3–4,* 24-29.

Kharchenko, V., Shmelova, T., Artemenko, O., & Otryazhiy, V. (2011). *A. r. Computer Program "Choosing the Pre-Flight Information and Decision Making about Departure for Automated System of Pre-Flight Information Preparation (AS PIP)"*. Kyiv: State Department of Intellectual Property.

Kharchenko, V., Shmelova, T., Sikirda, Yu., & Gerasimenko, O. (2011). *A. r. Computer Program "Optimizing the Choice of Alternative Variant of Aircraft Flight Completion in Flight Emergencies "Prompt"*. Kyiv: State Department of Intellectual Property.

Khorasani, A., Jalali Aghchai, A., & Khorram, A. (2010). Chatter prediction in turning process of conical workpieces by using case-based resoning (CBR) method and taguchi design of experiment. *International Journal of Advanced Manufacturing Technology*, *55*(5-8), 457–464. doi:10.100700170-010-3060-1

Khramov, P. (2016, March 5). *"Maps" section: Photos and taxa linking to the maps markers*. Retrieved from http://insecta.pro/community/60496

Kim, A., Wampler, B., Goppert, J., Hwang, I., & Aldridge, H. (2012) Cyber Attack Vulnerabilities Analysis for Unmanned Aerial Vehicles. American Institute of Aeronautics and Astronautics. doi:10.2514/6.2012-2438

Kim. (2011). The Evolution of Network Configuration: A Tale of Two Campuses. *Proc. 2011 ACM SIGCOMM Conf. Internet Measurement Conf.*, 499–514.

Kim, N. V., & Chevonenkis, M. A. (2015). Situation Control of Unmanned Aerial Vehicles for Road Traffic Monitoring. *Modern Applied Science*, *9*(5), 1–13.

Kociubiński, J. (2016). *Rozwój bezzałogowych statków powietrznych w świetle prawa do prywatności na gruncie Europejskiej Konwencji o Ochronie Praw Człowieka*. Available at: www.eumed.net.rev.rednma.33.koc.html

Kogure, S., & Kishimoto, M. (2012). *MICHIBIKI (QZS-1) and SSV*. Retrieved from: http://www.unoosa.org/pdf/icg/2012/icg-7/wg/wgb2-3.pdf

Komashynskiy, V., & Smirnov, D. (2002). *Neural Networks and Its Use in Systems of Management and Communication*. Moscow: Hotline-Telecom.

Kondratiev, A. (2012). Modern trends in the study of critical infrastructure in foreign countries. *Foreign Military Review*, *1*, 19–30.

Konin V. & Harchenko V. (2010) *Sistemy sputnikovoj radionavigacii* [Satellite radionavigation systems]. Kyiv: Holtech Reference.

Konin, V., Shyshkov, F., & Pogurelskiy, O. (2016). Estimation of coordinates on geostationary orbit by using GNSS signals. 2016 IEEE Radar Methods and Systems Workshop (RMSW), 32-35. doi:10.1109/RMSW.2016.7778544

Konin, V., Shyshkov, F., & Pogurelskiy, O. (2017). Differences in measurements with separate use of frequencies L1 and L2 for the application of satellite navigation in near-earth space. *2017 IEEE Microwaves, Radar and Remote Sensing Symposium (MRRS)*, 67-70. doi: 10.1109/MRRS.2017.8075028

Konin, V., & Harchenko, V. (2010). *Sistemy sputnikovoj radionavigacii* [Satellite radionavigation systems]. Kyiv: Holtech.

Konin, V., & Shyshkov, F. (2016). Autonomous navigation of service spacecrafts on geostationary orbit using GNSS signals. *Radioelectronics and Communications Systems*, *59*(12), 562–566. doi:10.3103/S0735272716120049

Koponen, Casado, Gude, Stribling, Poutievski, Zhu, … Hama. (2010). Onix: A distributed control platform for large-scale production networks. *OSDI, 10*, 1–6.

Korn, G. A., & Korn, T. M. (1968). *Mathematical handbook*. McGraw-Hill Book Company.

Korotonoshko, A.N., & Perunov, Y.M. (2006). *Stoykost' i radiotekhnicheskiy immunitet transportnykh sistem, osnovannykh na tochnoy sputnikovoy navigatsii* [Persistence and radio technical immunity of transport systems relying on precise satellite navigation]. Moscow: Novosti navigatsii.

Korotonoshko, A.N., & Perunov, Y.M. (2007). *Imitatsionniye radiotekhnicheskiye pomekhi sistemam spitnikovoy navigatsii* [Imitating radio technical interferences to the satellite navigation systems]. Moscow: Novosti navigatsii.

Korzhov, V. (2012). *Children's Diseases SDN*. Open Systems.

Krasnov, L. A., Liamtsev, S. E. (2016). Research of antijammingness of algorithms of the trajectory measuring on viewdatas [Issledovaniye pomekhoustoychivosti algoritmov traektornykh izmereniy po videodannym]. *Radioelectronic and Computer Systems, 4*(78), 113-117.

Kucherov, D. P., & Kozub, A. N. (2015). Control System Objects with Multiple Stream of Information. *Proceedings IEEE 3rd International Conference "Actual Problems of Unmanned Aerial Vehicles Developments (APUAVD)"*, 290–293. 10.1109/APUAVD.2015.7346623

Kucherov, D., Kozub, A., & Kostyna, O. (2016, October). Group behavior of UAVs in obstacles presence. In *Methods and Systems of Navigation and Motion Control (MSNMC), 2016 4th International Conference on* (pp. 51-54). IEEE.

Kucherov, D. P. (2007). The synthesis of adaptive terminal control algorithm for inertial secondary order system with bounded noises. *Journal of Automation and Information Sciences, 39*(9), 16–25. doi:10.1615/JAutomatInfScien.v39.i9.20

Kucherov, D. P. (2016). Reconfiguration multisensory system in conditions of impact of destabilizing factors. *Sensor Electronics and Microsystem Technologies, 2*(13), 101–112. doi:10.18524/1815-7459.2016.2.73659

Kułaga, Ł. (2016). Używanie dronów w celu zwalczania terroryzmu w świetle 'ius ad bellum' [The use of drones to combat terrorism in the light of 'ius ad bellum']. *Zeszyty Prawnicze, 16*(2), 141–162. doi:10.21697/zp.2016.16.2.07

Kulik, A. S. (1991, September). Fault diagnosis in dynamic Systems via signal-parametric approach. In *IFAC/IMACS Symp. Baden-Baden* (*Vol. 1*, pp. 157-162). Academic Press.

Kulik, A. S. (2014). Rational control of aerospace objects operability under the influence of the destabilizing actions. *Aerospace Engineering and Technology*, (1), 31-38.

Kuusniemi, H. (2012). *Effects of GNSS jammers and potential mitigation approaches*. Retrieved from: http://www.unoosa.org/pdf/sap/2012/un-latvia/ppt/3-14.pdf

Kuzmichov, A.I., & Dodonov, E.O. (2017). Optimization models of network structure. *Registration, Storage, and Processing of Data, 2*, 24 – 35.

Larin, V. (2009). *New test and development procedures for precision instruments and information-measurement systems design*. Donetsk, Ukraine: Weber. (in Russian)

Larin, V. (2012). Measurement of slewing angle of elevator of unmanned aircraft vehicle using ferrimagnetic transformer. [in Ukrainian]. *Radioelectronic and Computer Systems, 4*, 29–34.

Larin, V. J. (2009). Development of mathematical model of sensitive elements of ferrimagnetic transducer. *Eastern-European Journal of Enterprise Technologies, 2/6*(38), 52–56. (in Russian)

Lastovchenko, M. M., Zubareva, E. E., & Sachenko, V. O. (2009). Method for analyzing the efficiency of the reconfiguration of the topology of wireless multi-server networks with increased noise immunity. *USM, 6*, 79–86.

Latsys, A. O. (2003). *How to build and use a supercomputer*. Moscow: Bestseller, 240. Retrieved from http://www.myshared.ru/ slide/

Latsys, A. O. (2003). How to build and use supercomputer. Moscow: Academic Press.

Lawton, J., Beard, R., & Young, B. (2003). A decentralized approach to formation maneuvers. *IEEE Transactions on Robotics and Automation, 19*(6), 933–941. doi:10.1109/TRA.2003.819598

Lee, D., Lillywhite, K., Fowers, S., Nelson, B., & Archibald, J. (2006). An embedded vision system for an unmanned four-rotor helicopter. SPIE Optics East, Intelligent Robots and Computer Vision XXIV: Algorithms, Techniques, and Active Vision.

Lee, H., & Rhee, K. (2001). Developing Collision Avoidance System by Using Expert System and Search Algorithm. Int. *Shipbuilding Progress, 48*, 197–212.

Lesicar, J.C. (n.d.). Conceptual approach to utilization of the unmanned aerial vehicles in diverse activities. *Interdisciplinary Description of Complex Systems, 2*(15), 144-148.

Leychenko, S., Malishevskiy, A., & Mikhalic, N. (2006). *Human Factors in Aviation: monograph in two books*. Kirovograd: YMEKS.

Likhachev, M., & Arkin, R. C. (2001). Spatio-Temporal Case-Based Reasoning for Behavioral Selection. *Proc. of Int. Conf. of Robot Applications (ICRA)*, 1627–1634.

Littlepage, R. S. (1998). The Impact of Interference on Civil GPS. *Proceedings ION, GPS-98*, 821–828.

Liu, Z., Li, Z., Loannis, R., Liu, B., Fu, X., & Ren, Q. (2015). Rise of Mini-Drones: Applications and Issues. *Pamco '15. Proceedings of the 2015 Workshop of Privacy-Aware Mobile Computing*, 7-12.

Liu, Y., Li, S., Fu, Q., & Liu, Z. (2018). Impact assessment of GNSS spoofing attacks on INS/GNSS integrated navigation system. *Sensors (Switzerland), 18*(5), 1433. doi:10.339018051433 PMID:29734715

Liu, Y., Yang, C., & Du, X. (2008). A CBR-based Approach for Ship Collision Avoidance. In Lecture Notes In. *Artificial Intelligence, 5027*, 687–697.

Lopez, J., Markel, M., Siddiqi, N., Gebert, G., & Evers, J. (2003). Performance of passive ranging from image flow. *Proceedings of the IEEE International Conference on Image Processing*, 1, 929–932. 10.1109/ICIP.2003.1247116

Lorenc, W., Chodnicki, M., Kordowski, P., & Nowakowski, M. (2016). Drony transportowe – nowa era transportu towarów i ludzi [Transport drones - a new era of transporting goods and people]. *Autobusy, R17*(6), 627–631.

Losev, Y.I. (1988). *Adaptivnaya compensatsiya pomekh v kanalakh svyazi* [Adaptive compensation of interferences in communication channels]. Moscow: Radio i svyaz.

Lucas, B., & Kanade, T. (1981). An Iterative Image Registration Technique with an Application to Stereo Vision. *7th International Joint Conference on Artificial Intelligence (IJCAI): materials of the international seminar*, 123-129.

Luiz, G. B., Mirisola, & Dias, J. (2009). Exploiting Attitude Sensing in Vision-Based Navigation for an Airship. Journal of Robotics.

Luo, Z., Liu, Z., & Shi, J. (2017). A Two-Echelon Cooperated Routing Problem for a Ground Vehicle and Its Carried Unmanned Aerial Vehicle. *Sensors (Basel), 17*(5), 1144–1161. doi:10.339017051144 PMID:28513552

Lyadovska, V. (2014). Methods and criteria for the identification of objects of the state's critical infrastructure. *Integrated Intelligent Robotic Complexes*, 356-358.

Lyusin, S. V., Sazonov, L. B., Khazanov, I. G., & Komarov, A. S. (1998). Combined GPS/GLONASS Receiver with High Antijamming Performance. *Proceedings ION, GPS-98*(September), 775–782.

Magiera, J., & Katulski, R. (2015). Detection and mitigation of GPS spoofing based on antenna array processing. *Journal of Applied Research and Technology, 13*(1), 45–57. doi:10.1016/S1665-6423(15)30004-3

Ma, J., Tao, X.-P., Zhu, H.-H., & Lü, J.MA. (2014). Enhancing Multi-Agent System with Context-Awareness. *Journal of Software, 23*(11), 2905–2922. doi:10.3724/SP.J.1001.2012.04301

Maksimenko, Y. (2016). *Investigation of the parameters of ground and onboard receivers using the GPS system*. Academic Press.

Manathara, J., & Ghose, D. (2011). Reactive collision avoidance of multiple realistic UAVs. *Aircraft Engineering and Aerospace Technology, 83*(6), 388–396. doi:10.1108/00022661111173261

Marquis, W. (2014). *The GPS Block IIR/IIR-M Antenna Panel Pattern Appendix B – SV-Specific Patterns, Data*. LMCO.

Martin, P. G., Kwong, S., Smith, N. T., Ymashiki, Y., Payton, O. D., Russell-Pavier, F. S., ... Scott, T. B. (2016). 3D unmanned aerial vehicle radiation mapping for assessing contaminant distribution and mobility. *International Journal of Applied Earth Observation and Geoinformation, 52*(6), 12–19. doi:10.1016/j.jag.2016.05.007

Massive-Parallel Architecture. (n.d.). Retrieved from http://www.myshared.ru/ slide / 641118 /

Mataric, M. (1995). Designing and understanding adaptive group behaviors. *Adaptive Behavior, 4*(1), 51–80. doi:10.1177/105971239500400104

Matese, A., Toscano, P., Di Gennaro, S. F., Genesio, L., Vaccani, F. P., Primicerio, J., ... Gioli, B. (2015). Intercomparison of UAV, Aircraft and Satellite Remote Sensing Platforms for Precision Viticulture. *Remote Sensing, 7*(3), 2971–2990. doi:10.3390/rs70302971

Mathews, B. D. (2015). Potential Tort Liability for Personal Use of Drone Aircraft. *St. Mary's Law Journal, 46*, 573–602.

Matias, J. (2014). *FlowNAC: Flow-Based Network Access Control*. Proc. Euro. Wksp. Software Defined Networks.

Matthews, B., Das, S., Bhaduri, K., Das, K., Martin, R., & Oza, N. (2013). Discovering anomalous aviation safety events using scalable data mining algorithms. *Journal of Aerospace Information Systems, 10*(10), 467–475. doi:10.2514/1.I010080

Mattioli, R., & Levy-Bencheton, C. (2014). Methodologies for the identification of critical information infrastructure assets and services. ENISA Report.

McLain, T., & Beard, R. (2000). Cooperative rendezvous of multiple unmanned air vehicles. *Proceedings of the AIAA Guidance, Navigation and Control Conference, (Denver, CO), AIAA 2000-4369.*

Mengmeng, D., Noboru, N., Atsushi, I., & Yukinori, S. (2017). Multi-temporal monitoring of wheat growth by using images from satellite and unmanned aerial vehicle. *International Journal of Agriculture and Biology, 10*(5), 1–14.

Merino, L., Martínez de Dios, J., & Ollero, A. (2015). Cooperative Unmanned Aerial Systems for Fire Detection, Monitoring, and Extinguishing. In *Handbook of Unmanned Aerial Vehicles* (pp. 2693–2722). Springer Netherlands. doi:10.1007/978-90-481-9707-1_74

Merkisz, J., & Nykaza, A. (2016). Perspektywy rozwoju i wykorzystania bezzałogowych statków powietrznych w służbach ratowniczych [Prospects for the development and use of unmanned aerial vehicles in emergency services]. *Autobusy, R17*(6), 291–296.

Merkisz, J., & Nykaza, A. (2016a). Zastosowanie bezzałogowych statków powietrznych w kryminalistyce rozpoznawczej i wykrywawczej [The use of unmanned aerial vehicles in forensic and detection criminalistics]. *Autobusy, R17*(6), 297–301.

Mesas-Carrascosa, F. J., Santano, D. V., Porras, F. P., Merono-Larriva, J. E., & Garcia-Ferrer, A. (2017). The Development of an Open Hardware and Software System Onboard Unmanned Aerial Vehicles to Monitor Concentrated Solar Power Plants. *Sensors (Basel)*, *17*(6), 1329–1343. doi:10.339017061329 PMID:28594353

Mettler, B., Tischler, M. B., & Kande, T. (1999). *System Identification of Small-Size Unmanned Helicopter Dynamics.* American Helicopter Society 55th Forum, Montreal, Quebec, Canada.

Michael, N., & Kumar, V. (2011). Control of ensembles of aerial robots. *Proceedings of the IEEE*, *99*(9), 1587–1602. doi:10.1109/JPROC.2011.2157275

Milenin, A., Dyya, H., Muskalsky, Z., & Pylyarchik, Ya. (2006). Simulation with the finite element method of the wire drawing process in roller coils. *Metizy*, *2*(12), 30–33.

mininet.org. (2014). *Mininet. An Instant Virtual Network on your Laptop (or other PC).* Retrieved from http://mininet.org/

Ministry of Transport and Communications of Ukraine. (2007). *The Rules for Issuing Certificates of Aviation Personnel in Ukraine*. Kyiv: Author.

Ministry of Transport of Ukraine. (2003). *About Approving the Flight Rules of Aircraft and Air Traffic Services in the Classified Airspace of Ukraine*. Kyiv: Author.

Mojsilovic, A. (2005). A computational model for color naming and describing color composition of images. *IEEE Transactions on Image Processing*, *14*(5), 690–699. doi:10.1109/TIP.2004.841201 PMID:15887562

Molodtsov, D. (1999). Soft Set Theory – first results. *Computers & Mathematics with Applications (Oxford, England)*, *37*(4-5), 19–31. doi:10.1016/S0898-1221(99)00056-5

Monastyrskyi, D. (2017, February 19). *Geotag photos. Part one.* Retrieved from https://monastyrskiy.ru/?p=3354

Monzingo, R. A., & Miller, T. W. (2004). *Introduction to adaptive arrays.* SciTech Publishing, Inc.

Moreau, M. C., Davis, E., Carpenter, J. R., Kelbel, D., Davis, G., & Axelrad, P. (2002). *Results from the GPS Flight Experiment on the High Earth Orbit AMSAT OSCAR-40 Spacecraft*. Paper presented at the ION GPS 2002 Conference, Portland, OR.

Morris, D. Z. (2015). *Why BNSF Railway is using drones to inspect thousands of miles of rail lines.* Available at: http://fortune.com/2015/05/29/bnsf-drone-program/

Moving to a Balanced Computer Architecture. (n.d.). Retrieved from http://tslab.com.ua/en/publication/category/3/50

Mozaffari, M., Saad, W., Bennis, M., Nam, Y.-H., & Debbah, M. (2018). *A Tutorial on UAVs for Wireless Networks: Applications, Challenges, and Open Problems*. Retrieved from https://arxiv.org/abs/1803.00680

Munjulury, R., Staack, I., Berry, P., & Krus, P. (2015). A knowledge-based integrated aircraft conceptual design framework. *CEAS Aeronautical Journal*, *7*(1), 95–105. doi:10.100713272-015-0174-z

Nadobnik, J. (2016). The use of drones in organizing the Olympic Games. *Handel Wewnętrzny*, *365*(6), 288–299.

NagibinP.KrylosovD. (2014). *Ambiguity SDN.* Retrieved from http://nag.ru/articles/reviews/26333/neodnoznachnost-sdn.htm

Nathan, S., Janusz, M., Milind, T., Paul, S., Lewis, J. P., & Kasinadhuni, N. (2005). *The Future of Disaster Response: Humans Working with Multiagent Teams Using DEFACTO*. Published Articles & Papers. Paper 41. Retrieved from http://research.create.usc.edu/published_papers/41

National Aeronautics and Space Administration. (2016). *NASA Goddard Space Flight Center Navigator GPS receiver.* Retrieved from: https://technology.nasa.gov/patent/Navigator_GPS_Receiver

Nechausov, A., & Zamirec, O. (2005). Metodika sozdaniya mozaiki izobrazheniy na osnove dannykh bespilotnogo letatelnogo apparata [Technique of creation of a mosaic of images on the basis of data of an unmanned aerial vehicle] [in Russian]. *System Information Boxes*, *133*, 51–56.

Nehrey, M., & Hnot, T. (2017). Using recommendation approaches for ratings matrixes in online marketing. *Studia Ekonomiczne*, *342*, 115–130.

Ng, K., & Trivedy, M. (1998). A Neuro-Fuzzy Controller for Mobile Robot Navigation and Multirobot Convoying. IEEE Trans. on Systems, Manufacture and Cybernetics, Part B. *Cybernetics*, *28*(6), 829–840. PMID:18256001

Nicolescu, M., & Mataric, M. (2001). Expirience-Based Representation Construction: Learning from Human and Robot Teachers. *Proc. of the IEEE Int. Conf. on Intelligent Robots and Sys.*, 2, 740–745.

Niyato, D. (2005). Analysis of fair scheduling and connection admission control in differentiated services wireless networks. *Proceedings of the IEEE International Conference on Communications*, 3137–3141. 10.1109/ICC.2005.1494984

Nobahari, H., & Mohammadkarimi, H. (2017). Application of model aided inertial navigation for precise altimetry of Unmanned Aerial Vehicles in ground proximity. *Aerospace Science and Technology*, *69*(7), 650–658. doi:10.1016/j.ast.2017.07.033

Non-Uniform Memory Architecture (NUMA): Investigating the subsystem memory of AMD Opteron dual-processor platforms using the RightMark Memory Analyzer. (n.d.). Retrieved from http://www.ixbt.com/cpu/rmma-numa.shtml

Official Page of the LAM-MPI Package. (n.d.). Retrieved from http://www.lam-mpi.org

Official Page of the mpich1 Package. (n.d.). Retrieved from http://www-unix.mcs.anl.gov/mpi/mpich1

Official Site of the MPI Forum. (n.d.). Retrieved from http://www.mpi-forum.org

Official Website of the Laboratory. (n.d.). Retrieved from http://www.anl.gov

Omar, T., & Nehdi, M. L. (2017). Remote sensing of concrete bridge decks using unmanned aerial vehicle infrared thermography. *Automation in Construction*, *83*(6), 360–371. doi:10.1016/j.autcon.2017.06.024

On the identification and designation of European critical infrastructures and the assessment of the need to improve their protection: Council Directive 2008/114/EC.

Ong, W. H., Chiu, W. K., Kuen, T., & Kodikara, J. (2017). Determination of the State of Strain of Large Floating Covers Using Unmanned Aerial Vehicle (UAV) Aided Photogrammetry. *Sensors (Basel)*, *17*(6), 1731–1755. doi:10.339017081731 PMID:28788081

Open Networks Foundation (ONF). (2012). *Software-Defined Networking: The New Norm for Networks, white paper.* Retrieved from https://www.opennetworking.org

Open Street Map Project. (n.d.). Retrieved from http://wiki.openstreetmap.org/wiki/Main_Page

Orlov. (2014). SDN. *Journal of Network Solutions.*

Oumar, O. A., Siyau, M. F., & Sattar, T. P. (2012). Comparison between MUSIC and ESPRIT direction of arrival estimation algorithms for wireless communication systems. *The First International Conference on Future Generation Communication Technologies*, 99-103. 10.1109/FGCT.2012.6476563

Ovcharenko, L. A., & Poddybniy, V. N. (1992). *Pomekhoustoychivost priema fazomanipulirovannykh signalov na fone naiboleye neblagopriyatnykh pomekh* [Interference immunity of receiving phase-shift keying signals with the most unfavorable interferences]. Moscow: Radiotekhnika.

Overlay Tiled Images on a Map. (n.d.). Retrieved from https://docs.microsoft.com/en-us/windows/uwp/maps-and-location/overlay-tiled-images

Ozerov, S., & Kalinichenko, A. (n.d.). Parallel computations: clusters. *Computerra: The Site of Computer Magazine*. Retrieved from http://offline.computerra.ru/2005/618/244051

Pagacz, S. & Witczuk, J. (2016). Wykorzystanie samolotów bezzałogowych i termowizji do nocnej inwentaryzacji zwierząt kopytnych. *Studia i materiały CEPL w Rogowie, 49A*(4), 50-57.

Panchal, P. M., Panchal, S. R., & Shah, S. K. (2013). A Comparison of SIFT and SURF. *International Journal of Innovative Research in Computer and Communication Engineering, 1*(2).

Paneque-Galves, J., McCall, M. K., Napoletano, B. M., Wich, S. A., & Koh, L. P. (2014). Small Drones for Community-Based Forest Monitoring: An Assessment of Their Feasibility and Potential in Tropical Areas. *Forests, 5*(6), 1481–1507. doi:10.3390/f5061481

Parish, E. J., & Duraisamy, K. (2016). A paradigm for data-driven predictive modeling using field inversion and machine learning. *Journal of Computational Physics, 305*, 758–774. doi:10.1016/j.jcp.2015.11.012

Parker, J. J. K. (2017). *Refining the GPS Space Service Volume (SSV) and Building a Multi-GNSS SSV*. Presented at the SCaN Workshop on Emerging Technologies for Autonomous Space Navigation. Retrieved from https://www.nasa.gov/sites/default/files/atoms/files/session_1_-_3_refining_the_gps_space_service_volume_and_building_a_multi-gnss_ssv_joel_parker_1_0.pdf

Patent 57663 Ukraine, IPC G06F 15/16. (2011). *The module of the high-efficiency multiprocessor system of high readiness*. National Metallurgical Academy of Ukraine.

Patriarca, R., Di Gravio, G., & Costantino, F. (2017). A Monte Carlo evolution of the Functional Resonance Analysis Method (FRAM) to assess performance variability in complex systems. *Safety Science, 91*, 49-60.

Patruno, C., Nitti, M., Stella, E., & D'Orazino, T. (2017). Helipad detection for accurate UAV pose estimation by means of a visual sensor. *International Journal of Advanced Robotic Systems, 5*, 1–15.

Paukkeri, M., García-Plaza, A., Fresno, V., Unanue, R., & Honkela, T. (2012). Learning a taxonomy from a set of text documents. *Applied Soft Computing, 12*(3), 1138–1148. doi:10.1016/j.asoc.2011.11.009

Paul, T., Krogstad, T. R., & Gravdahl, J. T. (2008). Modeling of UAV Formation Flight Using 3D Potential Field. *Simulation Modelling Practice and Theory, 16*(9), 1453–1462. doi:10.1016/j.simpat.2008.08.005

Pavlyshenko, B. M. (2016, August). Linear, machine learning and probabilistic approaches for time series analysis. In *Data Stream Mining & Processing (DSMP), IEEE First International Conference on* (pp. 377-381). IEEE.

Pawlak, Z., Jerzy, W., Slowinski, R., & Ziarko, W. (1995). Rough Sets. *Communications of the ACM, 38*(11), 88–95. doi:10.1145/219717.219791

Pedrycz, W. (2011). Introducing WIREs Data Mining and Knowledge Discovery. *Wiley Interdisciplinary Reviews. Data Mining and Knowledge Discovery, 1*(1), 1–1. doi:10.1002/widm.17

Périaux, J., Chen, H. Q., Mantel, B., Sefrioui, M., & Sui, H. T. (2001). Combining game theory and genetic algorithms with application to DDM-nozzle optimization problems. *Finite Elements in Analysis and Design, 37*(5), 417–429. doi:10.1016/S0168-874X(00)00055-X

Pietrzykovski, Z., & Chomski, J. (2003). A Navigator Decision Support System in Planning a Safe Trajectory. In *Proc. of the 4th General Assembly Int* (pp. 4–10). Alexandria, VA: Ass. of Maritime Universities IAMU.

Pietrzykowski, Z., & Uriasz, J. (2009). The Ship Domain – A Criterion of Navigational Safety Assessment in an Open Sea Area. *Journal of Navigation, 62*(01), 93–108. doi:10.1017/S0373463308005018

Pinker, D., & Walker, C., (1998). Jamming the GPS signal. *Proceedings ION-98*, 829-837.

Piotrowski, P., Witkowski, T., & Piotrowski, R. (2015). Bezzałogowa zdalnie sterowana jednostka latająca [Unmanned remote controlled flying unit. Measurements Robotics Automation]. *Pomiary Automatyka Robotyka, R19*(1), 49–55. doi:10.14313/PAR_215/49

Piskunov, O., Yurchuk, I., & Bilyanska, L. (2017). Vyznachennja oblasti bachennja kamery pry aerofotozjomci [Determination of camera vision field at aerophotography] [in Ukrainian]. *Science-Based Technologies, 35*(3), 204–208. doi:10.18372/2410-5431.35.11839

Plaza, J. (2017). *Will Hydrogen Fuel Cells Help Drones Stay in the Air?* Available at: https://www.expouav.com/news/latest/hydrogen-fuel-cells-drones/

Polymaps. (n.d.). *Image overlay*. Retrieved from http://polymaps.org/ex/overlay.html

Porras, P. (2012). A Security Enforcement Kernel for OpenFlow Networks. Proc. 1st Wksp. *Hot Topics in Software Defined Networks, 2012*, 121–126.

Project Ryptide - A life Saving Drone Accessory. (2017). Available at: https://www.kickstarter.com/projects/flying-robots/project-ryptide-a-life-saving-drone-accessory

Prystavka, P., Chyrkov, A. & Sorokopud V. (2017) Eksperymentaljnyj zrazok avtomatyzovanoji systemy poshuku pidozrilykh ob'jektiv na video z bezpilotnogho povitrjanogho sudna [Experimental sample of automated search system for suspicious objects on unmanned aircraft video]. *Arms Systems and Military Equipment, 20*(2), 26–32.

Psiaki, M. L., & Humphreys, T. E. (2016). GNSS Spoofing and Detection. *Proceedings of the IEEE, 104*(6), 1258–1270. doi:10.1109/JPROC.2016.2526658

Qin, Dai, Huang, & Xu. (2017). Bandwidth-Aware Scheduling With SDN in Hadoop: A New Trend for Big Data. *IEEE Systems Journal, 11*(4).

Qin, C., Wu, X., Jiang, J., & Zhu, A. (2016). Case-based knowledge formalization and reasoning method for digital terrain analysis – application to extracting drainage networks. *Hydrology and Earth System Sciences, 20*(8), 3379–3392. doi:10.5194/hess-20-3379-2016

Radio Technical Commission for Aeronautics. (1998). *Minimum Aviation Performance Standards for the Automatic Dependent Surveillance Broadcast (ADS-B), RTCA DO-247*. Author.

Radio Technical Commission for Aeronautics. (2008). *Assessment of radio frequency interference relevant to the GNSS L1 frequency band RTCA/DO-235B*. Author.

Rahman, M. M., McDermid, G. J., Strack, M., & Lovitt, J. (2017). A New Method to Map Groundwater Table on Peatlands Using Unmanned Aerial Vehicles. *Remote Sensing, 9*(5), 1057–1071. doi:10.3390/rs9101057

Rajkumar, B. (1999). *High-Performance Cluster Computing* (Vol. 1). Architectures and Systems.

Rao, B., Gopi, A. G., & Maione, R. (2016). The social impact of commercial drones. *Technology in Society, 45*(5), 83–90. doi:10.1016/j.techsoc.2016.02.009

Ratynskiy, M.V. (2003). *Adaptatsiya i sverkhrazresheniye v antennikh reshetkakh* [Adaptation and superresolution in antenna arrays]. Moscow: Radio i svyaz.

Rausand, M. (2004). *System reliability theory: models, statistical methods and applications*. New York: John Wiley & Sons, Inc.

Red Hat Enterprise Linux 4 System Administration Guide. (n.d.). Retrieved from http://www.rhd.ru/docs/manuals/enterprise/RHEL-4-Manual/sysadmin-guide/ch-diskless.html

Resolution on approval of the procedure for the formation of the list of information and telecommunication systems of critical infrastructure objects of the state. Resolution of the Cabinet of Ministers of Ukraine of 23.08.2016, N°563.

Review of the new platform “1C Enterprise 8.3”. (n.d.). Retrieved from http://erp–project.com.ua/index.phpuk/novini/item/206–obzor–novoj–platformy–1s–predpriyatie–8–3

Rey, N., Volpi, M., Joost, S., & Tuia, D. (2017). Detecting animals in African Savanna with UAVs and the crowds. *Remote Sensing of Environment, 200*(8), 341–351. doi:10.1016/j.rse.2017.08.026

Reynolds, M. (2017). AI coach helps chatbots seem more human. *New Scientist, 235*(3135), 14. doi:10.1016/S0262-4079(17)31407-0

Rhudy, M., Yu, G., Haiyang, C., & Gross, J. (2015). Unmanned Aerial Vehicle Navigation Using Wide-Field Optical Flow and Inertial Sensors. Journal of Robotics.

Ricci, F., Rokach, L., & Shapira, B. (2015). Recommender systems: introduction and challenges. In Recommender systems handbook (pp. 1-34). Springer US. doi:10.1007/978-1-4899-7637-6_1

Rizun, N., & Shmelova, T. (2017). Decision-Making Models of the Human-Operator as an Element of the Socio-Technical Systems. In *Strategic Imperatives and Core Competencies in the Era of Robotics and Artificial Intelligence* (pp. 167-204). IGI Global. Retrieved from http://rpubs.com/tarashnot/orders_full

Robakowska, M., Tyranska-Fobke, A., Nowak, J., Slezak, D., Zuratynski, P., Robakowski, P., ... Ladny, J. R. (2017). The use of drones during mass events. *Disaster and Emergency Medicine Journal, 2*(3), 129–134.

Rodgers, C. (1991). Development of a Low Cost PC Controlled GPS Satellite Signal Simulator. In *Proceedings of the 15th Biennial Guidance Test Symposium.* Holloman AFB.

Rohde, S., & Wietfeld, C. (2012). Interference Aware Positioning of Aerial Relays for Cell Overload and Outage Compensation. *Proceedings IEEE Vehicular Technology Conference (VTC Fall)*, 1 – 5. 10.1109/VTCFall.2012.6399121

Roosjen, P. P. J., Suomalainen, J. M., Bartholomeus, H. M., Kooistra, L., & Clevers, J. G. P. W. (2017). Mapping Reflectance Anisotropy of a Potato Canopy Using Aerial Images Acquired with an Unmanned Aerial Vehicle. *Remote Sensing, 9*(5), 417–440. doi:10.3390/rs9050417

Roosta, S. H. (2000). *Parallel Processing and Parallel Algorithms: Theory and Computation*. Springer-Verlag. doi:10.1007/978-1-4612-1220-1

Ros, R., López De Màntaras, R., Sierra, C., & Arcos, J. L. (2005). A CBR system for autonomous robot navigation. In *Proc. of the 2005 Conf. on Artificial Intelligence Research and Development*, 131, 299–306.

Rügamer, A., & Kowalewski, D. (2015). *Jamming and Spoofing of GNSS Signals – An Underestimated Risk?!* Academic Press.

Rus, C., & Patrascoiu, N. (2016). Technical and legal aspects on the use of drones. *Annals of the University of Petroscani, 18*(6), 69–78.

Russell, S. J., & Norvig, P. (2016). *Artificial intelligence: a modern approach.* Pearson Education Limited.

Russian Institute of Space Device Engineering. (2008). *GLONASS Interface Control Document*, Edition 5.1. Author.

Sałaciński, T. (2015). *Pojazdy bezzałogowe - nowe wyzwanie dla materiałów wybuchowych. Przegląd.* Available at: http://yadda.icm.edu.pl/yadda/element/bwmeta1.element.baztech-244768f7-5471-4098-b4b9-298a2916e671/c/Salacinski1_t_6.pdf

Sandbrook, C. (2015). The social implications of using drones for biodiversity conservation. *Ambio, 44*(4), 636–647. doi:10.100713280-015-0714-0 PMID:26508350

Sandita, A., & Popirlan, C. (2015). Developing A Multi-Agent System in JADE for Information Management in Educational Competence Domains. *Procedia Economics and Finance, 23*, 478–486. doi:10.1016/S2212-5671(15)00404-9

Saripalli, S., Sukhatme, G. S., Mejias, L. O., & Cervera, P. C. (2005, April). Detection and tracking of external features in an urban environment using an autonomous helicopter. In *Proceedings of the 2005 IEEE International Conference on Robotics and Automation* (pp. 3972-3977). IEEE. 10.1109/ROBOT.2005.1570728

Saulnier, A., & Thompson, S. N. (2017). Police UAV use: Institutional realities and public perceptions. *Policing: An International Journal of Police Strategies & Management, 39*(4), 680–693. doi:10.1108/PIJPSM-11-2015-0136

Schmid, S., & Suomela, J. (2013). Exploiting locality in distributed sdn control. *Proceedings of the second ACM SIGCOMM workshop on Hot topics in software defined networking*, 121–126. 10.1145/2491185.2491198

Semenov, Y. A. (2014). *Network technology OpenFlow.* Moskow, Russia: SDN.

Seneff, S. (2002). Response planning and generation in the mercury flight reservation system. *Computer Speech & Language, 16*(3-4), 283–312. doi:10.1016/S0885-2308(02)00011-6

Sensor Design Application NoteL. D. C. (2015). Retrieved from: http://www.ti.com/lit/snoa930.pdf

Sensor products. Inductive sensing redefined. (2015). Texas Instruments. Retrieved from: http://www.ti.com/lsds/ti/sensors/inductive-sensing-multi-channel-ldc.page?DCMP=multichldcs&HQS=tlead-sensing-sva-psp-ssp-multichldcs-vanity-lp-en

Seok-Wun, H., & Yong-Ho, M. (2011). Multiple Object Tracking Using SIFT Features and Location Matching. *International Journal of Smart Home, 5*(4), 17–26.

Sharma, B., Vanualailai, J., & Chand, U. (2009). Flocking of Multi-agents in Constrained Environments. *Europen J. of Pure and Appl. Math., 2*(3), 401–425.

Shershakov, V., Trahtengerts, E., & Kamaev, D. (2015). *Network-centric methods of computer support for the management of emergency response.* LENAND.

Sherstjuk, V. (2013). The case-scenario approach to control the dynamic objects. Pressing issues and priorities in development of the scientific and technological complex, 17, 97–103.

Sherstjuk, V. (2015). Scenario-Case Coordinated Control of Heterogeneous Ensembles of Unmanned Aerial Vehicles. *Proc. of 2015 IEEE 3rd Int. Conf. Actual Problems of Unmanned Aerial Vehicles Developments*, 275–279. 10.1109/APUAVD.2015.7346620

Sherstjuk, V., Zharikova, M., & Sokol, I. (2018). Forest Fire-Fighting Monitoring System Based on UAV team and Remote Sensing. *Proc. of 2018 IEEE 38th Int. Conf. on Electronics and Nanotechnology (ELNANO-2018)*, 99-106. 10.1109/ELNANO.2018.8477527

Sherstjuk, V., Zharikova, M., Sokol, I., & Tarasenko, E. (2017). Onboard Joint Motion Coordination System for Heterogeneous Ensembles of Unmanned Vehicles. *Proc. of 2017 IEEE 37th Int. Conf. on Electronics and Nanotechnology (ELNANO-2017)*, 480-484.

Shevchuk, D. (2010). Fiber-optic intelligence systems for diagnostics contraction integrity of aircrafts. *Proceedings Of National Aviation University*, *42*(1). doi:10.18372/2306-1472.42.1816

Shilin, W., Jianli, S., Xiongkui, H., Le, S., Xiaonan, W., Changling, W., ... Yun, L. (2017). Performances evaluation of four typical unmanned aerial vehicles used for pesticide application in China. *International Journal of Agriculture and Biology*, *10*(4), 22–31.

Shim, D., Chung, H., Kim, H. J., & Sastry, S. (2005, August). Autonomous exploration in unknown urban environments for unmanned aerial vehicles. In *AIAA Guidance* (p. 6478). Navigation, and Control Conference and Exhibit. doi:10.2514/6.2005-6478

Shin, J., & Kim, H. J. (2009). Nonlinear model predictive formation flight. *IEEE Transactions on Systems, Man, and Cybernetics. Part A, Systems and Humans*, *39*(5), 1116–1125. doi:10.1109/TSMCA.2009.2021935

Shmelova, T., Sikirda, Yu., Zemlyanskiy, A., & Danilenko, O. (2015). Fuzzy Assessment of the Situation's Complexity during Simulation Training of Specialists in Air Traffic Services. In *Materials of International Scientific-Practical Conference "Problems of Energy Efficiency and automation in industry and agriculture"*. Kirovohrad: Kirovohrad National Technical University.

Shmelova, T., Lazorenko, V., & Bilko, A. (2014). Estimation of Pre-Simulating Training Tasks Complexity. *Proceedings of the National Aviation University, No.*, *1*(62), 17–22.

Shmelova, T., Sikirda, Yu., Zemlyanskiy, A., Danilenko, O., & Lazorenko, V. (2016). Artificial Neural Network for Air Traffic Controller's Pre-Simulator Training. *Proceedings of the National Aviation University, No.*, *3*(68), 13–23.

Shmelova, T., Yakunina, I., Moyseenko, V., & Grinchuk, M. (2014). *A. r. Computer Program "Network Analysis of Flight Emergency"*. Kyiv: State Intellectual Property Service of Ukraine.

Shmelova, T., Yakunina, I., & Yakunin, R. (2013). *A. r. Computer Program "Test Extraordinary Incident"*. Kyiv: State Intellectual Property Service of Ukraine.

Shoro, A. G., & Soomro, T. R. (2015). *Big data analysis: Apache spark perspective. Global Journal of Computer Science and Technology.*

Shvachych, G. G. (2007). On parallel computer technologies of cluster type solutions of multidimensional non-stationary problems. *Vědeckэ potenciбl světa - materiбly IV mezinбrodnн vědecko-praktickб conference. Technickův̆ědy. Mathematics Fyzika. Moderni informaun technologie. Vэstavbaaarchitektura.* Prague: Publishing House "Education and Science" s.r.o.

Shvachych, G. G., & Shmukin, A. A. (2014). Features of the design of parallel computing algorithms for a computer in problems of heat and mass transfer. *East-European Magazine of Advanced Technologies, 3,* 42–47.

Shvachych, G. G., Kholod, E. G., Ivaschenko, E. V., & Busygin, V. V. (2018). Visualization of the decisions of applied problems in multiprocessor system. Natural and Technical Sciences, 1(17), 65–69.

Shvachych, G. G., Kolpak, V. P., & Sobolenko, M. A. (2007). Mathematic simulation of the speed modes of thermo processing of the long products. *Theory and Practice of the Metallurgy, 4–5*(59–60), 61–67.

Shvachych, G. G., Moroz, B. I., Pobochii, I. A., Ivaschenko, E. V., & Busygin, V. V. (2018). Maximally parallel forms of distributed simulation of the dynamic system. World Science, 4(32), 12–20.

Shvachych, G. G., Sobolenko, A. V., & Sobolenko, M. A. (2003). Research of the temperature regime of operation of annealing furnaces of mine type of wires. *Theory and Practice of the Metallurgy, 1,* 59–62.

Shvachych, G. G., Tkach, M. A., & Shcherbyna, P. A. (2008). Supercomputers and high-efficient computing. *Proceedings from B″deshcheto problemyte na svetovnata nauka: materyaly za 4-a mezhdunarodna praktychna konferentsyya. V. 21. S″vremenyy tekhnolohyy na ynformatsyy,* 22–27.

Shvachych, G. G. (2006). On the design of parallel computing in the simulation of environmental identification problems. *Mathematical Simulation, 2*(14), 23–34.

Shvachych, G. G., Kolpak, V. P., & Sobolenko, M. A. (2007). Mathematical simulation of thermal processing high-speed modes of long-length articles. *Theory and Practice of Metallurgy: State Scientific and Technical Journal, 4-5*(59-60), 61–67.

Shvachych, G. G., & Shmukin, A. A. (2014). On the concept of unbounded parallelism in the problems of thermal conductivity. *East-European Journal of Advanced Technologies, 3*(9), 81–84.

Shvachych, G. G., Sobolenko, O. V., & Ivashchenko, O. V. (2017). Network Interface link aggregation of multiprocessor computing systems. *Strategy of Quality in Industry and Education: Proceedings from XII International Conference,* 236–241.

Shvets V.A. (2013). *Formuvannia vahovykh koefitsiientiv priamymy metodamy v antennykh reshitkakh system GPS* [Formation of weight coefficients by direct methods in antenna arrays of GPS systems]. Kyiv: Visnyk inzhenernoyi akademiyi Ukrayiny.

Shvets, V. A., & Kharchenko, V. P. (2017). Pulse characteristics of network satellite systems adaptive antenna for assessing correlation interference matrix. *Proceedings of the National Aviation University, 4*(73), 30 – 35. Retrieved from: http://er.nau.edu.ua:8080/handle/NAU/32567

Shvets, V.A. (2012). *Eksperymentalni doslidzhennya zavadostiykosti system GPS* [Experimental studies of impedance of GPS systems]. Kyiv: Visnyk inzhenernoyi akademiyi Ukrayiny.

Shvets, V.A., Kasperovych, N.O., & Ostrovskiy, Y.P. (2010). Certificate about copyright on work No 36059 from 14.12.2010. Computer program. *Modelyuvannya zavad suputnykovykh aeronavigatsiynykh system iz vyznachenoyu corelyatsiynoyu funktsiyeyu* [Modeling the interferences to satellite aeronavigation systems with defined correlation function].

Shvets, V.A., Kasperovych, N.O., Ostrovskiy, Y.P. (2011). Certificate about copyright on work No 36677 from 25.01.2011. Computer program. *Modelyuvannya zavad suputnykovykh aeronavigatsiynykh system* [Modeling the interferences to satellite aeronavigation systems].

Shvets, V.A., Shvets, O.V. (2013). *Formuvannya vahovykh koefitsiyentiv bez opornoho syhnalu v adaptyvnykh antennykh reshitkakh system GPS, HALILEO, GLONASS* [Formation of weight factors without reference signal in adaptive antenna arrays of GPS, GALILEO, GLONASS systems]. Kyiv: Visnyk inzhenernoyi akademiyi Ukrayiny.

Skoldstrom, P. (2015). *Towards Unified Programmability of Cloud and Carrier Networks.* Proc. Euro. Wksp. Software Defined Networks.

Sladek, J. (2013). Drones and their use in geovedical research. *Nové trendy v geovedách, 088UK(4),* 1-19.

Sładkowski, A. (Ed.). (2013). *Some Actual Issues of Traffic and Vehicle Safety. Faculty of Transport.* Silesian University of Technology.

Smaczyński, M. (2015). Wizualizacja dynamiki zmian liczby uczestników imprezy masowej z wykorzystaniem dronów. *Badania Fizjograficzne - Geografia Fizyczna, 66*(6), 157-172.

Smelyansky, R. (2012). *Program-configurable networks.* Ruslan Smelyansky.

Smith, D. J. (2005). *Reliability, Maintainability and Risk. Practical methods for engineers.* London: Elsevier.

Smith, K. W. (2015). Drone Technology: Benefits, Risks, and Legal Considerations. *Seattle Journal of Environmental Law, 5*(1), 290–302.

Snovidov, I. (2015). Factors affecting measurement accuracy. *Bulletin of the State Hydrography, 1*(48).

Soleymani, T., & Saghafi, F. (2010). Behavior-Based Acceleration Commanded Formation Flight Control. *International Conference on Control, Automation and Systems.*

Solomentsev, O. V., Melkumyan, V. H., Zaliskyi, M. Yu., & Asanov, M. M. (2015). UAV operation system designing. In *Proceedings of IEEE 3rd International Conference on Actual Problems of Unmanned Air Vehicles Developments (APUAVD 2015)* (pp. 95 – 98). Kyiv, Ukraine: IEEE. 10.1109/APUAVD.2015.7346570

Solomentsev, O., Zaliskyi, M., Kozhokhina, O., & Herasymenko, T. (2017). Reliability Parameters Estimation for Radioelectronic Equipment in Case of Change-point. In *Proceedings of Signal Processing Symposium 2017 (SPSympo 2017).* (pp. 1 – 4). Jachranka Village, Poland: Academic Press.

Solomentsev, O., Zaliskyi, M., Nemyrovets, Yu., & Asanov, M. (2015). Signal processing in case of radio equipment technical state deterioration. In *Proceedings of Signal Processing Symposium 2015 (SPS 2015).* (pp. 1 – 5). Debe, Poland: Academic Press. 10.1109/SPS.2015.7168312

Solomentsev, O., Zaliskyi, M., & Zuiev, O. (2013). Radioelectronic equipment availability factor models. In *Proceedings of Signal Processing Symposium 2013 (SPS 2013)* (pp. 1 – 4). Jachranka Village, Poland: Academic Press.

Solomentsev, O., Zaliskyi, M., & Zuiev, O. (2016). Estimation of quality parameters in the radio flight support operational system. *Aviation., 20*(3), 123–128. doi:10.3846/16487788.2016.1227541

Soloviev, Y. (2000). *Satellite navigation systems.* Eco-Trendz.

Sosnowicz, K. (2016). Dronem do wypadku [With drone to an accident]. *Geodeta Magazyn Informacyjny, 257*(10), 44–47.

Sotnik, M. (2003). Rabota s rastrom na nizkom urovne dlya nachinayushchikh [Working with a raster at a low level for beginners]. Retrieved from http:// habrahabr.ru/post/196578/ (in Russian)

Speed Control of a Train using Fuzzy Logic. (2017). *International Journal of Science and Research, 6*(6), 781-784. doi:10.21275/art20174396

Stallings, W. (2002). *High-Speed Networks and Internets. Performance and Quality of Service.* Prentice Hall PTR.

Statistics of the World's the Largest Aircraft Accidents for the Years 1974-2014. (2018). Foringshurer Insurance. Retrieved from https://forinsurer.com/public/17/01/10/3824

Stevens, L. B., & Lewis, L. F. (2003). *Aircraft Control and Simulation* (2nd ed.). Hoboken, NJ: John Wiley & Sons, Inc.

Stislavsky, A. (2009). Building a methodology for ensuring transport security based on categorization. *Bulletin of the National University of Water Management and Nature Management, 3*(47), 155-165.

Strzelczyk, P., & Macek-Kamińska, K. (2015). Kontroler lotu dla bezzałogowych obiektów latających [Flight controller for unmanned aerial vehicles. Measurements Robotics Automation]. *Pomiary Automatyka Robotyka, R19*(4), 69–73. doi:10.14313/PAR_218/69

Stull, R. (2015). *Practical Meteorology*. Retrieved April 19, 2018, from: https://www.eoas.ubc.ca/books/Practical_Meteorology/prmet/PracticalMet_WholeBook-v1_00b.pdf

Sumner, R. W., & Popovic, J. (2004). Deformation transfer for triangle meshes. *ACM Transactions on Graphics, 23*(3), 399–405. doi:10.1145/1015706.1015736

Surovka, D., Pertile, E., Dombek, V., Vastyl, M., & Leher, V. Monitoring of Thermal and Gas Activities in Mining Dump Hedvika, Czech Republic. *IOP Conference Series: Earth and Environmental Science*, *92*(1), 1-6. 10.1088/1755-1315/92/1/012060

Sushych, O.P. (2012). *Eksperymentalna otsinka vplyvu navmysnykh zavad na aparaturu spozhyvacha hlobalnoi navihatsiinoi suputnykovoi systemy* [Experimental assessment of the effect of deliberate interference on the equipment of the consumer of the global navigation satellite system]. Kyiv: Visnyk inzhenernoyi akademiyi Ukrayiny.

Suzuki, K. (2013). *Artificial Neural Networks – Architectures and Applications*. InTech. doi:10.5772/3409

Svaton, M. (2016). *Low-cost implementation of Differential GPS using Arduino* (PhD thesis). Prague: Czech Technical University in Prague.

Sydorenko, V., Aleksander, M., & Nagi, A. (2016). Modern approaches to the definition and identification of critical information infrastructure. *Current issues of cyber security and information security*, 134-137.

Sydorenko, V., Zhmurko, T., Polishchuk, Yu., & Gnatyuk, S. (2017). *Data model for forming critical infrastructure objects and determining its connectivity. In Inzynier XXI wieku* (pp. 329–350). Monografia, Poland: ATH.

Tate, A. (2016). Miley Cyrus and the Attack of the Drones: The Right of Publicity and Tabloid Use of Unmanned Aerial Vehicles. *Texas Review of Entertainment and Sports Law*, *17*(1), 73–99.

Tavallaee, M., Bagheri, E., Lu, W., & Ghorbani, A. (2009). A detailed analysis of the KDD CUP 99 data set. *Second IEEE Symposium on Computational Intelligence for Security and Defense Applications (CISDA)*. 10.1109/CISDA.2009.5356528

Taylor, S. J., & Letham, B. (2018). Forecasting at scale. *The American Statistician, 72*(1), 37–45. doi:10.1080/00031305.2017.1380080

The Blackford MultiCore cluster specification: Institute of systems dynamics and control theory of the SB RAS. (n.d.). Retrieved from http://www.mvs.icc.ru/cluster_info.html

The Commission. (2001). Report of the Commission to Assess United States National Security Space Management and Organization.

Thermodiagnostics in the power engineering sector. (2017). Available at: https://www.drone-thermal-camera.com/drone-uav-thermography-inspection-highvoltage/

Thiels, C. A., Aho, J. M., Zietlow, S. P., & Jenkins, D. H. (2015). Use of Unmanned Aerial Vehicles for Medical Product Transport. *Air Medical Journal, 34*(2), 104–108. doi:10.1016/j.amj.2014.10.011 PMID:25733117

Three ways that drones could change the railway industry. (2017). Available at: http://railwayinnovation.com/3-ways-that-drones-could-change-the-railway-industry/

Toner, J., & Yuhai, T. (1998). Flocks, herds, and schools: A quantitative theory of flocking. *Physical Review. E, 58*(4), 4828–4858. doi:10.1103/PhysRevE.58.4828

Tootoonchian, A., & Ganjali, Y. (2010). HyperFlow: A Distributed Control Plane for Open-Flow. *Proc. 2010 Internet Network Management Conf. Research on Enterprise Networking*.

Tootoonchian, A., & Ganjali, Y. (2010). Hyperflow: a distributed control plane for OpenFlow. *Proceedings of the 2010 internet network management conference on Research on enterprise networking*, 3–3.

Torenbeek, E. (2000). Aircraft design education in Europe. *Aircraft Design*, *3*(4), 205–206. doi:10.1016/S1369-8869(00)00022-7

Torrero, L., Seoli, L., Molino, A., Giordan, D., Manconi, A., Allasia, P., & Baldo, M. (2015). The Use of Micro-UAV to Monitor Active Landslide Scenarios. *Engineering Geology for Society and Territory*, *5*, 701–704.

Tošić, P., & Vilalta, R. (2010a). A Unified Framework for Reinforcement Learning, Co-Learning and Meta-Learning How to Coordinate in Collaborative Multi-Agent Systems. *Procedia Computer Science*, *1*(1), 2217–2226. doi:10.1016/j.procs.2010.04.248

Tošić, P., & Vilalta, R. (2010b). Learning and Meta-Learning for Coordination of Autonomous Unmanned Vehicles: A Preliminary Analysis. *Proc. of 19th European Conf. on Artificial Intelligence ECAI-2010*, 163–168.

Tripicchio, P., Satler, M., Stefanini, G., Dabisias, G., Ruffaldi, E., & Avizzano, C. A. (2015). Towards Smart Farming and Sustainable Agriculture with Drones. *Computers & Society*, *29*, 140–143.

Tsigichko, V., Smolyan, G. & Chereshkin, D. (2006). *Ensuring the safety of critical infrastructures in the USA*. Academic Press.

Tunstel, E., de Oliveira, M., & Berman, S. (2002). Fuzzy Behaviour Hierarchies for Multi-Robot Control. *International Journal of Intelligent Systems*, *17*(5), 449–470. doi:10.1002/int.10032

Tuśnio, N., Tuśnio, J., Wolny, P., & Nowak, A. (2016). Bezzałogowe statki powietrzne w działaniach Państwowej Straży Pożarnej - propozycja dedykowana Państwowej Straży Pożarnej [Unmanned aerial vehicles in the activities of the State Fire Service - a proposal dedicated to the State Fire Service]. *Zeszyty Naukowe SGPS*, *58*(2), 105–122.

Tuśnio, N., & Wolny, P. (2017). Nowoczesne narzędzia i sprzęt wykorzystywane do poszukiwań osób zaginionych [Modern tools and equipment used to search for missing persons]. *Zeszyty Naukowe SGPS*, *61*(1), 7–23.

Upadhyay, A., & Kashyap, R. (2016). Fast Segmentation Methods for Medical Images. *International Journal of Computers and Applications*, *156*(3), 18–23. doi:10.5120/ijca2016912399

Upadhyay, A., & Kashyap, R. (2017). Robust hybrid energy based method for accurate object boundary detection. *Journal of Advanced Research in Dynamical and Control Systems*, *1*(06), 1476–1490.

US Department of Defense. (1999). *US Department of Transportation*. Federal Radionavigation Plan.

US Department of Defense. (2000a). *GPS IIF Operational Requirements Document*. ORD.

US Department of Defense. (2000b). *GPS III Capability Development Document*. CDD.

US Department of Defense. (2008a). *Global Positioning System Standard Positioning Service Performance Standard (Ed.4)*. Retrieved from https://www.gps.gov/technical/ps/2008-SPS-performance-standard.pdf

US Department of Defense. (2008b). *SS-SYS-800 GPS III System Specification*. Author.

US Department of Defense. (2008c). *2008 Federal Radionavigation Plan*. Retrieved from https://www.navcen.uscg.gov/pdf/2008_Federal_Radionavigation_Plan.pdf

US Department of Defense. (2017). *2017 Federal Radionavigation Plan*. Retrieved from: https://www.navcen.uscg.gov/pdf/FederalRadioNavigationPlan2017.pdf

US military develops pilotless helicopter. (2012). *Physics Today*. doi:10.1063/pt.5.026359

Vahidov, R., & Kersten, G. (2004). Decision station: Situating decision support systems. *Decision Support Systems, 38*(2), 283–303. doi:10.1016/S0167-9236(03)00099-X

Ventcel, E. S. (1988). *Issledovanie operacii: zadachi, principi, metodologia. Moscow.* Nauka.

Villa, T. F., Jayaratne, E. R., Gonzalez, L. F., & Morawska, L. (2017). Determination of the vertical profile of participle number concentration adjacent to a motorway using an unmanned aerial vehicle. *Environmental Pollution, 230*(6), 134–142. doi:10.1016/j.envpol.2017.06.033 PMID:28649041

Voevodin, V. V., & Zhumaty, S. A. (2007). Computation and cluster systems. Moscow: Publishing House of Moscow University.

Voevodyn, V. V., & Zhumatyy, S. A. (2007). Computing and cluster systems. Moscow: Academic Press.

Waldmann, J. (2002). Line-of-sight rate estimation and linearizing control by an imaging seeker in a tactical missile guided by proportional navigation. *IEEE Transactions on Control Systems Technology, 10*(4), 556–567. doi:10.1109/TCST.2002.1014675

Wallis, S. (1999). GPS Open Air Testing – Jamming at Woomera. *Proceedings of 1999 Technical Meeting & 19th Biennial Guidance Test Symposium.*

Wanderer, J. (2014). *Case Study: The Google SDN WAN.* Retrieved from http://www.computing.co.uk/ctg/analysis/2235886/case-study-the-google-sdn-wan

Wang, L., Chen, F., & Yin, H. (2016). Detecting and tracking vehicles in traffic by unmanned aerial vehicles. *Automation in Construction, 72*(5), 294–308. doi:10.1016/j.autcon.2016.05.008

Wang, R., Xu, X., Jia, S., Ma, R., Ran, L., Deng, Z., ... Ma, Z. (2017). Lower tropospheric distributions of O3 and aerosol over Raoyang a rural site in the North China Plain. *Atmospheric Chemistry and Physics, 17*(3), 3891–3903. doi:10.5194/acp-17-3891-2017

Waoo, N., Kashyap, R., & Jaiswal, A. (2010). DNA nano array analysis using hierarchical quality threshold clustering. In *2010 2nd IEEE International Conference on Information Management and Engineering*. IEEE.

Ward, P. (1994). GPS Receiver RF Interference Monitoring, Mitigation and Analysis Technique. *Journal of the Institute of Navigation, 41*(4).

Waslander, S. (2013). Unmanned Aerial and Ground Vehicle Teams: Recent Work and Open Problems. Autonomous Control Systems and Vehicles, Intelligent Systems, Control and Automation: Science and Engineering, 65, 21–36.

Widrow, B., & Stearns, S. (1985). *Adaptive signal processing.* Englewood Cliffs, NJ: Prentice-Hall.

Wildemeersch, M. (2010). *Impact Study of Unintentional Interference on GNSS Receivers.* Tech. Rep. EUR-24742-EN. Publications Office of the European Union. doi:10.2788/57794

Wilson, R. L. (2014). Ethical Issues with Use of Drone Aircraft. *IEEE Explore, 56,* 1–4.

Winer, B. (1996). GPS Receiver Laboratory RFI Tests. *Proceedings of the Institute of Navigation National Technical Meeting.*

Wintemitz, L., Moreau, M., Boegner, G., & Sirotzky, S. (2004). *Navigator GPS Receiver for Fast Acquisition and Weak Signal Space Applications.* Retrieved from: https://ntrs.nasa.gov/archive/nasa/casi.ntrs.nasa.gov/20040171175.pdf

Wittenberg, H. (2001). Aircraft performance, theory and practice. *Aircraft Design, 4*(1), 79–80. doi:10.1016/S1369-8869(00)00026-4

Woytcehovskii, Ya. (1977). *Distance control of models. Manual for draftsmen and radio-amateur*. Moscow, USSR: Svyaz. (in Russian)

Wyszywacz, W. (2016). *Drony, budowa, loty, przepisy*. Łódź: Wydawnictwo Poligraf.

Xie, H., Tsou, T., Lopez, D., Yin, H., & Gurbani, V. (2012). *Use cases for alto with software defined networks*. IETF Internet-Draft.

Xiong, J., Yu, G., & Zhang, X. (2017). Research on Governance Structure of Big Data of Civil Aviation. *Journal of Computer and Communications*, *5*(05), 112–118. doi:10.4236/jcc.2017.55009

Yeganeh, H., Tootoonchian, A., & Ganjali, Y. (2013). On scalability of software-defined networking. *Communications Magazine, IEEE*, *51*(2), 136–141. doi:10.1109/MCOM.2013.6461198

Yoon, H., Hoskere, V., Park, J., & Spencer, B. Jr. (2017). Cross-Correlation-Based Structural System Identification Using Unmanned Aerial Vehicles. *Sensors (Basel)*, *17*(9), 1–12. doi:10.339017092075 PMID:28891985

Yoon, H., & Shahabi, C. (2009). Accurate Discovery of Valid Convoys from Moving Object Trajectories. *IEEE Int. Conf. on Data Mining (ICDMW'09)*, 636–643. 10.1109/ICDMW.2009.71

Yuan, C., Zhang, Y., & Liu, Z. (2015). A Survey on Technologies for Automatic Forest Fire Monitoring, Detection and Fighting Using UAVs and Remote Sensing Techniques. *Canadian Journal of Forest Research*, *45*(7), 783–792. doi:10.1139/cjfr-2014-0347

Yue, J., Yang, G., Li, C., Li, Z., Wang, Y., Feng, H., & Xu, B. (2017). Estimation of Winter Wheat Above-Ground Biomass Using Unmanned Aerial Vehicle-Based Snapshot Hyperspectral Sensor and Crop Height Improved Models. *Remote Sensing*, *9*(5), 708–727. doi:10.3390/rs9070708

Zade, L. (1965). Fuzzy sets. *Information and Control*, *8*(3), 338–353. doi:10.1016/S0019-9958(65)90241-X

Zahorski, T., & Żurek, Ł. (2016). Wykorzystanie druku przestrzennego w budowie bezzałogowego statku powietrznego [The use of 3D printing in the construction of unmanned aircraft]. *Autobusy*, *R17*(12), 1495–1499.

Zak, B. (2004). The problems of collision avoidance at sea in the formulation of complex motion principles. *International Journal of Applied Mathematics and Computer Science*, *14*, 503–514.

Zhan, X., & Jing, S. (2014). *Space Service Volume (SSV). Characteristics of BDS*. Retrieved from: http://www.unoosa.org/pdf/icg/2014/wg/wgb03.pdf

Zhang, L., Li, Y., & Nevatia, R. (2008). Global Data Association for Multi-Object Tracking Using Network Flows. *Proceedings IEEE Conference on Computer Vision and Pattern Recognition*, 1 – 8. 10.1109/CVPR.2008.4587584

Zhang, N., & Zhu, J. (2014). Fuzzy Analytic Hierarchy Process Method for Civil Aviation Airport Security Information Management. *Applied Mechanics and Materials*, *701-702*, 40–43. . doi:10.4028/www.scientific.net/AMM.701-702.40

Zhang, T., Li, P., & Wang, S. (2012). A personal air distribution system with air terminals embedded in chair armrests on commercial airplanes. *Building and Environment*, *47*, 89–99. doi:10.1016/j.buildenv.2011.04.035

Zhao, J., Wu, Z., & Wang, F. (1993). Comments of ship domains. *Journal of Navigation*, *46*(03), 422–436. doi:10.1017/S0373463300011875

Zharikova, M., & Sherstjuk, V. (2016). Case-based approach to intelligent safety domains assessment for joint motion of vehicles ensembles. *Proc. of the 4th International Conference on Methods and Systems of Navigation and Motion Control (MSNMC)*, 245-250. 10.1109/MSNMC.2016.7783153

Zheltov S., Veremeenko K., Kim N., Kozorez D. & Krasilshchikov N. (2009). *Sovremennye informatsionnye tekhnologii v zadachakh navigatsii i navedeniya bespilotnykh manevrennykh letatelnykh apparatov* [Modern information technologies in the tasks of navigation and guidance of unmanned maneuverable aircraft]. Moskow: House of Physical and Mathematical Literature Publishing. (in Russian)

Zhu. (2016). *MCTCP: Congestion-aware and robust multicast TCP in Software-Defined networks.* IEEE.

Zubair, M., Khan, M. J., & Awais, M. M. (2012). Prediction and analysis of air incidents and accidents using case-based reasoning. In *Intelligent Systems (GCIS), 2012 Third Global Congress on* (pp. 315-318). IEEE.

Бикоптер UH-144 из Halo. (2014). Available at: http://www.parkflyer.ru/static/user_files/2014/5/9/1.1402320683936.jpg

История развития дронов. (2018). Available at: https://dronomania.ru/faq/istoriya-razvitiya-dronov.html

About the Contributors

Tetiana Shmelova, PhD, DSc, Professor of Department of Air Navigation Systems in National Aviation University (Ukraine). Thesis in the field of Scientific and Methodological Basis of Decision-Making in the Air Navigation System (2012). Areas of Scientific Interests: Mathematical Models of Decision-Making by the Human operator, especially in Emergency Situation; Decision Support System; Research of Air Navigation system as a Socio-technical system; Aviation Sociometry and Socionics; Management and Marketing in aviation; Information technology and Informatics of Decision-Making; Problems of Human factors in aviation. Author of more than 250 scientific articles and guides (about 200 articles is in Ukrainian and Russian, about 35 articles in English, 14 methodical manuals, 12 copyright certificates for computer programs), 6 monographs in fields of aviation, economics, mathematics, the theory of system. Teaching courses: Theory of decision-making, Mathematical Programming, Effectiveness of Air Traffic Management. Has got two children: daughter and son, husband. Hobbies: traveling, music, reading, and swimming.

Yuliya Sikirda is Associate Professor of Management, Economy, Law and Tourism Department at the Flight Academy of National Aviation University. Has graduated from the Flight Academy of National Aviation University (former State Flight Academy of Ukraine) in 2001, received a Master degree on specialty "Air Traffic Services" specialization "Air Traffic Control". Has got a diploma of PhD in Technical Sciences on specialty "Automated Control Systems and Advanced Information Technologies" (2004). PhD thesis "Modelling of Decision Support System for an Air Traffic Controller in Flight Emergency Situations", the thesis has defended in Kyiv Automation Institute. In 2006 has received an attestat of Associate Professor of Management and Economy Department. Since 2014 is the member of the specialized scientific council for the defense of dissertations for PhD in Technical Sciences on the specialty "Navigation and Motion Control". The author and co-author of about 180 scientific works, including three manuscripts and more than 40 articles in specialized scientific publications, three copyright certificates on computer programs. Together with the scientific work has actively engaged in the methodological provision of the educational process, during the years of labor activity above 80 teaching and methodological works have been issued. Areas of scientific interests: increasing of the decision making efficiency by the human-operator of the Air Navigation System in flight emergencies; assessment of the influence of aviation enterprises' internal and external management environment factors on the activity of the Socio-technical System's operator. Teaching courses: "Introduction to the Specialty", "Transport and Transport Infrastructure", "Information Systems in Management", "Methodology of Scientific Researches", "Mathematical Modelling of Professional Tasks".

* * *

Volodymyr Busygin graduated Nobel Universit. His scientific interests relate to the multiprocessor computing systems and design of the complex dynamic systems.

Nina Chichikalo was born 1939. Nina Ivanovna Chichikalo is a professor of Department of Audio engineering and Information registration of National Technical University of Ukraine "Igor Sikorsky Kiev Polytechnic Institute". Doctor of Sciences (Engineering), Professor. Academician of Ukraine Metrology Academy (2018). He graduated from Kiev Polytechnic Institute in 1961. Since 2017 he has being worked in Kiev Polytechnic Institute as a professor.

Wojciech Kamiński, in October 2012, began studies at the Faculty of Transport in the Silesian University of Technology. As a specialty in 1st degree studies (engineering), he chose Railway Vehicles Operation. In January 2016 he passed the engineering diploma exam. After graduation 1st degree studies, he began the 2nd degree studies (master's) in the same faculty with specialization in the Logistics of Transport. In July 2017 he defended his master's thesis and passed the diploma exam. From October 2017 Wojciech Kamiński is a PhD student at the Faculty of Transport in the Department of Logistics and Aviation Technology.

Ramgopal Kashyap's areas of interest are image processing, pattern recognition and machine learning. He has published many research papers, and book chapters in international journals and conferences like Springer, Inderscience, Elsevier, ACM and IGI-Global indexed by Science Citation Index (SCI) and Scopus (Elsevier). He has Reviewed Research Papers in the Science Citation Index Expanded, Springer Journals and Editorial Board Member and conferences programme committee member of the IEEE, Springer international conferences and journals held in countries: Czech Republic, Switzerland, UAE, Australia, Hungary, Poland, Taiwan, Denmark, India, USA, UK, Austria, and Turkey. He has written many book chapters published by IGI Global, USA.

Olena Kholod is a professor of department of economy and design of business processes of University of the name of Alfred Нобеля, city Dnepr, Ukraine. In 1976 made off the Dnepropetrovsk state university, механико-математический faculty, speciality of loud «Speaker and durability of machines». In 1984 protected candidate's dissertation on speciality «Structural mechanics» on a theme the «Nonlinear dynamic regional tasks of »Theory of plastins and cylindrical shells«. _ To 1988 worked in НИИ of mechanization of ferrous metallurgy in the laboratory of durability, reliability and ремонтопригодности of metallurgical asms, building and building (senior staff scientist, managing a laboratory). From 1988 to 1996 is an associate professor of department of higher mathematics of the Dnepropetrovsk metallurgical institute. The rank of associate professor on the department of higher mathematics got in 1992. From 1996 for a present tense works in University of the name of Alfred Нобеля. A department provides экономико-математическую preparation of students of economic specialities. A cold Helen teaches disciplines: higher and applied mathematics, theory of chances, mathematical statistics, optimization methods and models, analysis of operations. Professional interests are concentrated on research of tasks of mechanics of the deformed solid, dynamics of the thin-walled constructions.

Vitaliy Y. Larin received his MS in Computer's Informational Technologies from the Donetsk State Technical University in 1994. Before 2005 he worked at Donetsk State Technical University (last name – Donetsk Narional Technical University) where he passed several positions: senior laboratory assistant;

second research assistant; post-graduate student; lecturer; associate professor. In 2003 V.Lain defended his PhD dissertation from informational-measuring systems. From 2005 to 2008 he was developing his Doctor of Science dissertation, which was successfully defended in 2010 in National Technical University of Ukraine "Kiev Politechnical Institute". From 2008 to present time V. Larin has being working in AirNavigation Systems department at National Aviation University (Kiev, Ukraine).

Ekaterina Larina was born in 1973. Katerina Y. Larina received her MS in Computer's Informational Technologies from the Donetsk State Technical University in 1995. Before 2013 she worked at Donetsk State Technical University (last name – Donetsk National Technical University) where she passed several positions: second research assistant; post-graduate student; lecturer. In 2012 Katerina defended her PhD dissertation. Since 2013 she has been worked at Department of Computer-Aided Management and Data Processing Systems of National Technical University of Ukraine "Igor Sikorsky Kiev Polytechnic Institute" as the senior lecturer.

Oleksander Marynoshenko is Associate Professor of Department of aircraft and rocket engineering at the National Technical University of Ukraine "Igor Sikorsky Kyiv Polytechnic Institute». He graduated from National Technical University of Ukraine «Kyiv Polytechnic Institute» in 2002, received a Master degree on specialty «Aircraft Control Systems». Received the degree of Ph.D. in Technical Sciences on the specialty «Dynamics and strength of machines» in 2010, the subject of the study was "Dynamic deformation of elongated beam systems with interacting air flow". In 2013 he was certified as an associated professor in the Department of Aircraft and Space System. His areas of scientific interests are the flight dynamics of aircraft, navigation and control systems.

Roman Odarchenko, Associate professor in the Department of Telecommunications systems at the National Aviation University, Deputy director of Educational and Research Institute of Air Navigation, Electronics and Telecommunications PhD in Telecommunications with broad experience in teaching, scientific work and managing. Extensive knowledge of electronics and telecommunications, especially 2G/3G/4G/5G cellular networks. Strengths in problem solving and project management. Current research interests is creation of new technological solutions for future networks, LTE, 5G cellular networks, raising the quality of service level and reliability. Dr. Odarchenko received PhD at the National Aviation University in 2013. Research experience: "5G-XCast project" (broadcast/multicast services for 5G networks) (Horizon 2020 project (EU)) - Expert; "Creating IMS platforms for government communications" (Research Institute of the State Special Communications and National Aviation University) – Project manager; "Development of methods for data transfer speed increasing in wireless fourth generation cellular networks channels" (National Aviation University) – project manager; "Technology of creation, operation and examination of complex information security systems" (National Aviation University) – Executor.

Olga Ogirko, born in 1973, graduated from Lvov National University named by Ivana Franko, Ph.D., Senior Lecturer, Lvov State University of Internal Affairs, Lvov. Research interests - information technology.

Igor Ohirko (born on 1952, the Zborivskogo district, Ternopol's area) is a Ukrainian mathematician, doctor of physic and mathematical sciences (1990) and Professor. After that, he worked in the Institute of Applied Problems of Mechanics and Mathematics of the Academy of Sciences of Ukraine as well as the Lvov State University. There he also headed the Department of Numeral Methods of Mechanics. In 1990, Ohirko oversaw dissertations at the Kazan's University. From 1992 onwards, he worked as a professor and was head of the Department of Applied Mathematics of the Ukrainian Polygraphy Institute.

Oleksii Pikenin in 2014 completed the full course of National Technical University of Ukraine «Kyiv Polytechnic Institute» and obtained a complete higher education in the specialty «Aircraft Control Systems» and obtained the qualification of Research Engineer. He currently is Senior Lecturer of Department of aircraft and rocket engineering at the National Technical University of Ukraine "Igor Sikorsky Kyiv Polytechnic Institute". His areas of scientific interests are pattern recognition, artificial intelligence, development and design unmanned aerial vehicles.

Georgii Rozorinov is a professor of Department of Audio engineering and Information registration of National Technical University of Ukraine "Igor Sikorsky Kiev Polytechnic Institute". Doctor of Technical Sciences, Professor. A Laureate of the State Award of Ukraine is in the Field of Science and Technology (1989). Academician of the International Bioenergy Technologies Academy (2006). Academician of Ukraine Metrology Academy (2018). He graduated from Kiev Polytechnic Institute Electro-acoustic faculty (now Electronic faculty) in the specialty "Electro-acoustics and ultrasonic equipment". At first, he worked in Kiev scientific and research institute of radio electronics and Kiev design bureau MARS. Since 1973 he worked in Kiev Polytechnic Institute as a senior engineer, senior research worker, associate professor, deputy of the dean of electro-acoustic faculty from the advanced study (1979-1989), by a professor. Head the Department of Information security systems, deputy of the director, director of Educational-scientific information security institute State University of Telecommunications (2012 - 2016). Heorhiy Rozorinov is a well-known specialist in the field of the applied acoustics and audio engineering. He made a substantial contribution to the development of theory and practices of digital magnetic and optical record. The new types of high– informative fault-tolerant signals and facilities of their treatment are offered to them provided steady development of this industry in Ukraine. Scientific interests: different types of communication, multimedia, and terahertz technologies, signals compression, information security.

Vladimir Sherstjuk is currently Full Professor of Information technologies department at Kherson National Technical University, Ukraine. He earned his Master of Software Engineering at Kherson Industrial Institute in 1987, then Ph.D. in Computer Science at Kherson State Technical University in 1986 and Doctor in Computer Science at Kherson National Technical University in 2014. His thesis dwelled on scenario-case control of dynamic objects. After completing his Ph.D., Vladimir Sherstjuk accepted an Associate Professor position of Information Technologies department, and then in 2015, he accepted a Full Professor position in the same department. Vladimir Sherstjuk has over 250 refereed publications.

Fedir Shyshkov graduated from National Aviation University in 2015. He currently is a postgraduate student of Air Navigation Department of National Aviation University. His areas of scientific interests are satellite radionavigation and computer modeling.

Yuliya Sikirda, Associate Professor of Management, Economy, Law and Tourism Department at the Flight Academy of National Aviation University. Has graduated from the Flight Academy of National Aviation University (former State Flight Academy of Ukraine) in 2001, received a Master degree on specialty "Air Traffic Services" specialization "Air Traffic Control". Has got a diploma of PhD in Technical Sciences on specialty "Automated Control Systems and Advanced Information Technologies" (2004). PhD thesis "Modelling of Decision Support System for an Air Traffic Controller in Flight Emergency Situations", the thesis has defended in Kyiv Automation Institute. In 2006 has received an attestat of Associate Professor of Management and Economy Department. Since 2014 is the member of the specialized scientific council for the defense of dissertations for PhD in Technical Sciences on the specialty "Navigation and Traffic Control". The author and co-author of about 180 scientific works, including three manuscripts and more than 40 articles in specialized scientific publications, three copyright certificates on computer programs. Together with the scientific work has actively engaged in the methodological provision of the educational process, during the years of labor activity above 80 teaching and methodological works have been issued. Areas of scientific interests: increasing of the decision-making efficiency by the human-operator of the Air Navigation System in flight emergencies; assessment of the influence of aviation enterprises' internal and external management environment factors on the activity of the Socio-technical System's operator. Teaching courses: "Introduction to the Specialty", "Transport and Transport Infrastructure", "Information Systems in Management", "Methodology of Scientific Researches", "Mathematical Modelling of Professional Tasks".

Konin Valeriy graduated from Rybinsk Aviatechnological Institute in 1969. He is a doctor of technical sciences, professor of Air Navigation Department of National Aviation University. His primary interests lie in satellite radionavigation, theory and technique of microwave systems, computer modeling.

Maryna Zharikova is currently Associate Professor of Information technologies department at Kherson National Technical University, Ukraine. She earned her Bachelor of Software at Kherson National Technical University in 1999, and her PhD in Mathematical Modeling and Computational Methods from Kharkiv National University of Radioelectronics in 2004. Her thesis dwelled on forest fire modeling. After completing her PhD, Maryna Zharikova accepted a position as Senior Lecturer of the same department. In 2008, she accepted an Associate Professor position at Information Technologies department. Maryna Zharikova has over 60 refereed publications.

Index

U

V

W

www.ingramcontent.com/pod-product-compliance
Lightning Source LLC
LaVergne TN
LVHW060256011026
849161LV00011B/2720
* 9 7 8 1 5 2 2 5 7 5 8 8 7 *